Correlated Electrons in Quantum Matter

Correlated Electrons in Quantum Matter

Peter Fulde
Max Planck Institute for the Physics of Complex Systems, Dresden, Germany
and
Department of Physics, POSTECH, Pohang, Korea

NEW JERSEY · LONDON · SINGAPORE · BEIJING · SHANGHAI · HONG KONG · TAIPEI · CHENNAI

Published by

World Scientific Publishing Co. Pte. Ltd.
5 Toh Tuck Link, Singapore 596224
USA office: 27 Warren Street, Suite 401-402, Hackensack, NJ 07601
UK office: 57 Shelton Street, Covent Garden, London WC2H 9HE

British Library Cataloguing-in-Publication Data
A catalogue record for this book is available from the British Library.

CORRELATED ELECTRONS IN QUANTUM MATTER

Copyright © 2012 by World Scientific Publishing Co. Pte. Ltd.

All rights reserved. This book, or parts thereof, may not be reproduced in any form or by any means, electronic or mechanical, including photocopying, recording or any information storage and retrieval system now known or to be invented, without written permission from the Publisher.

For photocopying of material in this volume, please pay a copying fee through the Copyright Clearance Center, Inc., 222 Rosewood Drive, Danvers, MA 01923, USA. In this case permission to photocopy is not required from the publisher.

ISBN 978-981-4390-91-0
ISBN 978-981-4390-92-7 (pbk)

Printed by FuIsland Offset Printing (S) Pte Ltd Singapore

Preface

The field of correlated electrons has considerably grown and matured over the last twenty years. One reason is that an increasing number of materials, which are interesting from a technological point of view, turn out to have fairly strong electronic correlations. The transition metal oxides, which find application in various memory and dedicated chip devices are one example, and superconductors with high transition temperatures are another. Excitations with fractional charges, which in the future might play a role in quantum computing, require strong electron correlations as well. Last, and certainly not least, the field is profiting from the increasing possibility to test various theoretical models experimentally for correlated electrons by studying ultracold fermionic atoms on optical lattices.

Twenty years ago the first edition of the book *Electron Correlations in Molecules and Solids* appeared; the third and last edition of this monograph was published in 1995. Its goal had been to present the problem of electron correlation in a unified form and to provide a framework, within which we could treat weakly as well as strongly correlated electron systems. The framework was intended to be applicable independent of whether the electrons are part of a solid or a molecule. The original plan for this monograph was to write a simplified version of the previous book, in order to help students with little experience in condensed matter and field theory. However, so many new developments had taken place that this intention was given up. Rather it seemed more justified to keep the previous level of presentation, but to instead shorten and condense some of the previous material in order to make room to describe some of the new developments. The present monograph no longer includes a special discussion of molecules. The change in the title of the book is an indication of this factor. The wavefunction based approach to electronic structure calculations for molecules has only minimally advanced as the field has been more and more taken over by density functional methods. So at the end it became a new book. However, there remains some overlap with the former text.

There are two special topics which are particularly stressed here. One is the application of wavefunction-based methods to electronic structure calculations of solids. Through wavefunction-based methods, the ground state as well as energy bands of electrons in solids can be determined. As is well known, density functional calculations have revolutionized the field of electronic structure calculations and are strongly dominating it. During this period the parallel development of wavefunction-based methods has been neglected. However, wavefunction-based methods are definitely needed in order to gain better insight into different correlation contributions to physical quantities. Remember that density functional theory avoids any statements about the many electron wavefunction.

The second topic to which special attention is drawn is the projection method. In order to describe quantitatively the correlation hole of an electron, a comparatively small number of operators needs to be applied to the wavefunction of uncorrelated electrons. They constitute a very small portion of the complete operator- or Liouville space. Therefore, properly chosen microscopic processes which contribute to the construction of the correlations hole are considered a projected part of the full operator space. This may sound rather technical, but examples will show that the method can be used for rather efficient calculations, e.g., of satellite structures in the excitation spectra of solids.

This book is supposed to be a textbook rather than a review. Thus, no attempts have been made to provide a list of references that reflects the historical development or significance of different contributions to any of the subjects discussed here. Instead, reference is only made when an outside source helps to better describe or deepen specific arguments given in the book.

Needless to say, not all aspects of electron correlations have been discussed here. Instead, any coverage of this very broad field will necessarily be a biased selection. It is the sincere hope that the book might help some graduate students gain access to this fascinating field of theoretical physics.

Dresden, February 2012 *Peter Fulde*

Contents

1 **Introduction** 1

2 **Independent Electrons** 9
   2.1 Many-Electron Hamiltonian 10
   2.2 Basis Sets 11
   2.3 Self-consistent Field Equations 12
   2.4 Unrestricted SCF Approximation 18
   2.5 Missing Features of the Independent-Electron Approximation . 20

3 **Homogeneous Electron Gas** 25
   3.1 Uncorrelated Electrons 26
   3.2 Random-Phase Approximation 31
   3.3 Wigner Crystal 34

4 **Density Functional Theory** 39
   4.1 Theory of Hohenberg, Kohn and Sham 40
   4.2 Local-Density Approximation and Extensions 43
   4.3 Strong Electron Correlations: LDA+U 47
   4.4 The Energy Gap Problem 50
   4.5 Time-Dependent DFT 53

5 **Wavefunction-Based Methods** 57
   5.1 Method of Configuration Interactions 59
   5.2 Cumulants and their Properties 63
   5.3 Ground-State Wavefunction and Energy 64
      5.3.1 Method of Increments 68
   5.4 Different Approximation Schemes 70
      5.4.1 Partitioning and Projection Methods 71
      5.4.2 Coupled Cluster Method 72
      5.4.3 Selection of Excitation Operators 75
      5.4.4 Trial Wavefunctions 78

Contents

6 Correlated Ground-State Wavefunctions 83
- 6.1 Semiconductors 84
  - 6.1.1 Model for Interatomic Correlations 84
  - 6.1.2 Estimates of Intra-Atomic Correlations 89
  - 6.1.3 Ab Initio Results 90
- 6.2 Ionic and van der Waals Solids 93
  - 6.2.1 Three Oxides: MgO, CaO and NiO 93
  - 6.2.2 Rare-Gas Solids 96
- 6.3 Simple Metals 97
- 6.4 Ground States with Strong Correlations: CASSCF 99

7 Quasiparticle Excitations 101
- 7.1 Single-particle Green's Function 102
  - 7.1.1 Perturbation Expansions 106
  - 7.1.2 Temperature Green's Function 111
- 7.2 Quasiparticles in Metals 117
- 7.3 Quasiparticles in Semiconductors and Insulators 123
  - 7.3.1 Quasiparticle Approximation 124
  - 7.3.2 A Simple Model: Bond-Orbital Approximation 126
  - 7.3.3 Wavefunction-Based Ab Inito Calculations 132

8 Incoherent Excitations 137
- 8.1 Projection Method 138
- 8.2 An Example: Hubbard Model 141

9 Coherent-Potential Approximations 145
- 9.1 Static Disorder 146
- 9.2 Dynamical Disorder: DMFT and Beyond 148

10 Strongly Correlated Electrons 157
- 10.1 Measure of Correlation Strengths 159
- 10.2 Indicators of Strong Correlations 162
  - 10.2.1 Low-Energy Scales: a Simple Model 163
  - 10.2.2 Effective Hamiltonians 167
- 10.3 Kondo Effect 168
- 10.4 The Hubbard Model Revisited 178
  - 10.4.1 Spin-Density Wave Ground State 178
  - 10.4.2 Gutzwiller's Ground-State Wavefunction 183
  - 10.4.3 Hubbard's Approximations and their Extensions 186
  - 10.4.4 Kanamori Limit 189
- 10.5 The t-J Model 191
- 10.6 Mean-Field Approximations 207
  - 10.6.1 Test of Different Approximation Schemes 212
- 10.7 Metal-Insulator Transitions 220
- 10.8 Numerical Studies 224

	10.9	Break-down of Fermi Liquid Description 231
		10.9.1 Marginal Fermi Liquid Behavior 232
		10.9.2 Charged and Neutral Quasiparticles 234
		10.9.3 Hubbard Chains 235
		10.9.4 Quantum Critical Point 239

11 Transition Metals 241
  11.1 Ground-State Wavefunction 242
  11.2 Satellite Structures 248
  11.3 Temperature-Dependent Magnetism 250
    11.3.1 Local Spin Fluctuations 251
    11.3.2 Long-Wavelength Spin Fluctuations 264

12 Transition-Metal Oxides 281
  12.1 Doped Charge-Transfer Systems: the Cuprates 282
    12.1.1 Quasiparticle–like Excitations 284
  12.2 Orbital Ordering 299
    12.2.1 Manganites: $LaMnO_3$ and related Compounds 307
    12.2.2 Vanadates: $LaVO_3$ 314
    12.2.3 Ladder Systems: α'–NaV_2O_5 314
    12.2.4 Other Oxides 319

13 Heavy Quasiparticles 321
  13.1 Kondo Lattice Systems 324
    13.1.1 Renormalized Band Theory 325
    13.1.2 Large Versus Small Fermi Surface 333
    13.1.3 Mean-Field Treatment 336
  13.2 Charge Ordering in Yb_4As_3: an Instructive Example 340
  13.3 Partial Localization: Dual Role of 5f Electrons 348
  13.4 Heavy d Electrons: LiV_2O_4 355

14 Excitations with Fractional Charges 363
  14.1 Trans-Polyacetylene 364
  14.2 Fractional Quantum Hall Effect 368
  14.3 Correlated Electrons on Frustrated Lattices 375
    14.3.1 Loop Models 378
    14.3.2 Dimer Models 386
    14.3.3 Mapping to a U(1) Gauge Theory 390
    14.3.4 Magnetic Monopoles 393

15 Superconductivity 399
  15.1 The Superconducting State 402
    15.1.1 Pair States 405
    15.1.2 BCS Ground State 410
  15.2 Cooper Pair Breaking 417

15.2.1 Ergodic vs. Nonergodic Perturbations 418
15.2.2 Pairing Electrons with Population Imbalance 423
15.3 Cooper Pairing without Phonons 435
15.3.1 Filled Skutterudite $PrOs_4Sb_{12}$ 437
15.3.2 UPd_2Al_3: Pairing and Time-Reversal Symmetry Breaking 440
15.4 Magnetic Resonances 441
15.5 High-T_c Superconductors 448
15.5.1 Suppression of Antiferromagnetic Order by Holes 450
15.5.2 Pseudogap Regime 451
15.5.3 Strange Metal....................................... 454
15.5.4 Optical Properties: Drude Peak 455
15.5.5 Pairing Interactions................................ 460
15.5.6 Stripe Formation 467

A Some Relations for Cumulants 473

B Scattering Matrix in Single-Centre and Two-Centre Approximation 475

C Intra-atomic Correlations in a C Atom 479

D Landau Parameter: Quasiparticle Mass 481

E Kondo Lattices: Quasiparticle Interactions 483

F Lanczos Method 485

G Density Matrix Renormalization Group 489

H Monte Carlo Methods 497
H.1 Sampling Techniques...................................... 498
H.2 Ground-State Energy 500

I Computing the Memory Function by Increments 505

J Kagome Lattice at 1/3 Filling 507

References.. 509

Index.. 525

List of Acronyms

AF	antiferromagnet	
ARPES	angle resolved photoemssion spectroscopy	
BCS	Bardeen-Cooper-Schrieffer	
BKZ	Berezinskii-Kosterlitz-Thouless	
BOA	bond orbital approximation	
BZ	Brillouin zone	
CASSCF	complete active space SCF	
CC	coupled cluster	
CCSD	CC with single + double excitations	
CCSD(T)	CCSD with triples (perturbat.)	
CDW	charge density wave	
CDMFT	cluster DMFT	
CEF	crystalline electric field	
CEPA	coupled electron pair approximation	
CMO	canonical molecular orbital	
CO	charge order	
CPA	coherent potential approximation	
DCPA	dynamical CPA	
DE	double exchange	
DFT	density functional theory	
DMFT	dynamical mean field theory	
DMRG	density matrix renormalization group	
DOS	density of states	
DZ	double zeta	
DZ+P	DZ + polarization function	
EELS	electron energy loss spectroscopy	

Acronyms

e-ph	electron-phonon	
FFLO	Fulde-Ferrell-Larkin-Ovchinnikov	
FL	Fermi liquid	
FLEX	fluctuating exchange	
FM	ferromagnet	
FQHE	fractional quantum Hall effect	
FS	Fermi surface	
FSCP	fully self-consistent projection	
GGA	generalized gradient approximation	
GTO	Gauss-type orbital	
HOMO	highest occupied molecular orbital	
IQHE	integer quantum Hall effect	
ISN	inelastic neutron scattering	
J-T	Jahn-Teller	
LDA	local-density approximation	
LHB	lower Hubbard band	
LSDA	local spin density approximation	
MC-SCF	multiconfigurational SCF	
MFA	mean-field approximation	
M-I	metal-insulator	
MO	molecular orbital	
NFL	non-Fermi liquid	
NMR	nuclear magnetic resonance	
n.n.	nearest neighbor	
n.n.n.	next-nearest neighbor	
ODLRO	off-diagonal long range order	
PM	paramagnet	
QCP	quantum critical point	
QMC	quantum Monte Carlo	
QPT	quantum phase transition	
RKKY	Ruderman-Kittel-Kasuya-Yoshida	
RPA	random phase approximation	
RPT	renormalized perturbation theory	

RVB	resonating valence bond
SCF	self-consistent field
SCR	self-consistent renormalized
SDW	spin-density wave
SIC	self-interaction corrections
STO	Slater-type orbital
TB	tight binding
TDDFT	time-dependent DFT
UHB	upper Hubbard band
1D (2D)	one (two) dimensional

1

Introduction

For a long time condensed matter physics was based on the notion that most elementary physical phenomena in solids can be understood in terms of a single-particle description. This has changed considerably over the last few decades. It has become increasingly obvious that electron correlations play a much larger role than originally thought. Accounting for them has developed into an active field of research. The aim of this book is to describe a number of the most important recent developments at a level which enables students to follow. Although the main emphasis is on the theory of correlated electrons, we have included here numerous examples concerning its applications.

The electron-correlation problem appeared for the first time when in 1927 *Heitler* and *London* aimed at describing chemical bonding of a H_2 molecule by using Schrödinger's equation. Their ansatz for the two-electron wavefunction (Heitler-London wavefunction) did not contain any ionic contribution, i.e., it assumed that there is always one electron centered at atom 1 while the other is centered at atom 2. The wavefunction is then written as

$$\psi_{HL}(\mathbf{r}_1, \mathbf{r}_2) = \frac{1}{2}[\chi_1(\mathbf{r}_1)\chi_2(\mathbf{r}_2) + \chi_2(\mathbf{r}_1)\chi_1(\mathbf{r}_2)](\alpha_1\beta_2 - \beta_1\alpha_2) \ , \qquad (1.1)$$

where the functions $\chi_{1,2}(\mathbf{r})$ are centered on atoms 1 and 2, and the spin functions α and β denote spin up and down states, respectively. Avoiding ionic configurations, i.e., those in which both electrons are centered at one atom, has the advantage that the Coulomb repulsion of the two electrons is kept low because they are well separated. However, this is at the expense of their kinetic energy, which is being lowered if the above restriction is dropped. Therefore, the implicit assumption of the Heitler-London approach is that the mutual Coulomb repulsion of the electrons is more important than their energy gain due to delocalization. In present day terminology, we speak in that case of strongly correlated electrons.

Quite an opposite point of view is taken by the molecular-orbital approach for which the names of *Hückel*, *Hund*, *Mulliken*, *Slater* and others stand. The molecular-orbital theory describes the H_2 molecule within the independent electron approximation. Within that scheme a bonding molecular orbital is determined for the H$_2$ molecule

$$\phi(\mathbf{r}) = \frac{1}{\sqrt{2}}(\chi_1(\mathbf{r}) + \chi_2(\mathbf{r})) \ . \tag{1.2}$$

This orbital is occupied by two electrons of opposite spin, i.e., the antisymmetric total wavefunction is

$$\psi_{MO}^S(\mathbf{r}_1, \mathbf{r}_2) = \frac{1}{2^{3/2}}[\chi_1(\mathbf{r}_1)\chi_1(\mathbf{r}_2) + \chi_1(\mathbf{r}_1)\chi_2(\mathbf{r}_2) + \chi_2(\mathbf{r}_1)\chi_1(\mathbf{r}_2)$$
$$+ \chi_2(\mathbf{r}_1)\chi_2(\mathbf{r}_2)](\alpha_1\beta_2 - \beta_1\alpha_2) \ . \tag{1.3}$$

The electrons move independently of each other and the chance to find both of them at the same atomic site (ionic configuration) is 50 %. Their kinetic energy is optimally lowered, but their Coulomb repulsion remains relatively large.

In reality, a H$_2$ molecule is between the two limits just described. It is closer though to an independent electron description than to the Heitler-London limit of strong correlations. Therefore, it is more suitable to start from an independent electron or molecular-orbital wavefunction and improve it than to start from a Heitler-London wavefunction. This changes when one artificially pulls apart the two protons. The larger the distance between them the more we have to suppress the ionic configurations in the ground-state wavefunction. In the limit of complete separation we end up with two independent hydrogen atoms. Expressed differently, with increasing bond length the electronic correlations become more and more important until we end up in the Heitler-London or strong correlation limit.

Why are we stressing so much the simple case of a H$_2$ molecule and what has this to do with solids? The same competition between kinetic energy gain and Coulomb repulsion energy observed in the case of a H$_2$ molecule is found in other molecules, small and large. It governs also the electronic properties of solids. Here we find a rich variation in the strength of electron correlations at equilibrium distance of the ions forming a solid. The kinetic energy gain due to delocalization depends on how strong is the overlap of electronic wavefunctions of neighboring atoms. Therefore we expect that in a solid s and p electrons of the valence shell are less correlated in their motions than, e.g., f electrons of an incomplete f shell because they are closer to the nuclei. In fact, $4f$ electrons are the strongest correlated electrons we have to deal with. Describing them within an independent electron approximation makes no sense. Instead, their behavior is like in an atom. The only modifications that arise are due to a weak hybridization with the electrons of the neighboring atoms.

Various methods and techniques have been applied to deal quantitatively with the electron correlation problem or, more generally, with the electronic structure of solids. The one most commonly used is density functional theory and approximations to it as developed by *Hohenberg*, *Kohn* and *Sham*.

Strictly speaking this theory applies to the ground state. It avoids calculating many-electron wavefunctions. Instead, ground-state properties such as the energy, the electron density distribution, magnetization etc. are directly calculated. Although originally not designed for it, the highly successful theory has also been applied to excited states, i.e., energy-band calculations. Yet the approximations are uncontrolled and therefore it is no surprise that there are a number of cases where they fail, in particular when correlations are strong. Even then, by taking a pragmatic attitude and giving up a strict ab initio treatment of correlations one can often make reasonable improvements. The LDA+U serves as an example here. It combines a local density approximation (LDA) to density functional theory with a heuristic separate treatment of a local Coulomb interaction U. In any case, this development has resulted in a good understanding of the electronic structure of many different materials which in former times were considered as not accessible to quantitative computations.

A rather different approach is pursued by *wavefunction methods*, which aim at determining the (many-electron) ground-state wavefunction of a solid similarly as done before for the H_2 molecule. They have the advantage that one can learn more about the different aspects of electron correlations. The reason is that the correlation hole of the electrons is explicitly constructed. This holds true for the ground state as well as for excited states if a system. Starting point is a self-consistent field (SCF) calculation. When the correlations are relatively weak, a Hartree-Fock calculation serves that purpose. Correlations are included with the help of *local* excitation operators. They take optimal advantage of the fact that the correlation hole of an electron is a rather local object. When correlations are strong the corresponding electrons, e.g., d electrons, define an active space and the SCF calculation is done by including all configurations of this active space (CASSCF). This way the strong correlations are accurately treated while the remaining weak correlations are included by standard methods. How a CASSCF calculation and a corresponding wavefunction for an infinite solid can be formulated and executed will be discussed in detail in this book. Cumulants prove very useful here in setting up a clean theoretical frame for calculations of this kind.

Although wavefunction-based electronic structure calculations are presently more time consuming and require more program development compared with density functional based computations, they deserve special attention in the future. In fact, both approaches should be developed in parallel; there will be always systems which are too large to be handled by wavefunction based methods, but for which density functional calculations are possible. For other systems, nevertheless, may it be the computation of an energy gap of a semiconductor or the energy dispersion of holes in a Cu-O plane, wavefunction-based methods are clearly preferable due to their controlled approximations.

When dealing with electronic correlations in solids, one finds that they often resemble those in corresponding molecules or clusters. Hence one would expect quantum chemistry and solid-state theory to be two areas of research with many links and cross fertilization. Regrettably this is not the case. The two fields have diverged to such an extent that it is frequently difficult to find even a common language, something we hope will change in the future. In particular it has become clear that the various methods applied in chemistry and in solid-state theory are simply different approximations to the same set of cumulant equations.

Usually the effects of electron correlations on excited states are even more important than those on the ground state. The excitation energy of a system is often the difference between two large energies, i.e., the one of the excited state and that of the ground state. When electron correlations influence those two energies differently, the excitation energy may be dominated by correlation effects. For example, the energy gap of a semiconductor is strongly influenced by correlations. Consider a semiconductor like Si with covalent bonding. While in the ground state correlations lead to van der Waals type of interactions between different bonds, the correlation hole of an added electron includes a long-ranged polarization cloud. The latter has no analogy in the ground state and contributes significantly to reducing the excitation energy, hence the energy gap, from its Hartree-Fock values. An ab initio calculation of the energy gap of a semiconductor must therefore account for both, the relaxation effects in the vicinity of the added electron or hole as well as the long-ranged polarization cloud.

It was *Wigner* who first posed the question regarding the ground state of an electron system in which the mutual Coulomb repulsion is more important than the kinetic-energy gain due to delocalization. The answer he gave was that in that case the electrons would form a lattice because it would reduce most efficiently their mutual repulsion. A lattice keeps electrons well apart from each other. What are the conditions for such a dominance of the Coulomb repulsion? For the homogeneous electron gas which he considered, the crucial condition is that the gas must have a very low density. Due to Pauli's principle electrons in their ground state fill momentum **k** states up to a maximum momentum. The latter depends on the electron density. Let $2r_S$ denote the average distance between electrons. Then it is easy to show that the Coulomb repulsion dominates the kinetic energy when r_S is sufficiently large. Sophisticated Monte Carlo calculations show that r_S must be at least of the order of $70a_B$ where a_B is Bohr's radius or 0.53 Å, in order for a Wigner crystal to form. This condition may be fulfilled in inversion layers of doped semiconductors and there are indications that Wigner crystallization may indeed take place.

The condition for electron crystallization is dramatically improved if, instead of a homogeneous electron gas, one considers electrons located near the center of atoms. When the overlap between atomic wavefunctions of neighboring sites is small, the electrons cannot gain much kinetic energy by delocalization. Therefore their Coulomb repulsion may dominate even at high densities. It was pointed out before that the valence electrons which are closest to the nuclei and therefore have small overlap with the surrounding atoms are the

$4f$ electrons of rare-earth ions. Provided that in a system there are more rare-earth sites than $4f$ electrons (or holes), we expect electron crystallization or charge ordering to take place. Indeed, Yb_4As_3 is a prominent example for this type of charge order. Here we have one quarter of a $4f$ hole per Yb site.

The theory of metals has been strongly influenced by *Sommerfeld* and *Bethe* who treated conduction electrons like free electrons. They successfully explained a number of thermodynamic and transport properties with the help of the Fermi distribution function and its temperature derivatives. It was *Landau* who put that theory on firmer ground by introducing the concept of quasiparticles for the low-energy excitations of a system of conduction electrons. The former behave like weakly interacting electrons but with renormalized mass, Fermi velocity etc. A quasiparticle can be thought of as a bare electron together with its correlation hole which keeps the other electrons away. So the largest part of the electron interactions is taken into account by working with renormalized quantities. A quasiparticle, i.e., a bare electron plus its correlation hole has internal degrees of freedom. When they are excited they may lead to satellite structures or peaks in the density of states. These excitations are obtained from the incoherent part of the one-particle Green's function.

The concept of Landau seems to work to a reasonable extent even when the electron correlations are so strong that quasiparticles become heavy. The latter involve in most cases rare earth or actinide ions and are characterized by extremely large renormalized masses at low temperatures which may become several hundred times the free electron mass. Only recently have systematic deviations from the quasiparticle concept become the subject of intense investigations. This development was in part initiated by work on the high-temperature superconducting cuprates. The claim has been made that in the normal state of these materials Landau's Fermi liquid theory may become inapplicable. However, this subject is far from being resolved. Yet, in one-dimensional systems the Fermi liquid description is breaking definitely down. We are dealing here with a *Tomonaga* and *Luttinger* theory instead. A characteristic feature of a Luttinger liquid is a separation of spin and charge degrees of freedom, a phenomenon which can also occur in trans-polyacetylene with kinks or solitons present. It has been found that not only the dimension is important for the occurrence of spin-charge separation but also the lattice type plays an important role. Geometrically frustrated lattices are particularly amenable to deviations from Landau's quasiparticle approach.

Strongly correlated electron systems are frequently studied by means of model Hamiltonians. The multiband Hubbard models play a prominent role here. They can be diagonalized exactly for small clusters or in case of one-dimensional systems treated very accurately by means of the density-matrix renormalization group method (DMRG). The availability of powerful computers has initiated much research interest in those *brute force* techniques. Their impact on many-body theory has been steadily increasing. The same holds true for Monte Carlo calculations which belong in the same category.

Hund's rule or intra-atomic correlations play a distinctive role in many materials. They are important not only in rare-earth systems with incomplete $4f$ shells, but also in transition metals, in their oxides as well as in actinide compounds. Often they compete with crystalline electric field effects. For example, in transition metal oxides, whether d electrons are in a high- or low-spin state is effected by this competition. To what extent Hund's rule correlations are operative depends on the degree of delocalization of the electrons. The larger the hybridization matrix elements between atomic orbitals of neighboring sites, the less intra-atomic spin alignment can be established. In some of the $5f$ systems the strong intra-atomic correlations cause a partial localization of f electrons. Their dual character shows up in a number of experiments.

A characteristic feature of strong correlations is the generation of new low energy scales, something that has led to a large number of new physical phenomena. Examples are metals with heavy quasiparticle masses. They can be of different physical origin. The Kondo effect is one of them. Equally important are Hund's rule correlations, partial charge ordering, frustrations, or the Zeeman effect, to mention a few. The behavior of doped Mott-Hubbard insulators is also strongly affected by the characteristic low-energy scales.

The kinetic energy of electrons can also be efficiently reduced by applying a magnetic field. This forces the electrons into cyclotron orbits. Particularly interesting is the case of a two-dimensional electron system at low density. It can be realized, e.g., by epitaxial growth of GaAs/GaAl/GaAs heterostructures with carrier densities as low as $\rho \simeq 10^{11}$ cm^{-2}. When a sufficiently high magnetic field is applied to the layers, electrons occupy only the lowest Landau orbital and their kinetic energy is reduced to zero-point fluctuations. In that case Coulomb repulsion becomes crucial, in particular when the Landau orbital is only partially filled. Yet instead of forming a Wigner crystal, a new quantum state described by the Laughlin wavefunction is established at appropriate filling factors. It turns out that this new state is a strongly correlated electronic liquid with an energy lower than that of the Wigner crystal. The fractional quantum Hall effect is a consequence of it. Its outstanding features are excitations with fractional charges, e.g., of $\pm$ e/3, $\pm$ e/5 etc. depending on the fractional filling factor of the lowest Landau level. Fractional charges are intimately connected here with fractional statistics. Interchanging two quasiparticles shows that they are neither fermions nor bosons but anyons instead. But we can also device a simple model Hamiltonian which leads to fractionally charged excitations in three dimensions. Therefore the connection between fractionally charged excitations and fractional statistics is limited to two dimensions and is not a general one. It is well known that in three dimensions there are only fermions and bosons possible when we deal with point-like particles.

Superconductivity is solely due to correlations, more specific to pair correlations. They lead to Cooper pair formation and to an instability of the normal state of the electronic system. It came as a surprise when *Bardeen, Cooper* and *Schrieffer* showed that merely pair correlations are required, which contribute only a small fraction to the total correlation energy. All the remaining ones simply renormalize parameters which enter the superconducting transition temperature, such as the Fermi velocity or an effective Coulomb repulsion contribution. This explains why it is so difficult to calculate T_c. The situation is different in superconductors with strong electron correlations. Here the bosonic excitations, which are responsible for a net electron-electron interaction may result from electronic correlations as well. Their role is similar to the one played by phonons in conventional superconductors like Al or Pb. There is compelling evidence that specific bosonic excitations of the correlated electron system are acting as glue for Cooper pair formation. Therefore, good insight into the correlation problem is a prerequisite in order to understand superconductors with strong electron correlations like cuprates or Fe pnictides not only in the superconducting but also in the normal state.

2

Independent Electrons

When in a material electron correlations are not too strong, a convenient starting point is to consider first the electrons as being independent of each other and to add afterwards correlation corrections. The assumption of independent electrons implies that the wavefunction of the N-electron system $\Phi(\mathbf{r}_1\sigma_1,\cdots,\mathbf{r}_N\sigma_N)$ can be written in form of an antisymmetrized product of single-electron wavefunctions $\psi_i(\mathbf{r}_i\sigma_i)$. In this case the self-consistent field (SCF) or Hartree-Fock (HF) equations provide for the optimal wavefunction, i.e., the one with the lowest energy. However, for solids these equations are much too complicated to be solved without further simplifications. The most important one is to perform all calculations with a limited set of basis functions and to determine the self-consistent solution within the space spanned by that basis. When the basis set is a complete one, we often speak of the Hartree-Fock limit of the SCF equations. An important point is to find out how large a basis set has to be in order to obtain SCF wavefunctions and eigenvalues with a required accuracy.

It has been known for a long time now that *unrestricted* or symmetry-broken SCF wavefunctions enable us to partially include effects of electron correlations even within the independent-electron approximation. For example, an antiferromagnetic SCF ground-state wavefunction keeps electrons better apart than a paramagnetic one because electrons of different spin are concentrated on different sublattices. Therefore, for a paramagnet with short-range antiferromagnetic interactions, an antiferromagnetic (i.e., symmetry broken) wavefunction yields often a lower energy than does a paramagnetic SCF ground-state. The price for the improvement of the energy is a poor, i.e., unphysical, symmetry broken form of the wavefunction. A better description is, of course, to determine a paramagnetic correlated wavefunction by going beyond the independent-electron approximation. Such a wavefunction preserves the spinsymmetry of the system and at the same time keeps the electrons even better apart than a symmetry breaking SCF function.

The larger the effect of mutual Coulomb repulsion of electrons as compared with the kinetic energy gain due to delocalization, the less is the independent electron approximation justified. The inclusion of correlation effects becomes crucial. It is important to understand in a simple manner how correlations modify the SCF wavefunction, e.g., for the ground state. That is done in Section 2.5 where we also discusse how the correlations strength can be quantified.

2.1 Many-Electron Hamiltonian

We start out by defining the Hamiltonian of a system of N electrons moving in an external potential $V(\mathbf{r})$ set up by the nuclei and the inner shells and interacting via the Coulomb repulsion. By introducing electron field operators satisfying anticommutation relations

$$[\psi_\sigma^+(\mathbf{r}), \psi_{\sigma'}(\mathbf{r}')]_+ = \delta_{\sigma\sigma'}\delta(\mathbf{r}-\mathbf{r}'),$$
$$[\psi_\sigma(\mathbf{r}), \psi_{\sigma'}(\mathbf{r}')]_+ = [\psi_\sigma^+(\mathbf{r}), \psi_{\sigma'}^+(\mathbf{r}')]_+ = 0 \quad, \tag{2.1}$$

we can express the Hamiltonian in the form

$$H = \sum_\sigma \int d^3r\, \psi_\sigma^+(\mathbf{r})\left(-\frac{1}{2m}\nabla^2 + V(\mathbf{r})\right)\psi_\sigma(\mathbf{r})$$
$$+ \frac{e^2}{2}\sum_{\sigma\sigma'}\int d^3r\, d^3r'\, \psi_\sigma^+(\mathbf{r})\psi_\sigma(\mathbf{r})\frac{1}{|\mathbf{r}-\mathbf{r}'|}\psi_{\sigma'}^+(\mathbf{r}')\psi_{\sigma'}(\mathbf{r}') \quad. \tag{2.2}$$

We try to find eigenstates of this Hamiltonian within a given set of L basis function $f_j(\mathbf{r})$. These functions are generally not orthogonal to each other and their overlap matrix is

$$S_{ij} = \int d^3r\, f_i^*(\mathbf{r}) f_j(\mathbf{r}) \quad. \tag{2.3}$$

The expansion of the field operators in terms of the basis set

$$\psi_\sigma(\mathbf{r}) = \sum_{i=1}^L a_{i\sigma} f_i(\mathbf{r}) \tag{2.4}$$

defines annihilation operators $a_{i\sigma}$ and similarly creation operators $a_{j\sigma}^+$, which satisfy the relations

$$[a_{i\sigma}^+, a_{j\sigma'}]_+ = S_{ji}^{-1}\delta_{\sigma\sigma'},$$
$$[a_{i\sigma}^+, a_{j\sigma'}^+]_+ = [a_{i\sigma}, a_{j\sigma'}]_+ = 0 \quad. \tag{2.5}$$

As proof, we write (2.4) in the form

$$a_{i\sigma} = \sum_j S_{ij}^{-1} \int d^3r\, f_j^*(\mathbf{r})\psi_\sigma(\mathbf{r}) \quad. \tag{2.6}$$

Expressed in terms of the operators $a_{i\sigma}, a_{i\sigma}^+$, the Hamiltonian (2.2) becomes

$$H = \sum_{ij\sigma} t_{ij} a_{i\sigma}^+ a_{j\sigma} + \frac{1}{2} \sum_{\substack{ijkl \\ \sigma\sigma'}} V_{ijkl} a_{i\sigma}^+ a_{k\sigma'}^+ a_{l\sigma'} a_{j\sigma} \quad, \tag{2.7}$$

with the matrices t_{ij} and V_{ijkl} given by

$$t_{ij} = \int d^3r f_i^*(\mathbf{r}) \left(-\frac{1}{2m} \nabla^2 + V(\mathbf{r}) \right) f_j(\mathbf{r}),$$

$$V_{ijkl} = e^2 \int d^3r d^3r' f_i^*(\mathbf{r}) f_j(\mathbf{r}) \frac{1}{|\mathbf{r} - \mathbf{r}'|} f_k^*(\mathbf{r}') f_l(\mathbf{r}') \quad. \tag{2.8}$$

This Hamiltonian will be used very frequently. Sometimes it is advantageous to introduce operators which create or annihilate electrons in states $f_i(\mathbf{r})$ with spin σ. We denote these operators by $\hat{a}_{i\sigma}^+$ and $\hat{a}_{i\sigma}$, i.e., $|f_{i\sigma}\rangle = \hat{a}_{i\sigma}^+ |0\rangle$, where $|0\rangle$ is the vacuum state. The $\hat{a}_{i\sigma}^+$ are related to the operators $a_{i\sigma}^+$ through

$$\hat{a}_{i\sigma}^+ = \sum_j S_{ji} a_{j\sigma}^+ \tag{2.9}$$

and fulfill the anticommutation relations

$$[\hat{a}_{i\sigma}^+, \hat{a}_{j\sigma'}]_+ = S_{ji} \delta_{\sigma\sigma'} \quad. \tag{2.10}$$

This should be compared with (2.5). The other relations remain unchanged. It is a simple matter to check that

$$[\hat{a}_{i\sigma}^+, a_{j\sigma'}]_+ = \delta_{ij} \delta_{\sigma\sigma'} \quad. \tag{2.11}$$

2.2 Basis Sets

In deciding on a particular set of basis functions $f_i(\mathbf{r})$, a compromise will have to be made between high-accuracy results, which require a large basis set, and computational costs, which favor small basis sets. Similar arguments hold for their functional form. Functions which are particularly suitable as far as numerical accuracy is concerned often lack convenience from a computational point of view.

The functions $f_i(\mathbf{r})$ are generally centered at different atoms. *Slater* was the first to suggest the use of exponential functions of the form

$$f_i(\mathbf{r}) = N_i r^{n-1} e^{-\zeta_i r} Y_{lm}(\theta, \phi) \quad. \tag{2.12}$$

The $Y_{lm}(\theta, \phi)$ are the spherical harmonics and the N_i's are normalization factors. The function $f_i(\mathbf{r})$ does not only depend on i, but also on the parameters n, l and m, i.e., on the principal quantum number, angular momentum and

z component of the angular momentum. Basis functions of the form (2.12) are called Slater-type orbitals (STOs). Their advantage lies in that they approximate well the electron wavefunctions for small values of r, i.e., near the nucleus of an atom. Their disadvantage is that integrals of the form (2.8) are not easy to obtain when a SCF calculation is performed. *Boys* suggested instead the use of Gaussian-type orbitals (GTOs) of the form

$$f_i(\mathbf{r}) = N_i x^l y^m z^n e^{-\zeta_i r^2} \ . \tag{2.13}$$

Here N_i is again a normalization factor. The advantage of using GTOs is that all three- and four-center integrals of the two-electron interaction integrals V_{ijkl} can be reduced to two-center integrals. This is so because the product of two GTOs at different centers can be written as a GTO centered between the two centers. The remaining two-center integrals can be easily calculated. Note that, independent of the principal quantum number, all p_x orbitals are of the form $xe^{-\zeta r^2}$ and similarly for the other angular momentum functions (e.g., the form $xye^{-\zeta r^2}$ is used for all d_{xy} functions). The GTOs are simpler to use than the STOs, but they are less suitable for finding accurate SCF functions. One typically needs 3-4 times as many GTOs as STOs to achieve the same accuracy in a SCF calculation.

It is obvious that the size of the basis set is of crucial importance if we want to obtain results which are reasonably close to the HF limit. The simplest form of a basis set includes as many basis functions as there are electrons. This is called a *minimal basis set* and STOs must be used for it in order to obtain sensible results. We can also resort to contracted GTOs rather than to STOs. If one contacts n different s-GTOs into ν s-orbitals and m different p-GTOs into μ p-orbitals, the following notation is used ($ns\ mp/\nu s\ \mu p$), e.g., ($8s4p/2s1p$). A minimal basis set often proves insufficiently accurate for electronic structure calculations. We may use instead a double-zeta (DZ) set, which includes twice as many basis functions as there are electrons. In order to come close to the HF limit, we have to include basis functions of higher angular momenta than those of the valence electrons (e.g., d-functions in the case of carbon). The latter are called polarization (P) functions because they enable us to describe the polarization of an atom in a molecular field. We often use basis sets of double-zeta plus polarization function (DZ+P) size.

2.3 Self-consistent Field Equations

We are now in a position to derive the self-consistent field equations by making the approximation of independent electrons. It implies that the total wavefunction of the electron system $\Phi(\mathbf{r}_1\sigma_1; \mathbf{r}_2\sigma_2; ...; \mathbf{r}_N\sigma_N)$ can be written in form of an antisymmetrized product of one-electron wavefunctions $\phi_\mu(\mathbf{r}\sigma)$. They are called spin orbitals and are a product of a spatial orbital $\chi_\mu(\mathbf{r})$ and a two-component spinor σ. The latter equals $\alpha = \binom{1}{0}$ for spin-up and $\beta = \binom{0}{1}$ for spin-down electrons with respect to a given axis. Therefore

2.3 Self-consistent Field Equations

$$\phi_\mu(\mathbf{r}\sigma) = \chi_\mu(\mathbf{r})\sigma \ . \tag{2.14}$$

The $\chi_\mu(\mathbf{r})$ are constructed from the basis functions $f_i(\mathbf{r})$.

In the following we consider *closed shell* systems. These are systems with the property that whenever $\phi_\mu(\mathbf{r}\alpha)$ is occupied so is $\phi_\mu(\mathbf{r}\beta)$. Thus there is complete symmetry with respect to spin-up and spin-down electrons. In that case the antisymmetric total wavefunction can be written in form of a single determinant

$$\Phi(\mathbf{r}_1\sigma_1;\cdots;\mathbf{r}_N\sigma_N) = \frac{1}{\sqrt{N!}} \begin{vmatrix} \phi_1(\mathbf{r}_1\sigma_1) & \cdots & \phi_N(\mathbf{r}_1\sigma_1) \\ \vdots & & \vdots \\ \phi_1(\mathbf{r}_N\sigma_N) & \cdots & \phi_N(\mathbf{r}_N\sigma_N) \end{vmatrix} \tag{2.15}$$

often referred to as *Slater* determinant. Without loss of generality we may assume the functions $\phi_\nu(\mathbf{r}\sigma)$ to be orthonormal.

In second quantized form (2.15) is written as

$$|\Phi\rangle = \prod_{\mu\sigma} c^+_{\mu\sigma}|0\rangle \ , \tag{2.16}$$

where the $c^+_{\mu\sigma}$ are the creation operators of electrons in the spin orbitals $\phi_\mu(\mathbf{r}\sigma)$. Since the latter are orthogonal, the $c^+_{\mu\sigma}$ satisfy the anticommutation relations

$$[c^+_{\mu\sigma}, c_{\nu\sigma'}]_+ = \delta_{\mu\nu}\delta_{\sigma\sigma'} \ ,$$
$$[c^+_{\mu\sigma}, c^+_{\nu\sigma'}]_+ = [c_{\mu\sigma}, c_{\nu\sigma'}]_+ = 0 \ . \tag{2.17}$$

An important question to be answered is the best approximate ground-state $|\Phi\rangle$ of the Hamiltonians (2.7) which can be written in the form of (2.16). It will be denoted by $|\Phi_{\text{SCF}}\rangle$ in the following. The best state $|\Phi\rangle$ must minimize the energy and therefore fulfill the requirement of stationarity

$$\delta\langle\Phi|H|\Phi\rangle = 0 \ . \tag{2.18}$$

This condition leads to the so-called self-consistent field or SCF equations, which we now want to derive.

The SCF ground state $|\Phi_{\text{SCF}}\rangle$ has an energy expectation value $\langle\Phi_{\text{SCF}}|H|\Phi_{\text{SCF}}\rangle$ of the form

$$E_0 = \sum_{ij}^{L}\sum_\sigma t_{ij}\langle\Phi_{\text{SCF}}|a^+_{i\sigma}a_{j\sigma}|\Phi_{\text{SCF}}\rangle$$
$$+ \frac{1}{2}\sum_{ijkl}\sum_{\sigma\sigma'} V_{ijkl}\langle\Phi_{\text{SCF}}|a^+_{i\sigma}a^+_{k\sigma'}a_{l\sigma'}a_{j\sigma}|\Phi_{\text{SCF}}\rangle \ . \tag{2.19}$$

The occupied spin orbitals $\phi_\mu(\mathbf{r}\sigma)$ contained in $|\Phi_{\text{SCF}}\rangle$ are expanded in terms of the basis functions $f_i(\mathbf{r})$ as

$$c_{\mu\sigma}^+ = \sum_{n=1}^{L} d_{\mu n} \hat{a}_{n\sigma}^+ \tag{2.20}$$

and we aim at determining the coefficients $d_{\mu n}$. In the following we use the abbreviation

$$\langle ... \rangle = \langle \Phi_{\text{SCF}} | ... | \Phi_{\text{SCF}} \rangle \tag{2.21}$$

for convenience. In order to compute the first term of (2.19) we introduce the bond-order matrix

$$P_{ij} = \sum_\sigma \langle a_{i\sigma}^+ a_{j\sigma} \rangle \tag{2.22}$$

so that we may write

$$\sum_{ij}^{L} \sum_\sigma t_{ij} \langle a_{i\sigma}^+ a_{j\sigma} \rangle = \sum_{ij}^{L} t_{ij} P_{ij} \ . \tag{2.23}$$

We want to relate the P_{ij} to the coefficients $d_{\mu n}$ and for that purpose we use (2.20) and (2.11) in order to write

$$a_{j\sigma} \prod_{\mu\sigma'} c_{\mu\sigma'}^+ |0\rangle = (\pm) \sum_\nu^{\text{occ}} d_{\nu j} \prod_{\mu\sigma' \neq \nu\sigma} c_{\mu\sigma'}^! |0\rangle \ . \tag{2.24}$$

The sum over ν refers to all occupied orbitals. The sign depends on the spin direction, but independent of it we find

$$P_{ij} = 2 \sum_\nu^{\text{occ}} d_{\nu i}^* d_{\nu j} \ . \tag{2.25}$$

The second term on the right hand side of (2.19) can be treated similarly. Applying the relation (2.24) together with the one where $a_{j\sigma}$ is replaced by $a_{i\sigma}^+$, we find that

$$\langle a_{i\sigma}^+ a_{k\sigma'}^+ a_{l\sigma'} a_{j\sigma} \rangle = \begin{cases} \langle a_{i\sigma}^+ a_{j\sigma} \rangle \langle a_{k\sigma'}^+ a_{l\sigma'} \rangle & \text{if } \sigma \neq \sigma', \\ \langle a_{i\sigma}^+ a_{j\sigma} \rangle \langle a_{k\sigma}^+ a_{l\sigma} \rangle - \langle a_{i\sigma}^+ a_{l\sigma} \rangle \langle a_{k\sigma}^+ a_{j\sigma} \rangle & \text{if } \sigma = \sigma'. \end{cases} \tag{2.26}$$

Therefore the energy (2.19) can be expressed as

$$E_0 = \sum_{ij} t_{ij} P_{ij} + \frac{1}{2} \sum_{ijkl} \left(V_{ijkl} - \frac{1}{2} V_{ilkj} \right) P_{ij} P_{kl} \ . \tag{2.27}$$

The coefficients $d_{\mu n}$ have to be varied so that E_0 is minimized. The variation must leave the normalization of the spin orbitals $\phi_\mu(\mathbf{r}\sigma)$ unchanged, i.e., $\langle \phi_\mu | \phi_\mu \rangle = 1$. This condition can be included in the variation of $|\phi_\mu\rangle$ by requiring that

2.3 Self-consistent Field Equations

$$\delta \left(E_0 - \sum_\mu \varepsilon_\mu \langle \phi_\mu | \phi_\mu \rangle \right) = 0 \ . \tag{2.28}$$

The ϵ_μ are Lagrange parameters. From this relation we find the following set of equations for the coefficients $d_{\mu j}$

$$\sum_{j=1}^{L} (f_{ij} - \varepsilon_\mu S_{ij}) d_{\mu j} = 0 \ . \tag{2.29}$$

They constitute one particular form of the Hartree-Fock (HF) equations for a given basis set [387]. The matrix elements f_{ij} define the Fock matrix which is given by

$$f_{ij} = t_{ij} + \sum_{kl} \left(V_{ijkl} - \frac{1}{2} V_{ilkj} \right) P_{kl} \ . \tag{2.30}$$

Note that the solutions of (2.29) enter the Fock matrix through the P_{kl}. Therefore self-consistency must be achieved between the two. We can satisfy (2.29) by introducing the Fock operator

$$F = \sum_{ij\sigma} f_{ij} (a_{i\sigma}^+ a_{j\sigma} - \langle a_{i\sigma}^+ a_{j\sigma} \rangle) \tag{2.31}$$

and requiring that the following relations hold

$$\begin{aligned} F\, c_{\mu\sigma} |\Phi_{\text{SCF}}\rangle &= -\varepsilon_\mu c_{\mu\sigma} |\Phi_{\text{SCF}}\rangle \\ F\, c_{i\sigma}^+ |\Phi_{\text{SCF}}\rangle &= \varepsilon_i c_{i\sigma}^+ |\Phi_{\text{SCF}}\rangle \ . \end{aligned} \tag{2.32}$$

They are another form of the Hartree-Fock equations and apply when the coefficients of occupied ($d_{\mu n}$) and unoccupied (d_{in}) one-electron orbitals are considered. The corresponding spin orbitals $\phi_\mu(\mathbf{r}\sigma)$ and $\phi_i(\mathbf{r}\sigma)$ are the *canonical* orbitals, i.e., molecular orbitals or Bloch states (in case of a solid).

Finally, by using (2.8), (2.14), (2.20) and (2.25) we can write the Hartree-Fock equations also in the form

$$\left(-\frac{1}{2m} \nabla^2 + V(\mathbf{r}) + e^2 \int d^3\mathbf{r}' \frac{\rho(\mathbf{r}') - \rho_\mu^{\text{HF}}(\mathbf{r}, \mathbf{r}')}{|\mathbf{r} - \mathbf{r}'|} \right) \chi_\mu(\mathbf{r}) = \varepsilon_\mu \chi_\mu(\mathbf{r}) \ , \tag{2.33}$$

where

$$\rho(\mathbf{r}') = 2 \sum_{\nu=1}^{N/2} |\chi_\nu(\mathbf{r}')|^2 \qquad \text{and} \tag{2.34a}$$

$$\rho_\mu^{\text{HF}}(\mathbf{r}, \mathbf{r}') = \sum_\nu^{N/2} \frac{\chi_\mu^*(\mathbf{r}) \chi_\nu(\mathbf{r})}{|\chi_\mu(\mathbf{r})|^2} \chi_\nu^*(\mathbf{r}') \chi_\mu(\mathbf{r}') \tag{2.34b}$$

denote the density and exchange density of the electrons, respectively. The latter ensures that the effective potential in (2.33) does not contain interactions of an electron with itself. A self-interaction contribution results from the $\rho(\mathbf{r}')$ term in the interaction potential because the orbital $\chi_\mu(\mathbf{r})$ appears in (2.34a) too. However, its contribution is canceled by the $\rho_\mu^{\mathrm{HF}}(\mathbf{r},\mathbf{r}')$ term. The exchange density, being vital for avoiding self-interactions, complicates considerably the finding of self-consistent solutions of the Hartree-Fock equations. It has a non-local character.

We proceed by introducing the self-consistent field Hamiltonian H_{SCF} according to

$$H_{\mathrm{SCF}} = F + E_0 \; . \tag{2.35}$$

From (2.31) we find that $\langle F \rangle = 0$ and therefore $\langle H_{\mathrm{SCF}} \rangle = \langle H \rangle = E_0$. This suggests to decompose H into

$$H = H_{\mathrm{SCF}} + H_{\mathrm{res}} \tag{2.36}$$

with H_{res} given by

$$H_{\mathrm{res}} = \frac{1}{2} \sum_{\substack{ijkl \\ \sigma\sigma'}} V_{ijkl} a_{i\sigma}^+ a_{k\sigma'}^+ a_{l\sigma'} a_{j\sigma} - \sum_{\substack{ijkl \\ \sigma}} \left(V_{ijkl} - \frac{1}{2} V_{ilkj} \right) P_{kl} a_{i\sigma}^+ a_{j\sigma}$$
$$+ \frac{1}{2} \sum_{ijkl} \left(V_{ijkl} - \frac{1}{2} V_{ilkj} \right) P_{ij} P_{kl} \; . \tag{2.37}$$

One can check easily that $\langle H_{\mathrm{res}} \rangle = 0$.

When a SCF calculation is supplemented by one, which includes correlations, it is advantageous to work with *localized* SCF orbitals instead of molecular- or Bloch orbitals. Therefore we want to reexpress $|\Phi_{\mathrm{SCF}}\rangle$ in terms of optimally localized orbitals with creation operators $\tilde{c}_{\nu\sigma}^+$ so that

$$|\Phi_{\mathrm{SCF}}\rangle = \prod_{\nu\sigma}^N \tilde{c}_{\nu\sigma}^+ |0\rangle \; . \tag{2.38}$$

We obtain the $\tilde{c}_{\nu\sigma}^+$ from the $c_{\mu\sigma}^+$ by a unitary transformation U within the space spanned by the *occupied* canonical spin orbitals, i.e.,

$$\tilde{c}_{\nu\sigma}^+ = \sum_{\mu=1}^{N/2} U_{\nu\mu} c_{\mu\sigma}^+ \; . \tag{2.39}$$

Several procedures have been proposed to find these optimally localized orbitals. We mention here the one of *Foster* and *Boys* which is based on the requirement that the sum of the quadratic repulsions of localized orbitals with themselves be minimized,

$$\sum_\mu^{\mathrm{occ}} \int d^3r d^3r' |\lambda_\mu(\mathbf{r})|^2 (\mathbf{r}-\mathbf{r}')^2 |\lambda_\mu(\mathbf{r}')|^2 = \mathrm{minimum} \; . \tag{2.40}$$

We can show that this condition maximizes the distances between different orbitals [121].

At this stage a comment is in order regarding *open-shell* systems. In contrast to the closed-shell systems considered until now, the latter are characterized by a ground state with partially filled molecular (or atomic) energy levels. For example, consider a C atom. Only two of the six possible $2p$ orbitals are occupied and it is generally not possible to represent a SCF state of such a system by a single Slater determinant. The SCF ground state of that atom is a triplet state ^3P. In this particular case it can be represented by a single determinant, symbolized by

$$\Phi_0 = (1s)^2(2s)^2(2p_{0\sigma})(2p_{1\sigma}) \quad , \qquad (2.41)$$

where we have characterized each atomic orbital by n, l, m, σ, i.e., the principal quantum number, the angular momentum and its z-component, and the spin, respectively. However, the lowest-lying singlet state ^1S with wavefunction Φ_1 can be written only in the form of a superposition of *three* Slater determinants, i.e.,

$$\Phi_1 = \frac{1}{\sqrt{3}}(1s)^2(2s)^2[(2p_0)^2 - (2p_{-1\sigma})(2p_{1-\sigma}) + (2p_{-1-\sigma})(2p_{1\sigma})] \quad . \qquad (2.42)$$

Even when several determinants have to be used for the construction of a SCF wavefunction of a given symmetry, there is no difficulty in setting up SCF equations.

Table 2.1. SCF energy for the ground state of a B atom using basis sets of different sites with optimized GTOs. (From [197, 484])

Basis set	SCF ground-state energy [a.u.][a]
$(2s1p)$	-20.7667
$(4s2p)$	-24.3359
$(5s3p)$	-24.4646
$(7s4p)$	-24.5185
$(9s4p)$	-24.5271
$(10s6p)$	-24.5283
$(11s7p)$	-24.5287
Estimated HF limit	-24.5291
Three STOs	-24.4984

[a] Atomic units: 1 a.u. = 27.2107 eV

What are the accuracies which can be achieved for various physical quantities within the SCF approximation? Of course, they depend to some extent on the size of the basis sets used. The deviations listed below refer to large basis sets close to the HF limit. So, they originate primarily in the insufficiencies of the SCF- or independent-electron approximation. The examples we give concern molecules, since systematic investigations are lacking for solids. We expect, however, that the deviations are practically the same in these cases.

Bond lengths and bond angles are obtained with an accuracy of 1 pm and $1^0 - 2^0$, respectively. The bond lengths usually come out too short, a minimal basis set typically producing deviations of 3 pm and 3^0 - 4^0.

As far as harmonic force constants are concerned, we usually overestimate the diagonal constants in SCF calculations. Typical stretching and bending force constants are accurate to an order of $0.1 \times 10^2 N/m$. This assumes the use of at least DZ basis sets.

Binding-energy calculations yield relatively poor results within the independent-electron approximation, even for large basis sets. For example, in the HF limit the binding energies for the molecules H_2 and N_2 are found to be 3.6 eV and 5.3 eV, respectively. These values must be compared with the experimental ones of 4.72 eV and 9.91 eV. The basis-set dependence of the SCF ground-state energy is presented in Table 2.1, where we have chosen a B atom as an example.

2.4 Unrestricted SCF Approximation

A wavefunction of a solid or a molecule must fulfill certain symmetry and equivalence requirements. For example, a Bloch- or a molecular orbital must be an eigenfunction of the different symmetry operators with which the Hamiltonian commutes, i.e., it must transform according to the irreducible representation of the point group of the crystal or molecule. Similarly, an atomic orbital must be an eigenfunction of the angular-momentum operator.

For an example of an equivalence restriction, consider the $^1\Sigma^+$ ground state of the CO molecule

$$\Phi_{\mathrm{SCF}} = (1\sigma)^2(2\sigma)^2(3\sigma)^2(4\sigma)^2(5\sigma)^2(1\pi_{-1})^2(1\pi_{+1})^2 \tag{2.43}$$

where different σ and π bonds are counted consecutively.

The orbital function of $4\sigma_\uparrow$ has to be the same as that of $4\sigma_\downarrow$. Another equivalence restriction is that the orbitals $1\pi_{+1}$ and $1\pi_{-1}$ are the same, except for a phase factor $\exp(i\phi)$ and $\exp(-i\phi)$ respectively, resulting from the azimuthal quantum number. Equivalence restrictions result from conservation laws. For example, the requirement that orbital parts be the same for different spin directions results from $[H, \mathbf{S}^2]_- = 0$. A Slater determinant with different orbitals for different spin directions would not be an eigenfunction of the total electron spin operator $\mathbf{S}^2$. Similarly, the condition on $1\pi_{+1}$ and

2.4 Unrestricted SCF Approximation

$1\pi_{-1}$ results from the conservation of the total orbital momentum operator, i.e., from $[H, \mathbf{L}^2]_- = 0$.

Unrestricted SCF (or HF) wavefunctions break both symmetry and equivalence requirements and often obtain lower energies this way. It is well known that the ground state of an infinite system like a solid can break a symmetry that the Hamiltonian obeys. Examples are ferromagnets or antiferromagnets which break rotational symmetry in spin space. Clearly, in this case an unrestricted SCF calculation is expected to yield a lower energy than one which is symmetry restricted, i.e., resulting in a non-magnetic ground state. However, a lower energy may be obtained erroneously for an unrestricted ground state even in cases when the system does *not* break a symmetry. The reason is that electronic correlations are partially simulated when an unrestricted SCF calculation is done. In order to demonstrate this consider the H$_2$ molecule. The SCF ground state is $\Phi_{\text{SCF}} = (1\sigma)^2$ or

$$|\Phi_{\text{SCF}}\rangle = c_\uparrow^+ c_\downarrow^+ |0\rangle \quad , \tag{2.44}$$

where c_τ^+ creates an electron with spin τ in the molecular orbital (MO) 1σ. In the simplest approximation this MO is of the form

$$c_\tau^+ = \frac{1}{\sqrt{2}}[a_{A\tau}^+ + a_{B\tau}^+] \quad , \tag{2.45}$$

where $a_{A\tau}^+, a_{B\tau}^+$ create electrons in atomic-like s wavefunctions centered on atom A and B, respectively. As is well known, the product state $c_\uparrow^+ c_\downarrow^+ |0\rangle$ overrates the probability amplitude of finding both electrons at the same atom. Stated differently, the ionic part of $|\Phi_{\text{SCF}}\rangle$ is too large, and therefore electron repulsion costs too much energy, a consequence of having left out electronic correlations. By breaking the symmetry and assigning different orbitals to different spins, the ionic part of the wavefunction can be reduced compared with that of $|\Phi_{\text{SCF}}\rangle = c_\uparrow^+ c_\downarrow^+ |0\rangle$. The spin-*unrestricted* SCF wavefunction Φ_{USCF} is of the form

$$|\Phi_{\text{USCF}}\rangle = \tilde{c}_{1\uparrow}^+ \tilde{c}_{2\downarrow}^+ |0\rangle \quad , \tag{2.46}$$

where

$$\tilde{c}_{1\uparrow}^+ = \frac{1}{\sqrt{1+\lambda^2}}[a_{A\uparrow}^+ + \lambda a_{B\uparrow}^+]$$

$$\tilde{c}_{2\downarrow}^+ = \frac{1}{\sqrt{1+\lambda^2}}[\lambda a_{A\downarrow}^+ + a_{B\downarrow}^+] \quad . \tag{2.47}$$

We notice that the ionic configurations $a_{A\uparrow}^+ a_{A\downarrow}^+ |0\rangle$ and $a_{B\uparrow}^+ a_{B\downarrow}^+ |0\rangle$ have here weight $\lambda/(1+\lambda^2)$. They are partially suppressed depending on the size of λ. This way we can reduce the Coulomb repulsion energy of the electrons. Clearly, $|\Phi_{\text{USCF}}\rangle$ is not an eigenstate of $\mathbf{S}^2$.

Returning to the solid, the above example of H$_2$ shows why a spin-unrestricted SCF or HF calculation for a nonmagnetic solid may predict erroneously, e.g., an antiferromagnetic ground state. The price paid for an energy gain due to a reduction of the ionic configurations is a wavefunction with incorrect symmetry properties.

2.5 Missing Features of the Independent-Electron Approximation

The SCF approximation is the best one when electrons are treated as independent of each other. This implies that an electron of the system interacts with the other ones according to their *average* location. In reality, however, the electronic motion occurs according to the other electrons' *actual* placement. [Side remark: imagine the effect on daily traffic, if cars interacting by hard core repulsions moved according to where the other cars are on *average* and not according to their actual position]. The Coulomb repulsion between electrons becomes sufficiently reduced only when a correlated motion of the electrons takes place. What is missing in the SCF approximation is the correlation hole which an electron carries with it in addition to the exchange hole discussed before. It prevents electrons, in particular those with opposite spins, to come too close to each other. The difference between the exact N-electron wavefunction and its Hartree-Fock counterpart is therefore related to the correlation aspect of the electron motion. Consequently, the correlation energy of a system is defined as the difference between the exact energy and the SCF or HF energy. The missing correlation hole shows up in the pair-distribution function $g(\mathbf{r}, \mathbf{r}')$. The latter is defined by the probability of finding an electron at point $\mathbf{r}'$ provided there is one at point $\mathbf{r}$ *relative* to the one without that constraint. An example is the homogeneous electron gas for which $g(\mathbf{r}, \mathbf{r}')$ is discussed in the next section.

If electrons are treated as being uncorrelated, charge fluctuations at an atomic site of the solid turn out too large. In order to illustrate this important point in more detail, it is instructive to consider first a molecule instead of a solid. The π electron system of the benzene molecule C_6H_6 is well suited in order to explain the point we want to make. The π molecular orbitals of that molecule have the simple form

$$\phi_\mu(\mathbf{r}\sigma) = \sum_{n=1}^{6} d_{\mu n} p_z(n) \sigma \qquad (2.48)$$

if we reduce the basis function for site n to one real function $p_z(n)$ (see Fig. 2.1). The six π electrons of C_6H_6 fill the lowest three energy levels with two electrons each. Imagine the ground-state wavefunction $|\Phi_{\text{SCF}}\rangle$ decomposed into a number of different terms (configurations), two of which are schematically shown in Fig. 2.2. They are obtained by inserting the six functions (2.48) into the Slater determinant (2.15) and decomposing it into different products of the functions $p_z(n)$. Notice that the Coulomb repulsions between electrons are quite different in the two configurations shown in that figure. The one in

2.5 Missing Features of the Independent-Electron Approximation

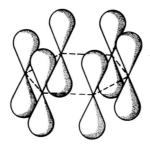

Fig. 2.1. p_z atomic-like orbitals for the description of valence electrons in C_6H_6.

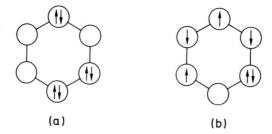

Fig. 2.2. Two configurations contained in the Slater determinant for the ground state of the π electrons in C_6H_6: (a) unfavorable and (b) favorable configurations as far as the Coulomb interactions are concerned.

Fig. 2.2a is energetically less favorable than the one in Fig. 2.2b because of the much larger Coulomb repulsions. The former contains large charge fluctuations, i.e., deviations from the average charge distribution. The carbon p_z orbitals are seen to be either empty or doubly occupied, while on the average there is one π electron per carbon atom. The important point is that unfavorable configurations (like the one in Fig. 2.2a) have too large a weight in $|\Phi_{\text{SCF}}\rangle$. Electron correlations suppress them partially, thereby lowering the energy of the system. The degree of suppression depends on the reduction of kinetic energy gain associated with it. The latter counterbalances the Coulomb-energy gain.

Very similar arguments hold true for a solid. There are again unfavorable and favorable configurations, two of which are shown in Fig. 2.3. One notices that the configuration in Fig. 2.3a is particularly favorable because the electron spins at the C sites are nearly arranged as required by Hund's rules.

The partial suppression of unfavorable configurations can be made more quantitative. For that purpose we assume that the bonds are well localized, and that they contain two electrons each. The probability of finding both electrons of a bond in the same sp^3 hybrid of an atom is 1/4 in SCF approx-

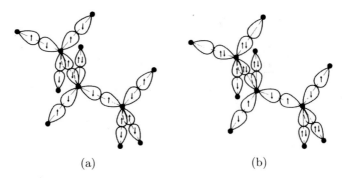

Fig. 2.3. Favorable (a) and unfavorable (b) configurations of electrons in a diamond-like structure. Only 10 bonds of the infinite system are showed. In the favorable configuration electrons are better separated from each other than in the unfavorable one.

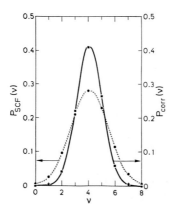

Fig. 2.4. Schematic plot of the probability $P(\nu)$ of finding ν valence electrons on a C atom in diamond or a hydrocarbon molecule with C-C bonds when the SCF ground state and when the correlated ground-state wavefunction are used. (From [367])

imation. Therefore the chance of finding at a C atom *eight* valence electrons is $1/256$ ($= (1/4)^4$). This value is too high in view of the Coulomb repulsions which are very large when eight instead of four valence electrons occupy a site. Inclusion of electron correlations reduces that chance by 85%. The same holds true when we calculate the probability of finding seven valence electrons or zero or one at a C site. All these probabilities are too high in the SCF approximation and they are reduced by including correlations. This is illustrated in Fig. 2.4. For well-localized bonds as in diamond the probability distribution $P_{\text{SCF}}(\nu)$ of finding ν valence electrons at a C site is nearly a Gaussian one, i.e.,

$$P_{\text{SCF}}(\nu) = \alpha \exp[-(\nu - \nu_0)^2/(2\Delta n^2)] \ . \tag{2.49}$$

2.5 Missing Features of the Independent-Electron Approximation

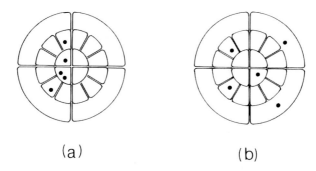

Fig. 2.5. Spatial segmentation of the atomic volume at a C site. The three rings indicate three different sets of hybrid functions with varying radial decrease. A four-fold angular segmentation can be obtained from hybridized $s-p$ functions, whereas a 12-fold subdivision requires also d functions. (a) An unfavorable configuration; (b) a favorable one.

The three parameters contained in it, i.e., α, ν_0 and Δn^2 are determined by the three moments

$$\sum_\nu P_{\text{SCF}}(\nu) = 1$$
$$\sum_\nu \nu P_{\text{SCF}}(\nu) = \bar{n}$$
$$\sum_\nu \nu^2 P_{\text{SCF}}(\nu) = \Delta n^2 + (\bar{n})^2 \ . \qquad (2.50)$$

The first condition normalizes the probability distribution, while the second and third equations determine the average valence number $\bar{n} = 4$ and the average of the squared number (note that $\Delta n^2 + (\bar{n})^2 = \overline{n^2}$). If we assume that the probability distribution $P_{\text{corr}}(\nu)$ for the correlated ground state can also be approximated by a Gaussian, we can use the reduction of the second moment, i.e.,

$$\frac{\overline{(n^2)}_{\text{SCF}} - \overline{(n^2)}_{\text{corr}}}{\overline{(n^2)}_{\text{SCF}}} < 1 \qquad (2.51)$$

in order to quantify the importance of electron correlations, for a particular system. The subscripts SCF and corr refer to $|\Phi_{\text{SCF}}\rangle$ and to the ground-state wavefunction $|\Psi_0\rangle$ containing correlations, respectively.

Those correlations which reduce charge fluctuations at an atomic site can be described by a minimal basis set and shall be called *interatomic* from now on. They are supplemented by *intra-atomic* correlations.

Consider a C site with ν valence electrons. They arrange themselves so as to reduce their Coulomb repulsion as much as possible, i.e., so as to minimize the sum of the kinetic and potential energy. Fig. 2.5 shows a segmentation of the atomic volume of a C atom with five electrons. Describing it requires a larger basis set than a minimal one. The segments can be constructed from sets of hybridized angular-momentum functions with different spatial extent. Again, in the independent-electron approximation, the weight of configurations in which the electrons are too close to each other proves too large. An example of an unfavorable configuration is shown in Fig. 2.5a. Correlations will decrease the weight of such configurations because they favor configurations in which the electrons are well separated as in Fig. 2.5b. In particular, when one electron is close to the nucleus, the others stay preferably further away from the latter (in-out correlations). Similarly, the electrons prefer to stay relatively uniformly distributed over the different angular segments of the atom (angular correlations). In order to describe accurately intra-atomic correlations, one needs a very fine spatial segmentation and hence large basis sets including high angular momenta. The above examples show the severe limitations and shortcomings of the independent-electron approximation.

3
Homogeneous Electron Gas

The homogeneous electron gas has served for a long time as a simple model for ordinary metals. In particular it has been used as a model system to study the effect of complete screening, an important hallmark of metallic behavior. It is interesting that the SCF approximation leads here to an unphysical density of states at the Fermi energy. It results from the long-range part of the Coulomb interaction. The response to it, i.e., the screening cloud is missing when uncorrelated electrons move through the system. The random phase approximation (RPA) discussed in Sect. 3.2 stands for its description. When applied to the ground-state wavefunction it includes the zero-point fluctuations of plasmons, the collective excitations of Coulomb interacting electrons. At low densities the Coulomb repulsion of the electrons is more important than the kinetic energy gain caused by delocalization. As a result the homogeneous gas becomes unstable against crystallization, because the latter strongly reduces the mutual Coulomb repulsions of electrons. The electrons form a Wigner crystal. This is discussed in Section 3.3. In between the liquid and the solid phase, there is a liquid phase with a broken rotational symmetry.

The kinetic energy gain due to delocalization can be also reduced by applying a magnetic field perpendicular to the plane of a two-dimensional homogeneous electron gas. When the field becomes large enough, we enter the strong correlation limit. The cyclotron orbitals of the electrons decrease to a minimum required by the uncertainty principle and the kinetic energy is reduced to zero point motions. All electrons are in the lowest Landau level in this case. It is found that at certain field strengths, which depend on the density of the two-dimensional electron gas, the system is an incompressible liquid – a special feature of the strong correlations. Instead of forming a Wigner crystal the system remains a liquid with highly unusual properties like excitations with fractional charges (fractional quantum Hall effect). Those unusual features are discussed in Sect. 14.2 together with other possible origins of fractional charges.

3.1 Uncorrelated Electrons

We consider a homogeneous electron gas with the compensating positive charges distributed uniformly over the volume (jellium model). This way charge neutrality is achieved without introducing any inhomogeneities. When we apply the HF approximation a number of interesting results are obtained which we will now discuss.

The normalized HF eigenfunctions of a homogeneous electron gas are plane-wave states $\Omega^{-1/2}\exp[i\mathbf{k}\mathbf{r}]$, where Ω is the total volume. Therefore, the ground state $|\Phi_{HF}\rangle$ can be expressed in terms of creation operators $c^+_{\mathbf{k}\sigma}$ of plane-wave states as follows

$$|\Phi_{\mathrm{HF}}\rangle = \prod_{|\mathbf{k}|\leq k_F} c^+_{\mathbf{k}\sigma}|0\rangle \ . \tag{3.1}$$

All plane-wave states with momentum less than the Fermi momentum k_F are occupied with two electrons each while all other states are empty. Let $\hat{n}_{\mathbf{k}\sigma} = c^+_{\mathbf{k}\sigma}c_{\mathbf{k}\sigma}$ denote the occupation-number operator with eigenvalue $n_{\mathbf{k}\sigma}$. Then it follows that

$$\hat{n}_{\mathbf{k}\sigma}|\Phi_{\mathrm{HF}}\rangle = n_{\mathbf{k}\sigma}|\Phi_{HF}\rangle \tag{3.2}$$

with

$$n_{\mathbf{k}\sigma} = \begin{cases} 1 & |\mathbf{k}| \leq k_F, \\ 0 & |\mathbf{k}| > k_F. \end{cases} \tag{3.3}$$

In order to calculate the ground-state energy we express the Hamiltonian in terms of plane-wave operators and split it into a kinetic energy- and an interaction part

$$\begin{aligned} H &= H_0 + H_{int}, \\ H_0 &= \sum_{\mathbf{p}\sigma} \varepsilon_{\mathbf{p}} c^+_{\mathbf{p}\sigma} c_{\mathbf{p}\sigma}, \\ H_{\mathrm{int}} &= \frac{1}{2\Omega} \sum_{\substack{\mathbf{pkq}\\ \sigma\sigma'}} v_{\mathbf{q}} c^+_{\mathbf{p}+\mathbf{q}\sigma} c^+_{\mathbf{k}-\mathbf{q}\sigma'} c_{\mathbf{k}\sigma'} c_{\mathbf{p}\sigma} \ . \end{aligned} \tag{3.4}$$

Here $\varepsilon_{\mathbf{p}} = p^2/(2m)$ is the kinetic energy of an electron. The interaction matrix element is

$$v_{\mathbf{q}} = \frac{4\pi e^2}{q^2}(1 - \delta_{\mathbf{q}0}) \ , \tag{3.5}$$

where the Kronecker $\delta_{\mathbf{q}0}$ ensures that $v_{\mathbf{q}=0} = 0$ because the homogeneous system is charge neutral.

The ground-state energy E_0 is obtained from (2.19). The direct Coulomb interaction does not contribute because of $v_{\mathbf{q}=0} = 0$. Therefore only the kinetic and exchange energy have to be calculated. Thus

3.1 Uncorrelated Electrons

$$E_0 = \sum_{\mathbf{p}\sigma} \varepsilon_{\mathbf{p}} n_{\mathbf{p}\sigma} + \frac{1}{2\Omega} \sum_{\substack{\mathbf{pq}\\\sigma}} v_{\mathbf{q}} \langle \Phi_{\mathrm{HF}} | c^+_{\mathbf{p}+\mathbf{q}\sigma} c^+_{\mathbf{p}\sigma} c_{\mathbf{p}+\mathbf{q}\sigma} c_{\mathbf{p}\sigma} | \Phi_{\mathrm{HF}} \rangle \quad . \tag{3.6}$$

When the kinetic-energy term is summed over the states with $|\mathbf{k}| < k_F$ we obtain the following contribution per electron

$$\frac{E_{\mathrm{kin}}}{N} = \left(\frac{3}{5}\right) \frac{k_F^2}{2m} \quad . \tag{3.7}$$

The exchange contribution is

$$\frac{E_{\mathrm{ex}}}{N} = -\frac{1}{2N\Omega} \sum_{\mathbf{pq}\sigma} v_{\mathbf{q}} n_{\mathbf{p}+\mathbf{q}\sigma} n_{\mathbf{p}\sigma}$$

$$= -\frac{1}{N\Omega} \sum_{|\mathbf{k}_1|<k_F} \sum_{|\mathbf{k}_2|<k_F} \frac{4\pi e^2}{|\mathbf{k}_1 - \mathbf{k}_2|^2} \quad . \tag{3.8}$$

The evaluation of the integral

$$I = \int_{k_1, k_2 < k_F} d^3k_1 d^3k_2 \frac{1}{|\mathbf{k}_1 - \mathbf{k}_2|^2}$$

$$= 4\pi^2 k_F^4 \tag{3.9}$$

is found in [242], for example. Expressed in terms of the electron density $\rho = N/\Omega$, the exchange energy per electron can be written as

$$\frac{E_{\mathrm{ex}}}{N} = -\frac{2e^2 k_F^4}{(2\pi)^3 \rho}$$

$$= -\frac{3e^2 k_F}{4\pi} \quad . \tag{3.10}$$

Here k_F and ρ are related through

$$\frac{2}{(2\pi)^3} \frac{4\pi}{3} k_F^3 = \rho \quad . \tag{3.11}$$

The prefactor results from the spin and the fact that the phase space is in units of h^3, instead of $\hbar^3$, where h is Planck's constant. Often a mean radius per electron r_0 is used as a characteristic length, which is related to the total electron number N through

$$\frac{4\pi}{3} r_0^3 N = \Omega \quad . \tag{3.12}$$

In units of Bohr's radius a_B (see Table 3.1) r_0 defines a dimensionless number

$$r_s = r_0 / a_B \tag{3.13}$$

in terms of which the Fermi momentum k_F is

$$k_F = \frac{1}{\alpha a_B r_s} , \quad \alpha = \left(\frac{4}{9\pi}\right)^{1/3} = 0.521 . \tag{3.14}$$

When expressed as function of r_s, the energy per electron takes the following form in atomic units

$$\begin{aligned}\frac{E_0}{N} &= \frac{3}{10} \frac{1}{\alpha^2 r_s^2} - \frac{3}{4\pi} \frac{1}{\alpha r_s} \quad a.u. \\ &= \frac{1.105}{r_s^2} - \frac{0.458}{r_s} \quad a.u. .\end{aligned} \tag{3.15}$$

Notice that for small values of r_s (high densities), the kinetic-energy term dominates. As r_s increases the exchange contribution increases too and eventually gives binding ($E_0 < 0$). In the low-concentration limit ($r_s \to \infty$) the exchange contribution dominates. Electron correlations, neglected in the HF theory, become very important. In the limit of large r_s electrons minimize their mutual Coulomb repulsions by forming a lattice (Wigner lattice). Metals have typically r_s values in the range of order unity: for Li, Cu and Al one finds $r_s = 3.2$, 2.7 and 2.1, respectively. Owing to the neglect of electron correlations, binding energies for metals generally turn out much too small in the independent-electron or HF approximation.

Another serious deficiency of the HF approximation appears when we calculate the eigenvalues $\omega_{\mathbf{k}\sigma}$ of the single-particle plane-wave states. They are computed according to (2.30–2.32) and we find that

$$\omega_{\mathbf{k}\sigma} = \frac{k^2}{2m} - \frac{1}{\Omega} \sum_{\mathbf{q}} \frac{4\pi e^2}{q^2} n_{\mathbf{k}+\mathbf{q},\sigma} . \tag{3.16}$$

The sum over $\mathbf{q}$ can be performed with the result that

Table 3.1. Atomic units and their equivalents. The three basic quantities are the electron mass and charge (e, m) and Planck's constant $(\hbar)$.

Quantity	Atomic unit	Equivalent
Length	$a_B = \frac{\hbar^2}{me^2}$	52.9 pm = 0.529 Å
Energy	$E_h = \frac{e^2}{a_B}$	4.3597×10^{-20} J = 27.211 eV
Time	$t = \frac{\hbar}{E_h}$	2.419×10^{-17} s
Probability density	$|\psi|^2 = \frac{1}{a_B^3}$	6.749×10^{30} m^{-3} = 6.749 Å^{-3}
		$[6.749 \times 10^{-6}$ pm$^{-3}]$

3.1 Uncorrelated Electrons

$$\omega_{\mathbf{k}\sigma} = \frac{k^2}{2m} - \frac{e^2 k_F}{2\pi}\left(2 + \frac{k_F^2 - k^2}{kk_F}\ln\left|\frac{k+k_F}{k-k_F}\right|\right)$$

$$= \frac{k^2}{2m} - \frac{e^2 k_F}{2\pi} F(k/k_F) \ . \qquad (3.17)$$

Due to the appearance of the logarithm we find a singularity in the derivative $d\omega/dk$ when $k = k_F$. It can be traced back to the long-range part of the Coulomb interaction, i.e., to the q^{-2} behavior of $v_\mathbf{q}$ as $\mathbf{q} \to 0$. According to (2.32) we can consider $\omega_{\mathbf{k}\sigma}$ an electron ionization potential or affinity, depending on whether an electron is removed or added to the system (Koopmans' theorem). Therefore $\omega_{\mathbf{k}\sigma} - \omega_{k_F\sigma}$ plays the role of an excitation energy of the electron gas. The density of states $\rho(\omega)$ is a measure of the number of available excited states N_e per energy interval and unit volume. It is defined as

$$\rho(\omega) = \frac{dN_e}{d\omega}\frac{1}{\Omega}$$

$$= \frac{1}{\Omega}\frac{dN_e}{dk}\frac{dk}{d\omega} \ . \qquad (3.18)$$

From phase-space counting we find that at the Fermi surface, i.e., for $|\mathbf{k}| = k_F$, the relation $\Omega^{-1}(dN_e/dk)\big|_{k=k_F} = k_F^2/\pi^2$ holds when both spin directions are taken into account. Since $d\omega/dk$ is singular for $k = k_F$ the density of states $\rho(\omega)$ of a homogeneous electron gas vanishes at the Fermi surface in HF approximation. This result is clearly in disagreement with experiments on metals like Na and K, which resemble closely homogeneous electron systems. Measurements of the low-temperature specific heat, spin susceptibility, etc. demonstrate that the density of states at the Fermi energy is nearly that of a noninteraction electron gas, i.e., mk_F/π^2. Obviously, correlation effects neglected here remedy the errors introduced by the independent-electron approximation. Indeed, it is the screening of the long-range part of the Coulomb interaction which modifies the density of states near the Fermi energy so that it is of order k_F/π^2 despite the fact that electron interactions are anything but small.

An important quantity is the pair-distribution function $g(\mathbf{r},\mathbf{r}')$. It is defined by the probability of finding an electron at point $\mathbf{r}'$, provided there is one at point $\mathbf{r}$ *relative* to the one without that constraint. The pair-distribution function is closely related to the equal-time density-density correlation function $S(\mathbf{r},\mathbf{r}')$ defined with respect to a given state $|\Phi\rangle$ by

$$S(\mathbf{r},\mathbf{r}') = \frac{1}{N}\langle\Phi|\hat{\rho}(\mathbf{r}')\hat{\rho}(\mathbf{r})|\Phi\rangle \ . \qquad (3.19)$$

The density operator $\hat{\rho}(\mathbf{r})$ is

$$\hat{\rho}(\mathbf{r}) = \sum_{i=1}^{N}\delta(\mathbf{r} - \mathbf{r}_i) \qquad (3.20)$$

where $\mathbf{r}_i$ are the positions of the N electrons. The density is $\rho(\mathbf{r}) = \langle \Phi | \hat{\rho}(\mathbf{r}) | \Phi \rangle$. In accordance with the above definition the pair-distribution function is given by

$$g(\mathbf{r}, \mathbf{r}') = \frac{1}{\rho(\mathbf{r})\rho(\mathbf{r}')} \langle \Phi | \sum_{i \neq j} \delta(\mathbf{r}' - \mathbf{r}_i) \delta(\mathbf{r} - \mathbf{r}_j) | \Phi \rangle \ . \tag{3.21}$$

The condition $i \neq j$ ensures that an electron does not correlate with itself.

For the homogeneous electron gas $g(\mathbf{r}, \mathbf{r}') = g(\mathbf{r} - \mathbf{r})$. Taking the Fourier transform and identifying $|\Phi\rangle$ with $|\Phi_{\text{HF}}\rangle$, we find that

$$\begin{aligned} g_{\text{HF}}(\mathbf{k}) &= \int d^3 r \, g_{\text{HF}}(\mathbf{r}) e^{-i\mathbf{k}\mathbf{r}} \\ &= \frac{1}{N} \left[\frac{1}{N} \langle \Phi_{\text{HF}} | \sum_{i,j} e^{i\mathbf{k}(\mathbf{r}_i - \mathbf{r}_j)} | \Phi_{\text{HF}} \rangle - 1 \right] \ , \end{aligned} \tag{3.22}$$

where we have assumed a unit volume for convenience. In second quantization the Fourier transform $\rho_{\mathbf{k}}$ of the density operator is

$$\begin{aligned} \rho_{\mathbf{k}} &= \sum_i e^{-i\mathbf{k}\cdot\mathbf{r}_i} \\ &= \sum_{\mathbf{p}\sigma} c^+_{\mathbf{p}-\mathbf{k}\sigma} c_{\mathbf{p}\sigma} \ . \end{aligned} \tag{3.23}$$

The pair-distribution function is then rewritten as

$$\begin{aligned} g_{\text{HF}}(\mathbf{k}) &= \frac{1}{N} \left(\frac{1}{N} \langle \Phi_{\text{HF}} | \rho^+_{\mathbf{k}} \rho_{\mathbf{k}} | \Phi_{\text{HF}} \rangle - 1 \right) \\ &= \frac{1}{N^2} \langle \Phi_{\text{HF}} | \sum_{\substack{\mathbf{p}\mathbf{q} \\ \sigma\sigma'}} c^+_{\mathbf{p}+\mathbf{k}\sigma} c^+_{\mathbf{q}-\mathbf{k}\sigma'} c_{\mathbf{q}\sigma'} c_{\mathbf{p}\sigma} | \Phi_{\text{HF}} \rangle \ . \end{aligned} \tag{3.24}$$

This expression is easily evaluated and gives

$$\begin{aligned} g_{\text{HF}}(\mathbf{k}) &= \frac{N-1}{N} \delta_{0\mathbf{k}} - \frac{1}{N^2} \sum_{\mathbf{p}\sigma} n_{\mathbf{p}\sigma} n_{\mathbf{p}+\mathbf{k}\sigma} (1 - \delta_{\mathbf{k}0}) \\ &= \delta_{0\mathbf{k}} - \frac{1}{N} + \frac{1}{N^2} \sum_{\mathbf{p}\sigma} n_{\mathbf{p}\sigma} (1 - n_{\mathbf{p}+\mathbf{k}\sigma})(1 - \delta_{\mathbf{k}0}) \ . \end{aligned} \tag{3.25}$$

Transforming back into $\mathbf{r}$ space, we find

$$g_{\text{HF}}(\mathbf{r}) = 1 - \frac{9}{2} \left(\frac{\sin(k_F r) - k_F r \cos(k_F r)}{(k_F r)^3} \right)^2 \ . \tag{3.26}$$

This function is displayed graphically in Fig. 3.1. The oscillations are too small to be seen in the figure. They result from the discontinuity in the $\mathbf{k}$ state occupation at the Fermi surface (see (3.1)). The pair-distribution function drops to

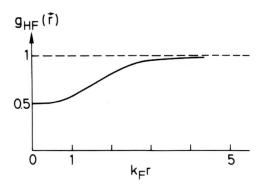

Fig. 3.1. Pair distribution function $g_{\mathrm{HF}}(r)$ for the homogeneous electron gas.

0.5, at $r = 0$ because Pauli's principle prevents two electrons of the same spin from occupying the same space volume. The drop in $g_{\mathrm{HF}}(\mathbf{r})$ is therefore called *exchange hole*. In the HF approximation electrons with antiparallel spin come arbitrarily close despite their Coulomb repulsions, which is unphysical. The energy associated with the exchange hole is given by the contribution (3.10). It is negative because the Coulomb repulsion of electrons with the same spin is reduced this way. The moving electron with the exchange hole around it is the simplest case of a quasiparticle.

3.2 Random-Phase Approximation

As pointed out before, the HF approximation gives unphysical results when applied to a homogeneous electron gas, because the density of states vanishes at the Fermi energy ϵ_F. As pointed out before, this is in disagreement with the measurements, e.g., of low temperature specific heat $C = \gamma T$ for metals like Na, which have an almost homogeneous conduction electron density. Those experiments require a constant density of states at ϵ_F like for noninteracting electrons. The failure of the HF approximation is closely related with the long-range part of the Coulomb interaction. Indeed, it is easy to show that a perturbation treatment of H_{res} (see (2.37)) yields divergent energy corrections in every order of the perturbation due to the r^{-1} behavior of the Coulomb interaction. This implies, that the contribution of H_{res} to the ground-state energy is non-analytic and can be obtained only by summing up the most divergent contribution in each order of v_q. This was done by *Gell-Mann* and *Brueckner* [145] based on earlier work by *Macke* [299]. The result which they obtained for the correlation energy is

$$\frac{E_{\mathrm{corr}}}{N} = 0.0311 \ln r_s - 0.048 + O(r_s) \ a.u. \qquad (3.27)$$

One notices the non-analytic logarithmic dependence of E_{corr} on the electron density here represented by r_s. In order to visualize the origin of the failure of perturbation theory, we imagine an electron put as a test charge into a metal. The conduction electrons respond to that test charge by screening it. Therefore the potential felt by another electron at distance r from the test charge is much smaller than $V(\mathbf{r}) \sim r^{-1}$. It turns out that, at sufficiently large r, it is of the form

$$V(\mathbf{r}) \sim \frac{1}{r^3} \cos(2p_F r) \qquad (3.28)$$

and oscillates. These are the well-known Friedel oscillations [124] and they result from the discontinuous change of the momentum distribution $n_{\mathbf{p}\sigma}$ at p_F.

A simple ansatz for the correlated ground-state wavefunction is due to *Jastrow* [216]. In first quantization its form is

$$\psi(\mathbf{r}_1, \ldots, \mathbf{r}_N) = \prod_{\langle i,j \rangle} \tilde{f}(\mathbf{r}_i - \mathbf{r}_j) \Phi_{\text{HF}}(\mathbf{r}_1, \ldots, \mathbf{r}_N)$$

$$= \exp\left(\sum_{i,j} f(\mathbf{r}_i - \mathbf{r}_j)\right) \Phi_{\text{HF}}(\mathbf{r}_1, \ldots, \mathbf{r}_N)$$

$$= \exp\left(\sum_{\mathbf{q}} \tau(\mathbf{q}) \rho_{\mathbf{q}}^\dagger \rho_{\mathbf{q}}\right) \Phi_{\text{HF}}(\mathbf{r}_1, \ldots, \mathbf{r}_N) \quad . \qquad (3.29)$$

By means of the function $\tilde{f}(\mathbf{r}_i - \mathbf{r}_j)$ it reduces the amplitude of finding two electrons close to each other. The function $\tau(\mathbf{q})$ is the Fourier transform of $f(\mathbf{r})$ and can be considered as a variational function. In writing the last equation we have used the form (3.23) for $\rho_{\mathbf{q}}$. The wavefunction (3.29) consists of a part describing independent electrons and a prefactor which has the form of a ground state of independent harmonic oscillators. The oscillator variables are proportional to the density fluctuations $\rho_{\mathbf{q}}$. The exponential prefactor in (3.29) describes zero-point motions of density fluctuations which may give rise to collective plasmon excitations. The determination of their energies is our next goal. For that we have to derive and solve the equations of motion for the $\rho_{\mathbf{q}}$. Thereby a Random Phase Approximation (RPA) is made.

We start from

$$\dot{\rho}_{\mathbf{q}} = i\left[H, \rho_{\mathbf{q}}\right]_- \qquad (3.30)$$

and write the Hamiltonian (2.12) in the form

$$H = \sum_i \frac{p_i^2}{2m} + \frac{1}{2\Omega} \sum_{\mathbf{q}} \frac{4\pi e^2}{q^2} \left(\rho_{\mathbf{q}}^\dagger \rho_{\mathbf{q}} - N\right) \quad . \qquad (3.31)$$

This leads immediately to

$$\dot{\rho}_{\mathbf{q}} = -i \sum_j e^{-i\mathbf{q}\mathbf{r}_j} \frac{\mathbf{q}}{m} \cdot \left(\mathbf{p}_j - \frac{\mathbf{q}}{2}\right) , \qquad (3.32)$$

where the relation

$$\left[\mathbf{p}_j, e^{-i\mathbf{q}\mathbf{r}_j}\right]_- = -\mathbf{q}\, e^{-i\mathbf{q}\mathbf{r}_j} \qquad (3.33)$$

has been used. By repeating these steps we find

$$\ddot{\rho}_{\mathbf{q}} = -\sum_j e^{-i\mathbf{q}\mathbf{r}_j} \frac{1}{m^2} \left[\mathbf{q}\cdot\left(\mathbf{p}_j - \frac{\mathbf{q}}{2}\right)\right]^2$$
$$-\frac{1}{\Omega}\sum_{\mathbf{q}'jn} \frac{4\pi e^2}{mq'^2}\, \mathbf{q}\cdot\mathbf{q}'\, e^{-i(\mathbf{q}-\mathbf{q}')\mathbf{r}_j} e^{-i\mathbf{q}'\mathbf{r}_n} . \qquad (3.34)$$

Since it is the long-range part of the Coulomb interaction which is screened, it suffices to consider small q values only. By taking an average with respect to the direction of $\mathbf{p}_j$, we approximate

$$\frac{1}{m^2}\sum_j e^{-i\mathbf{q}\mathbf{r}_j}\left[\mathbf{q}\cdot\left(\mathbf{p}_j - \frac{\mathbf{q}}{2}\right)\right]^2 \simeq \frac{q^2 p_F^2}{3m^2}\sum_j e^{-i\mathbf{q}\mathbf{r}_j}$$
$$= \frac{q^2 p_F^2}{3m^2}\rho_{\mathbf{q}} . \qquad (3.35)$$

The RPA consists in keeping in the second term on the right-hand side of (3.34) only the contribution $\mathbf{q}' = \mathbf{q}$. Note that the neglected terms $\sum_j e^{-i(\mathbf{q}-\mathbf{q}')\mathbf{r}_j}$ cancel to zero when $\mathbf{q}' \neq \mathbf{q}$ and the phases appear at random. With these approximations (3.34) becomes

$$\ddot{\rho}_{\mathbf{q}} = -\left(\frac{4\pi e^2 n}{m} + \frac{p_F^2}{3m^2}q^2\right)\rho_{\mathbf{q}} . \qquad (3.36)$$

This is the equation of a harmonic oscillator with eigenfrequency

$$\omega_{p\ell}(\mathbf{q}) = \sqrt{\frac{4\pi e^2 n}{m} + \frac{p_F^2}{3m^2}q^2} . \qquad (3.37)$$

In the long wavelength limit it reduces to the plasma frequency

$$\omega_{p\ell} = \sqrt{\frac{4\pi e^2 n}{m}} , \qquad (3.38)$$

which is the frequency of a uniform oscillation of the electron gas of density n against a positively charged background.

From (3.36) it is noticed that the eigenmodes characterized by a momentum $\mathbf{q}$ are independent harmonic oscillator modes. The ground-state wavefunction for any of these modes is a Gaussian. That results in the Jastrow-type variational ansatz (3.29). In Sect. 5.4.4 we will show explicitly how the zero-point fluctuations of the plasma oscillations enter the ground-state wavefunction (3.29).

3.3 Wigner Crystal

It was *Wigner* who first dealt with the ground-state problem in the dilute limit of a homogeneous electron gas [486]. It is easy to see that in this limit the mutual Coulomb repulsion of electrons is much more important than their kinetic energy. If $2r_s$ is the average distance between electrons (in atomic units) then in the dilute limit the average kinetic energy per electron scales like

$$E_{\text{kin}}(r_s) \simeq \frac{(\Delta p)^2}{2m} \sim \frac{1}{r_s^2} \qquad (3.39)$$

because of the uncertainly relation (note that $(\Delta p)r_s \simeq 1$). On the other hand, the average potential energy scales like

$$V(r_s) \simeq \frac{e^2}{2r_s} \quad . \qquad (3.40)$$

Thus when $r_s \to \infty$ the potential energy $V(r_s)$ dominates the kinetic one. Therefore, the electron system will minimize the repulsive energy and according to Wigner the way it does this is by forming a lattice. The change from an itinerant electron system with a Fermi surface to a Wigner crystal is a liquid to solid phase transition. It will take place when the average potential energy $\langle V \rangle$ and kinetic energy $\langle T \rangle$ are comparable. We want to draw attention to the fact that $\langle T \rangle$ can be quite different for a homogeneous electron gas and an inhomogeneous system of the same average density. For example, when the electrons are close to the nuclei like $3d$ or $5f$ electrons, the gain in kinetic energy due to delocalization is much less than it is for a homogeneous system. Therefore, for an inhomogeneous system $\langle T \rangle = \langle V \rangle$ is fulfilled at lower values of r_s than for a homogeneous one. This point is discussed in detail in Sect. 13.2. Here we limit ourselves to treating the homogeneous electron gas.

In order to determine the critical value of r_s^c at which the liquid to solid transition does occur, we have to compute accurately the energy of the two phases. The energy of the electron liquid is given by (3.15, 3.27). In the crystalline phase there is first of all the Madelung energy, which scales like r_s. The exchange contribution falls off exponentially at large distances like $\exp\left[\alpha_s^{1/2}\right]$ with $\alpha_s \approx 2$. In addition the zero-point energy of the lattice vibrations, i.e., phonons has to be accounted for. The best estimates yield a dependence proportional to $r_s^{-3/2}$ [54]. Anharmonic correlations scale like r_s^{-2}. This gives us the ground-state energy in the crystalline phase of the electron system in terms of an expansion for large r_s in the form [53]

$$\mathcal{E}_0 = \frac{a}{r_s} + \frac{b}{r_s^{3/2}} + \frac{c}{r_s^2} + \ldots \qquad (3.41)$$

with $a \simeq -1.8$, $b \simeq 2.7$, $c = -0.7$.

The momentum distribution function $n(\mathbf{p})$ for noninteracting electrons is the step function (3.3). For an interacting electron liquid the discontinuity at

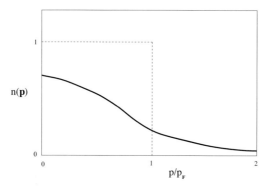

Fig. 3.2. Schematic plot of $n(\mathbf{p})$ for non-interacting electrons (dashed lines) and for a Wigner lattice (solid line). The size of $n(\mathbf{p}=0)$ depends on r_s in the latter case and becomes lower the larger r_s gets. (From [312])

the Fermi momentum p_F is reduced to $Z(p_F) < 1$ (see Fig. 7.3), where Z is the quasiparticle renormalization constant. A quasiparticle consists of a bare electron plus its correlation hole and $Z(p_F)$ is the weight of the bare electron in the quasiparticle. In the crystalline phase $Z=0$, i.e., there is no discontinuity in $n(\mathbf{p})$ and no Fermi surface. Neither are there quasiparticles. Instead $n(\mathbf{p})$ looks like shown in Fig. 3.2. The lower the electron density, the more the Coulomb repulsion dominates the kinetic energy. Hence the localization of electrons increases. Therefore, $n(\mathbf{p})$ approaches more and more a constant. This argument can be made more quantitative. In the low-density limit the only potential acting on an electron is the positive background charge within a sphere of radius r_s. Thus the potential energy is

$$V(r) = -\frac{e^2 r^2}{2r_s^3} + \text{const.} \qquad (3.42)$$

and the corresponding ground-state wavefunction is that of a harmonic oscillator, i.e.,

$$\psi(\mathbf{r}) = \left(\frac{\alpha}{\pi}\right)^{3/4} e^{-\frac{\alpha}{2} r^2} \qquad (3.43)$$

with $\alpha = r_s^{-3/2}$. With $\psi(\mathbf{r})$ known the momentum distribution function is obtained by a Fourier transformation of $n(\mathbf{r})$ [312], i.e.,

$$n(\mathbf{k}) = \frac{3\sqrt{\pi}}{r_s^{3/4}} \exp\left[-\left(\frac{9\pi}{4}\right)^{2/3} \frac{1}{\sqrt{r_s}} \left(\frac{k}{k_F}\right)^2\right]. \qquad (3.44)$$

Hereby it has been assumed that the orbitals (3.43) at different sites do not overlap with each other, which is justified when r_s is sufficiently large. Similarly one can determine the pair-distribution function (3.21) for a Wigner crystal. It differs considerably from that in Fig. 3.1 and resembles the one of an atomic lattice. A plot of $g(\mathbf{r}) = g(\mathbf{r}, 0)$ is shown in Fig. 3.3 for $r_s = 100$.

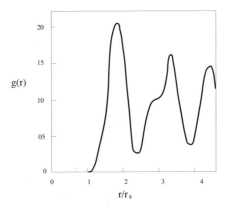

Fig. 3.3. Pair-distribution function $g(\mathbf{r})$ for a Wigner lattice with $r_s = 100$. (From [312])

An estimate of the critical value of r_s^c is obtained by applying Lindemann's criterion for melting. According to it a lattice becomes unstable when the zero-point fluctuations of the lattice vibrations have an averaged mean-square amplitude $\langle u^2 \rangle$ of the order of the lattice spacing c squared, i.e., $\langle u^2 \rangle^{1/2} = \gamma c$. Clearly r_s^c depends sensitively on the choice of γ. It also depends on the way $\langle u^2 \rangle$ is calculated, i.e., whether it is based on an harmonic approximation or includes anharmonicities. Therefore it is no surprise that estimates for r_s^c vary considerably, i.e., between 5 and 100. The average $\langle u^2 \rangle$ does not only depend on r_s but also an temperature. Here an estimate of the melting temperature T_m based on Lindemann's criterion gives

$$k_B T_m \simeq \frac{10^{-3}}{r_s} \text{a.u.} \quad . \tag{3.45}$$

The magnetic properties of a Wigner crystal depend on the value of r_s. For values $r_s \gtrsim 250$ a ferromagnetic ground state is expected caused by direct exchange. For smaller values of r_s antiferromagnetism is likely to occur as in the Hubbard model at half filling (see Sect. 8.2).

For an experimental observation of Wigner crystallization, two-dimensional (2D) systems like semiconducting heterojunctions or electrons on the surface of ^{4}He are particularly suitable. Therefore 2D Wigner lattices have been studied in special detail by numerical methods, in particular Monte-Carlo techniques. In the static limit, i.e., in the case of vanishing kinetic energy, a 2D electron system solidifies in a hexagonal (trigonal) lattice structure. Of all 2D lattices this one has the largest lattice spacing for a given electron density and therefore minimizes the mutual repulsions of the electrons. In the numerical calculations a variational wavefunction for electrons on a grid (L_x, L_y) is used of the form

$$\psi(\mathbf{r}_1, ..., \mathbf{r}_N) = \exp\left(\sum_{i<j} f(\mathbf{r}_i - \mathbf{r}_j)\right) \text{Det}\left[\phi_i(\mathbf{r}_j)\right] \quad . \tag{3.46}$$

3.3 Wigner Crystal

Here Det $[\phi_i(\mathbf{r}_j)]$ is a Slater determinant formed with plane-wave states in case of a liquid and with localized Gaussian orbitals in case of a solid. The prefactor is a Jastrow factor, which in the liquid phase keeps electrons better apart than does a state described by a Slater determinant. For example $f(\mathbf{r}_i - \mathbf{r}_j)$ can be chosen as

$$f(\mathbf{r}_i - \mathbf{r}_j) = -\alpha |\mathbf{r}_i - \mathbf{r}_j|^2 \quad . \tag{3.47}$$

One finds that the liquid becomes unstable for $r_s \gtrsim 40$. Yet the Wigner solid with localized wavefunctions $\phi_i(\mathbf{r}_j)$ is not the most stabile state either. This is seen by using in the liquid phase Bloch states on a hexagonal lattice instead of plane waves with $\mathbf{k}$ states limited to the first Brillouin zone. This implies a breaking of rotational symmetry. The static density-correlation function $S(\mathbf{r}, \mathbf{r}')$ (see (3.19)) has therefore no fully developed Bragg peaks yet. This symmetry broken liquid state may be termed a *Wigner liquid* and has in the regime $30 \lesssim r_s \lesssim 80$ a lower energy than both the electron liquid and the Wigner crystal. For $r_s \gtrsim 80$ the crystal has the lowest energy. The type of transition, i.e., true phase transition *vs.* simple cross-over from the symmetry broken hexadic phase to the crystalline one is unclear at present.

4
Density Functional Theory

Density functional theory has had a major impact on electronic-structure calculations. Thereby emphasis has been on ground-state properties of solids. It has given the calculations a sounder theoretical basis than they ever had before. Previously they depended to a considerable extent on model potentials. With density functional theory this is no longer the case. The theory has also been widely applied to energy-band calculations where its basis is, however, much less founded than for ground-state calculations. The combination of density functional theory with new and powerful linearized methods for solving self-consistent single-particle Schrödinger equations has led to an outburst of electronic structure work in condensed matter physics. The theory was developed by *Hohenberg, Kohn and Sham* [186, 246]. It was preceded by the work of *Slater*. His X_α method contained a number of important ideas which later entered density functional theory [418].

Density functional theory avoids the problem of calculating the many-electron ground-state wavefunction. Instead, ground-state properties – such as total energies, lattice constants and magnetic moments – are directly expressed in terms of the electronic density $\rho(\mathbf{r})$ or spin density $\rho_\sigma(\mathbf{r})$. A scheme is provided for calculating the latter from the solution of a single-particle Schrödinger equation with a self-consistent potential.

When the theory is applied to real solids, approximations to the general theoretical scheme are required. The most important one is the local-density approximation (LDA) which provides us with a simple, yet very successful potential. The Schrödinger-like equation (Kohn-Sham equation)) which has to be solved contains a *local* self-consistent potential instead of the *nonlocal* one entering the Hartree-Fock equations. So the numerical computations are much simpler than Hartree-Fock calculations. Yet the results are much better since correlation effects are partially included. The LDA potentials entering Schrödinger's equation are derived from a homogeneous electron gas. They can be improved by gradient corrections, i.e., corrections in which the gradients of the electronic density are included. When the LDA including gradient corrections fail, as it is the case when the electron correlations are strong, other kinds of improvements can be applied such as self-interaction corrections (SIC) or a LDA+U approximation scheme, where U denotes a local Coulomb interaction which must be estimated and put in by hand into the calculations. There exists also a time-dependent generalization of density functional theory, due to *Runge* and *Gross* [394], which has been applied in order to treat excited states.

Calculations based on the density functional have the advantage of a significant economy of computational expenses compared with calculations of a many-body wavefunction. At the same time we may expect only limited gain of insight into the electronic correlation problem. A detailed understanding of the correlated motion of electrons requires information which is contained in many-electron wavefunctions. For example, we would like to know to what extent electronic charge fluctuations are suppressed by correlations or how strongly spin fluctuations are enhanced. We would also like to understand to what extent Hund's rules are operative in a given system. This is relevant when the question is asked, in how good or bad an approximation the paramagnetic state, e.g., of iron may be described by a spin Hamiltonian in order to estimate the Curie temperature. There are also experimental techniques like Compton scattering which test the wavefunction of a system rather than the density. Therefore, despite all the successes of density functional theory, it is important not to neglect wavefunction-based methods.

4.1 Theory of Hohenberg, Kohn and Sham

Density functional theory is based on two theorems by *Hohenberg* and *Kohn*. The first one states that the ground state energy E of an interacting many-electron system in the presence of an external potential $V(\mathbf{r})$ is a functional of the electronic density $\rho(\mathbf{r})$. It can be written in the form

$$E_V[\rho] = \int d^3r \; V(\mathbf{r}) \; \rho(\mathbf{r}) + F[\rho] \quad , \tag{4.1}$$

where $F(\rho)$ is an unknown, yet universal functional of the density only, i.e., it does not depend on $V(\mathbf{r})$. The second theorem states that the ground-state density $\rho_0(\mathbf{r})$ minimizes $E_v[\rho]$. A particularly simple proof of the two theorems is due to *Levy* [280]. Consider a given density $\rho(\mathbf{r})$ and require that it be written as the expection value of an N-electron wavefunction $|\psi\rangle$. Furthermore, assume that there exists a set $S(\rho)$ of different wavefunctions, all yielding the same density $\rho(\mathbf{r})$. For any given operator A, a functional in terms of the density can be defined by requiring that

$$A[\rho] = \min_{|\psi\rangle \epsilon S(\rho)} \langle \psi | A | \psi \rangle \quad . \tag{4.2}$$

Stated differently, we select the particular wavefunction $|\psi\rangle$ contained in the set $S(\rho)$ which minimizes the expection value of A. This minimum value is then taken as the value of the observable A in the presence of the density $\rho(\mathbf{r})$. Equation (4.1) follows immediately, if A is identified with the sum of the kinetic and interaction energies of the electron, i.e.,

$$F = H_{\text{kin}} + H_{\text{int}}, \quad (4.3)$$
$$F[\rho] = \min_{|\psi\rangle \epsilon S(\rho)} \langle \psi | F | \psi \rangle \quad .$$

This relation refers neither to a specific system nor to any particular external potential $V(\mathbf{r})$ and is universal.

We will show now that the ground-state density $\rho_0(\mathbf{r})$ minimizes $E_V[\rho]$ and that this minimum value is the ground-state energy E_0. Let $|\psi_0\rangle$ denote the ground state wavefunction and $\rho_0(\mathbf{r})$ the density associated with it.

Furthermore, consider a density $\rho_1(\mathbf{r})$ and denote by $|\psi_1\rangle$ a wavefunction which yields $\rho_1(\mathbf{r})$ and is in addition that particular member of the set $S(\rho_1)$ which also gives $F[\rho_1]$, i.e.,

$$\langle \psi_1 | F | \psi_1 \rangle = F[\rho_1] \quad . \quad (4.4)$$

Then it follows that

$$\langle \psi_1 | F + V | \psi_1 \rangle \geq \langle \psi_0 | F + V | \psi_0 \rangle \quad (4.5)$$

since $|\psi_0\rangle$ is the ground state and $F + V = H$. From the definition (4.4) of F and $E_V[\rho]$, (see (4.1)), one obtains

$$E[\rho_1] \geq E_0 \quad (4.6)$$

because by definition $E_0 = \langle \psi_0 | H | \psi_0 \rangle$. From (4.6) it follows that for the particular density $\rho_1(\mathbf{r}) = \rho_0(\mathbf{r})$, the energy $E[\rho]$ is minimized and equal to the ground-state energy, i.e.,

$$E_V[\rho_0] = E_0(V) \quad . \quad (4.7)$$

The above arguments are easily generalized to spin-polarized systems, where the role of the density $\rho(\mathbf{r})$ is taken by the spin-density matrix

$$\rho_{\sigma\sigma'}(\mathbf{r}) = \langle \psi | \psi_\sigma^+(\mathbf{r}) \psi_{\sigma'}(\mathbf{r}) | \psi \rangle \quad . \quad (4.8)$$

Equation (4.18) is then replaced by

$$A[\rho_{\sigma\sigma'}] = \min_{|\psi\rangle \epsilon S(\rho_{\sigma\sigma'})} \langle \psi | A | \psi \rangle \quad , \quad (4.9)$$

where all wavefunctions $|\psi\rangle$ which yield a given spin-density matrix $\rho_{\sigma\sigma'}$ are included in $S(\rho_{\sigma\sigma'})$. Similarly, the ground-state energy is obtained from the ground-state spin-density matrix. The external potential may be spin dependent.

As pointed out before, $F[\rho]$ is a unique-though unknown-functional of the density $\rho(\mathbf{r})$. In order to apply the theory, approximations have to be made. Before describing them, we will show how $\rho(\mathbf{r})$ is obtained from the requirement that $E_V[\rho]$ be minimized. For that purpose $F[\rho]$ is divided into

$$F[\rho] = \frac{e^2}{2} \int d^3r\, d^3r'\, \frac{\rho(\mathbf{r})\rho(\mathbf{r}')}{|\mathbf{r}-\mathbf{r}'|} + T_0[\rho] + E_{xc}[\rho] \quad. \tag{4.10}$$

The first term describes the Coulomb repulsion of the electrons (Hartree-term). From the rest we single out the kinetic energy $T_0[\rho]$ of a system of *noninteracting* electrons with the same density $\rho(\mathbf{r})$ as the interacting one. What remains is $E_{xc}[\rho]$, which is usually called the exchange and correlation energy. It should be noted that $T_0[\rho]$ is *not* the true kinetic energy of the system, which would be hard to calculate owing to the many-body effects; instead, it is the kinetic energy of a fictitious, noninteracting system with the ground-state density $\rho(\mathbf{r})$. The part of the kinetic energy which is difficult to calculate is contained in $E_{xc}[\rho]$. It also includes the exchange and the remaining correlation energy. In order for $E_V[\rho]$ to be minimized, the density $\rho(\mathbf{r})$ must satisfy the variational equation

$$\int d^3r\, \delta\rho(\mathbf{r}) \left\{ V(\mathbf{r}) + e^2 \int d^3r'\, \frac{\rho(\mathbf{r}')}{|\mathbf{r}-\mathbf{r}'|} + \frac{\delta T_0[\rho]}{\delta \rho(\mathbf{r})} + \frac{\delta E_{xc}[\rho]}{\delta \rho(\mathbf{r})} \right\} = 0 \quad. \tag{4.11}$$

Since the total electron number is conserved, the variation $\delta\rho(\mathbf{r})$ has to fulfill the subsidiary condition

$$\int d^3r\, \delta\rho(\mathbf{r}) = 0 \quad. \tag{4.12}$$

The crucial point is that (4.11) is also the condition for a system of *noninteracting* electrons moving in an effective external potential

$$V_{\text{eff}}(\mathbf{r}) = V(\mathbf{r}) + e^2 \int d^3r'\, \frac{\rho(\mathbf{r}')}{|\mathbf{r}-\mathbf{r}'|} + v_{xc}(\mathbf{r}) \quad, \tag{4.13}$$

where the last contribution, the local exchange-correlation potential $v_{xc}(\mathbf{r})$ is defined through

$$v_{xc}(\mathbf{r}) = \frac{\delta E_{xc}[\rho]}{\delta \rho(\mathbf{r})} \quad. \tag{4.14}$$

The equivalence of (4.11) to that of a noninteracting electron system in an external potential $V_{\text{eff}}(\mathbf{r})$ has become possible because of the way the kinetic energy has been divided into $T_0[\rho]$ and a remaining part included in $E_{xc}[\rho]$. This division is a key point for making density functional theory a useful tool for practical calculations. It implies that $\rho(\mathbf{r})$ can be obtained by first solving a Schrödinger equation of the form

$$\left(-\frac{1}{2m}\nabla^2 + V_{\text{eff}}(\mathbf{r}) \right) \chi_\mu(\mathbf{r}) = \varepsilon_\mu \chi_\mu(\mathbf{r}) \quad, \tag{4.15}$$

and then determining it from

$$\rho(\mathbf{r}) = 2\sum_{\mu}^{N/2} |\chi_\mu(\mathbf{r})|^2 \quad . \tag{4.16}$$

The sum is over the eigenfunctions (Kohn-Sham orbitals) with the lowest eigenvalues. Since $V_{\text{eff}}(\mathbf{r})$ depends on $\rho(\mathbf{r})$, the equations (4.15), (4.16) must be solved self-consistently. They are often referred to as Kohn-Sham equations. Note that the $\chi_\mu(\mathbf{r})$ should *not* be used to construct a ground-state wavefunction, e.g., in form of a Slater determinant or else. Within the frame of density functional theory nothing can be said about the form of the total wavefunction.

The real eigenvalues ϵ_μ do *not* describe electronic excitation energies which are generally complex quantities due to lifetime effects. But for metallic infinite systems with extended states the energy of the highest occupied level $\epsilon_{N/2}$ turns out to be equal to the chemical potential μ. The Fermi energy is therefore correctly reproduced by density functional theory. A similar statement cannot be made as regards the shape of the Fermi surface. Despite this the ϵ_μ, or better the $\epsilon_\nu(\mathbf{k})$ where ν is a band index and $\mathbf{k}$ is the momentum of a Bloch state, are often successfully interpreted as energy bands of a solid. This has contributed significantly to the wide application of density functional theory.

The complexity of the many-electron problem is contained in the unknown exchange-correlation potential $v_{xc}(\mathbf{r})$. By making reasonable approximations for it we can hope to deal in a simple way with this highly complex problem. Indeed, the simplest possible approximation, the local-density approximation, has proven very successful. There are a number of good textbooks on density functional theory available see, e.g., [98, 110].

4.2 Local-Density Approximation and Extensions

In the local-density approximation (LDA) the exchange-correlation energy $E_{xc}[\rho]$ is replaced by

$$E_{xc}[\rho] = \int d^3 r \rho(\mathbf{r}) \varepsilon_{xc}(\rho(\mathbf{r})) \quad , \tag{4.17}$$

where $\epsilon_{xc}(\rho(\mathbf{r}))$ is the sum of the exchange and correlation energy per electron of a homogeneous electron gas of density ρ. This quantity can be calculated with good accuracy for a wide range of densities and therefore is considered to be known. Then

$$v_{xc}(\mathbf{r}) = \frac{d(\rho(\mathbf{r})\varepsilon_{xc}(\rho(\mathbf{r})))}{d\rho(\mathbf{r})} \tag{4.18}$$

is a direct function of $\rho(\mathbf{r})$ and the Schrödinger equation (4.15) becomes much easier to solve than the Hartree-Fock equation with the nonlocal exchange potential (see (2.33)). At the same time (4.15) contains correlations because of $v_{xc}(\mathbf{r})$ entering it. The results are therefore superior to those of an independent electron or Hartree-Fock approximation.

For magnetic ground states a spin-dependent generalization of (4.15, 4.16) is required. We obtain it by imposing a local spin-density approximation (LSDA). Starting from (4.8, 4.9) and repeating the steps which led to (4.15) we recover an analogous equation in (2 x 2) matrix form. The same holds true for (4.16). The LSDA employs the fact that exchange and correlation energy of a homogeneous electron system in the presence of an applied uniform magnetic field $\mathbf{H}$ depend only on the density and on the magnetization, i.e., on the spin density parallel to the field. Therefore a convenient choice for the two quantities are the spin densities $\rho_\uparrow$ and $\rho_\downarrow$ with $\rho = \rho_\uparrow + \rho_\downarrow$. The energy $\epsilon_{xc}(\rho(\mathbf{r}))$ in (4.17) is therefore replaced by $\epsilon_{xc}(\rho_\uparrow(\mathbf{r}), \rho_\downarrow(\mathbf{r}))$ and the 2 x 2 matrix equation reduces to the coupled equations

$$\left(-\frac{1}{2m}\nabla^2 - \mu_B \boldsymbol{\sigma} \cdot \mathbf{H}(\mathbf{r}) + V_\sigma^{\text{eff}}(\mathbf{r})\right) \phi_{\mu\sigma}(\mathbf{r}) = \varepsilon_{\mu\sigma} \phi_{\mu\sigma}(\mathbf{r}) \;, \quad (4.19)$$

which include a Zeeman term. The spin-dependent effective single-particle potential is given by

$$V_\sigma^{\text{eff}}(\mathbf{r}) = V(\mathbf{r}) + e^2 \int d^3 r' \frac{\rho(\mathbf{r}')}{|\mathbf{r} - \mathbf{r}'|} + v_\sigma^{xc}(\mathbf{r}) \quad (4.20)$$

with

$$v_\sigma^{xc}(\mathbf{r}) = \frac{d}{d\rho_\sigma(\mathbf{r})} \left\{ [\rho_\uparrow(\mathbf{r}) + \rho_\downarrow(\mathbf{r})] \, \varepsilon_{xc}(\rho_\uparrow(\mathbf{r}), \rho_\downarrow(\mathbf{r})) \right\} \;. \quad (4.21)$$

The spin densities are obtained from the functions $\phi_{\mu\epsilon}(\mathbf{r})$ via

$$\rho_\sigma(\mathbf{r}) = \sum_\mu^{\text{occ}} |\phi_{\mu\sigma}(\mathbf{r})|^2 \;, \quad (4.22)$$

where the sum is over all occupied orbitals with spin σ.

As pointed out above, we use for the exchange-correlation potential $v^{xc}(\mathbf{r})$ the one of a homogeneous electron gas for a spin density ρ_σ. In Sect. 3.1 we have discussed the exchange energy of a homogeneous system (see (3.10)), and in Sect. 3.2 the leading correlation contributions to the ground-state energy have been pointed out. Here it suffices to simply state the results for v_σ^{xc} in a parameterized form with a proper inclusion of the magnetization. For that purpose we introduce two dimensionless parameters, which specify the homogeneous gas. One is r_S defined in (3.12). It is related to the density through

$$\frac{1}{r_S} = a_B \left(\frac{4\pi}{3}\rho\right)^{\frac{1}{3}} \;. \quad (4.23)$$

4.2 Local-Density Approximation and Extensions

The second parameter is $m_S = (\rho_\uparrow - \rho_\downarrow)/\rho$ and characterizes the spin polarization of the system. In terms of r_S and m_S the potential v_σ^{xc} can be approximated by

$$v_\sigma^{xc} = -\frac{0.611}{r_S}\left(\beta(r_S) + \frac{1}{3}\frac{\delta(r_S)\,m_S\sigma}{1+0.297 m_S\sigma}\right) , \qquad (4.24)$$

where $\sigma = \pm 1$ and

$$\beta(r_S) = 1 + 0.0545 r_S \ln\left(1 + \frac{11.4}{r_S}\right)$$

$$\delta(r_S) = 1 - 0.036 r_S - \frac{1.36 r_S}{(1+10 r_S)} . \qquad (4.25)$$

This special parameterization goes back to [161]. Other parameterized forms have been used too, but the uncertainties caused by the different choices remain smaller than the ones generated by the LSD approximation itself.

From the above it seems clear that the LDA will work better, the less the electronic density changes in the unit cell. If the density changes are large, we must provide for additional density-gradient corrections [268, 269]. Instead of the density we may also use the Fermi momentum $k_F(\mathbf{r}) = (3\pi^2\rho(\mathbf{r}))^{\frac{1}{3}}$ as a variable for spin-unpolarized systems. The generalized gradient corrections are taken into account by writing

$$E_{xc}[k_F] = E_{xc}^{\text{LDA}}[k_F] + \delta E_{\text{GG}}[k_F] , \qquad (4.26)$$

where $E_{xc}^{\text{LDA}}[k_F]$ is the LDA contribution. In order to determine $\delta E_{\text{GG}}(k_F)$ it proves useful to introduce a dimensionless measure of the density gradient

$$\lambda(\mathbf{r}) = \frac{1}{2}\frac{|\nabla\rho(\mathbf{r})|}{\rho(\mathbf{r}) k_F(\mathbf{r})} . \qquad (4.27)$$

Even in metals $\lambda(\mathbf{r})$ can be appreciable. This is seen in Fig. 4.1, where the density gradient in bulk Cu is shown.

The gradient correction $\delta E_{\text{GG}}[k_F]$ is of the general form

$$\delta E_{\text{GG}}[k_F] = \int d^3 r\, F(k_F(\mathbf{r}), \lambda(\mathbf{r})) . \qquad (4.28)$$

Several expressions have been suggested for the function $F(k_F, \lambda)$ [144]. An especially successful one is

$$F(k_F, \lambda) = \frac{e^2}{18\pi^3}\lambda^2\left(e^{-b\lambda k_F} - \frac{7}{18}\right) , \qquad (4.29)$$

where $b \simeq 1.2$. For arguments why this form suggests itself we refer to the original literature.

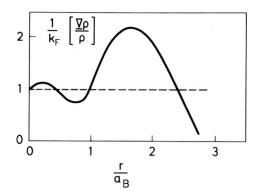

Fig. 4.1. Density gradient in bulk Cu. (From [160])

Another way of comparing different generalized gradient approximations (GGA) with respect to each other is by defining an exchange-correlation enhancement factor $F_{xc}(r_S(\mathbf{r}), \lambda(\mathbf{r}))$ through

$$E_{xc}^{GGA}[k_F] = \int d^3 r \rho(\mathbf{r}) \epsilon_x(\rho(\mathbf{r})) F_{xc}(r_S(\mathbf{r}), \lambda(\mathbf{r})) \;, \qquad (4.30)$$

where $\epsilon_x(\rho)$ is the exchange-energy per electron of a homogeneous electron gas of density ρ (see (3.10)). Successful functional dependences of $F_{xc}(r_S, \lambda)$ have been constructed by *Perdew, Wang* and *Becke* [24, 25, 361, 362]. An especially popular one is called PW 91 [361].

Generalized gradient corrections improve considerably the quality of ground-state calculations for solids. One noticeable improvement is a correct ferromagnetic ground state for bcc Fe which LDA does not reproduce. But some problems do remain, e.g., it is difficult to produce for FeO or CoO an insulating antiferromagnetic ground state. The differences of various forms of $E_{xc}^{GGA}[k_F]$ are often of similar size as those between $E_{xc}[\rho]$ and $E_{xc}^{GGA}[k_F]$. For a critical evaluation of gradient corrections see, e.g., Ref. [119]. The cohesive energy of silicon is found to be 4.64 eV per unit cell when the PW 91 version is used as compared with 5.35 eV within LDA. The experimental value is 4.63 eV. The almost exact agreement for silicon is somewhat fortuitous, though. The corresponding lattice constants are $a = 5.37 Å$ (LDA) and $a = 5.59 Å$ (PW 91) while the experimental value is $a_{exp} = 5.43 Å$.

As pointed out in Sect. 2.3, the Hartree-Fock equations do not contain any unphysical self-interactions because the contributions from the charge density and exchange density to the potential energy cancel each other in an orbit by orbital basis (see (2.33)). The same cancellation takes place in (4.10) if we use the correct energy $E_{xc}[\rho]$. However, when an approximation is made like the LDA or LSDA, the cancellation is incomplete. As a result an electron in a H atom provides an unwanted self-interaction contribution to the energy. Self-interactions can be avoided by a self-interaction correction (SIC) to the local spin-density approximation [363]. We write for the corrected exchange-correlation energy functional

$$E_{xc}^{SIC} = E_{xc}^{LSD}\left[\rho_\uparrow(\mathbf{r}), \rho_\downarrow(\mathbf{r})\right] - \sum_{i\sigma} \Delta_{i\sigma}$$

$$\Delta_{i\sigma} = \frac{e^2}{2} \sum_{i\sigma} \int d^3r\, d^3r'\, \frac{\rho_{i\sigma}(\mathbf{r})\rho_{i\sigma}(\mathbf{r'})}{|\mathbf{r}-\mathbf{r'}|} + \sum_{i\sigma} E_{xc}^{LSD}\left[\rho_{i\sigma}, 0\right] \quad . \quad (4.31)$$

The SIC consists of subtracting two terms from E^{LSD}. One is the Coulomb self-interaction of an electron in an orbital i with orbital density $\rho_{i\sigma}(\mathbf{r})$. The second term is the exchange-correlation energy of that electron when the fully polarized LSD expression for that orbital is used. Equation (4.31) ensures that a single electron, e.g., in the H atom does not interact with itself, since the exchange-correlation energy of a single, fully occupied spin orbital cancels exactly the self-direct Coulomb interaction.

The SIC is a useful concept only when the spin orbitals $\phi_{i\sigma}(\mathbf{r})$ are localized and are not Bloch states, for example. It shifts the Kohn-Sham orbital energies of localized orbitals downwards. This shift can be as large as 10 eV, e.g., for $4f$ electrons, which act then like core states.

4.3 Strong Electron Correlations: LDA+U

The LDA and its spin dependent version LSDA have many merits and had many successes. But as pointed out before, they can fail too, in particular when dealing with strongly correlated electrons. Examples are the insulators FeO and CoO which come out metallic when the LSDA is applied. Other examples are systems with heavy quasiparticles. The effective masses calculated within the LDA disagree with experiments sometimes by more than a factor of ten. For a molecular-field theory which the LDA is, this is not unexpected. Charge fluctuations are grossly overrated by a mean-field treatment when the Coulomb repulsion exceeds the kinetic-energy gain due to hybridization. We know this from the SCF approximations, the simplest version of a mean-field theory.

We want to point out the physical origin of these difficulties and discuss a phenomenological extension of the LDA, namely LDA+U, which partially avoids them. A second, quite different extension, i.e., renormalized band theory is discussed in Sect. 13.1.1.

In order to understand the difficulties just mentioned we need to get ahead of ourselves and use some of the results derived and discussed in later chapters. Consider again a homogeneous electron system. The exact electron excitation energies $\epsilon_\mathbf{k}^{ex}$ are given by the implicit equation

$$\varepsilon_\mathbf{k}^{ex} = \frac{k^2}{2m} + \Sigma(\mathbf{k}, \varepsilon_\mathbf{k}^{ex}) \quad , \quad (4.32)$$

where $\Sigma(\mathbf{k},\omega)$ is the wavenumber- and frequency-dependent electron self-energy discussed in Sect. 7.1.

In the LDA the eigenvalues of (4.15) are of the form $\epsilon_{\mathbf{k}} = k^2/2m + \text{const.}$ because for a homogeneous electron gas $V_{\text{eff}}(\mathbf{r}) = \text{const}$. Being a ground-state theory, density function theory and, hence, (4.15) must correctly describe the Fermi energy ϵ_F. The latter is just the difference between the ground-state energy of an (N+1)- and an N-electron system. This fixes the constant in the expression for $\epsilon_{\mathbf{k}}$, which becomes $\Sigma(k_F, \epsilon_F)$. The exact expression for the excitation energy differs therefore from the eigenvalues $\epsilon_{\mathbf{k}}$ of the Kohn-Sham equation (4.15) by

$$\varepsilon_{\mathbf{k}}^{ex} - \varepsilon_{\mathbf{k}} = \Sigma(\mathbf{k}, \varepsilon_{\mathbf{k}}^{ex}) - \Sigma(k_F, \varepsilon_F) \quad . \tag{4.33}$$

As long as the self-energy $\Sigma(\mathbf{k},\omega)$ varies sufficiently slowly with $\mathbf{k}$ and ω, one may identify the eigenvalues $\epsilon_{\mathbf{k}}$ with the excitation energies. This is seen as follows. By expanding $\Sigma(\mathbf{k},\omega)$ we may rewrite (4.32) in the form

$$\omega = \frac{k^2 - k_F^2}{2m} + \left(\frac{\partial \Sigma}{\partial k}\right)_{k=k_F}(k - k_F) + \left(\frac{\partial \Sigma}{\partial \omega}\right)_{\omega=0} \omega \quad , \tag{4.34}$$

where ω is counted from the Fermi energy ϵ_F. The requirement that $\Sigma(\mathbf{k},\omega)$ changes sufficient slowly with $\mathbf{k}$ and ω can be recast into a requirement for the effective mass m^* associated with the electronic excitations. In the effective mass approximation, the ansatz $\omega = (k^2 - k_F^2)/2m^*$ is made for the excitation energies. From (4.34) it follows that the effective mass is given by

$$\frac{m^*}{m} = \frac{1 - (\partial \Sigma/\partial \omega)_{\omega=0}}{1 + (m/k_F)(\partial \Sigma/\partial k)_{k=k_F}} \quad . \tag{4.35}$$

Therefore requiring that $\epsilon_{\mathbf{k}}^{ex} \simeq \epsilon_{\mathbf{k}}$ is well described by LDA is equivalent to the requirement that $m^*/m \simeq 1$. When the homogeneous electron system is replaced by an inhomogeneous one, a similar result can be derived except that m is replaced by the electron mass m_b in the presence of an external (periodic) potential, i.e., the band mass. In systems with heavy quasiparticles (see Chap. 13) $m^*/m_b \gg 1$ and therefore a LDA must fail.

One way of improving the situation for d- or f-electron systems is to extend the LDA method to one called LDA + U [11]. The central idea of that extension is to push the occupied part of the d or f shell downwards in energy, and to make sure that an additional electron added to the $d(f)$ shell has an energy much higher than ϵ_F. This shift is due to an effective repulsive energy U_{eff} with the other $d(f)$ electrons.

The basic idea behind LDA+U is to treat the strong correlations of d or f electrons more accurately than in LDA. This is done by simplifying the Coulomb repulsion between two electrons on an atomic site to a single Coulomb integral U.

4.3 Strong Electron Correlations: LDA+U

The intention of LDA + U is to combine LDA calculations with an improved treatment of the on-site interaction U. In order to derive an appropriate value for the model parameter U, we use that n electrons at a given site and repelling each other with energy U have a total repulsive energy $E(n) = (U/2)n(n-1)$. This implies that

$$U = E(n+1) + E(n-1) - 2E(n) \quad . \tag{4.36}$$

One may use the LDA in order to determine the $E(n)$ and from them the screened interaction U. For that purpose one sets, e.g., the hybridization matrix elements of the atomic-like d orbitals with the surroundings, equal to zero. By keeping the d electron number n_d fixed, one allows the other electrons to relax self-consistently. By varying n_d the ground-states energies $E(n_d)$ are calculated within the LDA and U is obtained from $U = d^2 E(n_d)/dn_d^2$. For transition metal ions typical values of U vary from 6 - 8 eV.

In a generalized version of the above model one includes the spin dependence of the on-site interactions. The repulsion energy of two electrons on a given site differs by the exchange energy J depending on whether the two spins are parallel or antiparallel. Note that J is usually of order 1 - 2 eV and therefore much smaller than U. The interaction energy for site ℓ reads

$$E_{\text{int}}(\ell) = \frac{1}{2}\sum_{i,j,\sigma} U n_{i\sigma}(\ell) n_{j-\sigma}(\ell) + \frac{1}{2}\sum_{i\neq j\sigma}(U-J)n_{i\sigma}(\ell)n_{j\sigma}(\ell) \quad . \tag{4.37}$$

The LDA is an orbital-independent molecular-field approximation. It replaces the occupation numbers $n_{i\sigma}(\ell)$ by the average occupation $n_0(\ell) = n_d(\ell)/10$, where

$$n_d(\ell) = \sum_{i\sigma} n_{i\sigma}(\ell) \quad . \tag{4.38}$$

By using (4.37) the total energy is written as

$$E = E_{\text{LDA}} + \frac{U}{2}\sum_{\ell i j \sigma}\delta n_{i\sigma}(\ell)\delta n_{j-\sigma}(\ell) + \frac{(U-J)}{2}\sum_{\ell i(\neq j)\sigma}\delta n_{i\sigma}(\ell)\delta n_{j\sigma}(\ell), \tag{4.39}$$

where E_{LDA} is the total energy in LDA and $\delta n_{i\sigma}(\ell) = n_{i\sigma}(\ell) - n_0(\ell)$. The potential which enters the Kohn-Sham equation is obtained from $\delta E/\delta n_{i\sigma}(\ell)$ as

$$V^{\text{eff}}_{i\sigma}(\ell) = V_{\text{LDA}} + U\sum_j \delta n_{j-\sigma}(\ell) + (U-J)\sum_{j\neq i}\delta n_{j\sigma}(\ell) \quad . \tag{4.40}$$

The LDA potential V_{LDA} refers to a charge density with $n_d(\ell)$ d electrons. The last two equations show that results different from LDA are expected only in the case of different spin- or orbital occupancies, i.e., when the $\delta n_{j\sigma} \neq 0$. Otherwise LDA+U in its present form reduces again to LDA. The above scheme can be improved by replacing $n_0(\ell)$ by its spin dependent components $n_{0\sigma}(\ell) = \frac{1}{5}\sum_i n_{i\sigma}(\ell)$, i.e., the average is taken separately for the two spin components. In that case one starts from the LSDA instead of the LDA in order to be consistent. One notices that the Coulomb energy $U \gg J$ enters the spin-dependent potential $V_{\text{eff}}^{\sigma}(\ell)$, i.e., the deviations from LDA.

The difference between LDA and LDA+U is seen most easily by restricting oneself to one orbital and one electron per site. In that case (4.40) becomes

$$V_{\sigma}^{\text{eff}}(\ell) = V_{\text{LDA}} + U\delta n_{-\sigma}(\ell)$$
$$= V_{\text{LDA}} + U\left(\frac{1}{2} - n_{\sigma}(\ell)\right) , \qquad (4.41)$$

where $n_{\sigma}(\ell) + n_{-\sigma}(\ell) = 1$ and $n_0(\ell) = 1/2$. Depending on the occupation number $n_{\sigma}(\ell)$, the potential $V_{\sigma}(\ell)$ is shifted by an amount, which is varying from $U/2$ for $n_{\sigma}(\ell) = 0$ to $-U/2$ for $n_{\sigma}(\ell) = 1$. Thus the Coulomb repulsion favors an unequal occupancy of the two spin orbitals. There is a potential barrier U for occupying the orbital with a spin $-\sigma$ electron when a spin σ electron is already present. Double occupancies of sites are therefore strongly suppressed when U is large as compared with the hybridization matrix elements. This provides an explanation for why CoO is an insulator and not a metal as the LDA suggests. For a review of the LDA+U see, e.g., Ref. [8].

A quite different way of coping with the strong correlation problem within density functional theory is provided by *renormalized band theory*. The approach differs fundamentally from the LDA+U one and is particularly useful for heavy quasiparticle systems. It is based on the experimental observation that systems like $CeRu_2Si_2$ or UPt_3 are Fermi liquids at low temperatures with the f electron participating in the formation of the Fermi surface. Therefore a calculation aimed at determining the heavy quasiparticle energy bands must ensure that f electron-like excitations remain close to ϵ_F. In order to prevent a strong mixing with the other electrons, the hybridization matrix elements of the f orbitals with the surrounding neighborhood must be strongly renormalized. This contrasts with the LDA+U approach which reduces the hybridization by downshifting the occupied f states. The renormalized band structure approach has been very successful. The method is discussed in Sect. 13.1.1 in connection with heavy quasiparticles.

4.4 The Energy Gap Problem

Although density functional theory has been designed for the ground-state properties of a system, the orbital energies of the Kohn-Sham equations (4.15) are often identified with the energy bands of a solid. This has worked remarkably well in many cases, but has led also to some serious problems. One of these is the energy gap problem. It is well known that energy gaps of semiconductors and insulators are considerably overestimated when the SCF or Hartree-Fock approximation is made. This is intuitively obvious because, when we add (or remove) an electron from a system, the immediate neighborhood

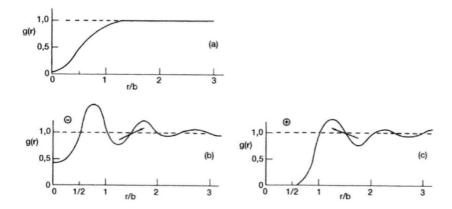

Fig. 4.2. Schematic comparison of the pair-distribution function $g(0,\mathbf{r})$ for a semiconductor or insulator (a) in the ground state of the system, (b) when an extra electron is added to the conduction band at point 0, and (c) when a hole is added at 0. The oscillatory behavior in (b) and (c) is a result of the polarization which the extra electron (hole) generates around itself. It decreases as $|\mathbf{r}|^{-2}$. The vector $\mathbf{r}$ follows a bond sequence in the semiconductor and b is the bond length. (From [187])

of that particle will respond to that change and lower the energy required for it. This is an effect of electron correlations and therefore not contained in a SCF approximation. As a result the energy which is necessary for moving an electron from the valence to the conduction band is overestimated in that approximation. On the other hand, when the eigenvalues of the Kohn-Sham equations are identified with the energy bands of a solid, energy gaps of semiconductors and insulators come out usually much too small. As discussed before, one may even find that within the LDA or LSD approximation an insulator or semiconductor shows metallic behavior, i.e., it does not have an energy gap in the excitation spectrum. In the following we want to offer a simple physical argument, why we cannot expect to obtain right energy gaps from the LDA or derivatives of it.

Consider the ground state of a semiconductor like silicon or germanium. The correlations are here of the van der Waals type. A charge fluctuation in a given bond results in a fluctuating electric dipole which induces dipoles in the other bonds. The corresponding correlation energy falls off rapidly at large distances (see Sect. 6.1). The pair-distribution function (3.21) looks in this case qualitatively as drawn in Fig. 4.2(a). Next let us add an electron to the system by putting it into the lowest-energy state of the conduction band. The resulting ground state of the (N+1)-electron system is charged and the added electron polarizes the bonds in its neighborhood. The pair-distribution function with the added electron at the origin must reflect this polarization and therefore differs considerably from the one shown in Fig. 4.2(a). In the neighboring bonds, charge is moving away from the added electron and with two electrons in each bond the pair-distribution function oscillates as indicated in Fig. 4.2(b). The oscillations fall off like $|\mathbf{r}|^{-2}$ because the electric field set up by the extra electron varies like $|\mathbf{E}(\mathbf{r})| = e/\epsilon r^2$, where ϵ is the dielectric constant. A similar situation prevails when an electron is removed from the top of the valence band, i.e., when we deal with the ground state of the (N-1)-electron system. The generated hole attracts charge from the neighboring bonds and the resulting pair-distribution function looks qualitatively as indicated in Fig. 4.2(c). The special features of the different pair-distribution functions are reflected in the ground-state energy of the (N-1)-, N-, and (N+1)-electron systems. Those findings have a strong effect on the size of the energy gap. The response of the system to the addition of an electron or a hole lowers the energy, which is required for that process as compared with the corresponding Hartree-Fock value. On the other hand, the *density* of an infinite system remains unchanged when one electron is added or removed, and so does $V_\sigma^{\text{eff}}(\mathbf{r})$ of the LDA, see (4.13). Therefore, it comes as no surprise, that the LDA fails to reproduce correctly energy gaps in semiconductors and insulators. The changes in the pair distribution function shown in Fig. 4.2 cannot be accounted for by the LDA. It is also apparent that there is no simple way of resolving this inherent difficulty of a method into which only the *density* of a system enters.

One way of improving the computed energy gaps is by using Green's functions and by applying the so-called $G * W$ approximation. Here $G(\mathbf{r}, \mathbf{r}', \omega)$ stands for Green's function and $W(\mathbf{r}, \mathbf{r}', \omega)$ for a dynamically screened Coulomb interaction. The $*$ indicates a convolution of the two functions. Finding simple forms for both is essential for applying that approximation. Constructing the Green's function, a subject dealt with in Chapter 7, we face no problems since the Kohn-Sham eigenfunctions may be used for that purpose. Yet the dynamically screened Coulomb interaction requires the knowledge of the nonlocal inverse dielectric function ϵ^{-1}, i.e.,

$$W(\mathbf{r}, \mathbf{r}', \omega) = \frac{e^2}{\Omega} \int d^3 r'' \epsilon^{-1}(\mathbf{r}, \mathbf{r}'', \omega) \cdot \frac{1}{|\mathbf{r}' - \mathbf{r}''|} \quad . \tag{4.42}$$

In momentum space W and ϵ^{-1} are matrices with respect to the reciprocal lattice vectors $\mathbf{G}, \mathbf{G}'$. The dielectric matrix $\epsilon_{\mathbf{G}\mathbf{G}'}(\mathbf{q}, \omega)$ can be calculated without too severe problems [117] and by numerical inversion of the matrix for each value of $\mathbf{q}$ and ω the screened interaction can be found. However the convolution of G and W requires an additional ω integration and that is difficult to achieve. Therefore one usually tries to approximate the matrix $\epsilon_{\mathbf{G}\mathbf{G}'}^{-1}(\mathbf{q}, \omega)$ by a form which allows for an analytic ω integration. When this is done we obtain much improved energy gaps for semiconductors, the reason being that the long-ranged polarization cloud of an extra electron or hole is described quite well by $1/(\epsilon r)$. Also the exchange part is improved as compared with LDA. By replacing $\epsilon_{\mathbf{G}\mathbf{G}}^{-1}(\mathbf{q}, \omega)$ with the unity matrix, all correlation effects are turned off and we obtain the non-local exchange. While this is very gratifying, there remains the problem of treating in a controlled way the short-range relaxation and polarization part. For that purpose one must work in r-space, instead of k-space, e.g., by working with GTO's. Like most of Green's function methods based on Feynman diagrams, the GW method uses k-space. That makes it almost impossible to make controlled approximations for the short range part of the correlations hole. Fortunately these deficiencies don't seem to play a major role when one compares the GW results with experiments.

The above is a sketchy outline of the essence of the method. A more detailed discussion is beyond the scope of this introductory book and we have to refer for it to the original literature [175, 176, 198, 199] or to [18].

4.5 Time-Dependent DFT

Time-dependent density function theory (TDDFT) is the endeavor to extend the successful, stationary DFT to time dependent processes. It was an important achievement of *Runge* and *Gross* that they were able to formulate a theorem considered to be the time-dependent analogue of the Hohenberg-Kohn theorem. The arguments involved are the following.

Instead of dealing with the external potential $V(\mathbf{r})$ in (4.1) we assume here a time-dependent external potential $V(\mathbf{r},t)$. It gives rise to a potential operator

$$V_{\text{ext}}(t) = \int d^3 r V(\mathbf{r},t)\rho(\mathbf{r}) \quad, \tag{4.43}$$

which enters the Schrödinger equation

$$i\frac{\partial \Phi(t)}{\partial t} = \tilde{H}(t)\Phi(t) \quad, \quad \Phi(t_0) = \Phi_0 \tag{4.44}$$

of the interacting electron system. Note that $\hat{\rho}(\mathbf{r})$ is given by (3.20). Except for the time-dependent external potential, the Hamiltonian is the same as the one used in (2.2), i.e., $\tilde{H}(t) = H + V_{\text{ext}}(t)$. Let us assume that by starting from a given initial state Φ_0, e.g., the ground state of H, equation (4.44) has been solved for different external potentials $V_{\text{ext}}(t)$. That provides for a map between $V(\mathbf{r},t)$ and $\Phi(t)$. For a given $\Phi(t)$ we may calculate the density

$$\rho(\mathbf{r},t) = \langle \Phi(t)|\hat{\rho}(\mathbf{r})|\Phi(t)\rangle \quad, \tag{4.45}$$

This gives us a map between $V(\mathbf{r},t)$ and $\rho(\mathbf{r},t)$. The Runge-Gross theorem specifies under which circumstances this map can be inverted. One notices immediately that two external potentials which differ by a function $C(t)$ map to the same density $\rho(\mathbf{r},t)$ as long as this function depends on time only and not on space. This is seen as follows. The effect of $C(t)$ is an additional multiplicative phase factor $e^{-i\alpha(t)}$ on the wavefunction with the result that $\tilde{\Phi}(t) = e^{-i\alpha(t)}\Phi(t)$. Here $\alpha(t)$ satisfies the equation $d\alpha/dt = C(t)$. Thus $\Phi(t)$

and $\tilde{\Phi}(t)$ give the same density and an inversion of the map is possible only up to a function $C(t)$. Note that $\langle \Phi(t)|i\frac{\partial}{\partial t} - \tilde{H}(t)|\Phi(t)\rangle$ is independent of $C(t)$. When $C(t)$ is added to $V_{\text{ext}}(t)$, it is canceled by the time derivative of $\alpha(t)$. A tacit assumption is that we are excluding from the beginning densities $\rho(\mathbf{r}, t)$ which do not correspond to any interacting electron system in an external potential $V(\mathbf{r}, t)$. An example is a discontinuous density. In other words, $\rho(\mathbf{r}, t)$ must be V-representable. It is also required that the potential $V(\mathbf{r}, t)$ can be Taylor expanded around $t = t_0$. We do not reproduce the proof of the map inversion here, but instead refer to the original literature [394]. The proof is restricted to finite systems.

While the ground state can be obtained by minimizing the total energy, this loses its meaning when we deal with time-dependent potentials. In that case, the energy is no longer conserved. Therefore, the determination of $\rho(\mathbf{r}, t)$ requires another starting point. One possibility is to consider the action

$$A = \int_{t_0}^{t} dt \left\langle \Phi(t) \left| i\frac{\partial}{\partial t} - \tilde{H}(t) \right| \Phi(t) \right\rangle \quad (4.46)$$

and to search for its stationary point. This is in analogy to classical mechanics, where the time evolution of a system is determined by the action $\int_{t_0}^{t} dt' L(t')$ with $L(t')$ denoting the Lagrangian. Note that the Schrödinger equation (4.44) follows from $\delta A/\delta \langle \Phi(t)|$. Indeed, it can be shown that the exact density $\rho(\mathbf{r}, t)$ is the one of a stationary point of (4.46) [394]. This proof is the equivalent of the one in DFT, which states that the ground-state density minimizes $E_V[\rho]$ (see Sect. 4.1).

The Runge-Gross theorem is far from being obvious. One should keep in mind that an electron system responds with retardation to an external time-dependent perturbation. Therefore, the potential at time t has an effect on $\rho(\mathbf{r}, t')$ at a later time, i.e., for $t' > t$. This effect must be correctly anticipated, since at t' the density $\rho(\mathbf{r}, t')$ is fixed by $V(t')$. Therefore it is not too surprising that response functions calculated from (4.46) suffer from violation of causality. However, the problem can be circumvented by making use of the Keldysh formalism [234]. To show this in detail is beyond the scope of this book as no additional insight is obtained into electron correlations.

The above-mentioned mapping, together with the fact that the matrix element $\langle \Phi|i\partial/\partial t - \tilde{H}|\Phi\rangle$ is independent of $C(t)$ implies that the action $A[\Phi]$ can be written as a functional of the density $A[\rho]$. In analogy to (4.1) we split the action into

$$A[\rho] = B[\rho] - \int_{t_0}^{t} dt' \int d^3 r \rho(\mathbf{r}, t') V(\mathbf{r}, t') \quad , \quad (4.47)$$

where

$$B[\rho] = \int_{t_0}^{t} dt' \left\langle \Phi([\rho], t') \left| i\frac{\partial}{\partial t'} - H \right| \Phi([\rho], t') \right\rangle \quad . \quad (4.48)$$

4.5 Time-Dependent DFT

The wavefunction $\Phi([\rho], t)$ is the one when $C(t) = 0$. This specification is necessary since H and not $\tilde{H}$ enters $B[\rho]$.

Note that as a consequence of the Runge-Gross theorem $B[\rho]$ is a universal functional of $\rho(\mathbf{r}, t)$, i.e., its form is independent of the external potential $V(\mathbf{r}, t)$. This is a simple consequence of the definition (4.48). The correspondence between $B[\rho]$ and $F[\rho]$ (see (4.1)) is apparent. The exact density $\rho(\mathbf{r}, t)$ is obtained from the stationary point of $A[\rho]$, i.e., from

$$\frac{\delta A[\rho]}{\delta \rho(\mathbf{r}, t)} = 0 \ . \tag{4.49}$$

For practical applications of the method we want to derive equations which are the time-dependent generalization of the Kohn-Sham equation. The role of $F[\rho]$ is here taken by $B[\rho]$. By splitting off a Hartree-like interaction term and a kinetic energy term of a noninteracting electron system with the exact density $\rho(\mathbf{r}, t)$, we can decompose $A[\rho]$ in analogy to (4.10) into the form

$$A[\rho] = \sum_{\mu=1}^{N} \int_{t_0}^{t} dt' \left\langle \chi_\mu(t') \left| i\frac{\partial}{\partial t'} + \frac{1}{2m}\nabla^2 - V_{\text{ext}}(t') \right| \chi_\mu(t') \right\rangle$$
$$- \frac{1}{2} \int d^3r d^3r' \, \frac{\rho(\mathbf{r}, t)\rho(\mathbf{r}', t)}{|\mathbf{r} - \mathbf{r}'|} - A_{xc}[\rho] \ , \tag{4.50}$$

with

$$\rho(\mathbf{r}, t) = \sum_{\mu=1}^{N} |\chi_\mu(t)|^2 \ . \tag{4.51}$$

The $\chi_\mu(t)$ are spin orbitals of a fictitious noninteracting electron system. The term $A_{xc}[\rho]$ contains the difference in kinetic energy of the noninteracting and the real, i.e., interacting system, as well as exchange and correlation contributions. Its form is unknown like $E_{xc}[\rho]$ in (4.10) is, and is subject to simple, yet reasonable, accurate approximations. The action $A[\rho]$ has a stationary point when the $\chi_\mu(t)$ are solutions of the generalized Kohn-Sham equations

$$\left(-\frac{1}{2m}\nabla^2 + V_{\text{eff}}(\mathbf{r}, t)\right)\chi_\mu(r, t) = i\frac{\partial}{\partial t}\chi_\mu(\mathbf{r}, t) \ , \tag{4.52}$$

with the effective potential given by

$$V_{\text{eff}}(\mathbf{r}, t) = V(\mathbf{r}, t) + e^2 \int d^3r' \, \frac{\rho(\mathbf{r}', t)}{|\mathbf{r} - \mathbf{r}'|} + v_{xc}(\mathbf{r}, t) \ , \tag{4.53}$$

and

$$v_{xc}(\mathbf{r}, t) = \frac{\delta A_{xc}[\rho]}{\delta \rho(\mathbf{r}, t)} \ . \tag{4.54}$$

The second functional derivative

$$f_{xc}(\boldsymbol{r},\boldsymbol{r}',t-t') = \frac{\delta v_{xc}(\mathbf{r},t)}{\delta\rho(\mathbf{r}',t')} \qquad (4.55)$$

is called exchange-correlation kernel. Its Fourier transform is the starting point for the linear response TDDFT, which is widely used in actual calculations. The simplest, yet quite successful approximation is the Adiabatic LDA (ALDA). The ALDA replaces the non-local, frequency-dependent density functional by the frequency-independent local response of the homogenous electron gas taken at zero frequency and zero momentum transfer and evaluated at the local density

$$f_{xc}^{ALDA}[\rho](\boldsymbol{r},\boldsymbol{r}',\omega) = \delta(\boldsymbol{r}-\boldsymbol{r}')f_{xc}^{hom}(\rho(\boldsymbol{r}),q=0,\omega=0)\ . \qquad (4.56)$$

The exchange correlation kernel f_{xc} is the appropriate starting point for recent developments to connect TDDFT with electronic structure calculations based on the Bethe-Salpeter equation.

5
Wavefunction-Based Methods

In the previous chapter we discussed density-functional theory and various approximations to it as a conceptually simple, yet effective scheme for going beyond the independent-electron approximation. A crucial feature of that method is that it avoids calculating the many-electron wavefunction. Instead, various ground-state properties such as the ground-state energy, magnetization, etc. are calculated directly, i.e., without determination of the ground-state wavefunction. On the other hand, it is also desirable to compute the effects of electron correlations on the wavefunction itself given that considerable insight into the correlation problem can be gained this way. For that reason wavefunction-based methods, traditionally used in quantum chemistry have not lost their importance. They have in addition the advantage that they are amenable to controlled approximations, a distinct advantage as compared to density functional theory. The computational efforts are generally much larger than those required by applying density functional theory, but often this disadvantage is not crucial.

As discussed extensively in Chap. 2, the independent-electron approximation neglects the correlation hole which every electron has attached to it. It prevents electrons from approaching each other too closely and reduces this way their mutual Coulomb repulsion. Therefore wavefunction-based methods must provide for a description of the correlation hole. The latter is a local object and therefore *local* operators are particularly suitable for generating it.

Consider a system of N electrons which we want to describe by using L basis functions and denote its ground state by $|\psi_0\rangle$. We can always expand $|\psi_0\rangle$ in terms of a complete basis of the Hilbert space $\mathcal{H}$ for which we choose the different N-electron configurations $|\Phi_I\rangle$. Thus

$$|\psi_0\rangle = \sum_I \alpha_I |\Phi_I\rangle \quad , \tag{5.1}$$

where the number of different terms is of order $\binom{2L}{N}$ and equals the dimension of $\mathcal{H}$. We may think of the $|\Phi_I\rangle$ as consisting of the SCF ground state

$|\Phi_{\mathrm{SCF}}\rangle$ and of states generated from the ground state by applying excitation operators. Except for very small systems the Hilbert space is much too large to be fully used. For practicable calculations it must be therefore drastically limited. The challenge consists in finding an appropriate subspace of dimensions as small as possible in order to achieve a given accuracy of the results. The different quantum-chemical methods vary in the way they do this.

The energy is a size-extensive quantity, i.e., doubling the size of a solid implies doubling the ground-state energy. Therefore, one requirement on wavefunction based methods is that they must yield size-extensive correlation-energy contributions, i.e., the correlation energy must be proportional to the electron number. This is not always the case, though. The popular *configuration-interaction* (CI) method is not size extensive and therefore it is applicable only to small molecules or clusters. Nevertheless, we will discuss it here in some detail. It is not only an important technique in quantum chemistry, but also in computations for solids. We may often determine matrix elements which are needed from calculations on small clusters. This will be demonstrated when the method of increments is discussed in Sect. (5.3.1). Then CI calculations are a useful tool. The Brillouin-Wigner perturbation theory also lacks size extensivity. The Rayleigh-Schrödinger perturbation theory, on the other hand, does not have this shortcoming. A simple way of ensuring size-extensivity of the correlation energy is to formulate the theory in terms of cumulants. They are well known from classical statistical mechanics and *Kubo* has been pivotal in emphasizing their usefulness in quantum statistical mechanics. Whenever an approximation is made within a theory based on cumulants, it will be size extensive. It guarantees also size consistency or separability of the energy, when a system is separated into independent parts.

Various different approximations can be made when the ground-state energy is formulated in terms of cumulants. Examples are many-body perturbation theory, coupled-electron pair approximations, coupled-cluster expansions, etc. A particularly useful one is the projection or partitioning technique. This method is a central theme of the book. Instead of expanding the ground-state energy in powers of a small quantity, like the residual interaction energy (see (2.36)), we partition the Liouville space $\mathfrak{R}$ of the excitation operators used to generate the correlation hole of the electrons. It is split into a relevant subspace $\mathfrak{R}_0$, which is kept, and a remaining part $(\mathfrak{R} - \mathfrak{R}_0)$, which is discarded. Another way of stating the same is by saying that the operators used to describe the correlation hole are projected onto $\mathfrak{R}_0$. The concept of *partitioning* $\mathfrak{R}$ goes back to *Löwdin* [292]. By combining it with the cumulant formalism, it is a tool whose efficiency will be demonstrated at numerous places. With its help and using local operators, we can compute the ground state of solids with high accuracy thereby establishing a connection to molecular calculations. Another distinct advantage of using cumulants is that they are also applicable in cases when, e.g., diagrammatic approaches are difficult to formulate as is the case for strongly correlated electrons. Diagrams represent in a pictorial way different terms of a perturbation expansion. This is easily done according to *Feynman* when the unperturbed system is one of noninteracting electrons. If, however, the unperturbed Hamiltonian contains already important correlation effects, as it is the case when we start from electrons on atoms instead of uncorrelated electrons, then a diagrammatic approach is difficult to set up and becomes rather complicated.

5.1 Method of Configuration Interactions

The configuration-interaction (CI) method is a very general theory for treating electron correlations [344]. It is a variational method and as such has the advantage of providing upper bounds for the correlation energy; it suffers, however, from the problem of size extensiveness and therefore it is not applicable to an infinite system like a solid.

Often it turns out that the SCF ground state of a finite system

$$|\Phi_{\text{SCF}}\rangle = \prod_{\mu=1}^{N} c_\mu^+ |0\rangle \qquad (5.2)$$

contributes dominantly to $|\psi_0\rangle$ when the expansion (5.1) is made (the spin index σ has been included in the label μ). In this case we may use $|\Phi_{\text{SCF}}\rangle$ as a reference state and label all other $|\Phi_I\rangle$'s according to the differences as compared to $|\Phi_{\text{SCF}}\rangle$. For example, they may differ from it by having one or several of the c_μ^+, c_ν^+ substituted by c_i^+, c_j^+. Here Greek indices μ, ν, etc. are used for occupied orbitals and Latin indices i, j etc. for unoccupied (or virtual) ones. Therefore we begin with the following expansion of the ground-state wavefunction

$$|\psi_0\rangle = \left(1 + \sum_{i\mu} \alpha_\mu^i c_i^+ c_\mu + \sum_{\substack{i<j \\ \mu<\nu}} \alpha_{\mu\nu}^{ij} c_i^+ c_j^+ c_\nu c_\mu + ...\right) |\Phi_{\text{SCF}}\rangle$$

$$= \left(1 + \sum_{i\mu} \alpha_\mu^i \omega_\mu^i + \sum_{\substack{i<j \\ \mu<\nu}} \alpha_{\mu\nu}^{ij} \omega_{\mu\nu}^{ij} + ...\right) |\Phi_{\text{SCF}}\rangle \quad , \qquad (5.3)$$

where we have set $\omega_\mu^i = c_i^+ c_\mu, \omega_{\mu\nu}^{ij} = c_i^+ c_j^+ c_\nu c_\mu$ and so on. The electrons are annihilated and created in *delocalized* canonical molecular (or Bloch-) orbitals (CMOs). We will later consider instead orthogonal localized as well as nonorthogonal *local* orbitals. The operators $\omega_\mu^i, \omega_{\mu\nu}^{ij}$ etc. span the full space $\Re$. When we terminate the expansion (5.3) we can obtain the $\alpha_\mu^i, \alpha_{\mu\nu}^{ij}$, etc. by diagonalizing H within a Hilbert space of given dimension spanned by the different configurations. Alternatively, one may consider the coefficients as variational parameters which are fixed by minimization of the energy $\mathcal{E} = \langle\psi_0|H|\psi_0\rangle/\langle\psi_0|\psi_0\rangle$.

From (5.1) we derive the following system of CI equations

$$\sum_J H_{IJ}\alpha_J = \mathcal{E}\alpha_I, \quad \text{with } H_{IJ} = \langle \Phi_I | H | \Phi_J \rangle, \tag{5.4}$$

provided the $|\Phi_I\rangle$ are orthogonal with respect to each other. Otherwise, on the right-hand side of (5.4), α_I has to be multiplied by the overlap matrix. Identifying $|\Phi_{I=0}\rangle$ with $|\Phi_{\text{SCF}}\rangle$ and using (5.3) for $|\psi_0\rangle$, we obtain from (5.4) for the ground-state energy $\mathcal{E}_0$

$$\langle H \rangle + \sum_{i\mu} \langle H\omega_\mu^i \rangle \alpha_\mu^i + \sum_{\substack{i<j \\ \mu<\nu}} \langle H\omega_{\mu\nu}^{ij} \rangle \alpha_{\mu\nu}^{ij} = \mathcal{E}_0 . \tag{5.5}$$

As before, $\langle ... \rangle = \langle \Phi_{\text{SCF}} | ... | \Phi_{\text{SCF}} \rangle$. The expansion terminates because H contains one- and two-particle terms only (see (2.7)). The CMOs from which $|\Phi_{\text{SCF}}\rangle$ is constructed follow from the stationarity condition (2.18) which implies for closed-shell systems

$$\langle H\omega_\mu^i \rangle = 0 \tag{5.6}$$

(Brillouin's theorem). The ground-state energy reduces therefore to

$$\mathcal{E}_0 = \langle H \rangle + \sum_{\mu<\nu} E_{\mu\nu},$$
$$E_{\mu\nu} = \sum_{i<j} \langle H\omega_{\mu\nu}^{ij} \rangle \alpha_{\mu\nu}^{ij} \tag{5.7}$$

and the $\alpha_{\mu\nu}^{ij}$ are obtained from (5.4).

When only single and double excitations are taken into account, (5.4) simplifies to the set of two equations

$$\sum_\nu \sum_j \langle (\omega_\mu^i)^+ H \omega_\nu^j \rangle \alpha_\nu^j + \sum_{\nu<\rho} \sum_{j<k} \langle (\omega_\mu^i)^+ H \omega_{\nu\rho}^{jk} \rangle \alpha_{\nu\rho}^{jk} = \mathcal{E}_0 \alpha_\mu^i$$
$$\langle (\omega_{\mu\nu}^{ij})^+ H \rangle + \sum_\rho \sum_k \langle (\omega_{\mu\nu}^{ij})^+ H \omega_\rho^k \rangle \alpha_\rho^k + \sum_{\rho<\tau} \sum_{k<l} \langle (\omega_{\mu\nu}^{ij})^+ H \omega_{\rho\tau}^{kl} \rangle \alpha_{\rho\tau}^{kl} = \mathcal{E}_0 \alpha_{\mu\nu}^{ij} . \tag{5.8}$$

Single substitutions are often only of secondary importance. They describe changes in the electronic charge distribution due to correlations, i.e., due to two-particle excitations. When only double excitations are kept, the ground state $|\psi_0\rangle$ has a form consisting of coupled electron-pair contributions, i.e.,

$$|\psi_0\rangle = |\Phi_{\text{SCF}}\rangle + \sum_{\mu<\nu} |\Phi_{\mu\nu}\rangle ,$$
$$|\Phi_{\mu\nu}\rangle = \sum_{i<j} \alpha_{\mu\nu}^{ij} \omega_{\mu\nu}^{ij} |\Phi_{\text{SCF}}\rangle . \tag{5.9}$$

Equation (5.8) simplifies accordingly to

$$\langle \Phi_{\text{SCF}} \mid (\omega_{\mu\nu}^{ij})^+ H \mid \Phi_{\text{SCF}} \rangle + \sum_{\rho<\tau} \langle \Phi_{\text{SCF}} \mid (\omega_{\mu\nu}^{ij})^+ H \mid \Phi_{\rho\tau} \rangle = \mathcal{E}_0 \alpha_{\mu\nu}^{ij} \ . \quad (5.10)$$

Even when we restrict ourselves to single- and double-excitations, i.e., configurations of the form $\omega_\mu^i|\Phi_{\text{SCF}}\rangle$ and $\omega_{\mu\nu}^{ij}|\Phi_{\text{SCF}}\rangle$ their number is of order $N^2(2L-N)^2/4$. Consider the molecule CH_4 as an example which has 10 electrons. A minimal basis set consists of 9 basis functions, i.e., $(1s, 2s, 3 \times 2p)$ for the C atom and four $(1s)$ functions for the H atoms, while a DZ + P basis set includes 35 functions. In the latter case the total number of double substitutions $\omega_{\mu\nu}^{ij}|\Phi_{\text{SCF}}\rangle$ is 79650 of which 22500 are singlets. With to-day's computing facilities, we can treat up to 10^8 configurations. Hence we can deal with molecules as large as C_2H_6 when the basis set is of $DZ + P$ quality and when only double substitutions are taken into account.

At this stage we want to comment on the convergence of the correlation energy with increasing size of the basis set. This issue is related to the description of the short-range part of the correlation hole. For $r \to 0$ the pair distribution function (3.21) has a cusp, usually referred to as *correlation cusp*. In order to describe it with sufficient accuracy, the atomic volume needs to be divided into very fine segments (see Fig. 2.5). Thus basis function with large angular momenta ℓ are required. An estimate of the convergence of the correlation energy with increasing values of ℓ is obtained by considering an He atom. A CI expansion is equivalent to an expansion of the two-electron wavefunction in terms of spherical harmonics in this case.

Thus we write

$$\psi(\mathbf{r}_1, \mathbf{r}_2) = \sum_{\ell=0}^{\infty} \left(\sum_i u_\ell^{(i)}(r_1) u_\ell^{(i)}(r_2) \right) P_\ell(\cos\vartheta_{12}) \quad (5.11)$$

where ϑ_{12} is the angle between $\mathbf{r}_1$ and $\mathbf{r}_2$. When we calculate the increments to the ground-state energy from different angular momenta ℓ, we find that they decrease like ℓ^{-4} [271]. Convergence is much faster if we add a term of the form $\frac{1}{2}|\mathbf{r}_1 - \mathbf{r}_2|u(\mathbf{r}_1, \mathbf{r}_2)$ to the right-hand side of (5.11), where $u(\mathbf{r}_1, \mathbf{r}_2)$ is an eigenfunction of the bare nuclear Hamiltonian. This improves considerably the modulation of the correlation cusp.

For small molecules, e.g., H_2O and basis sets up to triple zeta plus polarization functions for oxygen one can perform *full* CI calculations. Thereby all possible symmetry-adapted configurations which exist within that basis set are taken into account. Calculations of this type serve as benchmark for different approximation schemes which are later applied to solids.

An important generalization of the above CI equations are multireference configuration interactions. If some of the electrons in a molecule are strongly correlated, several configurations $|\Phi_n\rangle$ may contribute significantly to the expansion (5.1). The same holds true for a solid. In that case $|\Phi_{\text{SCF}}\rangle$ is no longer a good starting point for a correlation calculation. Consider a Li_2 molecule.

The SCF ground state is $(1\sigma_g)^2(1\sigma_u)^2(2\sigma_g)^2$. When the interatomic spacing is enlarged by pulling the atoms apart, one needs in addition the configuration $(1\sigma_g)^2(1\sigma_u)^2(2\sigma_u)^2$ in order to approach the limit of two separated Li atoms in their SCF ground state. In this case the correlation calculation has to start from two reference states.

The best way of taking M different reference configurations into account is by means of a multiconfiguration self-consistent field (MC-SCF) calculation onto which afterwards a CI calculation can be implemented. In MC-SCF calculations, not only the weighting factors of the included $|\Phi_n\rangle$ are optimized, but so are the molecular orbitals contained in them. The energy of the ground state is then of the form

$$\mathcal{E}_0 = \sum_{ij} A_i A_j \langle \Phi_i | H | \Phi_j \rangle \quad , \tag{5.12}$$

with real coefficients A_i and matrix elements $\langle \Phi_i|H|\Phi_j\rangle$. The A_i and the orbitals entering the $|\Phi_n\rangle$ are found as follows. The A_i are obtained by finding the eigenvector to the lowest eigenvalue of the secular equation

$$\det\left(\langle \Phi_i | H | \Phi_j \rangle - \mathcal{E}_0 \delta_{ij}\right) = 0 \quad . \tag{5.13}$$

We obtain the appropriate one-electron orbitals $\phi_\mu(\mathbf{r}, \sigma_\mu)$ if we require that

$$\delta\mathcal{E}_0 = \sum_{ij} A_i A_j \delta(\langle \Phi_i | H | \Phi_j \rangle) = 0 \tag{5.14}$$

The variation is done under the constraint that the orbitals are orthogonal to each other. The MC-SCF ground state $|\Phi_{\rm MC}\rangle$ is then written as

$$|\Phi_{\rm MC}\rangle = \sum_{n=1}^{M} A_n |\Phi_n\rangle \quad . \tag{5.15}$$

A special form of a MC-SCF calculation is one which includes *all* configurations of a defined active space. This active space must embrace the electrons which are strongly correlated. For example, it may be spanned by the d orbitals of a transition metal ion in a molecule or lattice. Since d electrons have usually strong on-site correlations, treating them by a simple SCF calculation is a poor starting point for improvement. Self-consistent field calculations which include all configurations of an active space are referred to us as Complete Active Space SCF (CASSCF). Supplementing a MC-SCF calculation by a CI calculation (MC-SCF-CI) we make the ansatz

$$|\psi_0\rangle = \left(1 + \sum_{i\mu} \alpha_\mu^i \omega_\mu^i + \sum_{\substack{i<j \\ \mu<\nu}} \alpha_{\mu\nu}^{ij} \omega_{\mu\nu}^{ij}\right) |\Phi_{\rm MC}\rangle \quad . \tag{5.16}$$

The state $|\Phi_{\text{MC}}\rangle$ is no longer an eigenstate of the occupation-number operators $n_\mu = c_\mu^+ c_\mu$ since it contains several different configurations. Therefore, for active orbitals the distinction of using Greek and Latin indices for occupied and virtual orbitals becomes obsolete. MC-SCF calculations have been performed with as many as 10^3 configurations and up to 10^8 single and double substitutions.

5.2 Cumulants and their Properties

For a unified discussion of various size-extensive correlation methods we first introduce cumulants and discuss their properties. Cumulants are closely related to matrix elements of products of operators. In distinction to ordinary matrix elements they do not contain any contributions from statistically independent processes. For example, consider the matrix element $\langle \Phi_1 | A_1 A_2 | \Phi_2 \rangle$ of two operators A_1 and A_2 with respect to the reference states $|\Phi_1\rangle$ and $|\Phi_2\rangle$. The latter must have a finite overlap $\langle \Phi_1 | \Phi_2 \rangle \neq 0$. The cumulant $\langle \Phi_1 | A_1 A_2 | \Phi_2 \rangle^c$ is then defined by

$$\langle \Phi_1 \mid A_1 A_2 \mid \Phi_2 \rangle^c = \frac{\langle \Phi_1 \mid A_1 A_2 \mid \Phi_2 \rangle}{\langle \Phi_1 \mid \Phi_2 \rangle} - \frac{\langle \Phi_1 \mid A_1 \mid \Phi_2 \rangle}{\langle \Phi_1 \mid \Phi_2 \rangle} \frac{\langle \Phi_1 \mid A_2 \mid \Phi_2 \rangle}{\langle \Phi_1 \mid \Phi_2 \rangle}, \tag{5.17}$$

i.e., it subtracts or eliminates that part of the matrix element which corresponds to statistically independent processes. A general definition of cumulants is found as follows. Consider the function

$$f(\lambda_1, \lambda_2, ..., \lambda_M) = \ln \langle \Phi_1 \mid \prod_{i=1}^{M} e^{\lambda_i A_i} \mid \Phi_2 \rangle \tag{5.18}$$

which depends on M parameters $\lambda_1, ..., \lambda_M$ and require that $\langle \Phi_1 | \Phi_2 \rangle \neq 0$. This function is analytic near $\lambda_1 = \lambda_2 = ... = \lambda_M = 0$ and therefore can be expanded in terms of the λ_i. The expansion coefficients define cumulants, i.e.,

$$\langle \Phi_1 \mid A_1 ... A_M \mid \Phi_2 \rangle^c = \frac{\partial}{\partial \lambda_1} ... \frac{\partial}{\partial \lambda_M} \ln \langle \Phi_1 \mid \prod_{i=1}^{M} e^{\lambda_i A_i} \mid \Phi_2 \rangle \bigg|_{\lambda_i = 0}. \tag{5.19}$$

One checks easily that (5.17) follows from (5.19). In the following we shall assume $\langle \Phi_1 | \Phi_2 \rangle = 1$, if not stated otherwise. In analogy to (5.17) one finds

$$\begin{aligned}\langle A_1 A_2 A_3 \rangle^c = &\langle A_1 A_2 A_3 \rangle - \langle A_1 \rangle \langle A_2 A_3 \rangle \\ &- \langle A_2 \rangle \langle A_1 A_3 \rangle - \langle A_3 \rangle \langle A_1 A_2 \rangle \\ &+ 2 \langle A_1 \rangle \langle A_2 \rangle \langle A_3 \rangle, \text{ etc. },\end{aligned} \tag{5.20}$$

where the abbreviation $\langle \Phi_1 | ... | \Phi_2 \rangle = \langle ... \rangle$ has been used. By setting in (5.18) $A_1 = ... = A_M = A$, multiplying with $\lambda^n/n!$ and summing over n, we obtain

$$\ln\langle e^{\lambda A}\rangle = \langle e^{\lambda A} - 1\rangle^c \quad . \tag{5.21}$$

Sometimes this expression is used to define cumulants instead of the more general form (5.19). It demonstrates that by using cumulants we can avoid working with the logarithm. This is of advantage, e.g., in statistical physics.

Cumulants have the property that

$$\langle A(\alpha B + \beta C)\rangle^c = \alpha\langle AB\rangle^c + \beta\langle AC\rangle^c$$
$$\langle \alpha_1 \Phi_1 \mid AB \mid \alpha_2 \Phi_2\rangle^c = \langle \Phi_1 \mid AB \mid \Phi_2\rangle^c \quad , \quad \alpha_1, \alpha_2 \neq 0 \quad . \tag{5.22}$$

When evaluating a cumulant, we must also distinguish between the number 1 and the unit operator 1_{op}. We find that $\langle 1 \cdot A\rangle^c = \langle A\rangle$ while $\langle 1_{op} \cdot A\rangle^c = 0$. Furthermore, by formally reducing in (5.19) the number of different λ_i to zero, we define

$$\langle \Phi_1 \mid 1 \mid \Phi_2\rangle^c = \ln\langle \Phi_1 \mid \Phi_2\rangle \quad . \tag{5.23}$$

We must also label special operator products, which are considered an entity when a cumulant is evaluated. For example, when the product $A_2 A_3$ is considered a unit with respect to a cumulant, we denote it by $(A_2 A_3)^{\bullet}$. Generally it is

$$\langle A_1 (A_2 A_3)^{\bullet}\rangle^c \neq \langle A_1 A_2 A_3\rangle^c \quad . \tag{5.24}$$

Sometimes we have to deal with cumulants of expectation values which vanish. Consider for example

$$\langle \Phi_{\text{SCF}} \mid (\omega_\mu^i)^+ (\omega_\mu^i)^+ \omega_\mu^i \omega_\mu^i \mid \Phi_{\text{SCF}}\rangle^c = -2\langle \Phi_{\text{SCF}} \mid (\omega_\mu^i)^+ \omega_\mu^i \mid \Phi_{\text{SCF}}\rangle^2 \quad . \tag{5.25}$$

Here ω_μ^i is defined as in Section 4.1. While the cumulant of that operator product obviously does not vanish, the expectation value does because $\omega_\mu^i \omega_\mu^i |\Phi_{\text{SCF}}\rangle = 0$. When the above rules are observed, calculations with cumulants are as simple as those with ordinary expectation values. A number of additional relations involving cumulants are to be found in Appendix A.

5.3 Ground-State Wavefunction and Energy

We want to derive a conceptually simple set of equations that allows for approximate yet accurate computations of the ground-state wavefunction and energy of a solid. The theory presented in the following is quite general and can be applied to weakly as well as strongly correlated electron systems. We start from a Hamiltonian H decomposed into

$$H = H_0 + H_1 \quad . \tag{5.26}$$

We assume the eigenstates and eigenvalues of H_0 to be known and the effect of H_1 on the ground-state energy to be relatively small. No further

5.3 Ground-State Wavefunction and Energy

assumptions about the above decompositions are being made. The ground state of H_0 is $|\Phi_0\rangle$ and for convenience is assumed to be nondegenerate, i.e.,

$$H_0 \,|\, \Phi_0\rangle = E_0 \,|\, \Phi_0\rangle \, . \tag{5.27}$$

We want to find the ground state $|\psi_0\rangle$ of H and its energy, i.e.,

$$H \,|\, \psi_0\rangle = \mathcal{E}_0 \,|\, \psi_0\rangle \tag{5.28}$$

by using the eigenstates of H_0. With the help of (5.19) we can write

$$\begin{aligned} \mathcal{E}_0 &= \frac{\langle \Phi_0 \,|\, H \,|\, \psi_0\rangle}{\langle \Phi_0 \,|\, \psi_0\rangle} \\ &= \langle \Phi_0 \,|\, H \,|\, \psi_0\rangle^c \end{aligned} \tag{5.29}$$

provided that there is a non-vanishing overlap $\langle \Phi_0|\psi_0\rangle \neq 0$. We also find the important relation

$$\langle \Phi_0 \,|\, AH \,|\, \psi_0\rangle^c = 0 \, , \tag{5.30}$$

where A is an arbitrary operator because $|\psi_0\rangle$ is an eigenstate of H and therefore the matrix element $\langle \Phi_0|AH|\psi_0\rangle$ factorizes and the cumulant vanishes (see (5.17)).

Next we want to express the exact ground state $|\psi_0\rangle$ in (5.29) and its energy in terms of $|\Phi_0\rangle$. Then we have to evaluate cumulants of the form $\langle \Phi_0|.....|\Phi_0\rangle^c$. This suggests a simplified notation of the form

$$(A \,|\, B) = \langle \Phi_0 \,|\, A^+ B \,|\, \Phi_0\rangle^c \, . \tag{5.31}$$

Note that this bilinear form is not a scalar product in the strict mathematical sense since $(A|A)$ need not be positive yet can vanish.

In order to find the desired relation between $|\psi_0\rangle$ and $|\Phi_0\rangle$ we use the identity

$$\lim_{t\to\infty} e^{-Ht} \,|\, \Phi_0\rangle = \lim_{t\to\infty} e^{-\mathcal{E}_0 t} \,|\, \psi_0\rangle\langle \psi_0 \,|\, \Phi_0\rangle \tag{5.32}$$

or

$$|\, \psi_0\rangle = \frac{1}{\langle \psi_0 \,|\, \Phi_0\rangle} \lim_{t\to\infty} e^{-(H-\mathcal{E}_0)t} \,|\, \Phi_0\rangle \, . \tag{5.33}$$

This implies the following form for the wave operator $\tilde{\Omega}$ which relates $|\psi_0\rangle$ and $|\Phi_0\rangle$ through $|\psi_0\rangle = \tilde{\Omega}|\Phi_0\rangle$,

$$\tilde{\Omega} = \frac{1}{\langle \psi_0 \,|\, \Phi_0\rangle} \lim_{t\to\infty} e^{-(H-\mathcal{E}_0)t} \, . \tag{5.34}$$

Because of the second equation (5.22) the cumulant wave operator $|\Omega)$ can be written as

$$|\, \Omega) = \lim_{t\to\infty} \,|\, e^{-Ht}) \, . \tag{5.35}$$

The round bracket implies that this expression has to be used together with the metric (5.31). Note that the effect of the operator $\exp(-Ht)$ remains finite even when the limit $t \to \infty$ is taken. In order to extract the remaining part we take the Laplace transform to which a constant term makes a z^{-1} contribution. By multiplying that transform by z and taking the limit $z \to 0$, we obtain the part we are looking for. Therefore

$$\lim_{z \to 0} \frac{1}{z} \mid \Omega) = -\lim_{z \to 0} \int_0^\infty dt \, e^{zt} \mid e^{-Ht}) \; ; \qquad \Re e\{z\} < 0$$

$$\mid \Omega) = \lim_{z \to 0} \mid \frac{1}{z - H} z) \; . \tag{5.36}$$

This expression is rewritten as

$$\mid \Omega) = \lim_{z \to 0} \mid 1 + \frac{1}{z - H} H)$$

$$= \lim_{z \to 0} \mid 1 + \frac{1}{z - H} H_1) \; . \tag{5.37}$$

The last equation results from the fact that $\mid \Phi_0\rangle$ is an eigenstate of H_0 and any cumulant with $\mid ...H_0)$ vanishes.

An equivalent expression for $\mid \Omega)$ is

$$\mid \Omega) = \lim_{z \to 0} \mid 1 + \frac{1}{z - L_0 - H_1} H_1) \; . \tag{5.38}$$

The Liouvillean L_0 which appears here, is a superoperator, i.e., it acts on operators and not on states. We call the space spanned by operators the Liouville space in distinction to the Hilbert space, i.e., the space spanned by states of a given particle number. The way L_0 acts on operators A is given by

$$L_0 A = [H_0, A]_- \; . \tag{5.39}$$

Equation (5.38) is obtained by starting from the decomposition

$$e^{-\lambda H} = e^{-\lambda(H_1 + L_0)} e^{-\lambda H_0} \tag{5.40}$$

and repeating the steps which lead to (5.37). The decomposition follows from integrating the equation of motion of

$$R(\lambda) = e^{-\lambda H} e^{\lambda H_0}, \qquad R(0) = 1 \; , \tag{5.41}$$

i.e.,

$$\frac{d}{d\lambda} R(\lambda) = -H R(\lambda) + R(\lambda) H_0$$

$$= -(H_1 + L_0) R(\lambda) \; . \tag{5.42}$$

At this stage we want to point out that cumulants containing L_0 or more generally a superoperator L have the property that

5.3 Ground-State Wavefunction and Energy

$$\langle BLA \rangle^c = \langle B(LA)^\bullet \rangle^c \quad . \tag{5.43}$$

A proof of that relation is found, e.g., in [243].

Within the cumulant formalism $|\Omega)$ characterizes the exact ground state $|\psi_0\rangle$. Therefore, when $|\Omega)$ is known we may claim that we know the ground-state wavefunction! We shall use repeatedly that identification despite the fact that $|\Omega)$ is not unique. Any other $|\Omega')$ with $(A|\Omega - \Omega') = 0$ for arbitrary operators A will correspond to the same ground state $|\psi_0\rangle$. For more details see Appendix A. With this in mind we may replace (5.29-5.30) by

$$\mathcal{E}_0 = (H \mid \Omega) \tag{5.44a}$$
$$0 = (A \mid H\Omega) \quad . \tag{5.44b}$$

The last two equations serve as a starting point for numerous different approximations and therefore are central for many applications. An important point is that size-extensivity is ensured independent of any approximation made for $|\Omega)$. Within the cumulant formulation the problem of size-extensivity has disappeared.

It is instructive to compare the expression for the ground-state energy (5.44a) with the corresponding one written in terms of $\tilde{\Omega}$, i.e.,

$$\mathcal{E}_0 = \frac{\langle \Phi_0 \mid H\tilde{\Omega} \mid \Phi_0 \rangle}{\langle \Phi_0 \mid \tilde{\Omega} \mid \Phi_0 \rangle} \quad . \tag{5.45}$$

In quantum-field theory where Feynman diagrams are used, the denominator can be eliminated by taking into consideration only *linked* or *connected* diagrams. When (5.44a) is used instead, there is no denominator which has to be canceled. Since cumulants eliminate any contribution from statistically independent processes, *unlinked* or *disconnected* diagrams do not appear a priori. Hence, energies written in terms of cumulants are size extensive. Instead of (5.44a) we may also write

$$\mathcal{E}_0 = (\Omega \mid H\Omega) \quad , \tag{5.46}$$

where use of (5.37) and (5.44b) has been made. This is a special case of a more general relation which holds for any operator A

$$\frac{\langle \psi_0 \mid A \mid \psi_0 \rangle}{\langle \psi_0 \mid \psi_0 \rangle} = (\Omega \mid A\Omega) \quad . \tag{5.47}$$

It is proven as follows. First we write

$$\frac{\langle \psi_0 \mid A \mid \psi_0 \rangle}{\langle \psi_0 \mid \psi_0 \rangle} = \langle \psi_0 \mid A \mid \psi_0 \rangle^c \quad . \tag{5.48}$$

By applying the transformation (A3) in going over from $|\psi_0\rangle$ to $|\Phi_0\rangle$ we find that

$$\langle \psi_0 \mid A \mid \psi_0 \rangle^c = \langle \Omega\Phi_0 \mid A \mid \Omega\Phi_0 \rangle^c$$
$$= (\Omega \mid A \, \Omega) \quad . \tag{5.49}$$

5.3.1 Method of Increments

In the following we discuss a practical way of calculating the correlation energy and ground-state wavefunction of a periodic solid. We assume that the SCF- or Hartree-Fock ground state has been determined within a given set of basis functions, e.g., by applying the program package CRYSTAL [371]. What needs to be done is the post-SCF part. By treating a solid we are dealing with an unlimited number of electrons. Yet in any practical calculation we can correlate only a finite, not too large number of electrons. Therefore, we first have to find out how we can reduce the correlation problem of a solid to that of a relatively small number of electrons. Afterwards we will discuss different approximate methods which enable us to deal with the problem of few electrons (see Sect. 5.4).

One expects that a reduction of the many-electron problem to that of a few electrons is possible in view of the small extent of the correlation hole of an electron. One should keep in mind, however, that there remain special correlations like the Cooper-pair correlation leading to superconductivity, which extend over several hundred lattice distances and require special treatment.

The simplest case of relating the correlation problem for a given number of particles to one of a smaller number of particles was treated by *Faddeev* [114]. He considered three particles interacting through a two-particle scattering potential. Assuming that the solution of the two-particle scattering problem is known, he used that information to set up an equation for the three-particle scattering problem. We want to proceed here in the same spirit, i.e., we want to use solutions of few-electron scattering problems in order to construct a solution for the N electron scattering problem.

For this purpose we introduce the scattering matrix S of the N electron system by writing

$$|\Omega) = |1 + S) \quad . \tag{5.50}$$

From (5.44a) it follows that

$$\mathcal{E}_0 = E_0 + (H \mid S) \quad . \tag{5.51}$$

When we identify H_0 with H_{SCF} and H_1 with the residual interactions H_{res}, the correlation energy of the system is

$$E_{\mathrm{corr}} = (H \mid S) \quad . \tag{5.52}$$

More generally $E_0 + \delta E_0 = (H|1) + (H|S)$ describes the change in the ground-state energy caused by H_1.

We decompose the scattering matrix into single-site (or bond), two-sites (bonds), three-sites etc. scattering matrices and write

$$S = \sum_I S_I + \sum_{\langle IJ \rangle} \delta S_{IJ} + \sum_{\langle IJK \rangle} \delta S_{IJK} + \ldots \quad , \tag{5.53}$$

5.3 Ground-State Wavefunction and Energy

where $\langle IJ \rangle$ and $\langle IJK \rangle$ denote pairs and triples of sites or bonds. Furthermore,

$$\delta S_{IJ} = S_{IJ} - S_I - S_J \tag{5.54}$$

is a two-sites increment, i.e., the scattering matrix S_{IJ} for sites I and J from which the single-site contributions have been subtracted. The higher order terms are defined accordingly. It is clear that when we continue up to the N-th order increment the exact scattering matrix of the N electron system is reproduced. A more detailed derivation of (5.53) starting from the form (5.37) is found in Appendix B. By using (5.52) we may write

$$\begin{aligned} E_{\text{corr}} &= \sum_I (H \mid S_I) + \sum_{\langle IJ \rangle} (H \mid \delta S_{IJ}) + \ldots \\ &= \sum_I \epsilon_I + \sum_{\langle IJ \rangle} \epsilon_{IJ} + \ldots \quad , \end{aligned} \tag{5.55}$$

We want to mention that this energy expansion resembles very much the Bethe-Goldstone expansion [33] known from nuclear physics. In distinction to that approach we determine here also the ground-state wavefunction by specifying $|\Omega\rangle = |1 + S\rangle$. This is done by computing the incremental contributions (5.53) to S step by step and is made possible by the use of cumulants. The approach enables us to determine ground-state wavefunctions with high accuracy (see Sect. 6.1.3). Note that this includes strongly correlated electron systems as well (see Sect. 6.4). Excluded are merely long-ranged pair correlations, which lead to superconductivity (see Chapter 15). In this case (5.53) is not manageable.

It turns out that in practice the incremental decomposition is rapidly convergent [432]. In most cases two-body increments are sufficient to achieve good accuracy for the correlation energy. This is discussed in more detail in Chapter 6 where this formalism is applied to semiconductors and insulators. There we also show how the single-site, two-sites etc. scattering matrices can be computed. For the benefit of the reader who wants to see right away how the formalism can be applied, we give here a brief sketch of the way this is done.

Starting point is the SCF ground-state wavefunction written in terms of localized, i.e., Wannier orbitals

$$|\Phi_{\text{SCF}}\rangle = \prod_{I,\nu\sigma} c^+_{\nu\sigma}(I) \mid 0 \rangle \quad , \tag{5.56}$$

where the creation operators refer to Wannier orbitals centered at site (or bond) I with additional orbital index ν and spin σ. Note that for a metal with partially filled bands, one cannot construct well-localized Wannier orbitals and therefore has to proceed somewhat differently. The single-site scattering matrix S_I is obtained by freezing all electrons in $|\Phi_{\text{SCF}}\rangle$ except those in $c^+_{\nu\sigma}(I)|0\rangle$, i.e., in Wannier orbitals centered at site I. They are of a small number and therefore the scattering matrix can be calculated by any of the methods discussed below. Similarly we can determine the S_{IJ} by freezing all electrons in $|\Phi_{\text{SCF}}\rangle$ except for those in Wannier orbitals at sites I and J. Again, that is still a small number and therefore poses no particular problems when the correlation energy is evaluated. As mentioned before and shown in Chapter 6, an extension to three sites is usually sufficient to obtain high-quality results for the total correlation energy.

The method of increments is not restricted to weakly correlated systems. It can be also applied when the electronic correlations are strong. If this is the case, different routes can be pursued. One consists in starting again from the SCF ground state $|\Phi_{\text{SCF}}\rangle$ despite the fact that the corrections to it become large. The one-center scattering operators S_I and to a lesser degree the two-center scattering operators S_{IJ} must provide for the strong modifications, e.g., by a MC-SCF calculation within a limited orbital space. An alternative way of treating such systems is by starting from a wavefunction $|\Phi_0\rangle$ which comes as close as possible to the exact wavefunction $|\psi_0\rangle$. A possible choice is to use an antisymmetrized product of the (correlated) wavefunctions of the different atoms. Once $|\Phi_0\rangle$ has been chosen the cumulant scattering operator $|S)$ is again determined as before, but this time with respect to the special form of $|\Phi_0\rangle$.

5.4 Different Approximation Schemes

After having reduced the correlation calculations for a solid to one of a few electrons, we still need approximation schemes in order to treat the latter. We use as starting point the set of equations (5.44a, 5.44b) with Ω given by (5.37) or (5.38).

The simplest approximation is to expand the change in the ground-state energy δE_0 due to H_1 in powers of H_1. From (5.38) and (5.37) we obtain

$$\delta E_0 = (H_1 \mid 1) + \lim_{z \to 0} \sum_{n=1}^{\infty} \left(H_1 \middle| \left(\frac{1}{z - H_0} H_1 \right)^n \right) \; . \tag{5.57}$$

Hereby the identity

$$\frac{1}{a+b} = \frac{1}{a} + \frac{1}{a} b \frac{1}{a} + \frac{1}{a} b \frac{1}{a+b} b \frac{1}{a} \tag{5.58}$$

has been used. This is nothing else but the Rayleigh-Schrödinger perturbation expansion in terms of cumulants. Note that in quantum chemistry a perturbation expansion based on H_{SCF} as unperturbed Hamiltonian is called Møller-Plesset expansion.

When H_0 describes noninteracting electrons and H_1 their interactions, then (5.57) is known as Goldstone's linked-cluster expansion. *Goldstone* developed a diagrammatic method to classify and compute the different terms

5.4 Different Approximation Schemes

of a Rayleigh-Schrödinger perturbation expansion (5.57) [148]. When those diagrams are analyzed, we find that only *linked* diagrams contribute to the changes δE_0 of the ground-state energy. Linked diagrams are those which do not separate into disconnected parts. Using cumulants amounts to dealing with linked diagrams only. Unlinked diagrams are eliminated by a cumulant because they correspond to statistically independent processes. Equation (5.57) can be considered a generalization of Goldstone's linked cluster theorem to arbitrary splittings of the Hamiltonian H into H_0 and H_1. No diagrams need to be considered here. By expanding (5.38) instead of (5.37) we can write as well

$$\delta E_0 = (H_1 \mid 1) + \lim_{z \to 0} \sum_{n=1}^{\infty} \left(H_1 \bigg| \left(\frac{1}{z - L_0} H_1 \right)^n \right) \quad . \tag{5.59}$$

The excitation energies in the denominators are obtained here through the Liouvillean L_0.

5.4.1 Partitioning and Projection Methods

The projection method provides for very useful and powerful approximations by limiting strongly the number of operators from which the cumulant wave operator is constructed. In practice this means that a relative small number of excitation operators acting on $|\Phi_0\rangle$, (see, e.g., (5.3)) are used when the effect of H_1 on the ground-state energy is accounted for. We divide the operator space $\Re$ into a *relevant* subspace $\Re_0$ spanned by a set of operators $\{A_\nu\}$ and a remaining *irrelevant* part $\Re_1 = \Re - \Re_0$ which we neglect. The operator Ω is projected onto $\Re_0$. By successively increasing the dimension of $\Re_0$, we can improve the quality of the approximation. We assume that the A_ν are orthonormal, i.e., $(A_\nu|A_\mu) = \delta_{\nu\mu}$. This suggests the ansatz

$$\mid \Omega) = \mid 1 + \sum_\nu \eta_\nu A_\nu) \quad . \tag{5.60}$$

We determine the parameters η_ν by making use of (5.44b), i.e., from

$$(A_\nu \mid H\Omega) = 0 \quad . \tag{5.61}$$

When $(A_\nu|H_1) \neq 0$ for all ν we obtain a particularly simple form for the energy change $\delta E(\Re_0)$ due to H_1. The last equation

$$(A_\mu \mid H\Omega) = (A_\mu \mid H_1) + \sum_\nu \eta_\nu (A_\mu \mid HA_\nu)$$
$$= 0 \tag{5.62}$$

is solved by

$$\eta_\nu = -\sum_\mu L_{\nu\mu}^{-1} (A_\mu \mid H_1) \quad , \tag{5.63}$$

where the matrix $L_{\rho\tau}$ is given by

$$L_{\rho\tau} = (A_\rho \mid HA_\tau) \ . \tag{5.64}$$

By virtue of (5.44a) and (5.60) the energy change is equal to

$$\delta E(\Re_0) = (H_1 \mid \Omega)$$
$$= \sum_\nu \eta_\nu (H_1 \mid A_\nu) \ , \tag{5.65}$$

showing that it consists of a sum of contributions from the different operators A_ν which span $\Re_0$. When some of the A_ν do not couple directly to H_1, i.e., when $(A_\rho \mid H_1) = 0$, these operators enter only indirectly $\delta E(\Re_0)$ by modifying the coefficients η_ν via the matrix $L_{\rho\tau}$. We want to point out that the projection method does not give bounds to the energy. This seems to be a general feature: size-extensive approximations give generally no bounds for the correlation energy.

A much used approximation in quantum chemistry is the coupled-electron pair approximation (CEPA). In combination with the method of increments it can also be used for calculations of the ground state of solids. In the CEPA we identify H_0 with H_{SCF} and choose for the set $\{A_\nu\}$ all single (S) and double (D) excitations. This suggests introducing operators with compound indices K and Γ. They stand for

$$A_\Gamma^K = \begin{cases} \omega_\mu^i \\ \omega_{\mu\nu}^{ij}; \ i < j \text{ and } \mu < \nu \end{cases} \ . \tag{5.66}$$

The ansatz

$$\mid \Omega) = \mid 1 + \sum_{K\Gamma} \eta_\Gamma^K A_\Gamma^K) \tag{5.67}$$

is called CEPA-0.

The accuracy of correlation calculations is improved if in addition operator products are included in the set which span $\Re_0$. For example, A_ν^2 or $A_\nu A_\mu$ are such products. An ansatz of the form

$$\mid \Omega) = \mid 1 + \sum_{K\Gamma} \eta_\Gamma^K A_\Gamma^K + \frac{1}{2} \sum_{KL\Gamma} \eta_\Gamma^K \eta_\Gamma^L A_\Gamma^K A_\Gamma^L) \tag{5.68}$$

is a variant termed CEPA-2 in quantum chemistry. Note that the prefactors $\frac{1}{2}\eta_\Gamma^K \eta_\Gamma^L$ of the product operators are not independent of those of the operators A_Γ^K. Instead they are products of the latter. There exist other CEPA variants for which we refer to the literature (see, e.g., [261]).

5.4.2 Coupled Cluster Method

Another powerful method in quantum chemistry is the coupled-cluster method (CC) [259] which is part of some of the available quantum chemistry program

5.4 Different Approximation Schemes

packages and has also been used for solid-state calculations. Coupled-cluster equations can be derived in a quite general form from the set of equations (5.44a, 5.44b), that is for any division of H into $H = H_0 + H_1$. In particular, H_0 need not be H_{SCF}, but can for example also be the Ising part $\sum_{\langle ij \rangle} J_{ij} S_i^z S_j^z$ of the Heisenberg Hamiltonian $H = \sum_{\langle ij \rangle} J_{ij} \boldsymbol{S}_i \boldsymbol{S}_j$. Here $\boldsymbol{S}_i, \boldsymbol{S}_j$ are spin operators for sites i and j, respectively.

We begin by deriving the CC equations in the most familiar form, i.e., in the form used mainly in quantum chemistry where H_0 is identified with H_{SCF}. Afterwards we generalize them so that they become more widely applicable.

Starting from the SCF ground state $|\Phi_{\text{SCF}}\rangle$ the CC ansatz for the exact ground state is

$$|\psi_0\rangle = \exp\left(\sum_{i\mu} \eta_\mu^i \omega_\mu^i + \sum_{i<j,\mu<\nu} \eta_{\mu\nu}^{ij} \omega_{\mu\nu}^{ij} + \ldots \right) |\Phi_{\text{SCF}}\rangle$$
$$= e^{\tilde{S}} |\Phi_{\text{SCF}}\rangle \; . \tag{5.69}$$

When compared with the CI ansatz (5.3) one notices that the excitation operators appear here in the exponent of the prefactor. This has the advantage that when the exponent is terminated, e.g., by stopping after the $\omega_{\mu\nu}^{ij}$ terms the corrections to $|\Phi_{\text{SCF}}\rangle$ are size extensive. This is seen best by considering an ensemble of N uncoupled atoms, e.g., He atoms. For a minimal description of intra-atomic correlations we take into account one excitation $1s^2 \to 2s^2$ out of each atomic SCF ground state. With the two-particle excitations in the exponent of (5.69) we obtain N-times the correlation energy of a single atom, a result which requires terms up to $\omega_{\mu_1\ldots\mu_{2N}}^{i_1\ldots i_{2N}}$ in the expansion (5.3). This is seen by expanding the prefactor.

From the ansatz (5.69) it follows immediately that

$$\mathcal{E}_0 = \langle \Phi_{\text{SCF}} | e^{-\tilde{S}} H e^{\tilde{S}} | \Phi_{\text{SCF}} \rangle \; . \tag{5.70}$$

For a determination of the coefficients $\eta_\mu^i, \eta_{\mu\nu}^{ij}, \ldots$ we rewrite $\tilde{S}$ in the form

$$\tilde{S} = \sum_\nu \eta_\nu \tilde{S}_\nu \; , \tag{5.71}$$

where ν is now a compact index of the form $\nu = (i;\mu), (i,j;\mu\nu)$ etc. Note that

$$\tilde{S}_\nu^+ | \Phi_{\text{SCF}} \rangle = 0 \tag{5.72}$$

because $\tilde{S}_\nu^+$ destroys electrons from virtual orbitals, i.e., orbitals which are unoccupied in $|\Phi_{\text{SCF}}\rangle$. This feature is used to set up the second CC equation

$$\langle \Phi_{\text{SCF}} | \tilde{S}_\nu^+ e^{-\tilde{S}} H e^{\tilde{S}} | \Phi_{\text{SCF}} \rangle = 0 \tag{5.73}$$

or
$$\left\langle \left(\omega^{i_1...i_n}_{\mu_1...\mu_n}\right)^+ e^{-\tilde{S}} H e^{\tilde{S}} \right\rangle = 0 \quad . \tag{5.74}$$

By expanding the exponential one obtains equations for the coefficients η_ν. In most calculations the excitation operators are limited the single- and double-excitations (CCSD). Sometimes triple excitations are added by perturbation theory (CCSD(T)). Program packages like MOLPRO [481] contain these CC options.

The coupled-cluster equations become particularly simple if we restrict $\tilde{S}$ to two-particle excitations (pair approximation). In that case

$$\tilde{S} = \tilde{S}_2$$
$$= \sum_{i_1 < i_2} \sum_{\mu_1 < \mu_2} \eta^{i_1 i_2}_{\mu_1 \mu_2} \omega^{i_1 i_2}_{\mu_1 \mu_2} \quad , \tag{5.75}$$

and the ground-state energy reduces to

$$\mathcal{E}_0 = E_{\text{SCF}} + \langle H \tilde{S}_2 \rangle \tag{5.76}$$

while (5.73) goes over into

$$\langle (\omega^{ij}_{\mu\nu})^+ (1 - \tilde{S}_2) H (1 + \tilde{S}_2 + \tilde{S}_2^2/2) \rangle = 0 \quad . \tag{5.77}$$

The expansion terminates as indicated, because H contains at most two creation and two annihilation operators. From the last equation the coefficients $\eta^{i_1 i_2}_{\mu_1 \mu_2}$ can be determined.

In a next step we generalize the coupled-cluster method to an arbitrary starting Hamiltonian H_0 with ground state $|\Phi_0\rangle$. We make for $|\Omega\rangle$ the ansatz

$$|\Omega\rangle = |e^S\rangle \tag{5.78}$$

and expand S similar to (5.71) into a basis $\{S_\mu\}$ of prime operators, i.e., operators which are treated as an entity when cumulants are evaluated. This implies that operators of the form $A_\nu A_\mu$ etc. are excluded. We also require that $S_\mu |\Phi_0\rangle \neq 0$. Thus

$$S = \sum_\mu \eta_\mu S_\mu \tag{5.79}$$

and the equations (5.44a) and (5.44b) become

$$\mathcal{E}_0 = (H \mid e^S)$$
$$0 = (S_\nu \mid H e^S) \quad . \tag{5.80}$$

These equations are the same as the ones (5.70, 5.73) when $H_0 = H_{\text{SCF}}$. This is seen by noticing that not only $(A|H\Omega) = 0$ but also $(AB|H\Omega) = 0$ for arbitrary operators B. Therefore

$$(S_\nu \mid He^S) = \left(e^{-S^+} S_\nu \mid He^S\right) = 0$$
$$\mathcal{E}_0 = (1 \mid He^S) = \left(e^{-S^+} \mid He^S\right) \quad , \tag{5.81}$$

which are evidently the CC equations. As pointed out before the advantage of (5.80) is that the equations hold independent of the form of H_0 and H_1.

5.4.3 Selection of Excitation Operators

As pointed out before, with to-day's computing facilities we can treat up to 10^8 configuration when a correlation calculation is performed. Therefore, we have to make a careful choice of the relevant operators A_ν from which the cumulant wave operator $|\Omega)$ or scattering operator $|S) = |\Omega - 1)$ are constructed.

In correlation calculations for small molecules the excitation operators w_μ^i and $w_{\mu\nu}^{ij}$ etc.[1] contain creation and annihilation operators of electrons in (delocalized) canonical molecular orbitals. As mentioned before, when referring to periodic solids the latter are called Bloch orbitals. For a correlation calculation they are not very useful. It is practically impossible to describe with them the local correlation hole of the electrons. Therefore we want to use *localized* or *local* orbitals instead of Bloch orbitals. From a computational point of view it is advantageous to work with *orthogonal* localized orbitals. In fact, almost all of the quantum chemical program packages are designed so that the annihilation operators in the $w_{\mu...\tau}^{i...l}$ refer to orthogonal orbitals; on the other hand, the creation operators may refer to nonorthogonal local orbitals. In solid-state physics orthogonal localized orbitals which are occupied in the SCF ground state are called Wannier orbitals. In quantum chemistry several localization procedures are used depending on the localization criteria. A popular one is named after *Foster* and *Boys*. The criterion for localization is here that the distance between different occupied orbitals is maximized. Another often used method is due to *Pipek* and *Mezey*. It requires that a localized orbital extends over as few atoms as possible. Generally we are facing the following problem. It is fairly easy to construct Wannier orbitals for a semiconductor or insulator, i.e., a system with a gap between the conduction and valence bands. However, this is generally not possible when we deal with metallic systems which are characterized by partially filled conduction bands. An extreme example is that of a homogeneous electron gas. The Wannier functions constructed from plane-wave states with momentum $\mid \boldsymbol{k} \mid < k_F$ fall off like r^{-2} as $r \to \infty$ rather than exponentially. Therefore, in this case one has to proceed differently, as will be explained in Sect. 6.3.

In order to be more specific let us consider single- and double excitations only, i.e., we assume that the operators A_ν in (5.60) are of the form (5.66). Since the correlation hole is a very local object, the creation operators in the A_Γ^K should create electrons in the immediate vicinity of the Wannier

---
[1] see Eq. (5.3)

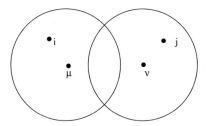

Fig. 5.1. Range of sites (bonds) i and j on which electrons are created by $w_{\mu\tau}^{ij}$ when the annihilation operators refer to Wannier orbitals centered at sites (bonds) μ and ν.

orbitals from which they were annihilated. This is schematically indicated in Fig. 5.1. As creation operators need not be mutually orthogonal we may use for them the $a_{i\sigma}^+$ which appear in the Hamiltonian[2]. However, they must be orthogonalized to the operators which create the occupied Wannier orbitals, i.e., they must act onto the virtual space. The above procedure is part of the package MOLPRO where it has been implemented according to a proposal by *Pulay* [380].

Discarding for a moment the requirements of presently available quantum chemical program packages, the local character of the correlation hole is accounted for best by giving up the requirement of double substituting orthogonal occupied orbitals and using instead non-orthogonal local orbitals [435]. Thus we define non-orthogonal local functions $g_i(\mathbf{r})$ and expand them in terms of the basis set, i.e.,

$$g_i(\mathbf{r}) = \sum_m \gamma_{in} f_m(\mathbf{r}) \quad . \tag{5.82}$$

For example, the $g_i(\mathbf{r})$ can be GTOs or combinations of them. The operators which create (destroy) electrons with spin σ in local orbitals $g_i(\mathbf{r})$ are denoted by $b_{i\sigma}^+$ ($b_{i\sigma}$). The $b_{i\sigma}^+$ can be expanded in terms of occupied (subscript μ) and unoccupied (subscript n) Wannier orbitals, in which case

$$b_{i\sigma}^+ = \sum_{\mu=1}^{N/2} u_{i\mu} c_{\mu\sigma}^+ + \sum_{n=N/2+1}^{L} v_{in} c_{n\sigma}^+ \quad . \tag{5.83}$$

Constructing the cumulant wave operator $|\,\Omega)$ with the help of the $b_{i\sigma}^+, b_{j\sigma}$ we have to take into account that a state $b_{i\sigma}^+ b_{j\sigma'}^+ b_{k\sigma'} b_{l\sigma} |\,\Phi_{\mathrm{SCF}}\rangle$ does not only contain double substitutions (D), but also single and zero substitutions. They must be eliminated when one wants to consider double substitutions only, i.e.,

$$b_{i\sigma}^+ b_{j\sigma'}^+ b_{k\sigma'} b_{l\sigma} |\,\Phi_{\mathrm{SCF}}\rangle \xrightarrow{D} \sum_{mn\mu\nu} v_{im} v_{jn} u_{k\mu} u_{l\nu} \omega_{\mu\nu}^{mn} |\,\Phi_{\mathrm{SCF}}\rangle \quad . \tag{5.84}$$

---

[2] see Eq. (2.7)

The spin indices are again included in the indices m, n, μ, ν. This shows that we can work with non-orthogonal local orbitals as well. However, sophisticated computer program packages, which would enable us to perform efficient calculations with them are still missing.

Working with local orbitals has the advantage that the number of, e.g., single and double substitutions can be kept minimal for a required accuracy of the ground-state or binding energy, for example. As pointed out before, the local orbitals k and l which are substituted and those which replace them (i.e., i and j) have to be in fairly close spatial neighborhood of each other. As an example consider a lattice with one orbital per site i only. Furthermore assume that electrons interact only when they are on the same lattice site. If so, the interaction Hamiltonian is

$$H_{\text{int}} = U \sum_i a^+_{i\uparrow} a_{i\uparrow} a^+_{i\downarrow} a_{i\downarrow}$$
$$= U \sum_i n_{i\uparrow} n_{i\downarrow} \ . \quad (5.85)$$

In order to treat the interactions better than in a mean-field approximation at least double substitutions are required. We choose for them operators of the form

$$O_{ij} = \begin{cases} n_{i\uparrow} n_{i\downarrow} \delta_{ij}, \\ n_i n_j, \\ \mathbf{s}_i \mathbf{s}_j \end{cases} \quad (5.86)$$

with number operators

$$n_i = \sum_\sigma n_{i\sigma} \quad , \quad n_{i\sigma} = a^+_{i\sigma} a_{i\sigma} \quad (5.87)$$

and spin operators

$$\mathbf{s}_i = \frac{1}{2} \sum_{\alpha\beta} a^+_{i\alpha} \boldsymbol{\sigma}_{\alpha\beta} a_{i\beta} \quad (5.88)$$

but with all zero- and single excitations removed. This is achieved by replacing the O_{ij} by δO_{ij} with, e.g., $\delta(n_{i\uparrow} n_{i\downarrow}) = \delta n_{i\uparrow} \delta n_{i\downarrow}$ where $\delta n_{i\sigma} = n_{i\sigma} - \langle n_{i\sigma} \rangle$ and similar for $n_i n_j$ and $\mathbf{s}_i \mathbf{s}_j$. Note that for the Hamiltonian considered here the $a^+_{i\sigma}$ take the role of the operators $b^+_{i\sigma}$. The number of operators O_{ij} per site remains small, since the index j need not go beyond n.n. or n.n.n. sites of site i.

The above choice of substitutions has been labeled Local Ansatz [435] and allows for a simple physical interpretation. As discussed in Chapter 2 the shortcomings of the independent electron approximation, i.e., of $|\Phi_{\text{SCF}}\rangle$, result from configurations with large deviations of the charges and spins from their local mean values. In these configurations, the Coulomb repulsions between electrons are large. Therefore they are partially suppressed by correlations. This is achieved by the Local Ansatz.

The δO_{ij} describe double substitutions which correlate charge and spin degrees of freedom between sites i and j. When δO_{ij} acts on $|\Phi_{\text{SCF}}\rangle$ all those configurations are picked out in which sites i and j are occupied by electrons. For an illustration, consider the operator $\delta O_i = \delta(n_{i\uparrow} n_{i\downarrow})$ when applied to $|\Phi_{\text{SCF}}\rangle$, i.e., $\delta(n_{i\uparrow} n_{i\downarrow})|\Phi_{\text{SCF}}\rangle$. All those configurations of $|\Phi_{\text{SCF}}\rangle$ are picked out, in which site i is doubly occupied. These particular configurations carry too large a weight in $|\Phi_{\text{SCF}}\rangle$ and therefore have to be partially suppressed. An ansatz of the form

$$|\psi_i\rangle = (1 + \eta \delta 0_i) | \Phi_{\text{SCF}}\rangle \quad (5.89)$$

with $\eta < 0$ shows explicitely the partial suppression of configurations with a doubly occupied site i contained in $|\Phi_{\text{SCF}}\rangle$. Similar arguments hold for $n_j n_j$ and $\mathbf{s}_i \mathbf{s}_j$. For example, an ansatz $[1 + \eta_{ij} \delta(\mathbf{s}_i \mathbf{s}_j)]|\Phi_{\text{SCF}}\rangle$ with $\eta_{ij} < 0$ enhances configurations with two electrons of opposite spin at site i and j, while configurations with parallel spin arrangement are partially suppressed. When i and j are nearest neighbor sites this describes antiferromagnetic correlations. For a simple extension of (5.86) leading to further improvements, see [434].

5.4.4 Trial Wavefunctions

Trial wavefunctions have played an important role in understanding electron correlations. They are often well suited when one is interested in a qualitative rather than quantitative treatment of correlation effects. Usually a trial wavefunction for the ground state of a correlated system contains parameters which are optimized by minimizing the ground-state energy. Examples are the Jastrow ansatz for the correlated ground state of a homogeneous electron gas or the trial wavefunction suggested by *Gaskell* which includes zero-point fluctuations of plasmons in an RPA description (compare with Sect. 3.2). For an understanding of strongly correlated electron systems the Gutzwiller trial wavefunction has played an important role. Another famous example is the BCS wavefunction, which includes Cooper-pair correlations. We want to show here how trial wavefunctions can be obtained from $|\Omega\rangle$, which according to Sect. 5.3 specifies the exact ground state. Special approximations are hereby required. In order to explain them we start from the decomposition $H = H_0 + H_1$ and assume that we know the ground state $|\Phi_0\rangle$ of H_0. We use for $|\Omega\rangle$ the form (5.38) and expand in powers of H_1, i.e., we perform a perturbation expansion

$$\Omega = \lim_{z \to 0} \sum_{\nu=0}^{\infty} \left(\frac{1}{z - L_0} H_1 \right)^\nu \ . \quad (5.90)$$

Furthermore we decompose H_1 in terms of eigenoperators A_ν of L_0. The A_ν satisfy the equations

$$L_0 A_\nu = a_\nu A_\nu \quad . \tag{5.91}$$

They generate transitions between eigenstates of the unperturbed system with energy differences a_ν. In terms of a complete operator set A_ν the Hamiltonian H_1 can be expanded as

$$H_1 = \sum_\nu \lambda_\nu A_\nu \quad . \tag{5.92}$$

In the *single-mode* approximation the different eigenvalues a_ν in (5.92) are approximated by a single, mean-excitation energy ω_0 so that

$$L_0 H_1 = \omega_0 H_1 \quad . \tag{5.93}$$

We use this relation in order to derive an approximate form for $\tilde{\Omega}$. By inserting (5.93) into (5.90) we find

$$\begin{aligned} \Omega &= \sum_{\nu=0}^{\infty} \frac{1}{\nu!} \left(-\frac{H_1}{\omega_0} \right)^\nu \\ &= \exp\left(-\frac{H_1}{\omega_0} \right) \quad . \end{aligned} \tag{5.94}$$

The parameter ω_0 is chosen so as to minimize the ground-state energy.

Thus, within the single-mode approximation, a very simple result is obtained. The trial wavefunction for the exact ground state is given by the unperturbed ground state $|\Phi_0\rangle$ of H_0 multiplied by the exponential function $\exp(-\frac{H_1}{\omega_0})$, i.e.,

$$|\psi_0\rangle = e^{-\frac{H_1}{\omega_0}} |\Phi_0\rangle \quad . \tag{5.95}$$

This is *not* a coupled-cluster ansatz, since H_1 destroys and creates electrons in occupied as well as virtual orbitals of $|\Phi_0\rangle$. The distinction becomes important when expansion terms of the exponent higher than linear order in H_1 are considered. We use that finding in order to derive the Gutzwiller wavefunction. It is a trial wavefunction for the ground state of the Hubbard Hamiltonian and is discussed in more detail in Sect. 10.4. The latter consists of a kinetic energy part H_0 of the form $\sum_{ij\sigma} t_{ij} c_{i\sigma}^+ c_{j\sigma}$ and an interaction part H_1 given by (5.85). By using (5.95) we find in single-mode approximation the following trial function

$$|\psi_0\rangle = e^{\eta \sum_i n_{i\uparrow} n_{i\downarrow}} |\Phi_0\rangle \tag{5.96}$$

first proposed by *Gutzwiller* [164]. An equivalent way of writing the above is

$$|\psi_0\rangle = \prod_i (1 - \tilde{\eta} n_{i\uparrow} n_{i\downarrow}) |\Phi_0\rangle \quad , \tag{5.97}$$

where the property $n_{i\sigma}^2 = n_{i\sigma}$ has been used. The parameter η is related to $\tilde{\eta}$ through

$$\eta = \ln(1 - \tilde{\eta}) \quad . \tag{5.98}$$

As a second application of the single-mode approximation we consider a homogeneous electron gas with H_1 given by

$$H_{\text{int}} = \frac{1}{2\Omega} \sum_{\mathbf{q}} \left(\frac{4\pi e^2}{q^2} \rho_{\mathbf{q}}^+ \rho_{\mathbf{q}} - N \right) \quad . \tag{5.99}$$

Applying (5.95) we find the following form for the ground-state wavefunction

$$|\psi_0\rangle = \exp\left(-\Sigma'_{\mathbf{q}} \frac{2\pi e^2}{\omega_0 q^2} |\rho_{\mathbf{q}}|^2 \right) |\phi_0\rangle \quad . \tag{5.100}$$

The dash on the sum means that q should remain sufficiently small. When we choose for ω_0 the plasma frequency ω_{pl} we arrive at a wavefunction suggested by *Gaskell* [142] which contains the zero-point fluctuations of plasmons. From (3.37) it is seen that for $(p_F^2/3m)q^2 \gtrsim \omega_{pl}^2$ the single-mode approximation becomes very poor. Therefore q should be summed up only to a critical value q_c.

In order to improve the limitation of the wavefunction (5.100) we extend the previous approximation to the *independent-mode* approximation. Here we form groups of eigenoperators of L_0 and replace the different eigenvalues within each group with a group dependent mean-excitation energy. More specifically, we write

$$H_1 = \sum_i B_i \quad , \tag{5.101}$$

where the groups B_i consist of sums of eigenoperators $A_{i\nu}$ of L_0, i.e.,

$$B_i = \sum_{\nu} \lambda_{i\nu} A_{i\nu} \tag{5.102}$$

with

$$L_0 A_{i\nu} = a_{i\nu} A_{i\nu} \quad . \tag{5.103}$$

The $a_{i\nu}$ are approximated by a mean-excitation energy ω_i. We also assume that the B_i are independent of each other, i.e., that they commute. The perturbing Hamiltonian then consists of a sum of independent excitation modes and in analogy to (5.95) we find

$$|\psi_0\rangle = e^{-\Sigma_i \left(\frac{1}{\omega_i}\right) B_i} |\Phi_0\rangle \quad . \tag{5.104}$$

Returning to the interaction Hamiltonian (5.99) we consider $(2\Omega)^{-1} v_{\mathbf{q}} \rho_{\mathbf{q}}^+ \rho_{\mathbf{q}}$ as independent excitation modes. They correspond to particle-hole excitations with a given momentum $\mathbf{q}$. We identify them with the B_i and the ω_i with $\omega_{\mathbf{q}}$. Within this approximation

$$|\psi_0\rangle = e^{\Sigma_{\mathbf{q}} \eta_{\mathbf{q}} |\rho_{\mathbf{q}}|^2} |\Phi_0\rangle \quad , \tag{5.105}$$

5.4 Different Approximation Schemes

where we have set $(2\Omega\omega_{\mathbf{q}})^{-1}v_{\mathbf{q}} = \eta_{\mathbf{q}}$. The $\eta_{\mathbf{q}}$ are variational parameters which are obtained by minimizing the energy. In distinction to (5.100) the relative weight of the density-fluctuation modes described by the prefactor of $|\Phi_0\rangle$ is optimized here. Therefore the results are improved when q increases.

After a Fourier transformation of the exponent in (5.105) the ground-state wavefunction is of the form

$$|\psi_0\rangle = \exp\left[\int d^3r d^3r' f(\mathbf{r}-\mathbf{r}')\rho(\mathbf{r})\rho(\mathbf{r}')\right] |\Phi_0\rangle \;, \qquad (5.106)$$

where $|\Phi_0\rangle$ is the ground state of noninteracting electrons. This is Jastrow's variational ansatz for a correlated ground state [216]. The function $f(\mathbf{r}-\mathbf{r}')$ is determined by minimizing the energy. When written in first instead of second quantized form the last equation goes over into

$$\psi_0(\mathbf{r}_1,...,\mathbf{r}_N) = \exp\left[\sum_{ij} f(\mathbf{r}_i - \mathbf{r}_j)\right] \Phi_0(\mathbf{r}_1,...,\mathbf{r}_N) \;. \qquad (5.107)$$

Here $\Phi_0(\mathbf{r}_1,...,\mathbf{r}_N)$ is the Slater determinant for the ground state of the non-interacting system.

The Jastrow prefactor depends on relative coordinates $\mathbf{r}_i - \mathbf{r}_j$ only and therefore requires improvements when inhomogeneous systems are considered. In that case an ansatz of the form

$$|\psi_0\rangle = \exp\left(\sum_i \eta_i \delta n_i + \sum_{ij} \eta_{ij} \delta O_{ij}\right) |\Phi_{\text{SCF}}\rangle \qquad (5.108)$$

proves appropriate. The single-particle excitations $\delta n_i = n_i - \langle \Phi_{\text{SCF}} | n_i | \Phi_{\text{SCF}}\rangle$ reoptimize the electron density distribution when correlations are taken into account. The latter are implemented by the double excitations δO_{ij} given, e.g., by (5.86) or by (5.3).

6
Correlated Ground-State Wavefunctions

Correlation effects play a significant role practically in all materials. Here we shall discuss the various correlation contributions to the ground-state wavefunction and energy. Hereby we have to distinguish between insulators, which have a gap and metals, which don't have one. In the former case the SCF ground-state wavefunction can be expressed in terms of well localized orthogonal Wannier orbitals. As pointed out in Sect. 5.4 this is not the case when dealing with a metal. Yet, well localized Wannier orbitals are very suitable for the construction of the local correlation hole of an electron. This explains why somewhat different approaches are required for the determination of the ground-state wavefunction and energy of an insulator and of a metal.

We begin with a discussion of semiconductors and insulators and continue by including metals. It may come as a surprise that correlations in semiconductors and insulators are significant in view of the fact that the physical properties of semiconductors are nearly always explained in terms of a one-electron theory. However, this merely demonstrates that the quasiparticle picture works very well in that case. Correlation effects are here incorporated in quasiparticle properties. In the following we are interested in the ground state, i.e., its wavefunction as well as its properties such as the binding energy or bulk modules. As it will turn out, electron correlations contribute to the binding energy, e.g., of elemental semiconductors as much as one third. Their role becomes even more pronounced when excitations such as the energy gap are considered. But this is the subject of Chapter 7 and therefore not discussed here.

It is instructive to consider the physics behind the various correlation contributions to the ground-state wavefunction and energy. We will do this first by means of a semi-empirical calculation for covalent semiconductors. An important advantage is that these calculations can be done analytically. They give good insight as to where the different correlation contributions come from and how big they approximately are. This is followed by a discussion of the ab initio results for various covalently bonded semiconductors and for systems with ionic bonding such as MgO, CaO and NiO. They are obtained by employing the program package CRYSTAL for the SCF part of the calculation and the package MOLPRO for the correlation part. The methods outlined in Chapter 5 find applications here. Finally we also discuss rare-gas solids because of the special van der Waals bonding in those systems.

As pointed out before, metals require a modified approach when the ground-state energy and wavefunction are calculated. The modifications are explicitly demonstrated for Li metal. The ab initio results obtained compare very favourably with the best DFT results.

Towards the end of the chapter, we show that the above concepts can be extended to ground states of strongly correlated electron systems. The method of choice is here the CASSCF. The obvious question is, how an active space may be defined for an infinite periodic system. It can be answered by using the cumulant wave operator for the characterization of the ground-state wavefunction.

6.1 Semiconductors

We start out by performing semi-empirical calculations for covalently bonded semiconductors with a diamond lattice structure. They have the advantage that one obtains good insight into the underlying physical processes. We shall distinguish between inter- and intraatomic correlations[1] and give simple estimates for both cases.

6.1.1 Model for Interatomic Correlations

Calculations of *interatomic* correlations become very simple when a bond-orbital approximation (BOA) is made. The starting Hamiltonian is written as

$$H = \sum_{ij\sigma} t_{ij} a^+_{i\sigma} a_{j\sigma} + \frac{1}{2} \sum_{\substack{ijkl \\ \sigma\sigma'}} V_{ijkl} a^+_{i\sigma} a^+_{k\sigma'} a_{l\sigma'} a_{j\sigma} \quad , \tag{6.1}$$

where the $a^+_{i\sigma}$ and $a_{i\sigma}$ are creation and annihilation operators for electrons in orthogonalized, tetrahedral atomic sp^3 hybrids. They fulfill the anticommutation relations

$$[a_{i\sigma}, a^+_{j\sigma'}]_+ = \delta_{ij}\delta_{\sigma\sigma'} \quad . \tag{6.2}$$

These hybrids take the place of the basis functions $f_i(\boldsymbol{r})$ in (2.8). The parameters t_{ij} can be obtained by fitting them, e.g., to more sophisticated band structure calculations. It is convenient to introduce a special notation for the most important of the matrix elements V_{ijkl}. We use the following abbreviation for matrix elements referring to hybrids i and j forming bond I

---

[1] see Sect. 2.5

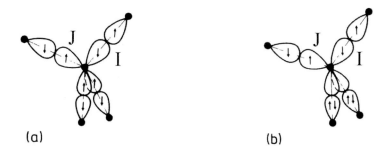

Fig. 6.1. Two configurations of a small segment of a diamond lattice structure. In (b) Coulomb repulsions are much larger than in (a). Note that the central C site has $\nu = 4$ valence electrons in (a) and $\nu = 2$ in (b).

$$\left. \begin{aligned} U &= V_{iiii} \\ J_1 &= V_{iijj} \\ K_1 &= V_{ijji} \end{aligned} \right\} \quad i \neq j, \quad i,j \in I \quad, \tag{6.3}$$

Calculations within the BOA are carried out best in terms of bonding and antibonding orbitals. The corresponding creation and annihilation operators are

$$B_{I\sigma}^+ = \frac{1}{\sqrt{2}}(a_{I1\sigma}^+ + a_{I2\sigma}^+) \quad, \tag{6.4a}$$

$$A_{I\sigma}^+ = \frac{1}{\sqrt{2}}(a_{I1\sigma}^+ - a_{I2\sigma}^+) \quad. \tag{6.4b}$$

Here we have indexed the two hybrids forming bond I by the subscripts 1 and 2. In this notation the SCF ground-state is approximated in the BOA by

$$|\Phi_{\text{BOA}}\rangle = \prod_{I\sigma} B_{I\sigma}^+ |0\rangle \quad. \tag{6.5}$$

There are two electrons in each bond.

In the correlated ground state $|\psi_0\rangle$ unfavorable configurations like the one in Fig. 6.1b must be partially suppressed. They have too large weights in $|\Phi_{\text{BOA}}\rangle$. Contrary to that, favorable configurations like the one in Fig. 6.1a become enhanced. We can do this by an ansatz of the form

$$|\psi_0\rangle = e^{\tilde{S}} |\Phi_{\text{BOA}}\rangle \tag{6.6}$$

like in (5.69) and applying subsequently the approximation of the local ansatz (5.86-5.88). Or, conversely, we can use the wave operator $|\Omega\rangle$ instead of the wavefunction $|\psi_0\rangle$ for characterizing the ground state and write

$$|\Omega\rangle = |1 + S\rangle \tag{6.7}$$

(see (5.50)). In the following we use the last equation as a starting point. For a construction of S we need to specify the correlations we want to describe. A plausible ansatz for S is of the form of (5.60), i.e.,

$$\begin{aligned} S &= \sum_{IJ} \eta_{IJ} S^\eta_{IJ} \\ &= \eta_0 \sum_I S^\eta_I + \eta_1 {\sum_{\langle IJ \rangle}}' S^\eta_{IJ} \quad , \end{aligned} \tag{6.8}$$

where S^η_I accounts for correlations of electrons in bond I and S^η_{IJ} for those within a pair of neighboring bonds $\langle IJ \rangle$ (indicated by a dash in the summation). The $S^\eta_I (= S^\eta_{II})$ and S^η_{IJ} are identified with the operators δO_{ij} of Sect. 5.4.3. The operator $S^\eta_I = \delta O_{ii}$ with $O_{ii} = n_{i\uparrow} n_{i\downarrow}$ and $n_{i\sigma} = a^+_{i\sigma} a_{i\sigma}$ is used to reduce configurations in which the hybrid 1 of bond I is doubly occupied. This reduces automatically also those configurations in which hybrid 2 of bond I is doubly occupied. In terms of the $A_{I\sigma}$ and $B_{I\sigma}$ operators, we may also write

$$S^\eta_I = \frac{1}{4} A^+_{I\uparrow} A^+_{I\downarrow} B_{I\downarrow} B_{I\uparrow} \quad . \tag{6.9}$$

A double excitation from bonding into antibonding states can be used to reduce configurations with two or zero electrons in a sp^3 hybrid orbital. This requires a negative value of η_0. Similarly we use $S^\eta_{IJ} = \delta O_{ij}$ with $O_{ij} = n_i n_j$ in order to describe correlations between electrons in neighboring bonds I and J. The S^η_{IJ} can be rewritten in the form

$$S^\eta_{IJ} = \frac{1}{4} \sum_{\sigma\sigma'} A^+_{J\sigma'} B_{J\sigma'} A^+_{I\sigma} B_{I\sigma} \quad . \tag{6.10}$$

The operator S^η_{IJ} generates dipoles in bonds I and J when acting on $|\Phi_{\text{BOA}}\rangle$. This is seen by expressing $A^+_{I\sigma} B_{I\sigma}$ in terms of $a^+_{I\nu\sigma}$ and $a_{I\nu\sigma}$ ($\nu = 1, 2$) and noticing that the operator contains parts which move within a bond an electron from one hybrid orbital to the other, resulting in a dipole. When a fluctuating dipole generates a dipole in another bond we speak of van der Waals interactions or correlations between the two bonds. The corresponding correlation energy falls off like R^{-6} with increasing bond separation R. With the above considerations we have specified the operators $\{A_\nu\}$ in (5.60) which span the relevant part $\mathfrak{R}_0$ of the Liouville space. We conclude that they are given by

$$\{A_\nu\} = \{S_I^\eta\} \oplus \{S_{I'J'}^\eta\} \quad . \tag{6.11}$$

What has been left out by the above choice for S_I^η and $S_{I'J'}^\eta$ are correlations between more distant bonds than nearest neighbors as well as spin-spin correlations. We use (5.52) in order to write for the (interatomic) correlation energy

$$E_{\text{corr}}^{\text{inter}} = \eta_0 \sum_I (H \mid S_I^\eta) + \eta_1 {\sum_{\langle IJ \rangle}}' (H \mid S_{IJ}^\eta) \quad . \tag{6.12}$$

The dash in the sum indicates that only nearest neighbor bonds are included. The parameters η_0 and η_1 are determined by applying (5.44b), i.e., from $(S_I^\eta | H\Omega) = 0$ and $(S_{IJ}^\eta | H\Omega) = 0$. More explicitly,

$$0 = (S_I^\eta \mid H) + \eta_0 \sum_K (S_I^\eta \mid HS_K^\eta) + \eta_1 {\sum_{\langle KL \rangle}}' (S_I^\eta \mid HS_{KL}^\eta)$$
$$0 = (S_{IJ}^\eta \mid H) + \eta_0 \sum_K (S_{IJ}^\eta \mid HS_K^\eta) + \eta_1 {\sum_{\langle KL \rangle}}' (S_{IJ}^\eta \mid HS_{KL}^\eta) \quad . \tag{6.13}$$

The different cumulants are readily evaluated. Thereby it is useful to express the electron interactions in terms of bonding and antibonding operators. These matrix elements of the interaction Hamiltonian are denoted by $\tilde{V}_{A_I B_J A_K B_L}$ etc. and are obtained from the V_{ijkl} by expressing the $a_{i\sigma}^+$ in terms of $A_{I\sigma}^+$ and $B_{I\sigma}^+$. With this notation we find

$$(S_I^\eta \mid H) = \frac{1}{2}\tilde{V}_{A_I B_I A_I B_I}$$
$$= \frac{1}{4}(U - J_1) = \frac{V_0^D}{2} \tag{6.14a}$$

$$(S_{IJ}^\eta \mid H) = \tilde{V}_{A_I B_I A_J B_J} - \frac{1}{2}\tilde{V}_{A_I B_J A_J B_I}$$
$$= V_1^D \quad . \tag{6.14b}$$

The matrix element V_1^D describes the van der Waals interaction between neighboring bonds I and J. The second term on the right-hand side of (6.14b) ensures that the interaction reduces to (6.14a) when $I = J$. Furthermore

$$(S_I^\eta \mid HS_K^\eta) = \delta_{IK} t_0 \quad , \tag{6.15}$$

where $t_0 = -t_{12} > 0$ denotes the (bare) hopping matrix element for the two sp^3 orbitals within a bond. It equals one-half of the energy splitting of bonding and antibonding states in the absence of the Coulomb interactions. In (6.14a) and (6.15) we have neglected all interaction matrix elements which are not of the form (6.3). We also find

$$(S_I^\eta \mid HS_{KL}^\eta) = \frac{1}{2}(\delta_{IK} + \delta_{IL})V_1^D \qquad (6.16)$$

$$\begin{aligned}(S_{IJ}^\eta \mid HS_{KL}^\eta) &= (\delta_{IK}\delta_{JL} + \delta_{IL}\delta_{JK})\left(t_0 + \frac{1}{2}\left(V_0^D + 4V_1^D\right)\right) \\ &= (\delta_{IK}\delta_{JL} + \delta_{IL}\delta_{JK})\alpha t_0 \;,\end{aligned} \qquad (6.17)$$

with

$$\alpha = 1 + \frac{V_0^D + 4V_1^D}{2t_0} \quad . \qquad (6.18)$$

As discussed before, only matrix elements involving nearest-neighbor bonds have been taken into account. The matrix element (6.16) corrects the formation of a dipole in bond I (see (6.14a)) in the presence of correlations in the neighboring bonds K or L, (local field corrections). It costs more energy to set up a dipole in the presence of these correlations than in their absence. The same feature shows up in (6.17). A dipolar fluctuation in bond I induces a dipole, e.g., in bond K and that dipole contributes to the electric field in bond J (via V_{ijkl}) and affects the size of the dipole which is formed there.

Inserting these matrix elements into (6.12) we obtain for the interatomic correlation energy per unit cell

$$E_{\text{corr}}^{\text{inter}} = -2V_0^D \eta_0 - 24V_1^D \eta_1 \;, \qquad (6.19)$$

while the set of equations (6.13) becomes

$$\begin{aligned} 0 &= \frac{V_0^D}{2} + t_0\eta_0 + 6V_1^D\eta_1 \\ 0 &= V_1^D + V_1^D\eta_0 + 2\alpha t_0\eta_1 \quad .\end{aligned} \qquad (6.20)$$

Table 6.1. Various matrix elements and the correlation parameters η_0 and η_1 for the elemental semiconductors. Also shown is the inter-atomic correlation energy per unit cell in eV. (From [42])

	Solid			
	C	Si	Ge	α-Sn
t_0	10.7	5.0	4.7	3.6
V_0^D	4.6	2.5	2.4	2.0
V_1^D	1.1	0.6	0.6	0.5
η_0	0.20	0.24	0.25	0.27
η_1	0.029	0.030	0.032	0.034
$-E_{\text{corr}}^{\text{inter}}$	2.6	1.6	1.6	1.5

By solving them we obtain

$$\eta_0 \simeq -\frac{V_0^D}{2t_0} \,, \quad \eta_1 \simeq -\frac{V_1^D}{2t_0\alpha}\left(1 - \frac{V_0^D}{2t_0}\right) \,. \tag{6.21}$$

Hereby we have neglected terms of order $(V_1^D/t_0)^2\alpha^{-1}$. The expression (6.19) shows in a simple form the energy contributions of correlations within a bond and between neighboring bonds. Table 6.1 lists for covalent semiconductors the parameters which enter (6.21). Before concluding we want to point out that interatomic correlations are reduced the more ionic a bond becomes, i.e., the larger the polarity α. The energy $E_{\text{corr}}^{\text{inter}}$ for homopolar bonds, which for given values of t_0, V_0^D and V_1^D is obtained from (6.19), is modified to

$$E_{\text{corr}}^{\text{inter}}(\alpha) = \left(1 - \alpha^2\right)^\nu E_{\text{corr}}^{\text{inter}}(\alpha = 0) \,. \tag{6.22}$$

The BOA predicts a value of $\nu = 2.5$ but a comparison with *ab initio* results using a minimal basis set suggest $\nu = 4.0 \pm 0.25$.

6.1.2 Estimates of Intra-Atomic Correlations

What remains to be discussed is the contribution of *intra*-atomic correlations to the binding energy. Their description requires large basis sets. Choosing diamond as an example, a simple estimate can be given as follows.

As pointed out in Sect. 2.4, in a chemical environment the number of valence electrons at a carbon site can vary between 0 and 8 with the average number given by 4. Therefore let us denote by $P_{\text{corr}}^C(\nu)$ the probability of finding n valence electrons at a C site. The total intra-atomic correlation energy (per unit cell) of diamond is then approximately given by

$$E_{\text{corr}}^{\text{intra}}(C_x) = 2\sum_\nu P_{\text{corr}}^C(\nu)\epsilon_\nu(C) \,, \tag{6.23}$$

where $\epsilon_\nu(C)$ is the correlation energy of a C atom in diamond with ν valence electrons. To be precise, we must exclude from it the contributions of $s^2p^{\nu-2} \to s^0p^\nu$ excitations because they are treated within a minimal basis set and therefore included in $E_{\text{corr}}^{\text{inter}}(C_x)$. The energy $\epsilon_\nu(C)$ is a weighted average of atomic correlation energies $\epsilon_\nu^{at}(i)$ belonging to different atomic terms or configurations i,

$$\epsilon_\nu(C) = \sum_i w_\nu(i)\epsilon_\nu^{at}(i) \,. \tag{6.24}$$

For the purpose of illustration we list in Appendix C the correlation energy of different configurations. Each configuration or term can give rise to different multiplets. For example, the configuration s^0p^2 can lead to a ^1S, ^3P or ^1D multiplet. In BOA the weights $w_\nu(i)$ are simply given by the sum of the degeneracies of those multiplets divided by 256 ($= 2^8$), i.e., the total number of possible states. Therefore we find for s^0p^2 that $\omega_\nu = 15/256$ (i.e., $(1+9+5)/256$ corresponding to the three multipletts). With their help $E_{\text{corr}}^{\text{intra}}(C_x)$ can be calculated provided we know the $P_{\text{corr}}^C(\nu)$. For the latter we may assume again a Gaussian distribution like in (2.49).

When estimating the intra-atomic correlation energy contribution to binding, one needs to know the correlation energy of a C atom in its ground state $(s^2p^2, {}^3P)$, which is $\epsilon_{\text{corr}}(C) = -4.27 eV$. By using (6.23) and subtracting the energy of the C atom, one obtains for the intra-atomic correlation contribution to binding of diamond a value of

$$E_B^{\text{corr}} = 1.2 \text{ eV/unit cell} \quad . \tag{6.25}$$

Together with the interatomic correlation energy contribution of 2.6 eV/unit cell (see Table 6.1) the total contribution of correlations to binding is

$$E_B^{\text{corr}}(C_x) = 3.8 \text{ eV/unit cell} \quad . \tag{6.26}$$

A comparison with the experimental cohesive energy for diamond of

$$E_B^{\text{exp}}(C_x) = 15.1 \text{ eV/unit cell} \tag{6.27}$$

reveals that after a correction for the zero-point energy of the atoms has been accounted for, correlations contribute approximately 25 % to it.

6.1.3 Ab Initio Results

After the above estimate of the correlation-energy contribution to binding we want to present some results based on the computational schemes discussed in Sect. 5. We consider again the elemental semiconductors. For the SCF part the program package CRYSTAL [372] is used. It contains a localization procedure for determining Wannier functions. An alternative approach to CRYSTAL is an embedded-cluster approach called Wannier [414], where the Wannier-Boys localization procedure is part of the SCF calculations. For C atoms a double-zeta basis set $(9s4p1d)/[3s2p1d]$ [99] is used to which $(1d1f)$ functions were added. For Si, Ge and Sn a four-valence-electron pseudopotential is employed for a simulation of atomic cores, while for the valence electrons, corresponding optimized basis sets $(4s4p2d1f)/[3s3p2d1f]$ are used. The use of pseudopotentials is not being discussed in this book, but a brief introduction of their use in the present context can be found, e.g., in [131] while profound reviews are, e.g., [91, 382].

When correlation calculations are performed the local character of the correlations hole is used extensively. For its description a finite cluster is completely sufficient as long as its size exceeds the extension of the correlations hole. This excludes long ranged pair correlations leading to Cooper pairing and superconductivity. Yet the cluster must be properly embedded in the infinite solid. The Wannier functions used in the cluster are these of the infinite system. We denote the cluster by C and subdivide it into an active central part C_A and an enclosing buffer C_B. The role of the buffer is to ensure that the tails of the Wannier functions centered at sites in C_A are properly taken into account. The occupied Wannier orbitals centered at sites in C_B are kept frozen. Yet, the unoccupied or virtual orbitals in C_B are included in the correlation calculations for C_A which involve all orbitals of that cluster. The effects of the solid outside the cluster C on the electrons inside are taken into account with the help of a special interface program connecting the cluster C and the remaining part of the infinite system. In its simplest form the infinite system is just modelled by point charges. When the electrons in C_A are correlated, their density distribution will slightly change. This is due to single-particle excitations which are part of the correlation calculations. These changes may be used to redefine the density distribution in C_B. This way self-consistency of the correlation calculations for the ground state is achieved.

The correlation contributions to the cohesive energy have been calculated by applying the methods described in Sect. 5, here in particular the CEPA-0 computational scheme. The package MOLPRO is used which has implemented a selection of excitation operators A_I^K (see (5.67)) in accordance with the discussion given in Sect. 5.4.3. Results are shown in Table 6.2 together with corresponding findings from density-functional calculations. The correlated ground-state wavefunction $|\psi_0\rangle$ is described by $|\Omega\rangle$ as given by (5.68). The numerical calculations include all excitation operators A_I^K which make significant contributions to the correlation energy and determine the corresponding parameters η_I^K. Therefore we may rightfully claim to have determined $|\psi_0\rangle$ with high accuracy. Instead of CEPA-0 we could also have applied the coupled-cluster method with marginal changes in the results. Not only the cohesive energy but also the lattice constant and the bulk modules have been calculated including correlation effects, both with good results.

Ground-state properties have also been calculated for III-V and II-VI compounds. Examples of III-V compounds are listed in Table 6.3 where the co-

Table 6.2. Cohesive energy per unit cell (in eV) as obtained from a SCF calculation and by including correlations by the method of increments ($E_{\text{incr}}^{\text{coh}}$) in percent of the experimental values. For comparison LDA results are shown. (From [358])

		C	Si	Ge	Sn
$E_{\text{SCF}}^{\text{coh}}$		10.74	6.18	4.25	3.65
$E_{\text{incr}}^{\text{coh}}$		14.36	8.83	6.98	6.08
		96 %	94 %	90 %	97 %
Expt.	[498]	15.10	9.39	7.75	6.23
LDA	[115], [154], [503]	17.25	10.58	9.06	–

92   6 Correlated Ground-State Wavefunctions

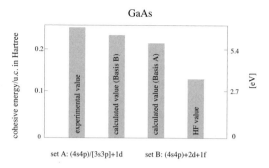

Fig. 6.2. Basis-set dependence of the correlation energy contributions to binding in GaAs. (From [359])

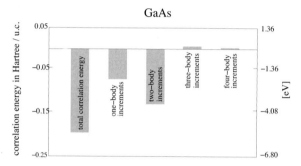

Fig. 6.3. Contributions of different many-body increments to the cohesive energy of GaAs. One notices a fast convergence with increasing number of correlated sites. Calculations are based on basis set A. (From [359])

hesive energy based on CCSD is compared with density functional results. As an example we consider GaAs. The basis-set dependence of the results is seen in Fig. 6.2 and gives an impression of how the size of the basis set influences the required accuracy. Fig. 6.3 demonstrates that the method of increments yields rapidly converging results. One- and two-body increments are essentially sufficient. Note that the sum of the two-body increments gives an almost twice as large contribution to cohesion than one-body increments

Table 6.3. Cohesive energy per unit cell (in eV) for selected III-V compounds. The results of wavefunction-based calculations are compared with LDA and GGA results based on density functional theory (from [359]).

	SCF	SCF+corr	LDA	GGA	Expt.
BN	9.12 (67 %)	12.38 (91 %)	16.57 (122 %)	13.74 (101 %)	13.60
AlP	5.39 (64 %)	7.94 (94 %)	10.09 (120 %)	8.38 (100 %)	8.41
GaAs	3.54 (53 %)	6.20 (93 %)	7.75 (116 %)	6.23 (93 %)	6.69

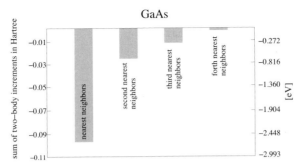

Fig. 6.4. Contributions to the cohesive energy of GaAs with increasing distance of the correlated electrons. Calculations were done with basis set A. (From [359])

do. It shows once more that single-site approximations for a treatment of electronic correlations are usually not sufficient. Correlation contributions fall off rather fast with increasing distance of the correlated electrons. This is seen from Fig. 6.4 which shows that one does not need to go beyond third-nearest neighbors in order to obtain rather accurate results for cohesion and for the correlations hole, as one might add. The above findings give a rather detailed picture of correlations in the ground state of that class of semiconductors. More details can be found, e.g., in the review [357].

6.2 Ionic and van der Waals Solids

Among the ionic solids oxides are of special interest. Consider for example MgO or alternatively $Mg^{2+}O^{2-}$, in order to stress the role of valency. There is no existing free O^{2-} ion. Evidently an O^{2-} ionic state in a solid is possible only because the chemical environment prevents electrons from leaving an O^{2-} site. This suggests a high polarizability of O^{2-} and therefore considerable correlation energy contributions to binding. Thus in addition to the ionic Coulomb repulsions, intersite correlations of van der Waals type will be important. This justifies their closer inspection, in particular since LDA type of calculations have problems with van der Waals interactions. The latter are crucial when binding in rare-gas solids is considered. Without van der Waals interactions there would be no binding in those systems. The method of increments is applied again for ionic as well as rare-gas solids, yet for the latter systems in a modified form. Since the rare-gas solids are weakly bound only, we calculate energy increments not of the correlation energy but instead of the *total* energy. This demonstrates the general character of the method.

6.2.1 Three Oxides: MgO, CaO and NiO

It is instructive to compare the correlation energy for the three oxides MgO, CaO and NiO, since Ca is in the Periodic Table just below Mg, and Ni has in distinction to the other two oxides in addition a nearly filled $3d$ shell. MgO is generally considered a nearly perfect ionic crystal consisting of Mg^{2+} and O^{2-} ions. The Wannier orbitals of the valence electrons as obtained, e.g., from CRYSTAL or elsewhere, have O $2p$, $2s$ character and are very compact and practicly limited to the oxygen sites. Therefore, when we decompose the scattering matrix S (see 5.53) the single-site contributions S_I involve excitations only from valence electrons on the oxygen sites. There are no valence electrons at the magnesium sites. In order to stick to a nearly perfect ionic description we include in S_I only excitations from Wannier orbitals at site I into virtual orbitals centered at the same site. All other electrons are kept frozen. Excitations into virtual orbitals centered at Mg sites are included in the matrices S_{IJ}. The different contributions to the S matrix are calculated by using the MOLPRO package and choosing either a CEPA-0 or a coupled-cluster approximation scheme, depending on the required accuracy. A basis set $[5s4p3d2f]$ is chosen for oxygens and a set $[6s6p5d2f1g]$ for Ca, combined with a small-core pseudopotential. For the Mg ions and their contributions to the S matrix a $[4s4p]$ valence basis set is used together with a large-core pseudopotential. The latter includes the effects of core-valence electron interactions on a SCF level together with a core-polarization potential accounting for correlation effects between the valence electrons and the core shell. For Ni the basis is $[6s5p4d3f]$ in combination with a Ne core pseudopotential. The correlation energy contributions for MgO and CaO were obtained by a CCSD computation. For NiO quasi-degenerate variational perturbation theory was used instead, because of the open d shell. Details are found in the original literature [93, 94].

The one-body correlation energy increments are listed in Table 6.4. Hereby O^{2-} ions were embedded by nearest neighbor X^{2+} pseudopotentials in order to simulate the repulsion due to Pauli's principle. The whole complex is embedded in a set of point charges arranged in a NaCl structure.

The two-body increments are calculated by correlating simultaneously electrons on two ions thereby keeping all other orbitals frozen. Results are to be multiplied with the number of such pairs of ions per primitive cell. Of interest is the decrease of the oxygen-oxygen increments with increasing distance d. For MgO this is shown in Table 6.5. Since these are van de Waals

Table 6.4. One-body increments of the correlation energy contributions to cohesion (in eV) for three oxides. (From [93, 94])

	MgO	CaO	NiO
O → O^{2-}	-2.62	-2.64	-2.74
X → X^{2+}	1.28	1.29	2.03
sum of 1-body increments	-1.35	-1.35	-0.72

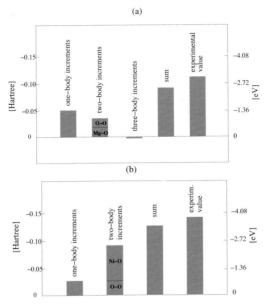

Fig. 6.5. One- to three-body correlation contributions to the cohesive energy per unit cell for (a): MgO and (b) NiO: three-body increments are omitted. (From [93])

interactions they decrease like d^{-6} but contribute significantly to the cohesive energy.

The one- to three-body increments to the correlation-energy contribution to cohesion for MgO and for NiO are shown in Fig. 6.5. The oxide CaO is very similar to MgO. The final results are listed in Table 6.6.

Table 6.5. Van der Waals like two-body increments to the correlation energy of oxygen-oxygen pairs at lattice sites (a, b, c) in MgO (in eV). (From [93])

pair of neighbors	2-body increm.	multiplic. factor	contribution to cohesive energy
O(0, 0, 0,) – O(0, 1, 1)	-0.069	6	-0.414
– O(2, 0, 0)	-0.006	3	-0.018
– O(2, 1, 1)	-0.002	12	-0.020
– O(2, 2, 0)	-0.001	6	-0.004
– O(3, 1, 0)	-0.000	12	-0.004
sum			-0.460

6.2.2 Rare-Gas Solids

Rare-gas crystals are particularly interesting since on a SCF level binding is not obtained. When second-order Møller-Plesset perturbation theory is applied we find for rare-gas dimers strong overbinding and the same holds true when the LDA is applied instead [252], [364]. Therefore rare-gas crystals are an ideal case for applying controlled approximations, i.e., wavefunction-based methods. In doing so we use an important modification: we apply the method of increments not to the computation of the correlation energy but instead to the total ground-state energy. Then H_0 in (5.26) describes a collection of free atoms while H_1 deals with the interactions between them. The decomposition of the S-matrix (5.53) begins with the S_{IJ} contributions and

Table 6.6. Sum of increments and experimental cohesive energy (in eV) for three oxides. The experimental correlation contribution is the difference between the measured cohesive energy and the Hartree-Fock value. (From [93, 94])

	MgO	CaO	NiO
experim. cohesive energy [73]	10.45	11.10	9.61
− Hartree Fock cohesive energy	-7.51	-7.59	-5.61
"experim." correl. contrib. to cohesive energy	= 2.93	= 3.51	= 4.00
calculated correl. contrib.	2.31 (79 %)	2.50 (71 %)	3.37 (84 %)
calculated cohesive energy	9.82	10.10	8.98
% of experim. value	94 %	91 %	93 %

Table 6.7. Sum of two-body increments and experimental cohesive energy (in eV) for rare-gas solids. (From [389])

including: E_{coh}	2-body contrib.	Expt.
Ne	-0.027	-0.027
Ar	-0.088	-0.089
Kr	-0.122	-0.122
Xe	-0.176	-0.170

yields the cohesive energy. Note that for large distances correlations are of van der Waals type. A fluctuating dipole on one atom induces a dipole on another atom as found before for bonds in the elemental semiconductors. For reliable results large basis sets are required. The following turn out being sufficient – for Ar, Kr, Xe a set: $(8s8p6d5f4g)/[7s7p6d5f4g]$ and for Ne: $(9s9p6d5f4g)/[7s7p6d5f4g]$. The core shells are described by pseudopotentials. Results for the cohesive energy are obtained by applying the CCSD(T) scheme. As seen from Table 6.7 the agreement with experiments is good. The same holds true for lattice constants, which are not shown here. More details are found in Ref. [389]. There it is also discussed that three-body contributions need to be included for a good agreement with experiments.

The above examples show that we can obtain detailed insight into the ground-state and the different correlation-energy contributions to cohesion by applying wavefunction-based methods and quantum-chemical techniques. Relative sizes of those contributions can be studied and compared when the chemical environment changes. We can also study differences in the correlation energy, e.g., of a bond when it is part of a molecule and of a solid. Until now only solids with small unit cells could be studied. In order to be able to treat solids with large unit cells and large basis sets considerable work is required in optimizing the available program packages.

6.3 Simple Metals

As we have seen in the preceding section, wavefunction-based methods are very efficient in van der Waals bonded rare-gas systems where DFT based approaches fail. On the other hand, they face some difficulties when the systems are metallic, i.e., where local density functionals are expected to be most accurate by construction. Thus from a practical point of view both approaches are complementary. When we consider bulk Li and imagine that the lattice constant is artificially enlarged, the system goes over from metallic to van der Waals bonding. Therefore, it is of interest to include metals in local abinitio wavefunction-based schemes. With this in mind, we turn to the cohesive energy of Li metal.

Metallic systems are difficult to treat with wavefunction-based correlation calculations because of the absence of well-localized Wannier orbitals. The latter fall off only algebraically and not exponentially like in insulators. Yet when we consider pairs of Li atoms, the pair orbitals are doubly occupied and act similarly as closed shell atoms. Therefore we determine localized orbitals ϕ_i, here according to the procedure proposed by *Pipek* and *Mezey* [370], associate them with single Li_2 units and compute the different increments to the S matrix for a single localized orbital, pairs of orbitals etc.

For a pair Li_2 of neighboring sites the wavefunction ϕ_i differs considerably from that of a free Li_2 molecule because of its chemical environment.

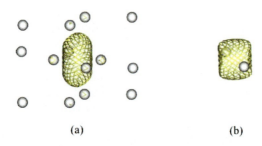

Fig. 6.6. (a) Localized Pipek-Mezey orbital of a Li$_2$ unit as compared with the valence orbital of a free Li$_2$ molecule (b). (From [433])

The differences are seen in Fig. 6.6. The correlation energy for the different increments is determined by using the CCSD computational scheme. The calculation can be done by considering a cluster of merely Li$_{14}$ sites with a basis set of triple-zeta quality. The experimental cohesive energy is $E_{\text{coh}}^{\text{exp}} = 1.66\text{eV/atom}$. When the Hartree Fock value of $E_{\text{coh}}^{\text{HF}} = 0.54\text{eV/atom}$ is subtracted one obtains for the experimental correlation-energy contribution to cohesion $E_{\text{coh}}^{\text{corr}} = 1.12\text{eV/atom}$. Results with the inclusion of second-nearest neighbor orbital pairs are shown in Fig. 6.7a. In order to see, how the quality of the basis influences the results, we show them also for a single- and a double-zeta basis set. The difference to the experimental value is less than 10 %. We may even go a step beyond and include pair energies up to the 6th-nearest neighbor and inter-pair correlations beyond those contained in the Li$_{14}$ sites cluster. For those contributions we increase the cluster to Li$_{50}$. For a cluster of that size only a single-zeta basis set can be used for computations. The final results for $E_{\text{coh}}^{\text{corr}}$ differ from the exact ones only by less than 3 % and compare very favorably with the best DFT results. The nice feature of the present wavefunction-based calculation is that one has a precise picture of where the correlation contributions come from. Similar computations can be done, e.g., for Na or K metal.

Before finishing we want to mention briefly mercury, which crystallizes below 233 K in a rhombohedral structure. Contrary to common belief, cohesion is solely due to electronic correlations and not to the single-particle kinetic energy. The cohesive energy is 0.79 eV/atom. A Hartree-Fock calculation does not give cohesion. It was considered a success when wavefunction-based calculations led to good results for the cohesive energy. The main findings are that binding results from two-site correlations, while single-site correlations

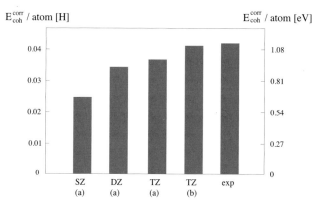

Fig. 6.7. Correlation energy contributions to the cohesive energy of Li metal, in Hartree and in eV. (a) Results from a Li_{14} cluster for different basis sets (single- to triple-zeta) up to 2nd-nearest neighbors (b) correlations for increments up to 6th-nearest neighbors. (From [433])

counteract binding here. Furthermore, it is crucial to include the d shell in the calculations. Relativistic effects are also important since without them we would find significant overbinding. For further details we refer, e.g., to the review [357].

6.4 Ground States with Strong Correlations: CASSCF

When electron correlations are strong like in d electron systems, which we assume in the following, an ordinary SCF calculation is not a proper starting point. The modifications that have to be made in order to account for correlations are very large. A much better starting point would be a CASSCF calculation in which the active space includes, e.g., the strongly correlated d electrons. But how do we formulate that computational scheme for an infinite solid, when the active space involves a finite number of orbitals per unit cell? In particular, how can we write down a ground-state wavefunction for a solid in that case? Here the advantages of the cumulant formulation of the correlation problem outlined in Sect. 5.3 come fully into play.

We characterize the many-body ground state by the cumulant wave operator $|\Omega\rangle$ or scattering operator $|S\rangle$ of the system. Consider the decomposition (5.53) of the S matrix. Furthermore, assume that a conventional SCF calculation has been done. The effect of S_I on $|\Phi_{\text{SCF}}\rangle$ should here be identified with the result of a CASSCF calculation with an active space restricted to orbitals on site I. Assume that the active space at site I is spanned by the atomic d orbitals $\phi_n(I)$ of that site. Then the projections $\{P_{\text{occ}}\phi_n(I)\}$ and $\{P_{\text{unocc}}\phi_n(I)\}$ onto the occupied and unoccupied orbitals in $|\Phi_{\text{SCF}}\rangle$ define localized orbitals in the two subspaces. A CASSCF calculation yields the optimal superposition of all possible configurations formed from the localized orbitals of the active space, while the remaining occupied orbitals are kept doubly occupied. By including in addition one-particle excitations from the active space to the remaining unoccupied localized orbitals, the ones in the active space are optimized. When 2-sites contributions S_{IJ} to the scattering operator are included, we can enlarge the active space so that it is the sum of the active spaces of sites I and J. However, that will generally not be necessary since interatomic correlations are usually not strong when I and J are some distance apart. Therefore, most of the remaining correlations can be computed by using one of the standard methods described in Sect. 5. CASSCF-based calculations have been performed for a number of strongly correlated systems. They enlarge considerably the range of application of wavefunction-based electronic structure calculations. Examples based on CASSCF are discussed in Chapter 12.

7
Quasiparticle Excitations

A determination of excited states of correlated electron systems would seem at first sight a formidable task, especially when the correlations are strong. Fortunately it turns out that the problem, though difficult is not unmanageable. The concept of quasiparticles is here of great help.

The traditional theories of electrons in metals, of which the theory of *Sommerfeld* and *Bethe* [420] is the most prominent, treat the electrons as an ideal gas of fermions. They move in an external potential set up by the nuclei and core electrons, however their mutual interactions are discarded. The Sommerfeld-Bethe theory has been very successful in describing qualitatively and - in its more elaborate form - even quantitatively the physical properties of systems like the alkali or earth-alkali metals. These findings became much better understood after *Landau* introduced the concept of quasiparticle and quasihole excitations [265, 266]. These excitations are restricted to an energy regime close to the Fermi surface and are indeed much weaker interacting than bare electrons. Instead of trying to calculate the quasiparticle interactions microscopically, indeed a difficult task, they are parameterized. These parameters enter different measurable quantities and can therefore be determined - at least in principle - by experiments. Landau's Fermi-liquid theory was originally devised for isotropic fermionic systems like ^{3}He, rather than realistic metals; if extended to anisotropic systems, it looses much of its simplicity and it becomes difficult to make predictions from it. Nevertheless, it had and still has a great influence on research in the field of metallic systems. In this Chapter we discuss only basis features of quasiparticles. The subject will come up again in a number of other chapters. In particular in Chapter 13 the usefulness and virtues of the concept of quasiparticles will become apparent.

One may speak of quasiparticle excitations not only in metals but also in semiconductors and insulators. When we add, e.g., an electron to a semiconductor its surroundings responds to this local perturbation by polarization and relaxation. The added electron plus its polarization and relaxation cloud forms a quasiparticle. It may move through the system in form of a Bloch wave giving rise to a dispersive conduction band. The same holds true when instead a hole is added to the system, i.e., when an electron is removed. It seems obvious that the energy gap between the conduction and valence bands will be reduced in comparison with the one in the absence of the response of the surroundings, i.e., without polarization and relaxation like in a SCF approximation. The reduction of energy gaps by correlation effects is an important subject in view of the very important role, which semiconductors play in research and development. Therefore, we have to be able to determine those gaps with high accuracy by ab initio methods, i.e., without adjustable parameters. Wavefunction-based methods are very useful here. Due attention will be paid to that topic in the later part of this chapter.

Quasiparticle excitations as well as incoherent excitations are defined by the poles of a single-particle Green's function. Therefore, we start by recalling basic facts about that function. At the beginning the Green's function formulation for zero temperature is considered. Then we go on to describe the extension to finite temperatures.

7.1 Single-particle Green's Function

In order to calculate the dispersion of quasiparticles and the spectral function $A(\mathbf{p}, \omega)$ we must determine the single-particle Green's function. This function will also enable us to calculate the incoherent contributions to the spectral density, which result from the internal degrees of freedom of the correlation hole and are the subject of the next chapter.

Let c_j^+ and c_j denote creation and annihilation operators for electrons in state j. The index j stands for any set of four quantum numbers that characterize the electron. For example, it may stand for the momentum vector $\mathbf{p}$ and spin σ, for a site index i and spin σ, or for a MO and spin index, depending on the problem. In the Heisenberg representation the time evolution of these operators is given by $idc_j^+/dt = [c_j^+, H]_-$ or

$$\begin{aligned} c_j^+(t) &= e^{iHt} c_j^+ e^{-iHt} \\ &= e^{iLt} c_j^+ \quad . \end{aligned} \tag{7.1}$$

Here H is the Hamiltonian of the system and L is the Liouville operator associated with H; see (5.39). The causal Green's function $G_{ij}(t-t')$ is defined as

$$G_{ij}(t-t') = -i \langle \psi_0^N | T\left(c_i(t) c_j^+(t')\right) | \psi_0^N \rangle \quad , \tag{7.2}$$

where ψ_0^N denotes the (exact) ground state of the N-electron system. T is a time-ordering operator; it orders products of time-dependent operators by placing operators with the larger time argument to the left of those with the smaller time argument. The overall sign depends on the number of permutations required to achieve time order. For the operator product in (7.2), this implies

7.1 Single-particle Green's Function

$$T\left(c_i(t)c_j^+(t')\right) = \begin{cases} c_i(t)c_j^+(t') & , \quad t > t' \\ -c_j^+(t')c_i(t) & , \quad t < t' \end{cases} \tag{7.3}$$

One notices that the Green's function is the probability amplitude of finding an electron at time t in state i, after at time t' an electron in state j has been added to the ground state. The Green's function describes electron propagation for $t > t'$ and hole propagation for $t < t'$. When we identify the quantum numbers i with $\mathbf{p}, \sigma$ it follows from (7.1,7.2) that

$$G(\mathbf{p}, t) = \begin{cases} -i e^{i\mathcal{E}_0^N t} \langle \psi_0^N | c_{\mathbf{p}\sigma} e^{-iHt} c_{\mathbf{p}\sigma}^+ | \psi_0^N \rangle & , \quad t > 0 \\ i e^{-i\mathcal{E}_0^N t} \langle \psi_0^N | c_{\mathbf{p}\sigma}^+ e^{iHt} c_{\mathbf{p}\sigma} | \psi_0^N \rangle & , \quad t < 0 \end{cases} \tag{7.4}$$

As before $\mathcal{E}_0^N$ denotes the ground-state energy of the interacting N electron system.

The *Fourier* transform of the Green's function

$$G(\mathbf{p}, \omega) = \int_{-\infty}^{+\infty} dt\, e^{i\omega t} G(\mathbf{p}, t) \tag{7.5}$$

can be written in form of a spectral representation

$$G(\mathbf{p}, \omega) = \int_0^\infty d\omega' \left(\frac{A(\mathbf{p}, \omega')}{\omega - \omega' + i\eta} + \frac{B(\mathbf{p}, \omega')}{\omega + \omega' - i\eta} \right) \tag{7.6}$$

with

$$A(\mathbf{p}, \omega) = \sum_n |\langle \psi_n^{N+1} | c_{\mathbf{p}\sigma}^+ | \psi_0^N \rangle|^2 \delta\left(\omega - \delta\mathcal{E}_n^{N+1}\right) ,$$

$$B(\mathbf{p}, \omega) = \sum_m |\langle \psi_m^{N-1} | c_{\mathbf{p}\sigma} | \psi_0^N \rangle|^2 \delta\left(\omega - \delta\mathcal{E}_m^{N-1}\right) . \tag{7.7}$$

The expressions for $A(\mathbf{p}, \omega)$ and $B(\mathbf{p}, \omega)$ follows from (7.4,7.5) by inserting into (7.4) a complete set of eigenstates of the $(N+1)$ and $(N-1)$ particle system with eigenvalues $\mathcal{E}_n^{N+1}$ and $\mathcal{E}_m^{N-1}$, respectively. The $\delta\mathcal{E}_n^{N\pm1}$ are the excitation energies of the $N \pm 1$ particle system.

The spectral representation of $G(\mathbf{p}, \omega)$, or more generally of $G_{ij}(\omega)$ shows that the Green's function represents an analytic function in the ω plane, except near the real axis. Therefore it can be constructed from two functions $G_{ij}^R(\omega)$ (retarded Green's function) and $G_{ij}^A(\omega)$ (advanced Green's function) which are analytic in the upper and lower ω half plane, respectively.

The relation is

$$G_{ij}(\omega) = \begin{cases} G_{ij}^R(\omega) & , \quad \mathrm{Re}\{\omega\} > \mu \\ G_{ij}^A(\omega) & , \quad \mathrm{Re}\{\omega\} < \mu \end{cases} \tag{7.8}$$

Because of their analytic properties the retarded and advanced Green's functions have for a translationally invariant system the following simple spectral representation:

$$G^R(\mathbf{p},\omega) = \int_{-\infty}^{+\infty} d\omega' \frac{\rho(\mathbf{p},\omega')}{\omega - \omega' + i\eta}$$
$$G^A(\mathbf{p},\omega) = \int_{-\infty}^{+\infty} d\omega' \frac{\rho(\mathbf{p},\omega')}{\omega - \omega' - i\eta} \quad , \qquad (7.9)$$

where the spectral function $\rho(\mathbf{p},\omega)$ is analytic in the whole ω plane. One notices that for real frequencies $G^R(\omega) = G^A(\omega)^*$. The corresponding time-dependent functions are

$$G^R_{ij}(t) = -i\Theta(t)\langle\psi_0^N|\left[c_i(t),c_j^+(0)\right]_+\left|\psi_0^N\right\rangle$$
$$G^A_{ij}(t) = i\Theta(-t)\langle\psi_0^N|\left[c_i(t),c_j^+(0)\right]_+\left|\psi_0^N\right\rangle \quad . \qquad (7.10)$$

Here $\Theta(t)$ is the step function, i.e., $\Theta(t) = 1$ for $t \geq 0$ and $\Theta(t) = 0$ for $t < 0$.

The equation of motion for the Green's function $G_{ij}(\omega)$ follows from (7.2) and is

$$i\frac{d}{dt}G_{ij}(t-t') = \frac{d}{dt}\left\{\Theta(t-t')\left\langle c_i(t)c_j^+(t')\right\rangle - \Theta(t'-t)\left\langle c_j^+(t')c_i(t)\right\rangle\right\}$$
$$= \delta(t-t')\left\langle\left[c_i(t),c_j^+(t')\right]_+\right\rangle + \left\langle T\left[c_i(t),H\right]_- c_j^+(t')\right\rangle. \quad (7.11)$$

From this equation of motion we can obtain another very useful representation, which looks similar for the retarded, advanced and full Green's function. Moreover, it applies not only to Green's functions formed with operators $c_i^+(t)$ and $c_j(t')$ but to Green' functions formed with arbitrary operators $A(t)$, $B(t')$ as well. We label the Fourier transforms of those Green's functions by $G_{AB}(\omega) = \ll A; B \gg_\omega$ a much-used notation in the literature. For such a function the equation of motion has the generalized form

$$\omega \ll A; B \gg_\omega = \langle\psi_0^N|[A,B]_+|\psi_0^N\rangle + \ll [A,H]_-; B \gg_\omega \qquad (7.12)$$

showing that the Green's function $\ll A; B \gg_\omega$ is coupled to another one, i.e., $\ll [A,H]_-; B \gg_\omega$. Setting up the equation for the new Green's function shows that it is coupled to yet another one, and so on. When we apply this to $G(\mathbf{p},\omega)$ and decompose the Hamiltonian into $H = H_0 + H_1$, with H_0 given by

$$H_0 = \sum_{\mathbf{p}\sigma}(\epsilon_\mathbf{p} - \mu)c_{\mathbf{p}\sigma}^+ c_{\mathbf{p}\sigma} \quad , \qquad (7.13)$$

we obtain the relation

$$\omega G(\mathbf{p},\omega) = 1 + (\epsilon_\mathbf{p} - \mu)G(\mathbf{p},\omega) + \ll [c_{\mathbf{p}\sigma}, H_1]_-; c_{\mathbf{p}\sigma}^+ \gg_\omega \quad . \qquad (7.14)$$

7.1 Single-particle Green's Function

We can define a mass operator $\Sigma(\mathbf{p}, \omega)$ by formally setting

$$\ll [c_{\mathbf{p}\sigma}, H_1]_-; c^+_{\mathbf{p}\sigma} \gg_\omega = \Sigma(\mathbf{p}, \omega) G(\mathbf{p}, \omega) \quad . \tag{7.15}$$

When set into (7.14) this yields the form

$$G(\mathbf{p}, \omega) = \frac{1}{\omega - (\epsilon_\mathbf{p} - \mu) - \Sigma(\mathbf{p}, \omega)} \quad . \tag{7.16}$$

The function $\Sigma(\mathbf{p}, \omega)$ is called self-energy. The poles of the Green's function give the excitations of the system. In the present case they are obtained from

$$\omega = \frac{p^2}{2m} - \mu + \Sigma(\mathbf{p}, \omega) \quad . \tag{7.17}$$

For "normal" Fermi-liquid systems, which exclude for example superconductors, one can expand the self-energy around the energy $\omega = 0$, i.e.,

$$\Sigma(\mathbf{p}, \omega) = \Sigma(\mathbf{p}, 0) + \left.\frac{\partial \Sigma(\mathbf{p}, \omega)}{\partial \omega}\right|_{\omega=0} \omega \quad . \tag{7.18}$$

When we set this expansion into (7.16) we obtain a pole with the real part given by

$$\begin{aligned}\epsilon_\mathbf{p} &= Z \left(p^2/2m - \mu + \Sigma(\mathbf{p}, 0)\right) \\ &\simeq v_F^* (p - p_F) \quad ,\end{aligned} \tag{7.19}$$

where the renormalization constant Z is

$$Z = \frac{1}{1 - \partial \Sigma(\mathbf{p}, \omega)/ \partial \omega|_{\omega=0}} \quad . \tag{7.20}$$

From (7.18) the quasiparticle mass is obtained according to (4.35).

It turns out that for a normal Fermi liquid the imaginary part of the pole is $\gamma_\mathbf{p} \sim (p - p_F)^2$ (see (7.82)). The self-energy of a Fermi liquid has therefore in the low-frequency limit the general form

$$\Sigma(\mathbf{p}, \omega) = A(\mathbf{p}) + B(\mathbf{p})\omega + iC(\mathbf{p})\omega^2 \quad . \tag{7.21}$$

As the Fermi energy is approached, the imaginary part vanishes faster than the real part and therefore quasiparticle excitations are well defined. For small excitation energies the Green's function can be split into the form

$$G(\mathbf{p}, \omega) = \frac{Z}{\omega - \epsilon_\mathbf{p} - i\gamma_\mathbf{p} \operatorname{sgn} \omega} + G_{\text{inc}}(\mathbf{p}, \omega) \quad , \tag{7.22}$$

where $G_{\text{inc}}(\mathbf{p}, \omega)$ is the incoherent part. It is well-behaved near the Fermi energy, having often branch cuts as a function of ω, instead of poles. By taking the Fourier transform of (7.22) we find for $t > 0$

$$G(\mathbf{p}, t) = -iZ(\mathbf{p})e^{-i\epsilon_\mathbf{p} t - \gamma_\mathbf{p} t} + G_{\text{inc}}(\mathbf{p}, t) \tag{7.23}$$

provided $t \gg 1/\epsilon_\mathbf{p}$ and $|\mathbf{p}| > p_F$. From the definition (7.2) of the Green's function it follows that $Z(\mathbf{p})$ can be interpreted as the weight of the bare electron within the quasiparticle. The part $G_{\text{inc}}(\mathbf{p}, t)$ describes the dynamics of the modified surroundings due to Coulomb repulsions or alternatively of the correlation hole of an added electron. The renormalization constant $Z(p_F)$ describes the discontinuity in the momentum distribution $n_\sigma(\mathbf{p})$ of the electrons at the Fermi surface. This is seen as follows. By making use of (7.4) we can write the momentum distribution $n_\sigma(\mathbf{p})$ of the electrons (not quasiparticles) in the ground state of the system as

$$\begin{aligned} n_\sigma(\mathbf{p}) &= \langle \psi_0^N | c_{\mathbf{p}\sigma}^+ c_{\mathbf{p}\sigma} | \psi_0^N \rangle \\ &= -i\, G_\sigma(\mathbf{p}, t)|_{t \to -0} \end{aligned} \tag{7.24}$$

In the discussion below we suppress the index σ. The last equation is written as

$$\begin{aligned} n(\mathbf{p}) &= -\frac{i}{2\pi} \lim_{t \to -0} \int d\omega\, e^{-i\omega t}\, G(\mathbf{p}, \omega) \\ &= -\frac{i}{2\pi} \int_C d\omega\, G(\mathbf{p}, \omega) \quad. \end{aligned} \tag{7.25}$$

The closed contour C extends along the real axis and includes a semicircle in the upper ω half-plane at $|\omega| \to \infty$. When using the quasiparticle representation (7.22) for $G(\mathbf{p}, \omega)$, the quasiparticle poles contribute for $p < p_F$ but not for $p > p_F$ because in the latter case they are in the lower ω half-plane. This causes a discontinuity in $n(\mathbf{p})$ at p_F given by

$$\lim_{\eta \to 0} [n(p_F - \eta) - n(p_F + \eta)] = Z(p_F) \quad. \tag{7.26}$$

We show this schematically in Fig. 7.1.

7.1.1 Perturbation Expansions

We demonstrate in what follows how the Green's function can be evaluated with the help of perturbation expansions. For that purpose, we consider the Green's function matrix $G_{ij}(t)$, where the operators c_i, c_j^+ destroy and create electrons in SCF spin orbitals i and j, respectively. The aim is to compute this matrix in powers of the residual interactions (2.37), for which one must express the expectation value (7.2) in terms of $|\Phi_{\text{SCF}}\rangle$. We achieve this by going over from the Heisenberg to the interaction representation.

When we start from the Schrödinger equation

$$i\frac{\partial \phi}{\partial t} = H\phi \tag{7.27}$$

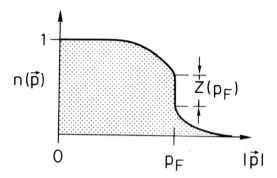

Fig. 7.1. Discontinuity $Z(p_F)$ in the momentum distribution $n(\mathbf{p})$. It determines the residue of the quasiparticle pole in the Green's function $G(\mathbf{p}, \omega)$.

the state $\phi(t)$ depends on time according to

$$\phi_S(t) = e^{-iHt}\phi_H \quad . \tag{7.28}$$

This is called the Schrödinger representation. In the Heisenberg representation, the state ϕ (i.e., $\phi = \phi_H$) remains time independent and the time dependence is shifted to the operators. Now consider (7.28) in order to establish the advantage of a third representation, i.e., the interaction representation. Dividing H into $H = H_0 + H_1$, we would like to split off a factor $\exp(-itH_0)$ so as to treat the remaining part of the time evolution by perturbation theory. For H_0 we can either choose H_{SCF} or, alternatively, the part without the electron-electron interactions. The following decomposition holds:

$$\begin{aligned} e^{-iHt} &= e^{-iH_0 t} T e^{-i\int_0^t d\tau H_1(\tau)} \\ &= e^{-iH_0 t} \mathbf{U}(t, 0) \quad , \end{aligned} \tag{7.29}$$

where the time dependence of $H_1(\tau)$ is according to

$$H_1(\tau) = e^{iH_0\tau} H_1 e^{-iH_0\tau} \quad . \tag{7.30}$$

Equation (7.29) is proven by differentiating both sides. Setting

$$y(t) = e^{-iH_0 t} T e^{-i\int_0^t d\tau H_1(\tau)} \quad , \tag{7.31}$$

we find by differentiation

$$\frac{dy(t)}{dt} = -iH_0 y(t) - i e^{-iH_0 t} H_1(t) T e^{-i\int_0^t d\tau H_1(\tau)} \quad . \tag{7.32}$$

In the last term $H_1(\tau)$ appears in front of the T product because t is larger than any of the values of τ. Using (7.30, 7.31), we write (7.32) as

$$\frac{dy(t)}{dt} = -i(H_0 + H_1)y(t)$$
$$= -iHy(t) \ . \qquad (7.33)$$

By integration we obtain the left-hand side of (7.29).

We introduce still another representation with the requirement that the states of the system $\phi_I(t)$ depend on time according to

$$\phi_I(t) = e^{iH_0 t}\phi_S(t) \ . \qquad (7.34)$$

This is the interaction representation. From (7.28, 7.29) it follows that

$$\phi_I(t) = \mathbf{U}(t,0)\phi_H \ . \qquad (7.35)$$

Note that $\phi_H = \phi_S(0) = \phi_I(0)$ at $t = 0$.

In the interaction representation, operators $A(t)_I$ evolve in time according to $idA(t)_I/dt = [A(t)_I, H_0]_-$, or

$$A(t)_I = e^{iH_0 t} A_S e^{-iH_0 t}$$
$$= \mathbf{U}(t,0)A(t)_H \mathbf{U}^+(t,0) \ . \qquad (7.36)$$

The indices S and H refer to operators in the Schrödinger and Heisenberg representation, respectively. As a reminder, the time dependence of operators in the Heisenberg representation is given by (7.1).

The matrix $\mathbf{U}(t,0)$ has the following properties:

$$\begin{aligned}\mathbf{U}\mathbf{U}^+ &= 1 \ , \\ \mathbf{U}(t,t') &= \mathbf{U}^+(t',t) = \mathbf{U}^{-1}(t',t) \ , \\ \mathbf{U}(t,t') &= \mathbf{U}(t,t'')\mathbf{U}(t'',t') \ . \end{aligned} \qquad (7.37)$$

The interaction representation allows for the required connection between the exact ground state $|\psi_0\rangle$ and the ground state $|\Phi_0\rangle$ of H_0, provided the adiabatic hypothesis is made. This assumes that $|\psi_0\rangle$ is obtained from $|\Phi_0\rangle$ by adiabatically switching on the interaction H_1 at time $t = -\infty$, so that the full interaction is present at $t = 0$. This implies that

$$|\psi_0\rangle = |\psi_0(0)\rangle = \mathbf{U}(0,-\infty)|\Phi_0\rangle \ . \qquad (7.38)$$

If we use the last of the relations (7.37), we can also write

$$|\psi_0\rangle = \mathbf{U}(0,\infty)\mathbf{U}(\infty,-\infty)|\Phi_0\rangle$$
$$= \mathbf{U}(0,\infty)\mathbf{S}|\Phi_0\rangle \ , \qquad (7.39)$$

where $\mathbf{S}$ is called the scattering matrix. It differs from the scattering matrix discussed in Chapter 5, yet both are interrelated. Starting from $|\Phi_0\rangle$ and slowly turning on and off the interaction H_1, we see that the final state can differ from the initial one only by a phase α. Thus, we consider $|\Phi_0\rangle$ an eigenstate of $S = \mathbf{U}(\infty, -\infty)$, i.e.,

$$S|\Phi_0\rangle = e^{i\alpha}|\Phi_0\rangle \quad . \tag{7.40}$$

It is not difficult to show that in the interaction representation the Green's function (7.2) can be written as

$$G_{ij}(t) = -i \frac{\langle T\left(c_i(t)c_j^+(0)S\right)\rangle}{\langle S \rangle} \quad , \tag{7.41}$$

where $\langle \ldots \rangle = \langle \Phi_0 | \ldots | \Phi_0 \rangle$ and the operators c_i, c_j^+ depend on time according to (7.36). In the course of this the assumption is that $t > 0$. The proof for $t < 0$ is completely analogous. With (7.38, 7.39) and (7.36), we can write (7.2) in the form

$$G_{ij}(t) = -i\langle\Phi_0|S^+\mathbf{U}^+(0,\infty)\mathbf{U}^+(t,0)c_i(t)\mathbf{U}(t,0)c_j^+(0)\mathbf{U}(0,-\infty)|\Phi_0\rangle$$
$$= -ie^{-i\alpha}\langle\Phi_0|\mathbf{U}(\infty,t)c_i(t)\mathbf{U}(t,0)c_j^+(0)\mathbf{U}(0,-\infty)|\Phi_0\rangle \quad . \tag{7.42}$$

After introducing the time-ordering operator, the operators can be reshuffled, resulting in

$$G_{ij}(t) = -ie^{-i\alpha} \left\langle T\left[c_i(t)c_j^+(0)S\right]\right\rangle \quad . \tag{7.43}$$

The phase factor $e^{-i\alpha}$ can be replaced by $\langle S \rangle^{-1}$, yielding (7.41). If we expand

$$S = Te^{-i\int_{-\infty}^{+\infty} d\tau H_1(\tau)} \quad , \tag{7.44}$$

we obtain the Green's function $G_{ij}(t)$ in the form of a perturbation expansion

$$G_{ij}(t) = \frac{-i}{\langle S \rangle} \sum_{n=0}^{\infty} \frac{(-i)^n}{n!} \int_{-\infty}^{+\infty} dt_1 \ldots \int_{-\infty}^{+\infty} dt_n$$
$$\times \left\langle T\left[c_i(t)c_j^+(0)H_1(t_1)\ldots H_1(t_n)\right]\right\rangle \quad . \tag{7.45}$$

The expectation values in this expansion can be evaluated if we apply Wick's theorem [485]. They are all of the form

$$\langle \Phi_0 | T[A_1 \ldots A_n] | \Phi_0 \rangle \quad , \tag{7.46}$$

where the A_n are either electron creation or annihilation operators $c_i^+(t)$, $c_i(t)$ in the interaction representation.

Wick's theorem describes how to break up such expectation values into products of expectation values, each involving one creation and one annihilation operator only. To state the theorem we first introduce a *normal order* of operators $N[A_1 \ldots A_n]$ defined by moving all "creation" operators to the left of the "annihilation" operators and associating a minus sign with each commutation. The quotation marks highlight the convention used here for c_i, which is considered to be an annihilation operator when the subscript i refers to a virtual orbital (i.e., an orbital which is unoccupied in $|\Phi_0\rangle$) and a creation operator (creation of a hole) when i refers to an occupied orbital. The advantage of introducing a normal order of operators is that

$$\langle \Phi_0 | N[A_1 \ldots A_n] | \Phi_0 \rangle = 0 \tag{7.47}$$

by definition. Having defined the normal order, a *contraction* between a creation operator A_1 and an annihilation operator A_2 is introduced by

$$\overrightarrow{A_1 A_2} = T[A_1 A_2] - N[A_1 A_2] \tag{7.48}$$

Wick's theorem states that a T product of creation and annihilation operators can be uniquely decomposed into a sum of normal ordered products according to the rule

$$T[A_1 \ldots A_n] = N[A_1 \ldots A_n] + N[\overrightarrow{A_1 A_2} \ldots A_n] + \cdots$$
$$+ N[\overrightarrow{A_1 A_2} \ldots A_\nu \ldots A_n] + N[\overrightarrow{A_1 A_2} \ldots A_\nu \ldots A_n] + \cdots$$
$$+ N[\overrightarrow{A_1 A_2 A_3} \ldots A_n] + \cdots$$
$$+ N[\overrightarrow{A_1 A_2} \ldots A_\mu \ldots A_\tau \ldots A_\rho \ldots A_n] + \cdots \quad . \tag{7.49}$$

The following notation is used:

$$N[A_1 \ldots \overrightarrow{A_\nu} \ldots \overrightarrow{A_\mu} \ldots A_n] = (-1)^{\mu-\nu-1} \overrightarrow{A_\nu A_\mu} N[A_1 \ldots A_{\nu-1} A_{\nu+1} \ldots$$
$$\times \ldots A_{\mu-1} A_{\mu+1} \ldots A_n] \quad . \tag{7.50}$$

The number of commutations determines the sign of this expression. Wick's theorem requires that the operators, which appear in the unperturbed Hamiltonian H_0, fulfill simple fermionic (or bosonic) commutation relations among themselves and the operators in H_1. We find this to be true in the case that H_0 is a one-particle Hamiltonian like H_{SCF}. Yet, when the electron correlations are strong, we should include most parts of the strong Coulomb repulsion in H_0. The remaining H_1 contains the weak resonance interactions (hybridizations) in powers of which we would like to expand. The operators, which diagonalize H_0 no longer obey a simple commutation algebra and therefore converting a time-ordered product into a normal-ordered one becomes much more complex. From (7.48) we conclude that

7.1 Single-particle Green's Function

$$\langle \overline{c_i(t) c_j^+(0)} \rangle = \langle T c_i(t) c_j^+(0) \rangle = i G_{ij}^{(0)}(t) \ . \tag{7.51}$$

The Green's function $G_{ij}^{(0)}(t)$ is that of a system treated in SCF approximation. Because of (7.47), the only terms which are left when the decomposition (7.49) is inserted into (7.46) are those in which all operators A_ν are contracted in pairs. Therefore, the perturbation expression (7.45) for the Green's function $G_{ij}(t)$ can be decomposed into a sum of products of Green's functions $G_{mn}^{(0)}(t)$ of independent electrons; it should be remembered, however, that while H_{SCF} contains electron interactions it does so only in a mean-field or independent-electron approximation. This decomposition provides the basis for associating Feynman diagrams with different orders of perturbation theory. These diagrams are obtained by representing each Green's function (7.51) by a directed line and help keeping track of the different contributions to the expansion (7.45). There are rules which specify the form of the diagrams to be associated with a given term of the expansion; these rules are found in numerous textbooks[1].

An important notion is that of connected versus disconnected diagrams. The latter are diagrams which divide into different unconnected pieces. They belong to parts of the expansion (7.45) for which the expectation values factorize into at least two independent products, a point we illustrate in Fig. 7.2. Since the disconnected diagrams just cancel the factor $\langle S \rangle$ in the denominator of (7.45), only connected or linked Feynman diagrams need be taken into account (linked-cluster theorem). Thus $G_{ij}(t)$ may also be written as

$$G_{ij}(t) = -i \sum_{n=0}^{\infty} \frac{(-1)^n}{n!} \int_{-\infty}^{+\infty} dt_1 \ldots \int_{-\infty}^{+\infty} dt_n$$
$$\times \langle T[c_i(t) c_j^+(0) H_1(t_1) \ldots H_1(t_n)] \rangle^c \ . \tag{7.52}$$

The superscript c refers to taking only connected diagrams as discussed above, which is equivalent to taking the cumulant of that expectation value. For a discussion of cumulants see Sect. 5.2.

7.1.2 Temperature Green's Function

An understanding of the effect of electronic correlations at finite temperatures requires a generalization of the single-particle Green's function to $T \neq 0$. Because of thermal excitations, effects of correlations on excited states come into play here.

When thermodynamic quantities like the temperature Green's function are determined the statistical operator ρ and the partition function Z enter the

---
[1] see, e.g., [4, 117, 158, 301, 403]

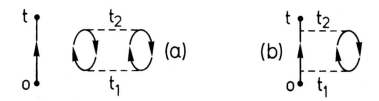

Fig. 7.2. Example of a disconnected diagram (a) and a connected diagram (b), which corresponds to particular contractions of operators in $\langle T[c_i(t)c_j^+(0)H_1(t_1)H_1(t_2)]\rangle$, i.e., a term in the expansion (7.45). Solid lines denote unperturbed functions $G_{nm}^{(0)}(t)$ or $G_{nm}^{(0)}(-t)$, depending on the direction of the arrows associated with them. The dashed lines represent matrix elements of $H_1(t_{1(2)})$. For a more detailed description of the rules governing the construction of diagrams, see, e.g., [117].

calculations. Therefore we start by introducing both. The statistical operator is defined by

$$\rho = \frac{e^{-\beta H}}{Tr\{e^{-\beta H}\}} \quad . \tag{7.53}$$

The symbol Tr stands for taking the trace of the operator. As usual $\beta = (k_B T)^{-1}$ where k_B is Boltzmann's constant. The definition (7.53) assumes that the temperature T, the volume V and the number N of electrons are independent thermodynamic variables. The partition function Z is defined by

$$\begin{aligned} Z &= Tr\{e^{-\beta H}\} \\ &= e^{-\beta F} \quad , \end{aligned} \tag{7.54}$$

where F is the free energy of the system. When dealing with a grand canonical ensemble, i.e., when instead of N the chemical potential μ is used as a thermodynamic variable, we have to make the substitution

$$H \to H - \mu N_{\text{el}}$$
$$\rho \to \rho_G = \frac{e^{-\beta(H-\mu N_{\text{el}})}}{Tr\{e^{-\beta(H-\mu N_{\text{el}})}\}} \quad . \tag{7.55}$$

We use the notation N_{el} for the electron-number operator in order to distinguish it from the electron number N. Then the grand canonical partition function is written in analogy to (7.54) as

$$Z_G = e^{-\beta \Omega} \quad , \tag{7.56}$$

where Ω is the thermodynamical potential.

Thermodynamic expectation values of operators A with respect to a grand canonical ensemble are defined through

7.1 Single-particle Green's Function

$$\langle A \rangle_H = Tr\{\rho_G A\} \quad . \tag{7.57}$$

Assume that H can be divided into

$$H = H_0 + H_1 \tag{7.58}$$

and that the eigenstates of H_0 are known. It is useful at this point to introduce the grand canonical partition function Z_0 corresponding to H_0, i.e.,

$$Z_0 = Tr\{e^{-\beta(H_0 - \mu N_{el})}\} \quad . \tag{7.59}$$

We define the expectation value $\langle A \rangle_{H_0}$ of an operator A with respect to H_0 in analogy to (7.57) as

$$\langle A \rangle_{H_0} = \frac{Tr\{e^{-\beta(H_0 - \mu N_{el})} A\}}{Z_0} \quad . \tag{7.60}$$

Clearly,

$$Z_G = Z_0 \left\langle e^{-\beta H} e^{\beta H_0} \right\rangle_{H_0} \quad , \tag{7.61}$$

and the thermodynamical potential may be written as

$$\beta \Omega = -\ln Z_G \quad . \tag{7.62}$$

The form of (7.61) suggests extracting a factor $\exp(-\lambda H_0)$ from $\exp(-\lambda H)$. Since H_0 and H_1 do not commute, this requires special care. Here (7.29) is of help. One notices that λ can be considered an imaginary time. Therefore we may use this analogy in order to write

$$e^{-\lambda(H_0 + H_1 - \mu N_{el})} = e^{-\lambda(H_0 - \mu N_{el})} T_\tau e^{-\int_0^\lambda d\tau H_1(\tau)}$$
$$= e^{-\lambda(H_0 - \mu N_{el})} \tilde{U}(\lambda) \quad . \tag{7.63}$$

The derivation of this equation is the same as for (7.29) except that t is replaced by the "imaginary time" λ. The operator $\tilde{U}(\lambda)$ is the analogue of the one in (7.29). We define the τ-dependent operator $H_1(\tau)$ by the relation

$$H_1(\tau) = e^{\tau(H_0 - \mu N_{el})} H_1 e^{-\tau(H_0 - \mu N_{el})} \quad . \tag{7.64}$$

The τ-ordering operator T_τ is defined in analogy to (7.3) as

$$T_\tau (A(\tau_1) B(\tau_2)) = \begin{cases} A(\tau_1) B(\tau_2) \quad , & \tau_1 > \tau_2 \quad , \\ (-) B(\tau_2) A(\tau_1) \quad , & \tau_2 > \tau_1 \quad . \end{cases} \tag{7.65}$$

The minus sign in brackets refers to the case where A and B are fermionic operators, e.g., electronic creation or annihilation operators.

If (7.63) is inserted into (7.60) we obtain

$$Z_G = Z_0 \left\langle T_\tau e^{-\int_0^\beta d\tau H_1(\tau)} \right\rangle_{H_0} . \tag{7.66}$$

Using (7.62) we may write the thermodynamic potential Ω in the presence of H_1 as

$$-\beta\Omega = -\beta\Omega_0 + \ln \left\langle T_\tau e^{-\int_0^\beta d\tau H_1(\tau)} \right\rangle_{H_0} . \tag{7.67}$$

The term Ω_0 denotes the thermodynamic potential in the absence of H_1. From the definition of cumulants, in particular from (5.21), it follows that the logarithmic term on the right-hand side can be expressed in terms of those. Thus

$$-\beta\delta\Omega = \left\langle T_\tau e^{-\int_0^\beta d\tau H_1(\tau)} - 1 \right\rangle_{H_0}^c , \tag{7.68}$$

with the obvious notation $\delta\Omega = \Omega - \Omega_0$. The superscript c has the same meaning as in Sect. 5.3. It is important that the cumulant can be taken despite the "time" ordering represented by T_τ. When computing the cumulants, the operators $H_1(\tau)$ as given by (7.64) must be treated as an entity. For a detailed discussion of these points, the original work of *Kubo* [253] should be consulted. The above form for $\delta\Omega$ can be expanded in powers of H_1 as

$$-\beta\delta\Omega = \sum_{n=1}^\infty \frac{(-1)^n}{n!} \int_0^\beta d\tau_1 \ldots \int_0^\beta d\tau_n \langle T_\tau H_1(\tau_1)\ldots H_1(\tau_n)\rangle_{H_0}^c . \tag{7.69}$$

When H_0 is a one-electron Hamiltonian and H_1 contains the electron-electron interactions, we can evaluate expectation values of the form $\langle T_\tau H_1(\tau_1)\ldots H_1(\tau_n)\rangle_{H_0}^c$ by means of finite-temperature Green's functions provided they are properly defined (see below). Wick's theorem is applicable and Feynman diagrams can be associated with each expectation value. Taking cumulants ensures that only connected diagrams have to be considered. The finite-temperature Green's function technique, standard by now, is described in great detail in a number of textbooks[2]; therefore, we include here only a summary of the main results. It is important to repeat that H_0 ought to be a one-electron Hamiltonian.

Next we discuss a proper definition of a finite temperature Green's function which allows for an expansion analogous to that of (7.45). One might think that we simply have to generalize $G_{ij}(t - t')$ given by (7.2) or, alternatively, the retarded Green's function $G_{ij}^R(t - t')$ (see (7.10)) to $T \neq 0$. That implies that the expectation value in (7.2) with respect to the ground state $|\psi_0^N\rangle$ has to be replaced by a thermodynamic ensemble average. When T, V and μ are chosen as thermodynamic variables, an assumption commonly made when

---

[2] see, e.g., [4, 158, 301, 403]

studying the effects of finite temperatures, the retarded Green's function (and similarly the advanced one) becomes

$$\begin{aligned} G_{ij}^R(t-t') &= -i\theta(t-t')\left\langle\left[c_i(t), c_j^+(t')\right]_+\right\rangle_H \\ &= -i\theta(t-t')\sum_{N,n} w_n^N \left\langle\Psi_n^N \left|\left[c_i(t), c_j^+(t')\right]_+\right|\Psi_n^N\right\rangle \end{aligned} \quad , \quad (7.70)$$

with the w_n^N given by

$$w_n^N = \frac{1}{Z_G} e^{-\beta(\mathcal{E}_n^N - \mu N)} \quad . \tag{7.71}$$

The partition function Z_G is defined by (7.56) and the sum in (7.70) is over all eigenstates $|\psi_n^N\rangle$ of H for different electron numbers N.

However, using the Green's function (7.70) has a serious drawback. This becomes apparent when we try to calculate it explicitly, e.g., in order to study the effects of finite temperatures on the excitation spectrum of the system. Then one realizes that the perturbation expansion described in the last section cannot be carried over the finite temperatures, the reason being that an equivalent of (7.41) does not hold. This is due to the fact that (7.40), a prerequisite for (7.41), does no longer apply when the ground state $|\Phi_0^N\rangle$ of H_0 is replaced by excited states $|\Phi_n^N\rangle (n \neq 0)$. Instead, the scattering matrix S when acting on $|\Phi_n^N\rangle$ transforms that state into the different scattering states resulting from the electron interactions. Without a form analogous to (7.42) or (7.43), one cannot generalize (7.52) to finite temperatures.

The difficulty can be circumvented by introducing a modified Green's function $\mathcal{G}_{ij}(t-t')$, which is related to, but not identical with $G_{ij}(t-t')$ and has the advantage that an expansion similar to (7.45) can be derived for it. It goes back to *Matsubara* and is called the temperature or Matsubara Green's function. Its definition is based on the observation that (7.63) and (7.29) are the same when t is replaced by $-i\lambda$ and H by $(H - \mu N_{\text{el}})$. This suggests the introduction of τ-dependent operators of the form

$$c_j^+(\tau) = e^{\tau(H - \mu N_{\text{el}})} c_j^+ e^{-\tau(H - \mu N_{\text{el}})} \quad , \tag{7.72}$$

i.e., with a τ evolution corresponding to (7.1) (Heisenberg representation). The temperature Green's function is then defined by

$$\mathcal{G}_{ij}(\tau, \tau') = -\left\langle T_\tau \left(c_i(\tau) c_j^+(\tau')\right)\right\rangle_H \quad . \tag{7.73}$$

The τ-ordering operator T_τ is the one given by (7.65). As in the preceding section, we shall go over to the analog of the interaction representation. By considering a Green's function which depends on *imaginary* times τ, τ' instead of *real* times, one avoids the obstacle of the scattering matrix $S = U(\infty, -\infty)$ transforming an excited state $|\Phi_n^N\rangle$ of H_0 into different scattering states. Before going into that, we would like to state briefly some important properties of $\mathcal{G}_{ij}(\tau,\tau')$, without necessarily explaining the details how they can be derived. As mentioned before, numerous fine textbooks are available on that topic.

First, it can be shown that $\mathcal{G}_{ij}(\tau,\tau') = \mathcal{G}_{ij}(\tau - \tau')$ and that for $\tau' = 0$ the variable τ is restricted to a range $-\beta \leq \tau \leq \beta$. Furthermore, the values of $\mathcal{G}_{ij}(\tau)$ for $\tau < 0$ are related to those for $\tau > 0$ by

$$\mathcal{G}_{ij}(\tau) = -\mathcal{G}_{ij}(\tau + \beta) \ , \quad -\beta \leq \tau \leq 0 \ . \tag{7.74}$$

The Fourier expansion of $\mathcal{G}_{ij}(\tau)$ is therefore of the form

$$\mathcal{G}_{ij}(\tau) = k_B T \sum_{n=-\infty}^{+\infty} \mathcal{G}_{ij}(\omega_n) e^{-i\omega_n \tau} \ , \tag{7.75}$$

where $\omega_n = \pi k_B T (2n+1)$ (Matsubara frequencies). The relation between $\mathcal{G}_{ij}(\tau)$ and the retarded (advanced) Green's function $G_{ij}^R(t)(G_{ij}^A)$, or better between their Fourier transforms, is

$$\mathcal{G}_{ij}(\omega_n) = \begin{cases} G_{ij}^R(i\omega_n) \ , & \text{for } \omega_n > 0 \\ G_{ij}^A(i\omega_n) \ , & \text{for } \omega_n < 0 \ . \end{cases} \tag{7.76}$$

On discrete points along the imaginary axis, the Fourier transform of the temperature Green's function agrees with that of the retarded and advanced Green's function. Provided that $\mathcal{G}_{ij}(\omega_n)$ is known, one can thus obtain $G_{ij}^R(\omega)$ or $G_{ij}^A(\omega)$ by analytic continuation from the points $i\omega_n$, to the real frequency axis.

As mentioned before, the main advantage of $\mathcal{G}_{ij}(\tau)$ is that an expression analogous to (7.45) can be derived for it. For that purpose, we go over to an *interaction representation*, in which, in analogy to (7.36), the τ evolution of an operator A is given by

$$A_I(\tau) = e^{\tau(H_0 - \mu N_{\text{el}})} A e^{-\tau(H_0 - \mu N_{\text{el}})} \tag{7.77}$$

[compare with (7.64)]. Within that representation, $\mathcal{G}_{ij}(\tau)$ takes the form [286, 343]

$$\mathcal{G}_{ij}(\tau) = \frac{\langle T_\tau \left(c_i(\tau) c_j^+(0) \mathscr{S}(\beta) \right) \rangle_{H_0}}{\langle \mathscr{S}(\beta) \rangle_{H_0}} \ , \tag{7.78}$$

with $\mathscr{S}(\beta)$ given by the equivalent of (7.44), i.e., $\mathscr{S}(\beta) = \tilde{U}(\beta)$ which here is

$$\tilde{U}(\beta) = T_\tau e^{-\int\limits_0^\beta d\tau H_1(\tau)} \ . \tag{7.79}$$

By expanding $\mathscr{S}(\beta)$ we obtain the desired series expansion of $\mathcal{G}_{ij}(\tau)$

$$\mathcal{G}_{ij}(\tau) = -\frac{1}{\langle \mathscr{S} \rangle_{H_0}} \sum_{n=0}^{\infty} \frac{(-1)^n}{n!} \int_0^\beta d\tau_i \ldots \int_0^\beta d\tau_n$$
$$\langle T_\tau \left[c_i(\tau) c_j^+(0) H_1(\tau_1) \ldots H_1(\tau_n) \right] \rangle_{H_0} \quad . \tag{7.80}$$

All operators should carry an additional index I as a reminder that their τ evolution is according to the interaction representation; however, for simplicity we have left out that index. In evaluating the matrix elements in (7.80), we can again introduce contractions as done above but, instead of (7.51), we have here

$$-\langle T_\tau \left(c_i(\tau) c_j^+(0) \right) \rangle_{H_0} = \mathcal{G}_{ij}^{(0)}(\tau) \quad , \tag{7.81}$$

where $\mathcal{G}_{ij}^{(0)}(\tau)$ is the temperature Green's function in the absence of H_1. Like for $T=0$, Feynman diagrams can be introduced and used to evaluate the temperature Green's function in various approximations. The factor $\langle \mathscr{S} \rangle_{H_0}$ in the denominator of (7.80) is cancelled by the *disconnected* diagrams. By considering *connected* diagrams only, we may replace $\langle \mathscr{S} \rangle_{H_0}$ by unity and $\langle \ldots \rangle_{H_0}$ by the cumulant $\langle \ldots \rangle_{H_0}^c$ (compare with (7.69)).

7.2 Quasiparticles in Metals

The concept of Fermi liquids was developed by *Landau* and has been extended by *Abrikosov, Khalatnikov, Nozieres, Pines, Silin* and others[3]. It explains why the low-temperature properties of metals resemble so much those of a free electron system, i.e., with neglected electron-electron repulsions, provided we renormalize properly some of the electronic parameters like the effective mass or density of states. A crucial part of that concept is the notion of quasiparticles and quasiholes. They describe the low-energy excitations of a metallic system. The basic assumption of Fermi liquid theory is a one-to-one mapping between those quasiparticles and the excitations of a corresponding noninteracting electronic system. The distribution function $n_{\mathbf{p}\sigma}$ helps to characterize the latter. This function depends on the energy $\epsilon_{\mathbf{p}\sigma}$ of the excitations and on temperature. When $n_{\mathbf{p}\sigma}$ is known, we can easily determine the energy of the system. In order to allow for the one-to-one mapping mentioned above, excitation-energy levels should not cross when the interactions are turned on. This implies that the energy of the interacting system must again be a functional of the distribution function $n_{\mathbf{p}\sigma}$, which now describes the distribution of the quasiparticles. Of course, $\epsilon_{\mathbf{p}\sigma}$ need not be the same in the presence of interactions as it is in their absence. Instead it will in general be renormalized. Many physical quantities like the specific heat, different susceptibilities, and in particular transport properties, involve electronic excitations with energies of the order of $k_B T$. These energies are usually on the order of or less than

---

[3] see, e.g., [277, 369]

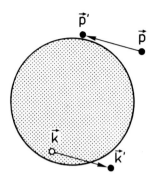

Fig. 7.3. Scattering of an electron with momentum **p** and $\epsilon_\mathbf{p} > \epsilon_F$ by an electron with momentum **k** inside the Fermi sphere. The final states with momentum **p'** and **k'** must be outside the Fermi sphere.

10^{-2} eV and therefore much smaller than the Fermi energy, which is on the order of few eV. This implies that the excited electrons are close to the Fermi surface, which makes their effective scattering rate τ^{-1} by interactions with other electrons small (Fig. 7.3).

Consider a filled Fermi sphere and an additional electron with momentum **p** and energy $\epsilon_\mathbf{p} \geq \epsilon_F$. In order to estimate its scattering rate, we look at the scattering of an electron with momentum **p** and energy $\epsilon_\mathbf{p} \geq \epsilon_F$ by an electron with momentum **k** and energy $\epsilon_\mathbf{k} < \epsilon_F$. After the scattering process the two electrons are in the final states **p'** and **k'** with $\epsilon'_\mathbf{p} > \epsilon_F$, $\epsilon'_\mathbf{k} > \epsilon_F$. The latter conditions result from the Pauli principle, which requires that the final states be empty before scattering. Energy conservation requires that $\epsilon_\mathbf{p} + \epsilon_\mathbf{k} = \epsilon'_\mathbf{p} + \epsilon'_\mathbf{k}$. If $\epsilon_\mathbf{p} = \epsilon_F$, it follows that $\epsilon_\mathbf{k} = \epsilon'_\mathbf{p} = \epsilon'_\mathbf{k} = \epsilon_F$, i.e., the two initial and the two final momenta are all on the Fermi surface. The available phase space has zero volume and the scattering rate is zero. When $\epsilon_\mathbf{p}$ is slightly larger than ϵ_F, the energies $\epsilon_\mathbf{k}$ and $\epsilon'_\mathbf{p}$ must be within a shell of thickness $(\epsilon_\mathbf{p}-\epsilon_F)$ around the Fermi surface. The fourth energy $\epsilon'_\mathbf{k}$ is not an independent variable because of energy conservation. There is now phase space available for scattering to take place and the scattering rate is $\tau^{-1} = a(\epsilon_\mathbf{p} - \epsilon_F)^2$ where a is a proportionality constant. At finite temperatures, i.e., for $T \neq 0$, an additional term proportional to T^2 enters the scattering rate because the Fermi surface is smeared out over an energy interval of order $k_B T$. Therefore we have

$$\frac{1}{\tau} = a\left(\epsilon - \epsilon_F\right)^2 + bT^2 \quad . \tag{7.82}$$

The electron mean free path due to electron-electron interactions is $\ell_{e-e} = \tau v_F$, where v_F is the velocity of the electrons at the Fermi surface and with τ given by the last equation. In order to estimate its actual value in a metal, we relate it to the effective electron scattering cross section $\sigma(T)$ through $\ell_{e-e} = 1/[n\sigma(T)]$. According to (3.11), the electron density n depends on the

Fermi energy ϵ_F. At sufficiently high temperatures, i.e., when $k_B T \gg \epsilon_F$ so that the effect of Pauli's exclusion principle can be neglected, the cross section is σ_0. It can be computed from Coulomb's law and is found to be in metals on the order $\sigma_0 \simeq 10^{-15} - 10^{-16} \mathrm{cm}^2$. Note that it corresponds to a disk with a radius of order r_0, see (3.13). At low temperatures, $\sigma(T)$ is given according to (7.82) by

$$\sigma(T) = \left(\frac{k_B T}{\epsilon_F}\right)^2 \sigma_0 \quad . \tag{7.83}$$

For $T = 4$ K and $\epsilon_F \simeq 5$ eV, we obtain a mean-free path of order $\ell_{e-e} \simeq 1$ cm. This demonstrates that at low temperatures the electronic excitations have long mean-free paths.

At finite temperatures T we want to describe the excitations by a distribution function depending on the energy ϵ of the excitations and on temperature. According to the uncertainty principle, the energy uncertainty caused by a mean-free time τ between electron collisions is $\Delta\epsilon = \tau^{-1}(= v_F/\ell_{e-e})$. This energy must be much less than the thermal broadening, i.e., $\Delta\epsilon \ll k_B T$, in order that the excitations be described by a thermal distribution function. Because of $(\ell_{e-e})^{-1} \propto T^2$, this condition is fulfilled at low temperatures for electrons within an energy interval $k_B T$ of the Fermi surface.

At this stage a comment concerning the spin index should be made. Since the spin is a quantum mechanical quantity, the distribution function is defined as a 2×2 density matrix with elements $n_{\mathbf{p},\alpha\beta}$. This becomes important when studying, for example, the effect of a homogeneous or inhomogeneous magnetic field on the electron system. Only when the locally defined quantization axis of quasiparticle excitations agrees with the z axis everywhere does $n_{\mathbf{p},\alpha\beta}$ reduce to a diagonal matrix $n_{\mathbf{p},\alpha\beta}\delta_{\alpha\beta}$. In order to simplify the notation, we assume here that the spin index σ stands for the matrix. When required, one replaces the sum over σ by a trace.

Assume that a given distribution $n_{\mathbf{p}\sigma}$ of quasiparticles is changed by an infinitesimal amount $\delta n_{\mathbf{p}\sigma}$. If the system is homogeneous, we may start from a step function of the form

$$n_{\mathbf{p}\sigma}^{(0)} = \Theta(|\mathbf{p}| - p_F) \quad , \tag{7.84}$$

where p_F is the magnitude of the Fermi momentum fixed by the electron density, and may consider the deviations from it. The change in the energy δE caused by the change in the distribution function is given by

$$\delta E = \sum_{\mathbf{p}\sigma} \epsilon_{\mathbf{p}\sigma} \delta n_{\mathbf{p}\sigma} + O\left(\delta n^2\right) \quad . \tag{7.85}$$

This serves as a definition of the quasiparticle energy matrix $\epsilon_{\mathbf{p}\sigma}$, i.e., the latter is the functional derivative of the energy with respect to the distribution function $\epsilon_{\mathbf{p}\sigma}(\{n_{\mathbf{p}\sigma}\}) = \delta E/\delta n_{\mathbf{p}\sigma}$. The entropy S of the quasiparticles is the same as that of noninteracting electrons, because of the requirement that the

energy levels correspond to each other in both cases. This implies the following form:
$$S = -k_B \sum_{\mathbf{p}\sigma} [n_{\mathbf{p}\sigma} + (1 - n_{\mathbf{p}\sigma}) \ln(1 - n_{\mathbf{p}\sigma})] \quad . \tag{7.86}$$

Consider a grand canonical ensemble, so that the electron number N is not fixed and may fluctuate. The chemical potential is denoted by μ. The quasiparticle distribution function is determined by the requirement that the free energy F remains stationary with respect to changes $\delta n_{\mathbf{p}\sigma}$ in the quasiparticle distribution, i.e.,
$$\delta F = \delta E - T \delta S - \mu \delta N = 0 \quad . \tag{7.87}$$
With the help of (7.85, 7.86) we find, in close analogy to the case of a noninteracting Fermi gas, that $n_{\mathbf{p}\sigma}$ is given by the Fermi distribution function
$$n_{\mathbf{p}\sigma} = \frac{1}{1 + e^{\beta(\epsilon_{\mathbf{p}\sigma} - \mu)}} \quad . \tag{7.88}$$

The energy $\epsilon_{\mathbf{p}\sigma}$ of a quasiparticle results from the motion of an electron in the self-consistent field of all the other electrons or quasiparticle excitations. When their distribution changes by $\delta n_{\mathbf{p}\sigma}$, the quasiparticle energy changes, too. The following ansatz is made for this change:
$$\delta \epsilon_{\mathbf{p}\sigma} = \sum_{\mathbf{p}'\sigma'} f_{\sigma\sigma'}(\mathbf{p},\mathbf{p}') \delta n_{\mathbf{p}'\sigma'} \quad . \tag{7.89}$$

The function $f_{\sigma\sigma'}(\mathbf{p},\mathbf{p}')$, introduced by *Landau*, characterizes the electron-electron interactions, although its microscopic calculation is generally not possible. Consequently, in Fermi-liquid theory no attempt is made to calculate it. Instead, we relate the interaction function to measurable physical quantities and determine it experimentally as accurately as possible. The information obtained proves useful in the prediction of the results of other experiments.

The relation (7.89) may be applied to write the quasiparticle energy $\epsilon_{\mathbf{p}\sigma}$ in the form
$$\epsilon_{\mathbf{p}\sigma} = \epsilon_{\mathbf{p}\sigma}^{(0)} + \sum_{\mathbf{p}'\sigma'} f_{\sigma\sigma'}(\mathbf{p},\mathbf{p}') \delta n_{\mathbf{p}'\sigma'} \quad . \tag{7.90}$$

Here $\epsilon_{\mathbf{p}\sigma}^{(0)}$ is the energy when a single quasiparticle is present, i.e., when $n_{\mathbf{p}\sigma} = n_{\mathbf{p}\sigma}^{(0)}$. One may expand $\epsilon_{\mathbf{p}\sigma}^{(0)}$ in terms of $|p - p_F|$, i.e., the distance in momentum space to the Fermi surface. For homogeneous systems, we obtain the simple form
$$\epsilon_{\mathbf{p}\sigma}^{(0)} = \mu + \frac{p_F}{m^*} |p - p_F| \quad . \tag{7.91}$$
The effective mass m^* of the quasiparticles is in this case given by
$$m^* = p_F \bigg/ \left(\frac{\partial \epsilon_{\mathbf{p}\sigma}^{(0)}}{\partial p}\right)_{p = p_F} \quad . \tag{7.92}$$

The change in the total energy resulting from the deviations $\delta n_{\mathbf{p}\sigma}$ of the quasiparticle distribution function from a step function becomes

$$\delta(E - \mu N) = \sum_{\mathbf{p}\sigma}\left(\epsilon_{\mathbf{p}\sigma}^{(0)} - \mu\right)\delta n_{\mathbf{p}\sigma} + \frac{1}{2}\sum_{\mathbf{pp}'\sigma\sigma'} f_{\sigma\sigma'}(\mathbf{p},\mathbf{p}')\delta n_{\mathbf{p}\sigma}\delta n_{\mathbf{p}'\sigma'} \ . \quad (7.93)$$

This equation proves basic to the theory of Fermi liquids. We notice that $f_{\sigma\sigma'}(\mathbf{p},\mathbf{p}')$ has to be symmetric under the permutation $\mathbf{p}, \sigma \rightleftarrows \mathbf{p}', \sigma'$. In regions where $\delta n_{\mathbf{p}\sigma} \neq 0$, i.e., close to the Fermi energy, the energy difference $\epsilon_{\mathbf{p}\sigma}^{(0)} - \mu$ is also small and therefore the two terms on the right-hand side are generally of comparable size. If the changes $\delta n_{\mathbf{p}\sigma}$ result solely from finite temperatures, then, in the limit $T \to 0$, the quasiparticle interactions may be neglected. Their contribution is proportional to T^4, in distinction to the first term on the right-hand side of (7.93), which is of order T^2. When in the second term of (7.90) the sum over $\mathbf{p}'$ is taken, the positive and negative contributions of $\delta n_{\mathbf{p}'\sigma'}$ cancel up to terms of order T^2. Since the interaction term in (7.93) contains a double summation, it is of order T^4 as stated above.

When the system is homogeneous, the $f_{\sigma\sigma'}(\mathbf{p},\mathbf{p}')$ depend only on the angle θ between $\mathbf{p}$ and $\mathbf{p}'$. If the spin-dependent interactions are of the exchange type, they are proportional to $\boldsymbol{\sigma}\cdot\boldsymbol{\sigma}'$. In that case, the function $f_{\sigma\sigma'}(\theta)$ is a tensor of the form

$$f(\theta) = f^s(\theta)\mathbf{1}\cdot\mathbf{1}' + \boldsymbol{\sigma}\cdot\boldsymbol{\sigma}' f^a(\theta) \quad (7.94)$$

or, alternatively,

$$f_{\alpha\beta,\gamma\delta}(\theta) = f^s(\theta)\delta_{\alpha\beta}\delta_{\gamma\delta} + \boldsymbol{\sigma}_{\alpha\beta}\cdot\boldsymbol{\sigma}'_{\gamma\delta}f^a(\theta) \ . \quad (7.95)$$

The Pauli matrices $\boldsymbol{\sigma}$ and $\boldsymbol{\sigma}'$ act on the spins of quasiparticles with momentum $\mathbf{p}$ and $\mathbf{p}'$, respectively.

The two functions f^s and f^a can be expanded in terms of Legendre polynomials

$$f^\lambda(\theta) = \frac{\pi^2}{\Omega m^* p_F}\sum_{l=0}^\infty F_l^\lambda P_l(\cos\theta), \qquad \lambda = s, a \ . \quad (7.96)$$

We can write the prefactor $\pi^2/\Omega m^* p_F$ as $[2N^*(0)]^{-1}$ where $N^*(0)$ denotes the density of states per spin of the quasiparticles close to the Fermi energy. The F_l^λ are the Landau parameters.

One immediate consequence of (7.87) together with (7.86, 7.91 - 7.93) is that, in the limit of low temperatures, the free energy of the interacting system is that of a noninteracting electron gas, its only modification being that the quasiparticle mass m^* is substituted for the free electron mass.

We will now give examples of how different physical quantities depend on the interaction parameters F_l^λ [267]. For homogeneous systems, the ratio of the quasiparticle mass m^* to the free electron mass m is

$$m^* = m\left(1 + \frac{F_1^s}{3}\right) \ , \quad (7.97)$$

i.e., it depends on the symmetric $l = 1$ Landau parameter only. Furthermore, the ratio of the spin susceptibility of an interacting system χ_s to that of a noninteracting system $\chi_s^{(0)}$ is

$$\frac{\chi_s}{\chi_s^{(0)}} = \frac{m^*/m}{1 + F_0^a} \quad . \tag{7.98}$$

The factor m^*/m results from the change in the quasiparticle density of states. The remaining factor $S = (1 + F_0^a)^{-1}$ is the Stoner enhancement factor, so called because it plays an important role, for example, in exchange-enhanced metals (Sect. 11.3). Similarly, the charge compressibility κ of an electron system is

$$\frac{\kappa}{\kappa^{(0)}} = \frac{m^*/m}{1 + F_0^s} \quad , \tag{7.99}$$

where $\kappa^{(0)}$ is the compressibility of a noninteracting electron gas. This relation becomes important in the theory of metals with heavy quasiparticles (see Chapter 13). The specific heat $C = -T(\partial^2 F/\partial T^2)$ takes the form

$$C(T) = \gamma T + \delta T^3 \ln T + O\left(T^3\right) \quad . \tag{7.100}$$

The Sommerfeld coefficient

$$\gamma = \frac{m^* p_F}{3} k_B^2 \tag{7.101}$$

remains the same as for free electrons except for the substitution of m^* for m. The contribution $T^3 \ln T$, solely due to electron interactions, is purely a Fermi-liquid effect and does not require a particular microscopic model for its derivation. Instead, it follows from general properties of the inverse lifetime of the quasiparticles. In order to derive it one must include in the scattering rate $1/\tau(\epsilon)$ (see (7.82)) the next higher order term, i.e., $1/\tau(\omega) = a\omega^2 + b|\omega|^3 + ...$ where $\omega = (\epsilon - \epsilon_F)$. Through the Kramers-Kronig relations, the term $a\omega^2$ makes a contribution to the real part of the quasiparticle energy which is proportional to ω, while the term $b|\omega|^3$ leads to one of the form $\omega^2 \ln \omega$. The last term results eventually in the $T^3 \ln T$ contribution to the specific heat. The form (7.100) has been widely used, in particular in the theory of almost ferromagnetic alloys and of metals with strongly correlated electrons.

As an example we show in Appendix D how the dependence (7.97) of the effective mass m^* on the Landau parameter F_1^s is obtained. It follows from Galilean invariance. The relation therefore does not strictly apply to electrons in a periodic lattice potential, because these systems are invariant only with respect to displacements by a lattice vector. In anisotropic systems the scattering amplitudes $f^\lambda(\mathbf{p}, \mathbf{p}')$ depend on the directions of $\mathbf{p}$ and $\mathbf{p}'$ separately, and not only on the angle θ between the two vectors. Yet, one may still use the property that $f^\lambda(\mathbf{p}, \mathbf{p}')$ remains unchanged under the operations $\Re$ of the symmetry group of the system, i.e.,

$$f^\lambda(\mathbf{p}, \mathbf{p}') = f^\lambda\left(\Re^{-1}\mathbf{p}, \Re^{-1}\mathbf{p}'\right) \quad . \tag{7.102}$$

Consequently one may expand $f^\lambda(\mathbf{p},\mathbf{p}')$ in terms of the basis functions belonging to the irreducible representations of the symmetry group. As is already obvious, the number of independent parameters increases considerably and it becomes very difficult to determine them experimentally.

7.3 Quasiparticles in Semiconductors and Insulators

Reliable calculations of energy-bands and in particular of energy gaps are important topics of semiconductor physics. Most phenomena in semiconductors and insulators are explained within a single-electron picture. This has led occasionally to the opinion that electron correlations are unimportant in those systems. That this is not the case has been demonstrated in Sect. 6.1, where it was shown that correlations contribute approximately one third to the binding energy. As regards excited states, correlation effects are even more important. Hartree-Fock energy gaps are usually reduced approximately by a factor of two by electronic correlations. The fact that a single-electron description of many phenomena seems to work so well proves that a quasiparticle description is highly appropriate. Correlation effects are hidden this way in renormalized parameters such as the energy gap between conduction and valence bands, quasiparticle mass, etc. In this section we show how quasiparticle excitations can be calculated by quantum chemical methods. Before discussing ab initio calculations and results for a specific example, i.e., MgO, we consider a simple model. It applies to elemental semiconductors and serves to demonstrate qualitatively different correlation contributions to excitation energies.

Consider what happens when we add an electron (or hole) to the system. The added particle polarizes the bonds of its neighborhood because the system is locally no longer charge neutral. The polarized bonds form a polarization cloud which moves with the extra electron (hole) and together they form a quasiparticle. It costs much less energy to create an electron-hole pair *with* a polarization cloud than without it. The generation of such a polarization cloud is a correlation effect not taken into account in the independent-electron approximation. It is reflected in the pair-distribution function $g(\mathbf{r},\mathbf{r}')$ introduced in Sect. 3.1 (see Fig. 4.2b). An analogous argument holds when a hole instead of an electron is considered. The correlation cloud around the added electron is quite distinct from that of electrons in the ground state of the system. In the latter case, the correlation hole is due to van der Waals interactions, which decrease rapidly with bond distance (Fig. 4.2a).

In the LDA we cannot distinguish between the correlation holes around the added and the remaining electrons. The density-dependent exchange-correlation potential remains unchanged when an extra electron is added to the infinite system. The *energy gap problem* described in Sect. 4.4 finds its origin here. One can circumvent it by calculating explicitly the relaxation and the polarization cloud around the added particle by quantum chemical techniques. This is the approach which is adopted here. Alternatively one may make use of Green's functions and of Feynman diagrammatic techniques. The GW approximation for the electron self-energy $\Sigma(\mathbf{p}, \omega)$ discussed in Sect. 4.4 proves an important computational scheme in this context [198, 199]. The conceptual simplicity of this method is definitely an advantage, yet it is difficult to free it from uncontrolled approximations.

7.3.1 Quasiparticle Approximation

In the following we outline the computation of energy bands when correlation effects are included and the quasiparticle approximation is being made.

As discussed in Section 7.1 excitations are generally determined from the poles of the single-particle Green's function. For the present purpose we write the latter in the form

$$G_{ij}(\omega) = \sum_{mm'} \langle \psi_0^N | c_i | \psi_m^{N+1} \rangle \left\langle \psi_m^{N+1} \left| \frac{1}{\omega - H + \mathcal{E}_0^N} \right| \psi_{m'}^{N+1} \right\rangle \langle \psi_{m'}^{N+1} | c_j^+ | \psi_0^N \rangle$$
$$+ \sum_{mm'} \langle \psi_0^N | c_j^+ | \psi_m^{N-1} \rangle \left\langle \psi_m^{N-1} \left| \frac{1}{\omega + H - \mathcal{E}_0^N} \right| \psi_{m'}^{N-1} \right\rangle \langle \psi_{m'}^{N-1} | c_i | \psi_0^N \rangle ,$$
(7.103)

where $|\psi_0^N\rangle$ is the ground-state of the N particle system with energy $\mathcal{E}_0^N$ and ψ_m^{N+1}, ψ_m^{N-1} are eigenstates of the $(N+1)$ and $(N-1)$ electron system. The valence bands v and conduction bands c are obtained from the poles of the matrices

$$R_{mm'}^v(\omega) = \left\langle \psi_m^{N-1} \left| \frac{1}{\omega + H - \mathcal{E}_0^N} \right| \psi_{m'}^{N-1} \right\rangle$$
$$R_{mm'}^c(\omega) = \left\langle \psi_m^{N+1} \left| \frac{1}{\omega - H + \mathcal{E}_0^N} \right| \psi_{m'}^{N+1} \right\rangle . \quad (7.104)$$

The indices m, m' comprise the momentum $\mathbf{k}$, a band index ν and spin σ.

Before we treat the effects of correlations, let us recall briefly how the energy bands of a semiconductor or insulator are calculated in SCF approximation. For that purpose we consider the valence bands with band index ν. Starting from the SCF ground state

$$|\Phi_{\text{SCF}}\rangle = \prod_{\mathbf{k}\nu\sigma} c_{\mathbf{k}\nu\sigma}^+ |0\rangle \quad (7.105)$$

with energy E_0^{SCF}, we denote the SCF states with one electron annihilated in a Bloch state with quantum numbers $\mathbf{k}, \nu, \sigma$ by

$$|\mathbf{k}\nu\sigma\rangle = c_{\mathbf{k}\nu\sigma} |\Phi_{\text{SCF}}\rangle \quad (7.106)$$

and write for the energies of the valence bands

7.3 Quasiparticles in Semiconductors and Insulators

$$\epsilon_{\mathbf{k}\nu}^{\mathrm{SCF}} = \langle \mathbf{k}\nu\sigma | H | \mathbf{k}\nu\sigma \rangle - E_0^{\mathrm{SCF}} \quad . \tag{7.107}$$

We want to express it in terms of localized orbitals since we later treat correlations by using those. Therefore we define localized hole states

$$|\mathbf{R}n\sigma\rangle = c_{\mathbf{R}n\sigma} |\Phi_{\mathrm{SCF}}\rangle \tag{7.108}$$

by a unitary transformation

$$|\mathbf{k}\nu\sigma\rangle = \frac{1}{\sqrt{N_0}} \sum_{\mathbf{R}n} \alpha_{\nu n}(\mathbf{k}) e^{i\mathbf{k}\mathbf{R}} |\mathbf{R}n\sigma\rangle \quad . \tag{7.109}$$

Here $c_{\mathbf{R}n\sigma}$ destroys an electron in the localized spin-orbital $(\mathbf{R}, n, \sigma)$, where n refers to the different localized states within a unit cell centered at a lattice vector $\mathbf{R}$. Furthermore, N_0 is the total number of cells. The choice of $|\mathbf{R}n\sigma\rangle$ and hence of the matrix $\alpha_{\nu n}$ is to some extent ambiguous. It depends on the special localization procedure. In solid-state physics the localized orbitals are called Wannier orbitals while in quantum chemistry one refers to them as Foster-Boys orbitals. We require that within the space of occupied SCF orbitals these orbitals are optimally localized.

Next we want to express the energies of the valence bands in terms of the $|\mathbf{R}n\sigma\rangle$. For this purpose we introduce the matrix

$$H_{\mathbf{R}-\mathbf{R}',nn'}^{\mathrm{SCF}} = \langle \mathbf{R}'n'\sigma | H | \mathbf{R}n\sigma \rangle - \delta_{\mathbf{R}\mathbf{R}'} \delta_{nn'} E_0^{\mathrm{SCF}}$$

$$= (c_{\mathbf{R}'n'\sigma} | H | c_{\mathbf{R}n\sigma}) \quad . \tag{7.110}$$

The round brackets were introduced in (5.31). The valence band energies are given by

$$\epsilon_{\mathbf{k}\nu}^{\mathrm{SCF}} = \sum_{\mathbf{R}} \sum_{nn'} \alpha_{\nu n}(\mathbf{k}) \alpha_{\nu n'}^*(\mathbf{k}) e^{i\mathbf{k}\mathbf{R}} H_{\mathbf{R},nn'}^{\mathrm{SCF}} \quad . \tag{7.111}$$

It is seen that we have reduced the computation of the valence band energies to that of a relatively small number of matrix elements $H_{\mathbf{R},nn'}^{\mathrm{SCF}}$ [155]. The matrix $\alpha_{\nu n}(\mathbf{k})$ is determined by the lattice structure.

We want to generalize the above calculations by including electronic correlations. In the course of this we use the identity

$$\frac{1}{\omega + H - \mathcal{E}_0^N} = \frac{1}{\omega - \mathcal{E}_0^N} - \frac{1}{\omega - \mathcal{E}_0^N} H \frac{1}{\omega + H - \mathcal{E}_0^N} \tag{7.112}$$

and write for the matrix $R_{mm'}^{\nu\sigma}(\omega)$ the matrix equation

$$\left[(\omega - \mathcal{E}_0^N) \mathbf{1} + \mathcal{H} \right] \mathcal{R} = \mathbf{1} \quad , \tag{7.113}$$

tacitly assuming that the $|\psi_m^{N-1}\rangle$ are orthogonal to each other. The quasiparticle assumption implies that to each state $|\mathbf{k}\nu\sigma\rangle$ there exists a corresponding state in the interacting $(N-1)$ electron system which we denote by $|\psi_{\mathbf{k}\nu\sigma}^{N-1}\rangle$ or $|\mathbf{k}\nu\sigma\}$. In analogy to (7.109), we write

$$|\mathbf{k}\nu\sigma\} = \frac{1}{\sqrt{N_0}} \sum_{\mathbf{R}n} \alpha_{\nu n}(\mathbf{k}) e^{i\mathbf{k}\mathbf{R}} |\mathbf{R}n\sigma\} \quad . \tag{7.114}$$

The state $|\mathbf{R}n\sigma\}$ contains a hole in Wannier orbital $(\mathbf{R}, n)$ together with its polarization- and relaxation cloud. In the quasiparticle approximation only the matrix element $\langle \psi_{\mathbf{k}\nu\sigma}^{N-1} | c_{\mathbf{k}\nu\sigma} | \psi_0^N \rangle \neq 0$ is kept in (7.103). We write the frequencies at which the bracket in (7.113) vanishes by setting $\omega = -\epsilon_{\mathbf{k}\nu\sigma}$ in the form

$$\mathcal{E}_{\mathbf{k}\nu\sigma}^{N-1} = \epsilon_{\mathbf{k}\nu\sigma} + \mathcal{E}_0^N \quad . \tag{7.115}$$

From (7.113) and (7.114) we find for the valence band of the correlated system

$$\epsilon_{\mathbf{k}\nu} = \{\mathbf{k}\nu\sigma | H | \mathbf{k}\nu\sigma\} - \mathcal{E}_0^N$$

$$= \sum_{\mathbf{R}} \sum_{nn'} \alpha_{\nu n}(\mathbf{k}) \alpha_{\nu n'}^*(\mathbf{k}) e^{i\mathbf{k}\mathbf{R}} \{\mathbf{0}n'\sigma | H | \mathbf{R}n\sigma\} - \mathcal{E}_0^N \quad . \tag{7.116}$$

A corresponding expression is obtained for the conduction band. Here we start from $|\mathbf{k}\nu\sigma\rangle = c_{\mathbf{k}\nu\sigma}^+ |\Phi_{\mathrm{SCF}}\rangle$ and after repeating the steps from (7.107) to (7.116) we end up again with matrix elements $\{\mathbf{0}n'\sigma | H | \mathbf{R}n\sigma\}$. They refer to states of the $(N+1)$-particle system. In passing we note that the correlation contribution to $\epsilon_{\mathbf{k}\nu}$ can be written in an elegant, condensed form if we make use of the cumulant formulation discussed in Sect. 5.2. We find

$$\epsilon_{\mathbf{k}\nu\sigma}^{\mathrm{corr}} = \sum_{\mathbf{R}} \sum_{nn'} \alpha_{\nu n}(\mathbf{k}) \alpha_{\nu n'}^*(\mathbf{k}) e^{i\mathbf{k}\mathbf{R}} \left(c_{\mathbf{0}n'\sigma} | H_{\mathrm{res}} S \, c_{\mathbf{R}n\sigma} \right) \quad , \tag{7.117}$$

where $S = \Omega - 1$ (see (5.50)).

In order to obtain the energy bands, the matrix elements in (7.116) have to be evaluated. For that one needs to know the states $|\mathbf{R}n\sigma\}$. They can be obtained in a fairly simple way as will be demonstrated later.

7.3.2 A Simple Model: Bond-Orbital Approximation

In Sect. 6.1 we introduced the bond-orbital approximation (BOA) in order to study qualitatively the effects of correlations on the ground-state wavefunction of elemental semiconductors. Here we use the same approximation in order to discuss the different physical processes which are contained in (7.117) and affect the band structure. Main emphasis is on the energy gap. For that reason it is important to understand first, why the gap is so large in SCF approximation. The exchange plays an important role here. We want to make this transparent. Starting Hamiltonian is (6.1) which in SCF approximation is

7.3 Quasiparticles in Semiconductors and Insulators

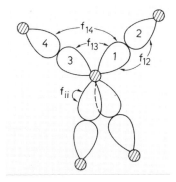

Fig. 7.4. The different matrix elements f_{ij} of the Fock operator relate to atomic hybrids as indicated.

$$H_{\text{SCF}} = \sum_{ij\sigma} f_{ij} \left(a^+_{i\sigma} a_{j\sigma} - \langle a^+_{i\sigma} a_{j\sigma} \rangle \right) + E_{\text{SCF}} \quad . \tag{7.118}$$

For a closer study it is useful to separate the Fock matrix defined in (2.30) into two parts, i.e.,

$$f_{ij} = f^H_{ij} + f^x_{ij} \quad , \tag{7.119}$$

where

$$f^H_{ij} = t_{ij} + \sum_{kl\sigma'} V_{ijkl} \langle a^+_{k\sigma'} a_{l\sigma'} \rangle \quad , \tag{7.120}$$

is the Hartree part and

$$f^x_{ij} = -\frac{1}{2} \sum_{kl\sigma'} V_{ilkj} \langle a^+_{k\sigma'} a_{l\sigma'} \rangle \quad , \tag{7.121}$$

is the exchange part. The labeling of the different Fock matrix elements is according to Fig. 7.4.

The SCF eigenstates of the (N + 1)-electron system are written in the form

$$| \Phi_{\mathbf{k}c\sigma} \rangle = c^+_{\mathbf{k}c\sigma} | \Phi_{\text{SCF}} \rangle \quad , \tag{7.122}$$

where $c^+_{\mathbf{k}c\sigma}$ creates an extra electron in conduction band c with momentum $\mathbf{k}$ and spin σ. The corresponding SCF energy is

$$E^{\text{SCF}}_{\mathbf{k}c\sigma}(N+1) = E_{\text{SCF}}(N) + \epsilon^{\text{SCF}}_{c\sigma}(\mathbf{k}) \quad , \tag{7.123}$$

with $E_{\text{SCF}}(N)$ denoting the ground-state energy of the semiconductor or insulator. This expression for $E^{\text{SCF}}_{\mathbf{k}c\sigma}$ is equivalent to the relation

$$\left[H_{\text{SCF}}, c^+_{\mathbf{k}c\sigma} \right]_- = \epsilon^{\text{SCF}}_{c\sigma}(\mathbf{k}) c^+_{\mathbf{k}c\sigma} \quad . \tag{7.124}$$

A corresponding relation holds for the valence band $\epsilon^{\text{SCF}}_{v\sigma}(\mathbf{k})$. Since we want to study the effect of the exchange, we divide

$$\epsilon_{n\sigma}^{\mathrm{SCF}}(\mathbf{k}) = \epsilon_{n\sigma}^{H}(\mathbf{k}) + \epsilon_{n\sigma}^{x}(\mathbf{k}) \ . \tag{7.125}$$

The first term $\epsilon_{n\sigma}^{H}(\mathbf{k})$ is the Hartree part determined by f_{ij}^{H} while $\epsilon_{n\sigma}^{x}$ is the exchange part originating from f_{ij}^{x}.

Consider first the Hartree part f_{ij}^{H} of the Fock matrix. It leads to Hartree energy bands. They are not self-consistent because the density matrix, which is used in (7.120) is that of a calculation including exchange. The four Hartree conduction (c) and four valence (v) bands are determined from

$$F_{\mathrm{H}} = \sum_{ij\sigma} f_{ij}^{H} a_{i\sigma}^{+} a_{j\sigma} \tag{7.126}$$

with help of the relations

$$\left[F_{\mathrm{H}}, c_{\mathbf{k}c\sigma}^{+}\right]_{-} = \epsilon_{c}^{H}(\mathbf{k}) c_{\mathbf{k}c\sigma}^{+} \ ,$$

$$\left[F_{\mathrm{H}}, c_{\mathbf{k}v\sigma}\right]_{-} = -\epsilon_{v}^{H}(\mathbf{k}) c_{\mathbf{k}v\sigma} \ . \tag{7.127}$$

We have omitted the spin index σ, because magnetic effects are not part of this analysis. When evaluating $\epsilon_{c(v)}^{H}(\mathbf{k})$, it is advantageous to single out matrix elements f_{ij}^{H} with i,j referring to the same bond because in BOA they do not contribute to the band dispersion. Thus with i,j equal to 1 or 2

$$\epsilon_{c}^{H}(\mathbf{k}) = f_{ii}^{H} - f_{12}^{H} + \tilde{\epsilon}_{c}^{H}(\mathbf{k}) \ , \quad c = 1,\ldots,4$$

$$\epsilon_{v}^{H}(\mathbf{k}) = f_{ii}^{H} + f_{12}^{H} + \tilde{\epsilon}_{v}^{H}(\mathbf{k}) \ , \quad v = 1,\ldots,4 \tag{7.128}$$

with $f_{12}^{H} < 0$. These expressions are still very general.

We apply the BOA in order to evaluate the effect of exchange on the energy bands. Now the exchange can be evaluated analytically because the one-particle density matrix is of the simple form

$$\langle a_{i\sigma}^{+} a_{j\sigma'} \rangle = \begin{cases} \delta_{\sigma\sigma'}/2 \ , & \text{for } i,j \epsilon I \\ 0 \ , & \text{otherwise} \end{cases} \ . \tag{7.129}$$

This follows from the form (6.5) for the ground-state wavefunction together with relation (6.4a). The inclusion of f_{ij}^{x} in (7.119) leads in (7.128) to the replacements

$$f_{ii}^{H} \to f_{ii}^{H} - \frac{U}{2} \ , \quad f_{12}^{H} \to f_{12}^{H} - \frac{J_1}{2} - \frac{K_1}{2} \ . \tag{7.130}$$

Here we have used the notation (6.3) for the most important matrix elements. The energies U and J are large while K is small. Estimates for diamond yield the following order of magnitude values: $U \simeq 22$ eV, $J_1 \simeq 13$ eV, $K_1 \simeq 0.3$ eV [239]. Thus there is a downward shift of all bands by $U/2$ which is due to the exchange. In addition there is an increase by $(J_1 + K_1)/2$ in the

7.3 Quasiparticles in Semiconductors and Insulators

effective hopping matrix elements between the two hybrids forming a bond. Accordingly, a dramatic increase in the bonding-antibonding splitting takes place and affects the direct gap at the Γ point. We find for the contribution of the exchange to the gap in SCF approximation

$$\Delta^{\mathrm{x}}(\Gamma) = \Delta^{\mathrm{SCF}}(\Gamma) - \Delta^{\mathrm{H}}(\Gamma) \simeq J_1 \quad , \tag{7.131}$$

where $\Delta^{\mathrm{H}}(\Gamma)$ and $\Delta^{\mathrm{SCF}}(\Gamma)$ are the gap in Hartree- and SCF (or HF) approximation, respectively. The contribution from K_1, may be neglected here. The exchange contribution J_1 enters with a different sign in the conduction and valence bands and therefore is discontinuous across the band gap. Exchange contributions from more distant hybrids are not significant for diamond in BOA considered here. The discontinuity is easily understood if we consider a single bond. When one of the two electrons within the bond is moved from a bonding into an antibonding state (particle-hole excitation), the charge distribution within that bond changes. Consequently, also the exchange energy of the two electrons changes by a finite amount, causing a discontinuity in the exchange energy across the gap.

The main contribution $-J_1/2$ to the exchange matrix elements f^{x}_{12} can be considered a classical Coulomb interaction energy of two electrons in a given bond because $J_1 = V_{1122}$. This suggests setting

$$\Delta^{\mathrm{SCF}} - \Delta^{\mathrm{H}} = \frac{e^2}{l^{\mathrm{x}}} \quad , \tag{7.132}$$

where l^{x} is a characteristic exchange length smaller than the interatomic distance d. In order to obtain agreement with ab initio results for diamond we have to use $l^{\mathrm{x}} = 0.93d$.

The effect of the exchange on the valence bandwidth is easily calculated when the BOA is used and when interactions extending beyond nearest neighbor bonds are neglected. In that case we find for the valence bandwidth

$$W^{\mathrm{SCF}} = W^{\mathrm{H}} + 4V_{1113} + 8V_{1114} \quad , \tag{7.133}$$

where W^{H} is the valence bandwidth when the energies (7.128) are used. Exchange therefore increases the valence bandwidth.

After having discussed the influence of the exchange, we consider the effects of correlations by assuming that one electron has been added to the system. Before a more detailed discussion of various correlation effects is given, it is worthwhile presenting a simple estimate for the change of the energy gap. As pointed out before, the dominant effect of correlations in the presence of the extra electron is the generation of a long-ranged polarization cloud which moves with the extra particle. For an estimate of the energy-gap correction, we simply calculate the classical polarization-energy gain in a continuum approximation. It is given by

$$\Delta\epsilon_{\mathrm{pol}} = \frac{1}{2}\int d^3\mathbf{r}\mathbf{P}\cdot\mathbf{E} = -\frac{\epsilon_0 - 1}{2\epsilon_0}\frac{e^2}{l^{\mathrm{c}}} \quad , \tag{7.134}$$

where **P** is the macroscopic polarization of the medium, and **E** is the electric field of the extra electron. Furthermore, ϵ_0 is the dielectric constant of the semiconductor or insulator, and l^c is an effective correlation length, which serves as a cutoff. We may fix it by requiring that it be equal to the length at which the dielectric function $\epsilon_0(r)$ reaches its asymptotic value ϵ_0 as a function of distance r; at this length, the correlation induced screening is fully developed. A more detailed analysis shows that this implies $l^c \simeq a/2 = 1.16d$, where a is the lattice constant and d is the interatomic distance. The correlation-energy reduction $\Delta^c(\Gamma)$ of the gap at the Γ point is twice the value of $\Delta\epsilon_{\text{pol}}$ because a hole in the valence bands also forms a polarization cloud. The contribution of exchange plus correlations to the direct gap is obtained by adding up (7.132) and (7.134), i.e.,

$$\Delta^{\text{xc}}(\Gamma) = \frac{e^2}{d}\left(\frac{1}{0.93} - \frac{1}{1.16}\frac{\epsilon_0 - 1}{\epsilon_0}\right) \ . \tag{7.135}$$

Some numerical results are found in Table 7.1. One notices a considerable cancellation of the exchange and correlation contributions. While the exchange enlarges the gap, correlations reduce it. This demonstrates the vital importance of electron correlations in semiconductors. The LDA treats both nonlocal exchange and correlations very approximately and the errors cancel to some extent when the gap is computed. But the computed gaps are unreliable and usually too small.

The GW approximation discussed in Sect. 4.4 includes the two effects on the gap pointed out here, i.e., the nonlocal exchange as well as the polarization cloud around an electron or hole. Therefore in a number of cases very good results are obtained [198]. That is not self-evident since the short-range part of the correlation hole is not well described in RPA on which the method is based. A proper reduction of the mutual Coulomb repulsion of electrons at

Table 7.1. Various contributions to the energy gap at the Γ point for diamond, silicon, and germanium when the BOA is made. Δ^{H}, Δ^{x}, and Δ^c are the Hartree-, exchange-, and the correlation contributions, respectively. The parameter d is adjusted slightly so that $\Delta = \Delta^{\text{SCF}} + \Delta^c$ agrees with the experimental values. Also shown are the gaps within the LDA.

	Solid		
	C	Si	Ge
Δ^{H}	3.9	1.6	1.6
Δ^{x}	10.1	6.6	6.4
Δ^{SCF}	14.0	8.2	8.0
Δ^c	-6.6	-4.8	-4.8
Δ (exp. value)	7.4	3.4	3.2
Δ^{LDA}	5.6	2.6	2.6

short distances is particularly important. Therefore one would have thought that the short-range part of the correlation hole must be particularly well described in order to obtain the correct gap. One reason might be that it changes little when in covalent systems one is going over from a crystal to the atomic limit. In that case the effect on the gap is small.

Next we apply the BOA in order to discriminate between the different correlation contributions to the energy bands. We consider valence bands with band index ν. In BOA the one-hole state in Bloch representation is written as

$$B_{\mathbf{k}\nu\sigma}|\Phi_{\text{BOA}}\rangle = \frac{1}{\sqrt{N_0}} \sum_I \alpha_{\mathbf{k}\nu}(I) B_{I\sigma} |\Phi_{\text{BOA}}\rangle \tag{7.136}$$

where $|\Phi_{\text{BOA}}\rangle$ is given by (6.5) and I is a bond index. In BOA the expression (7.117) for $\epsilon_{\mathbf{k}\nu\sigma}^{\text{corr}}$ becomes

$$\epsilon_{\mathbf{k}\nu\sigma}^{\text{corr}} = \sum_{II'} \alpha_{\nu I}(\mathbf{k}) \alpha_{\nu I'}^*(\mathbf{k}) \left(B_I | H_{\text{res}} S B_{I'}\right) \quad . \tag{7.137}$$

For the cumulant scattering operator S we make the ansatz

$$S = \sum_{I,J} \pi_{IJ} S_{IJ}^\pi + \sum_{I,J} \eta_{IJ} S_{IJ}^\eta \quad , \tag{7.138}$$

where

$$S_{IJ}^\pi = \sum_{\sigma\sigma'} A_{J\sigma'}^+ B_{J\sigma'} B_{I\sigma} B_{I\sigma}^+ \tag{7.139}$$

and

$$S_{IJ}^\eta = \sum_{\sigma\sigma'} A_{J\sigma'}^+ B_{J\sigma'} A_{I\sigma}^+ B_{I\sigma} \quad . \tag{7.140}$$

The parameters η_{IJ} are fixed according to (6.13) while the π_{IJ} are obtained from an analogue of (5.44b) but here for the (N-1)-particle state.

In order to understand the implications of the form (7.138) for S, or alternatively Ω, let us assume that the hole would not be present. In that case $S_{IJ}^\pi|\Phi_{\text{BOA}}\rangle = 0$ since $B_{I\sigma}^+|\Phi_{\text{BOA}}\rangle = 0$ (due to Pauli's principle). The ansatz (7.138) is then identical with (6.7 - 6.8) for the correlated ground state of the N particle system. When the S_{KL}^η act on $B_{I\sigma}|\Phi_{\text{BOA}}\rangle$ they generate again correlations between bonds I and J like in the N-particle ground state, but with one modification. The hole in bond I with spin σ blocks some processes which contribute to the N-particle ground state. This results in a *loss of ground-state correlations*.

When an operator S_{KL}^π is applied on $B_{I\sigma}|\Phi_{\text{BOA}}\rangle$ it is non-vanishing only when $K = L$, i.e.,

$$\sum_{K,L} \pi_{KL} S_{KL}^\pi B_{I\sigma}|\Phi_{\text{BOA}}\rangle = \sum_{L\sigma'} \pi_{IL} A_{L\sigma'}^+ B_{L\sigma'} B_{I\sigma} |\Phi_{\text{BOA}}\rangle \quad . \tag{7.141}$$

The hole state $B_{I\sigma}|\Phi_{\text{BOA}}\rangle$ is generating particle-hole excitations (or dipoles) $A^+_{L\sigma'}B_{L\sigma'}$ in bonds L. Thus a long-range polarization cloud is created around the hole. As pointed out before, it reduces considerably the energy required to remove an electron from the valence band. Very similar considerations apply when an electron is added to the system, i.e., to conduction bands. The simple model calculations employing the BOA pave the way for more elaborate ab initio calculations.

7.3.3 Wavefunction-Based Ab Inito Calculations

As an example for the application of the quasiparticle approximation we want to present the results of an ab initio calculation for the correlation contributions to the energy bands of MgO. Choosing MgO has the advantage that we have already discussed the correlation contributions to the ground state in Sect. 6.2. An extension to quasiparticle excitations complements our insight into correlation effects in that material. From (7.138) we know that when an electron or hole is added to the system we have to distinguish between correlation effects due to relaxation and polarization on the one hand and loss of ground-state correlations on the other. The operators S^π_{IJ} and S^η_{IJ} stand for the two different processes. They must reduce the band gap from the SCF value of 16.2 eV to the experimental value of 7.8 eV.

Starting point is a SCF bandstructure calculation using the CRYSTAL package. For Mg a triple-zeta basis set is used and for the highly polarizable O a triple-zeta set supplemented by polarization functions. The conduction band Wannier orbitals which are Mg $3s$- and Mg $3p$-like are shown in Fig. 7.5. There is substantial weight on the nearest-neighbor oxygen sites. The valence band O $2p$ and $2s$ Wannier orbitals are very compact with nearly vanishing contributions at the nearest Mg sites. Note that the O $2s$ valence bands are much lower in energy than the O $2p$ bands. Relaxation and polarization around an extra electron or hole can be determined by freezing the added electron or hole and performing a new SCF calculation in the presence of the frozen particle.

This corresponds precisely to (7.141) were the surroundings of the hole in $B_{I\sigma}|\Phi_{\text{BOA}}\rangle$ is modified by one-particle excitations. In practice the SCF calculation is done for a MgO cluster with the hole at the center. In order to obtain good results we choose a large cluster C which is divided into an active region C_A and a buffer region C_B, i.e., $C = C_A + C_B$. All Wannier orbitals centered at a site belonging to C_B are kept frozen. The buffer provides for a good representation of the tails of the Wannier orbitals centered on sites belonging to C_A. The changes in the diagonal matrix elements due to correlations, i.e., $\Delta H_{nn} = \{\mathbf{R}n\sigma|H|\mathbf{R}n\sigma\} - \langle\mathbf{R}n\sigma|H|\mathbf{R}n\sigma\rangle$ are listed in Table 7.2. One notices that the one-site and nearest-neighbor relaxation and polarization reduce the SCF band gap by more than 4 eV, which is more than 45 % of the required value.

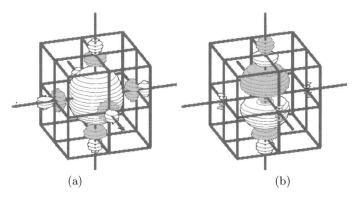

Fig. 7.5. (a): Mg $3s$-like conduction band Wannier orbital on a MgO lattice. Note that there is considerable weight on the nearest neighbor oxygen sites; (b): Mg $3p$-like Wannier orbital. (From [190])

The calculations do not yet account for the long-range part of the polarization cloud. In order to include it we make a continuum approximation and apply (7.134). The polarization energy beyond a radius R is

$$\Delta \epsilon_{\text{pol}} = -\frac{(\epsilon_0 - 1)}{2\epsilon_0} \frac{e^2}{R} \;, \qquad (7.142)$$

where $\epsilon_0 = 9.7$ is the dielectric constant of MgO. A more accurate estimate is obtained by choosing two different clusters $C_A^{(1)}$ and $C_A^{(2)}$ with radii R_1 and R_2 and extrapolating the corresponding $\Delta H_{nn}^{(1)}$ and $\Delta H_{nn}^{(2)}$. The radius R in (7.142) is determined by taking into account that the core-like electrons of Mg^{2+} ions contribute nearly nothing to the polarization energy. Therefore R is the average of the radii of the first and second oxygen coordination shells of a localized $2p$ ($2s$) hole, when valence bands are considered or of the

Table 7.2. Correlation induced corrections to the diagonal Hamiltonian matrix elements for the valence-band O $2s$, $2p$ and conduction-band Mg $3s$, $3p$ states. All numbers are in eV. Negative corrections induce upward shifts of the valence bands and shifts to lower energies for the conduction bands. Remember that the O $2s$ bands are far below the O $2p$ bands. (From [190])

	ΔH_{nn}			
	O $2s$	O $2p$	Mg $3s$	Mg $3p$
On-site orbital relaxation	-2.64	-2.04		
n.n. orbital relaxation	-1.23	-1.20	-0.81	-0.84
Long-range polarization	-1.80	-1.80	-2.25	-2.25
Total	-5.67	-5.04	-3.06	-3.09

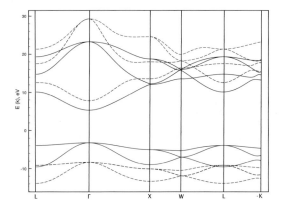

Fig. 7.6. Correlated valence and conduction bands of MgO (solid lines). For comparison the SCF bands are shown by dashed lines. The valence bands have predominantly O 2p and the conduction bands Mg 3s and 3p character. The O 2s bands are not shown (Courtesy of A. Stoyanova).

corresponding coordination shells of a localized 3s/3p electron when we treat conduction bands. The resulting energies are listed in Table 7.2.

Before turning to the off-diagonal matrix elements we have to discuss the loss of ground-state correlations. Here we find two competing effects. As discussed before, removing an electron results in a hole state localized on the oxygen sites. In its presence, i.e., for a frozen hole on an oxygen ion, 0.9 eV of the ground-state correlations are missing. The energy gap increases accordingly. On the other hand, when an electron is added, it resides on the Mg sites. Without it Mg^{2+} has closed electronic shells. They are polarized when an electron is added. The correlation energy gain of 0.3 eV reduces the gap and partially compensates the loss of 0.9 eV in the presence of a hole. The net result is an increase in the gap by 0.5 eV. Together with the results shown in Table 7.2 we find a gap reduction due to correlations of order 8.1 eV which is somewhat accidentally close, i.e., 95 % to the experimental value. Improved basis sets will certainly lead to some modifications.

Off-diagonal matrix elements determine the shape and width of the bands and therefore also effect the energy gap. Here we determine the effective hopping matrix elements

$$t_{\nu\nu'} = \frac{(H_{\nu\nu'} - S_{\nu\nu'} H_{\nu\nu})}{1 - S_{\nu\nu'}^2} \quad (7.143)$$

where ν is a compact index comprising $\mathbf{R}, n$ and σ. Thus $H_{\nu\nu'} = \{\mathbf{R}n\sigma|H|\mathbf{R}'n'\sigma\}$. Furthermore, $S_{\nu\nu'}$ is the overlap matrix. The latter is non-diagonal since states consisting of Wannier orbitals plus their polarization clouds are no longer orthogonal to each other. It turns out that not only nearest-neighbor but also next-nearest-neighbor hopping matrix elements are important in this rocksalt structure. They are of order 0.7 – 0.4 eV and change little when the SCF Wannier orbitals are supplemented by their relaxation and polarization cloud. The shapes and widths of the energy bands remain therefore practically unaffected by electron correlations. A plot of the SCF bands and of the quasiparticle bands are shown in Fig. 7.6. For further details we refer to the original literature [190]. Similar calculations were done for c-ZnS [437] and TiO_2 [85]. Here the inclusion of d electrons is a new feature.

We have discussed the example of MgO in some detail in order to show the accuracy, which presently can be achieved with wavefunction based calculations of energy bands. This remains a field with considerable potential. There seems to be no other way to obtain insight with respect to which microscopic processes affect to what extend the gap of an insulator or semiconductor. The approximations which are required are well controlled and can be improved step by step. It is still an open problem to generalize the wavefunction based approach to Green's functions. That would allow for determining the incoherent excitations with comparable accuracy as it is the case for the coherent excitations.

8

Incoherent Excitations

Very often excitations in solids can be described by quasiparticles. They consist of a bare electron (or hole) and a relaxation and polarization cloud around it. This cloud is what we have termed *correlation hole* before. Associated with the latter are internal degrees of freedom. When they are excited they give raise to incoherent excitations which are represented by the incoherent part $G_{\rm inc}(\mathbf{p},\omega)$ of the Green's function (see (7.22)). One may think of the excitation modes of a drumhead symbolizing the correlation hole. The latter is moving together with the bare electron in form of a quasiparticle through the system.

In a metal with the quasiparticle concept limited to low-energy excitations, the incoherent part of the excitation spectrum appears in the spectral density at much higher energies than the characteristic quasiparticle peak at $\omega = \epsilon_{\mathbf{k}\sigma}$. For strongly correlated electrons the incoherent contributions are generally large to the extent that the quasiparticle picture can break down completely.

In order to obtain the incoherent excitation spectrum of a system one has to determine its single-particle Green's function. The incoherent excitations are contained in the self-energy $\Sigma(\mathbf{p},\omega)$ and the computation of the latter is a central problem in solid-state theory. We can tackle it by applying either perturbation expansions with respect to the interaction part of the Hamiltonian, i.e., using the equations of motion (7.12) or by applying projection techniques. The two methods are related. The first method, i.e., perturbation theory, is usually formulated in terms of Feynman diagrams. It will not be discussed here, since there are numerous good textbooks available on the subject (see, e.g., [117]). But we want to consider the projection method in more detail. By identifying the most important degrees of freedom of the correlation hole and choosing them as relevant dynamical variables, one can determine the incoherent part of the spectrum.

8.1 Projection Method

The projection method was developed by *Mori* [330] and *Zwanzig* [507]. Even before it has been used in quantum chemistry by *Löwdin* and coworkers though in a somewhat modified form. There it has been called superoperator method[1]. That name relates to the use of the Liouville operator L which acts on operators and not on states.

Consider the operator space or Liouville space as it is called. There are various ways of defining an operator product in it. One particular choice was made in Chapter 5 and specified by (5.31). A different form is

$$(A|B)_+ = \langle \psi_0 | [A^+, B]_+ | \psi_0 \rangle \quad , \tag{8.1}$$

where the expectation value is with respect to the normalized *exact* ground state $|\psi_0\rangle$ of a Hamiltonian H. In distinction to (5.31) cumulants are not used here. Which is the most appropriate choice for the inner product depends on the problem we want to consider. The above choice is particularly suitable for calculating Green's functions, see (7.10). Most of the following considerations are independent of the particular choice of the inner product; they apply also when we use the form (5.31) instead. The Liouville operator L corresponding to H is defined in analogy to (5.39) by

$$LA = [H, A]_- \quad , \tag{8.2}$$

where A is an arbitrary operator. Since in addition $[H, A]_- = -idA/dt$, it follows that

$$LA = -i\frac{dA}{dt} \quad . \tag{8.3}$$

This equation has the formal solution

$$A(t) = e^{iLt} A(0) \quad . \tag{8.4}$$

We introduce in the text that follows a set of operators $|A_i\rangle$ called *dynamic variables*. We aim at evaluating the general Green's function matrix $\mathbf{R}(z)$ with matrix elements

$$R_{ij}(z) = \left(A_i \left| \frac{1}{z - L} \right| A_j \right)_+ \quad , \quad z = \omega + i\eta \quad . \tag{8.5}$$

The term $i\eta$ serves to specify the analytic properties of $R_{ij}(\omega)$. For example, when the $|A_i\rangle$ are the electron-creation operators $|c^+_{\mathbf{p}\sigma}\rangle$, then it can be shown that $R^\sigma_{\mathbf{pp}}$ is the retarded Green's function $G^R_\sigma(\mathbf{p}, \omega)$ of the electrons (see Sect. 7.1).

In order to evaluate (8.5) a projection operator

---

[1] see [292–294]

8.1 Projection Method

$$P = \sum_{ij} |A_i)_+ \chi_{ij}^{-1} (A_j| ,$$

$$\chi_{ij} = (A_i | A_j)_+ \qquad (8.6)$$

is introduced. It has the property that when it acts on any operator $|B)_+$ it extracts from it all those components proportional to the variables $|A_i)_+$. The operator

$$Q = 1 - P \qquad (8.7)$$

projects onto that part of the Liouville space, which is orthogonal to the one spanned by the $|A_i)_+$. The matrix χ_{ij} is often called susceptibility matrix.

We consider again the matrix $\mathbf{R}$ with matrix elements (8.5) and make use of the identity

$$\frac{1}{a+b} = \frac{1}{a} - \frac{1}{a} b \frac{1}{a+b} \quad . \qquad (8.8)$$

By using that $L = PL + QL$ we can write R_{ij} in the form

$$R_{ij} = \left(A_i \left| \left(\frac{1}{z - LQ} + \frac{1}{z - LQ} LP \frac{1}{z - L} \right) A_j \right)_+ \quad . \qquad (8.9)$$

Since $Q|A_j)_+ = 0$ it follows that

$$\left(A_i \left| \frac{1}{z - LQ} A_j \right)_+ = \frac{1}{z} (A_i | A_j)_+$$

$$= \frac{1}{z} \chi_{ij} \quad . \qquad (8.10)$$

Therefore (8.9) goes over into

$$R_{ij} = \frac{1}{z} \chi_{ij} + \sum_{lm} \left(A_i \left| \frac{1}{z - LQ} LA_l \right)_+ \chi_{lm}^{-1} R_{mj} \quad . \qquad (8.11)$$

When both sides are multiplied by z and when χ_{ij} is diagonal, this equation resembles the equations of motion (7.12,7.14,7.15). The terms, however, are ordered differently. We rewrite (8.11) in matrix notation as

$$\left(z\mathbb{1} - \boldsymbol{K}\boldsymbol{\chi}^{-1}\right) \boldsymbol{R} = \boldsymbol{\chi} \quad . \qquad (8.12)$$

The matrix $\boldsymbol{K}$ has matrix elements

$$K_{il} = \left(A_i \left| \frac{z}{z - LQ} LA_l \right)_+ \qquad (8.13)$$

and can be decomposed into

$$K_{il} = (A_i | LA_l)_+ + \left(A_i \left| LQ \frac{1}{z - LQ} LA_l \right)_+ \quad . \qquad (8.14)$$

One part consists of the matrix elements L_{ij} of the *frequency* matrix

$$L_{il} = (A_i \mid LA_l)_+ \quad , \tag{8.15}$$

while the remaining part defines the *memory* matrix

$$M_{il}(z) = \left(A_i \left| LQ \frac{1}{z - QLQ} QLA_l\right.\right)_+ \quad , \tag{8.16}$$

where $Q^2 = Q$ has been used in order to write the expression in a symmetric form. It should be noticed that $M_{il}(z)$ is again of the form of (8.5) but with $|A_j)_+$ replaced by $QL|A_j)_+$ and L replaced by QLQ. Therefore we can repeat the same procedure for evaluating the newly generated Green's functions over and over again. At each step the Liouville space considered is perpendicular to the former one.

When (8.15,8.16) are set into (8.12) it is of the form

$$\boldsymbol{R}(z) = \frac{1}{z\mathbf{1} - [\boldsymbol{L} + \boldsymbol{M}(z)]\,\chi^{-1}} \chi \quad , \tag{8.17}$$

or alternatively

$$\sum_l \left(z\delta_{il} - \sum_s [L_{is} + M_{is}(z)]\,\chi_{sl}^{-1}\right) R_{lj}(z) = \chi_{ij} \quad . \tag{8.18}$$

We see that when $\boldsymbol{M}(z)$ is written in the same form as $\boldsymbol{R}(z)$ a continued fraction is obtained. In each order a new frequency and memory matrix are generated.

There are two different ways in which one can use the projection method for the computation of the retarded Green's function (7.10). The first one uses only a single operator (or dynamic variable) $A = c_{\mathbf{p}\sigma}^+$. For the homogeneous electron gas the frequency matrix reduces in that case to the energy

$$\begin{aligned}L_{\mathbf{p}} &= \left(c_{\mathbf{p}\sigma}^+ \mid L\, c_{\mathbf{p}\sigma}^+\right)_+ \\ &= \left(\frac{p^2}{2m} - \mu\right) + \left(c_{\mathbf{p}\sigma}^+ \mid L_{\text{int}}\, c_{\mathbf{p}\sigma}^+\right)_+ \quad , \end{aligned} \tag{8.19}$$

where L_{int} refers to the interaction Hamiltonian H_{int} (see 3.4), i.e, it replaces H in (8.2). The memory matrix reduces to

$$M(\mathbf{p}, z) = \left(c_{\mathbf{p}\sigma}^+ \left| LQ \frac{1}{z - QLQ} QL\, c_{\mathbf{p}\sigma}^+\right.\right)_+ \tag{8.20}$$

and therefore, from the Fourier transform of (8.18), the self-energy is obtained as

$$\Sigma(\mathbf{p}, z) = \left(c_{\mathbf{p}\sigma}^+ \mid L_{\text{int}}\, c_{\mathbf{p}\sigma}^+\right)_+ + M(\mathbf{p}, z) \quad . \tag{8.21}$$

As pointed out above, we may continue by deriving a similar equation for $M(\mathbf{p}, z)$ and so on, thereby obtaining $\Sigma(\mathbf{p}, \omega)$ in form of a continued fraction. The latter has to be terminated at one point. With an increase in the number of fractions the number of poles of the Green's function increases. These poles with their residual model the incoherent part of the excitation spectrum. They give the coherent excitations a line width and model possible satellite structures. The latter show up in particular when correlations are strong.

The second way to determine the retarded Green's function (7.10) is by identifying again one of the dynamic variables A_i with $c_{\mathbf{p}\sigma}^+$ (i.e., $A_0 = c_{\mathbf{p}\sigma}^+$) but complementing this choice with a number of additional operators A_i. As previously discussed they describe the internal degrees of freedom of the correlation hole. Exciting them leads to the incoherent part of the spectral density $A(\mathbf{p}, \omega)$. We will demonstrate this below by giving a specific example. The projection method consists in neglecting all operators which are orthogonal to the space spanned by the $|A_i)_+$. Thus the memory matrix $M_{ij}(z) = 0$. We are dealing here with a generalization to excited states of the partitioning and projection method discussed before in Sect. 5.4.1, when we computed the ground state. This leaves us with the diagonalization of a Green's function matrix of dimension N, which equals the number of operators A_i. The diagonalization must be done for each vector $\mathbf{k}$, but is usually simple.

When $\boldsymbol{R}(z)$ is evaluated this way, it contains a number of static quantities, e.g., the susceptibility matrix χ_{ij} and the frequency matrix L_{il}. Those matrices cannot be computed directly since the exact ground state $|\psi_0\rangle$ is not known. Ways of approximating those quantities have been described in Chapt. 5.

We find that the conventional projection method of *Mori* and *Zwanzig* expresses *dynamic* correlation functions in terms of *static* quantities. The latter have to be determined separately. But as shown in Sect. 5.3, the projection method together with the method of increments can also be applied to the approximate calculation of the exact ground state. This opens the way for calculating the static quantities as well. We consider this a generalization of the Mori-Zwanzig method.

8.2 An Example: Hubbard Model

In order to demonstrate how the projection method works in practice, we consider a simple Hamiltonian which has been much applied to study strongly correlated electron systems. It is of the form

$$H = \sum_{i,j,\sigma} t_{ij} a_{i\sigma}^+ a_{j\sigma} + U \sum_i n_{i\uparrow} n_{i\downarrow}$$
$$= H_0 + H_{\text{int}} \tag{8.22}$$

and describes electrons on a lattice with one orbital per site. There is a hopping matrix element t_{ij} between different sites. We will mainly consider nearest neighbor hopping. Interactions between electrons take place only when they are located on the same site. As before $n_{i\sigma} = a_{i\sigma}^+ a_{j\sigma}$ are the occupation number operators. Strong correlations are present when the interaction U is large in comparison with the band width W of the system. The model Hamiltonian (8.22) was independently proposed and applied by *Gutzwiller, Hubbard* and *Kanamori*. It is commonly referred to as Hubbard Hamiltonian and the same convention will be adopted here. The Hubbard model is discussed in detail in Sect. 10.4. Here we limit ourselves to a demonstration of how incoherent excitations may be calculated by applying the projection technique.

Of particular interest is the case when $U \gg |t_{ij}|$. For less than one electron per site, the system will avoid configurations with doubly occupied sites since they cost an additional energy U. Similarly, for more than one electron per site the system tries to avoid empty sites.

In the limit of large U the case of half filling is special. On average there is one electron located at each site. The electrons take advantage of H_0 by hopping forth and back between neighboring sites. Because of the Pauli principle, this process can take place only if electrons on neighboring sites have opposite spins, thus resulting in antiferromagnetic correlations between sites.

Hubbard's approximate solutions have contributed significantly to our understanding of strongly correlated electrons [194, 195]. They have further had an influence on our understanding of another important problem of solid-state physics, namely that of electrons moving in a nonperiodic potential (disordered system). Now we turn to the so-called Hubbard I approximation and show how it is obtained by applying the projection method. For that purpose we determine the retarded Green's function matrix

$$R_{ij}(\omega) = \left(A_i \left| \frac{1}{\omega - L + i\eta} \right| A_j \right)_+ . \qquad (8.23)$$

For the $\{A_n\}$ we choose the following set of variables: $a_{i\sigma}^+$ and in addition $a_{i\sigma}^+ \delta n_{i-\sigma}$ with $\delta n_{i-\sigma} = n_{i-\sigma} - \langle n_{i-\sigma} \rangle$. The latter variables are orthogonal to the $a_{i\sigma}^+$ and the operator $a_{i\sigma}^+ \delta n_{i-\sigma}$ is used to describe in the simplest approximation the correlation hole of an electron on site i with spin σ. With its help we can modify double occupancies of site i because it filters out those configurations when it is applied to a state. When it acts on a configuration in which site i is singly occupied it gives zero. This is equivalent to the following choice of variables in **k**-space.

$$A_1(\mathbf{k}) = a_{\mathbf{k}\sigma}^+$$
$$A_2(\mathbf{k}) = \frac{1}{\sqrt{N_0}} \sum_i e^{i\mathbf{k}\cdot\mathbf{R}_i} a_{i\sigma}^+ \delta n_{i-\sigma} . \qquad (8.24)$$

We shall assume a paramagnetic state in which case $\langle \psi_0 | n_{i\sigma} | \psi_0 \rangle = n/2$, where n denotes the band-filling. The retarded Green's function $G^R(\mathbf{k}, z) = (a_{\mathbf{k}\sigma}^+ | (z - L)^{-1} a_{\mathbf{k}\sigma}^+)_+$ is given by R_{11}. For simplicity we omit in the following its upper script R. In order to determine the matrix $\boldsymbol{R}$, we apply (8.12) with the memory matrix set equal to zero, i.e.,

$$\left(\omega \mathbb{1} - \boldsymbol{L}\boldsymbol{\chi}^{-1}\right)\boldsymbol{R}(\mathbf{k},\omega) = \boldsymbol{\chi} \quad . \tag{8.25}$$

We find that

$$\boldsymbol{\chi} = \begin{pmatrix} 1 & 0 \\ 0 & \frac{n}{2}\left(1 - \frac{n}{2}\right) \end{pmatrix} \quad . \tag{8.26}$$

Similarly

$$\boldsymbol{L} = \begin{pmatrix} \epsilon(\mathbf{k}) + \frac{n}{2}U & \frac{n}{2}\left(1 - \frac{n}{2}\right)U \\ \frac{n}{2}\left(1 - \frac{n}{2}\right)U & \frac{n}{2}\left(1 - \frac{n}{2}\right)^2 U \end{pmatrix} \quad . \tag{8.27}$$

Hereby an additional term $W(\mathbf{k}) \propto t$ in L_{22} which describes a band shift has been neglected. This shift is partially caused by the $\mathbf{k}$-independent difference of the kinetic energy in the upper and lower band (see below) and partially by an effect of spin correlations on the band dispersion. Yet within the Hubbard I approximation those terms are not considered. From (8.25-8.27) and (8.17) we obtain for the retarded Green's function R_{11}

$$G(\mathbf{k},z) = \left[z - \epsilon(\mathbf{k}) - \frac{U}{2}n\left(1 + \frac{\hat{U}}{z - \hat{U}}\right)\right]^{-1} \quad , \tag{8.28}$$

where $\hat{U} = U(1 - (n/2))$. It has two poles centered around $\omega \simeq 0$ and $\omega \simeq U$, which result in two bands of excitations. Up to terms of order U^{-1}, we can rewrite $G(\mathbf{k},\omega)$ in the form

$$G(\mathbf{k},z) = \frac{1 - n/2}{z - \epsilon(\mathbf{k})(1 - n/2)} + \frac{n/2}{z - U - \epsilon(\mathbf{k})n/2} \quad . \tag{8.29}$$

We notice that the widths of the two bands differ except when $n = 1$, which is the case of half filling. An unusual feature - at least from the point of view of conventional band theory - is that the number of states per site in the two subbands is different, namely $(2-n)$ for the lower and n for the upper band. A simple physical argument for the n dependent number of states in a Hubbard band is given in Sect. 10.4.3. If we want to show that formally, we need to know the relation between the density of states per spin, $N(\omega)$, and Green's function. This relation is found in many textbooks and is given by

$$N(\omega) = -\frac{1}{\pi}\int d\mathbf{k}\, Im\left\{G(\mathbf{k},z)\right\} \quad . \tag{8.30}$$

Integrating $2N(\omega)$ over ω, we obtain the number of states in a subband. Its dependence on electron concentration is discussed below in more detail.

The dynamics of the correlation hole shows up as a satellite in the spectral function $A(\mathbf{k},\omega)$ (see (7.7)). This is shown in Fig. 8.1 for a square lattice with large ratio U/t at half filling $n = 1$ and when $n = 2/5$. The satellite structure

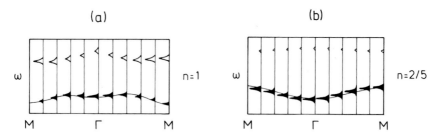

Fig. 8.1. Schematic representation of $A(\mathbf{k},\omega)$ for the Hubbard model on a square lattice with filling factors (a) $n = 1$ and (b) $n = 2/5$ in Hubbard I approximation. Only the centers of gravity of the spectral structure are shown. When the band filling is decreased to $n = 2/5$, one notices a change in the dispersion and a transfer of spectral density from the upper to the lower Hubbard band.

leads also to a band. Thus we are dealing with a lower and an upper Hubbard band. There are several things to notice when the bands are compared for different filling factors. For $n = 1$ both bands have equal spectral weight. The lower band has its minimum value at the M point, while for $n = 2/5$ the minimum is at the Γ point. The change in the band dispersion is obviously related to a decrease of antiferromagnetic correlations with decreasing band filling. They are present in the half-filled case as pointed out above. We mention here that the exact diagonalization of clusters with small and intermediate ratios U/t yields a dispersion for the lower Hubbard band at half-filling which is proportional to $(\cos k_x + \cos k_y)^2$ like in an itinerant antiferromagnet. For $n = 2/5$ the band resembles that of free electrons. At that band filling most of the spectral density of the upper Hubbard has transferred to the lower band and what remains looks like a small satellite structure. It can be considered an excitation of the correlation hole.

9

Coherent-Potential Approximations

In correlated electron systems as well as in disordered systems coherent-potential approximations (CPAs) play an important role[1]. The close relationship between these two different fields of condensed matter physics was pointed out by *Hubbard*, one of the inventors of the CPA [195]. When we speak of disorder it is important to distinguish between static and dynamical disorder. The first case is given when we deal with systems which deviate in an irregular fashion from a regular lattice structure. A glass is an extreme example here. But even on a regular lattice structure static disorder may exist. Think of an alloy with equal components A and B which are distributed irregularly over the lattice sites. The result is again static disorder. Dynamic disorder is different. Consider an electron moving on a lattice under the condition that it cannot hop onto a site which is already occupied by another electron. Thus it is experiencing a disordered surrounding because some of the sites are empty while others are occupied. However, this surrounding is changing with time. Not only does the original electron move, but also the surroundings of a given site are changing too, since the electrons hop on and off sites. The original CPA was done for static disorder. Its generalization to dynamical disorder is therefore an important step. There are several equivalent generalizations of which the Dynamical Mean-Field Theory (DMFT) is one that has obtained widespread application. It involves taking time dependent disorder into account in a local approximation. This reduces the problem to that of an impurity with on-site correlations embedded self-consistently in an environment. The success of DMFT is based in large part on the availability of efficient impurity solver programs. However, the neglected intersite dynamical disorder effects are not small. Therefore the DMFT needs generalization.

We discuss first the CPA for the case of static disorder before we turn to the DMFT and to generalizations of it.

---

[1] see [422, 447, 466, 500]

9.1 Static Disorder

Consider a disordered system with noninteracting electrons (alloy problem). Let us describe it by a Hamiltonian

$$H = \sum_{i,j\sigma} t_{ij} a_{i\sigma}^+ a_{j\sigma} + \sum_i \epsilon_i n_i \quad . \tag{9.1}$$

We speak of diagonal disorder when the ϵ_i vary irregularly from site to site and of off-diagonal disorder when the t_{ij} depend explicitly on i and j and not only on the relative positions of sites i and j. Consider for a moment an alloy with diagonal disorder only. An electron traveling through the system experiences at each site a different potential ϵ_i. In order to describe this system we start from the equation for the retarded Green's function matrix $G_{ij}(z)$ in site representation

$$(z\mathbb{1} - \boldsymbol{L})\,\boldsymbol{G}(\omega) = \mathbb{1} \quad , \tag{9.2}$$

where the Liouvillean $\boldsymbol{L}$ refers to H given by (9.1), i.e., $LA = [H, A]_-$. We find

$$L_{ij} = \left(a_{i\sigma}^+ \mid L\, a_{j\sigma}^+\right)_+ = t_{ij} + \epsilon_i \delta_{ij} \quad . \tag{9.3}$$

In the simplest approximation L_{ij} is replaced by the ensemble average

$$\overline{L_{ij}} = \langle L_{ij} \rangle \quad . \tag{9.4}$$

The disordered system is replaced here by a *virtual* periodic system. Equation (9.2) is then trivially solved after Fourier transformation. With the notations

$$\epsilon_{\mathbf{p}} = \frac{1}{N_0} \sum_{i \neq j} \langle t_{ij} \rangle \, e^{i\mathbf{p}(\mathbf{R}_i - \mathbf{R}_j)}$$

$$\bar{\epsilon} = \langle \epsilon_i \rangle \quad , \tag{9.5}$$

we obtain

$$\overline{G}\left(\mathbf{p}, z\right) = \frac{1}{z - \epsilon_{\mathbf{p}} - \bar{\epsilon}} \quad , \tag{9.6}$$

which is the Green's function of a virtual crystal.

The CPA improves considerably this approximation. In order to demonstrate the essence of it, assume again that there is diagonal disorder only, i.e., $t_{ij} = \langle t_{ij} \rangle$. In that case fluctuations of the frequency-matrix elements are given by

$$\delta L_{ij} = L_{ij} - \overline{L_{ij}}$$
$$= (\epsilon_i - \bar{\epsilon})\, \delta_{ij} = \Delta \epsilon_i \delta_{ij} \quad . \tag{9.7}$$

The system can therefore be considered as one of impurities with scattering potentials $\Delta \epsilon_i$ embedded in a virtual crystal. The respective scattering of an electron by site i can be described by the t-matrix

$$t_i = \frac{\Delta\epsilon_i}{1 - \Delta\epsilon_i \overline{G}_{ii}(z)} \quad . \tag{9.8}$$

It has the form of a geometric series. The scattering of an electron which repeatedly leaves and returns to site i enters through the Green's function $\overline{G}_{ii}(z)$ of the virtual crystal multiplied by the scattering potential $\Delta\epsilon_i$. In the CPA the effect of the scattering (or t-matrix) is approximately taken into account by a modified Green's function $\tilde{G}_{ij}(z)$ with a yet unspecified self-energy $\tilde{\Sigma}(z)$.

$$\tilde{G}_{ij}(z) = \overline{G}_{ij}\left(z - \tilde{\Sigma}(z)\right) \quad . \tag{9.9}$$

The system resembles therefore that of electrons in a virtual crystal but with an additional coherent potential $\tilde{\Sigma}(z)$ added to it. With respect to a medium described by $\tilde{G}_{ij}(z)$ the different sites behave like impurities with a modified, frequency dependent scattering potential

$$\Delta\tilde{\epsilon}_i = \Delta\epsilon_i - \tilde{\Sigma}(z) \quad . \tag{9.10}$$

Therefore they give raise to a $\tilde{t}$-matrix

$$\tilde{t}_i = \frac{\Delta\tilde{\epsilon}_i}{1 - \Delta\tilde{\epsilon}_i \tilde{G}_{ii}(z)} \quad . \tag{9.11}$$

The CPA requires as a self-consistency condition for the self-energy $\tilde{\Sigma}(z)$ that the averaged $\tilde{t}$-matrix in that effective medium vanishes, i.e., that

$$\langle \tilde{t}_i \rangle = 0 \quad . \tag{9.12}$$

For the purpose of demonstration consider a random binary alloy for which the ϵ_i take the two values ϵ_A and ϵ_B only. The concentrations of the two types of sites are c_A and $c_B = 1 - c_A$, respectively. The condition (9.12) together with (9.11) results in

$$\frac{c_A\left[\Delta\epsilon_A - \tilde{\Sigma}(z)\right]}{1 - \left[\Delta\epsilon_A - \tilde{\Sigma}(z)\right]\tilde{G}_{00}(z)} + \frac{c_B\left[\Delta\epsilon_B - \tilde{\Sigma}(z)\right]}{1 - \left[\Delta\epsilon_B - \tilde{\Sigma}(z)\right]\tilde{G}_{00}(z)} = 0 \quad , \tag{9.13}$$

where $\Delta\epsilon_{A(B)} = \epsilon_{A(B)} - \bar{\epsilon}$. Since $\tilde{G}_{00}(z) = [z - t_{00} - \tilde{\Sigma}(z)]^{-1}$ we can determine $\tilde{\Sigma}(z)$ from that equation. It is important to realize that the CPA is a *single-site* theory. Scattering off two sites or larger clusters is neglected here.

Equation (9.13) can be used to demonstrate the alloy analogy of strongly correlated electrons by means of the Hubbard model (8.22). Consider an electron with spin ↑ moving through the system. Let us assume for a moment that all electrons with spin ↓ are kept frozen. In that case the motion of the spin ↑ electron resembles the one in a disordered system with diagonal disorder. When a site is occupied by a frozen spin ↓ electron the acting potential at that site is U while when a site is empty the associated potential is zero. Thus we are dealing with a disordered system to which the CPA applies. We use (9.13) with $\Delta\epsilon_A = 0$ and $\Delta\epsilon_B = U$ and furthermore $c_A = (1 - \langle n_{-\sigma}\rangle)$ and $c_B = \langle n_{-\sigma}\rangle$. Here $\langle n_{-\sigma}\rangle$ is the probability of site being occupied by a spin $\downarrow$ electron. This gives us

$$\frac{-\tilde{\Sigma}(z)(1 - \langle n_{-\sigma}\rangle)}{1 + \tilde{\Sigma}(z)\tilde{G}_{00}(z)} + \frac{\left(U - \tilde{\Sigma}(z)\right)\langle n_{-\sigma}\rangle}{1 - \left(U - \tilde{\Sigma}(z)\right)\tilde{G}_{00}(z)} = 0 \quad . \tag{9.14}$$

In order to find a solution to that equation we consider the large U limit and neglect hopping processes in $\tilde{G}_{00}(z)$ (atomic limit). Therefore the latter reduces to

$$\tilde{G}_{00}(z) = \frac{1}{z - t_{00} - \Sigma(z)} \quad , \tag{9.15}$$

where t_{00} is the atomic orbital energy. When this expression is set into (9.14) we obtain for the self-energy

$$\tilde{\Sigma}(z) = \frac{(z - t_{00})\langle n_{-\sigma}\rangle U}{z - t_{00} - (1 - \langle n_{-\sigma}\rangle)U} \quad . \tag{9.16}$$

By setting this relation into (9.15) we obtain in the atomic limit

$$\tilde{G}_{00}^{\text{at}}(z) = \frac{1 - \langle n_{-\sigma}\rangle}{z - t_{00}} + \frac{\langle n_{-\sigma}\rangle}{z - t_{00} - U} \quad , \tag{9.17}$$

with two poles at energies t_{00} and $(t_{00} + U)$. In order to improve the result we include in the Fourier transform of $\tilde{G}_{ij}(\omega)$ the kinetic energy but still use for $\tilde{G}_{00}(\omega)$ the atomic limit.

$$\tilde{G}(\mathbf{k}, z) = \frac{1}{z - \epsilon(\mathbf{k}) - \tilde{\Sigma}(\omega)}$$

$$\simeq \frac{1}{\left[\tilde{G}_{00}^{\text{at}}(z)\right]^{-1} - \epsilon(\mathbf{k}) - t_{00}} \quad . \tag{9.18}$$

Because of the pole structure of $\tilde{G}_{00}^{\text{at}}(z)$ (see (9.17)), we find that also $\tilde{G}(\mathbf{k}, z)$ has two singularities for each $\mathbf{k}$ point. Neglecting terms of order t_{ij}/U, setting $t_{00} = 0$ and furthermore assuming $\langle n_{-\sigma}\rangle = n/2$ we reproduce (8.29), i.e., the Hubbard I approximation.

9.2 Dynamical Disorder: DMFT and Beyond

The alloy analogy discussed at the end of the last section shows that there is a close relation between strongly interacting electrons and noninteracting electrons in disordered systems. This relation is most transparent when formulated in terms of functional integrals [76]. As briefly pointed out before, the main idea is here the following. Consider an interacting electron moving through a lattice. When the electron hops onto a site, its energy will depend on the number of electrons which are already there. Since that number is fluctuating in space and time, the electron experiences a space- and time dependent potential. The conversion of interactions into fluctuating external potentials is achieved best by going over to finite temperatures and applying a Hubbard-Stratonovich transformation (see Sect. 11.3). The finite temperature version of the CPA discussed in that section is obtained by treating the potential in a single-site approximation and replacing it self-consistently by an effective medium. Thereby its time dependence is neglected (static disorder). Self-consistency implies here that the average scattering- or t-matrix of the spatially fluctuating potential vanishes in the effective medium.

Neglecting the time dependence of fluctuations by making the static approximation implies that at $T = 0$ the theory reduces to a mean-field theory. The ground state is the one obtained in a SCF approximation and all the correlation effects discussed in Chapter 6 are missing. This makes it clearly desirable to go beyond the static approximation and to treat dynamical disorder. An early attempt to generalize the coherent potential approximation to a many-body CPA was made by *Hirooka* and *Shimuzu* but did not get wide application. Other attempts started from a variational ansatz for the thermodynamic potential or from an expansion of the dynamical scattering matrix with respect to the frequency modes of the dynamical potential. However, a breakthrough was only obtained with the development of the dynamical mean-field theory (DMFT) for infinite dimensions by *Georges* and *Kotliar* [146] and by *Jarrell* [214]. They cast the theory into the form of a self-consistent quantum impurity problem, for which efficient numerical impurity solver became available. The work was stimulated by the findings of *Metzner* and *Vollhardt* proving that in infinite dimensions the Hubbard model is solvable [323]. Mean-field treatments become exact in that limit. The self-energy is a function of frequency only and the problem reduces to a single-site or local one [335]. This explains why DMFT reduces to an impurity problem. At the same time a dynamical CPA (DCPA) was developed by *Kakehashi* [223]. Subsequently it was shown that both theories are equivalent to each other and to the approach of *Hirooka* and *Shimizu* [225]. Impurity solvers, which use Monte Carlo techniques, have been particularly successful[2]. While the DMFT is based on temperature Green's functions, one can also formulate a dynamical CPA in terms of retarded Green's functions. The latter in its most advanced form is named fully self-consistent projection operator approach [227]. It avoids the problem of analytic continuation from imaginary to real frequencies that the DMFT approach faces.

---

[2] see, e.g., [214, 215]

Numerous works have successfully used DMFT in combination with the LDA. Reference to infinite dimensions is abandoned here. Instead, we simply limit ourselves to a single-site approximation, in which case the self-energy $\Sigma(\mathbf{p},\omega)$ is replaced by a function of frequency only. Correlations between different sites are neglected. By treating a small atomic cluster as a single "site", the self-energy $\Sigma(\mathbf{p},\omega)$ can be computed, at least for some $\mathbf{p}$ points (cluster dynamical mean-field theory (CDMFT) [302]). One of the remaining problems can be seen as follows.

Assume that we deal with a system with appreciable short-range antiferromagnetic correlations. A 4×4 sites cluster diagonalization will reflect these correlations by an antiferromagnetic arrangement of the spins in the cluster. A CDMFT transfers this result self-consistently to the environment, which in this case also consists of 4×4 sites cluster units. Since all these units become identical by the self-consistency process, we obtain long-range antiferromagnetic order. Antiferromagnetic correlations and long-range order are therefore strongly overemphasized in CDMFT. Whether a system is an antiferromagnet or not can be found out only when the dimension of the cluster exceeds the AF coherence length. In practice this is very difficult though. A special case is 2D. Here the Mermin–Wagner theorem excludes long-range AF order, except at $T = 0$. A CDMFT for the Hubbard model at half-filling on a square lattice shows a logarithmic decrease of the Néel temperature T_N with increasing clusters up to 26 sites in agreement with that theorem [303]. The field has been extensively reviewed in Ref. [147] and in a monograph [9].

We use here retarded Green's functions in order to describe the extension of the CPA to a dynamical CPA and beyond. Thereby we make use of the projection technique. When combined with the method of increments of Sect. 5.3.1 it allows for accurate calculations of the self-energy $\Sigma(\mathbf{k},\omega)$ for model systems like Hubbard's Hamiltonian. The method is not restricted to a single-site approximation but extends instead to far neighbors. We formulate the self-consistent projection method by using the Hubbard model (8.22). Starting point is the retarded Green's function

$$G_{mn}(z) = \left(a^+_{m\sigma} \left| \frac{1}{z-L} \right. a^+_{n\sigma} \right)_+ \quad , \quad z = \omega + i\eta \qquad (9.19)$$

and its Fourier transform

$$G(\mathbf{k},z) = \frac{1}{z - \epsilon(\mathbf{k}) - M(\mathbf{k},z)} \quad . \qquad (9.20)$$

The frequency matrix L_{il} (see (8.15)) has been included in $\epsilon(\mathbf{k})$. For the Hubbard model under consideration it leads to an energy shift only, which can be ignored. We assume a paramagnetic system so that the spin dependence of the Green's function can be omitted. In site representation the memory function is of the form (see (8.16) and (8.24)),

$$M_{ij}(z) = \left(a_{i\sigma}^+ \delta n_{i-\sigma} \left| \frac{1}{z - QLQ} \; a_{j\sigma}^+ \delta n_{j-\sigma} \right. \right)_+$$
$$= \left(A_{i\sigma} \left| \frac{1}{z - QLQ} \; A_{j\sigma} \right. \right)_+ , \qquad (9.21)$$

with $A_{j\sigma} = a_{j\sigma}^+ \delta n_{j-\sigma}$ and $Q = 1 - \sum_{i\sigma} |a_{i\sigma}^+)_+ (a_{i\sigma}^+|$. The function $M(\mathbf{k}, z)$ is obtained from
$$M(\mathbf{k}, z) = \sum_j M_{j0}(z) e^{i\mathbf{k}\mathbf{R}_j} , \qquad (9.22)$$

where $\mathbf{R}_j$ denotes the lattice vectors. Stopping at this stage, i.e., neglecting the new memory function in the denominator of $M_{ij}(z)$, would bring us back to the Hubbard I approximation.

In the spirit of the CPA we define an effective single-electron Hamiltonian with a nonlocal time- or frequency-dependent potential
$$\tilde{H}(z) = H_0 + \sum_{ij\sigma} \tilde{\Sigma}_{ij\sigma}(z) a_{i\sigma}^+ a_{j\sigma} . \qquad (9.23)$$

The operator H_0 is that of (8.22) with the slight modification that the frequency matrix has been included in $\epsilon(\mathbf{k})$ (see above).

The corresponding Liouvillean $\tilde{L}(z)$ has the property $\tilde{L}(z) A = \left[\tilde{H}(z), A \right]_-$ for arbitrary operators A. When we choose
$$\tilde{\Sigma}_{ij\sigma}(z) = M_{ij}(z) , \qquad (9.24)$$

then the Green's function
$$F_{ij}(z) = \left(a_{i\sigma}^+ \left| \frac{1}{z - \tilde{L}(z)} \; a_{j\sigma}^+ \right. \right)_+ \qquad (9.25)$$

is identical with (9.20). Equation (9.25) serves as a self-consistency condition, i.e.,
$$G_{ij}(z) = F_{ij}(z) . \qquad (9.26)$$

The goal is to determine $M_{ij}(z)$ as accurately as possible. Before we pursue this, we draw attention to a simplification which arises when we neglect nonlocality in $\tilde{\Sigma}_{ij\sigma}(z)$ and replace it by a local self-energy $\tilde{\Sigma}_{ij\sigma} = \tilde{\Sigma}(z) \delta_{ij}$. In that case the self-consistency condition becomes
$$\tilde{\Sigma}(z) = \frac{1}{N_0} \sum_{\mathbf{k}} M(\mathbf{k}, z) , \qquad (9.27)$$

where N_0 is the number of lattice sites. Here the medium acts like a time- or frequency dependent external- or molecular field, which is applied to an impurity site with on-site interaction $\left(U \delta n_{i\uparrow} \delta n_{i\downarrow} - \tilde{\Sigma}(z) n_i \right)$. Self-consistency

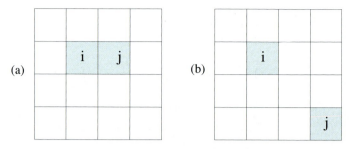

Fig. 9.1. 2-site clusters: (a) nearest-neighbor sites ij; (b) more distant sites i and j.

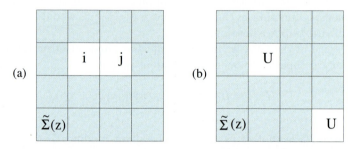

Fig. 9.2. (a) Medium with cavities at sites i and j. (b) the Hubbard two-particle interaction is substituted at the cavity sites. In the shaded areas the self-energy matrix $\tilde{\Sigma}(z)$ applies.

is attained when $\tilde{\Sigma}(z)$ ensures that the scattering matrix of the impurity vanishes. The theory is then essentially identical with the DMFT. In the text that follow we want to keep the nonlocal $\tilde{\Sigma}_{ij\sigma}(z)$ and go beyond the DMFT and equivalent theories.

We compute $M_{ij}(z)$ by increments. For that purpose we define groups of sites, i.e., clusters c which contain one-, two-, three- etc.. sites ($c = i, ij, ijk$, etc.). The sites need not be nearest neighbors but instead can be far apart (see Fig. 9.1). Furthermore we define an interaction Hamiltonian for the cluster

$$H_I^{(c)}(z) = U \sum_{i \in c} \delta n_{i\uparrow} \delta n_{i\downarrow} - \sum_{ij\sigma}{}' \tilde{\Sigma}_{ij\sigma}(z) n_{i\sigma} \quad , \tag{9.28}$$

where the dash at the summation of the second term indicates that the sites i and j must belong to the cluster c. The difference to the interaction part $\tilde{H}(z) - H_0$ (see (9.23)) is that at sites i of the cluster the coherent potential is replaced by the two-particle interaction of the Hubbard model. This is illustrated in Fig. 9.2. The corresponding Liouvillean is $L_I^{(c)}$ so that we may define a cluster Liouvillean

$$L^{(c)}(z) = \tilde{L}(z) + L_I^{(c)}(z) \tag{9.29}$$

9.2 Dynamical Disorder: DMFT and Beyond

corresponding to the cluster Hamiltonian $H^{(c)}(z) = \tilde{H}(z) + H_I^{(c)}(z)$. For the clusters we define cluster memory functions

$$M_{ij}^{(c)}(z) = \left(A_{i\sigma} \left| \frac{1}{z - QL^{(c)}(z)Q} \right| A_{j\sigma} \right)_+ \quad . \tag{9.30}$$

Our aim is to express the memory function (9.21) in form of increments. Thereby we have to distinguish between diagonal matrix elements

$$M_{ii}(z) = M_{ii}^{(i)}(z) + \sum_{\ell \neq i} \delta M_{ii}^{(i\ell)}(z) + \ldots \tag{9.31}$$

and off-diagonal matrix elements

$$M_{ij}(z) = M_{ij}^{(ij)}(z) + \sum_{\ell \neq ij} \delta M_{ij}^{(ij\ell)}(z) + \ldots \quad . \tag{9.32}$$

The $\delta M^{(c)}$ denote the changes in the matrix elements when the cluster includes an increasing number of sites. The DMFT corresponds to restricting oneself to

$$M_{ii}(z) = M_{ii}^{(i)}(z) \quad , \tag{9.33}$$

which reduces to an impurity problem (see Appendix I). Note that for the off-diagonal matrix elements one has to start from a two-site cluster.

In order to apply the theory in practice, we need to have explicit expressions for the cluster memory matrices $M^{(c)}(z)$. They can be obtained by making use of renormalized perturbation theory (RPT). Within that scheme, one is expanding around a suitably chosen Hamiltonian, respective Liouvillean. The latter is here $\tilde{L}(z)$. Therefore we expand $(z - QLQ)^{-1}$ in (9.21) by starting from

$$g_0(z) = \frac{1}{z - Q\tilde{L}(z)Q} \tag{9.34}$$

and including perturbationally $L_I^{(c)}(z)$ (see (9.29)). This defines a cluster T-matrix operator $T(z)$ through

$$\frac{1}{z - QLQ} = g_0 + g_0 T(z) g_0 \quad ,$$

so that the self-consistency condition (9.24) takes here the form

$$(A_{i\sigma} |g_0(z)T(z)g_0(z)| A_{j\sigma}) = 0 \quad . \tag{9.35}$$

This constitutes a generalization of the CPA to dynamical disorder. This way the different matrix elements of the memory-function increments can be determined (see Appendix I). After the cluster expansion is terminated at a stage, the potential $\tilde{\Sigma}_{ij}(z)$ is determined self-consistently by using (9.26). It enters the different matrix elements $M_{ij}(z)$ via $g_0(z)$.

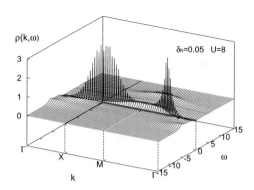

Fig. 9.3. Spectral density for the nearly half-filled Hubbard model with 5 % hole doping on a square lattice for $U = 8$ in units of $|t|$. Calculations employ the fully self-consistent projection (FSCP) operator technique. (Courtesy of Y. Kakehashi)

Actually, within RPT we can make one more simplification. We know from Sect. 8.2 that the variables $A_{i\sigma} = a_{i\sigma}^{+} \delta n_{i-\sigma}$ are sufficient in order to obtain the correct atomic limit. Therefore we introduce a projector $\bar{P}$ onto the subspace spanned by the $\{A_{i\sigma}\}$, i.e.,

$$\bar{P} = \sum_{i\sigma} |A_{i\sigma})_{+} \chi_{i\sigma}^{-1} (A_{i\sigma}| \quad , \tag{9.36}$$

where $\chi_{i\sigma} = \langle n_{i\sigma} \rangle (1 - \langle n_{i-\sigma} \rangle)$.

When we express $L_I^{(c)}(z)$ in (9.29) in terms of $\bar{P}$ in the form

$$L_I^{(c)}(z) = \bar{P} L_I^{(c)} \bar{P} + L_{IQ}^{(c)} \tag{9.37}$$

with

$$L_{IQ}^{(c)} = \bar{Q} L_I^{(c)} \bar{P} + \bar{L}_I^{(c)} \bar{Q} \tag{9.38}$$

and $\bar{Q} = 1 - \bar{P}$ we can neglect $L_{IQ}^{(c)}$ in lowest approximation. In that case the correct atomic limit is ensured and so is the weak interaction limit for small U (see (9.20)). The present scheme is therefore able to provide for an interpolation between the two limits.

When we require self-consistency with respect to the embedding potential only for the diagonal matrix elements $\tilde{\Sigma}_{ii\sigma}(z)$ we speak of the self-consistent projection method (SCPM). The fully self-consistent projection (FSCP) method includes also self-consistency for the off-diagonal contributions $\tilde{\Sigma}_{ij\sigma}(z)$.

The theory has been applied to the Hubbard Hamiltonian on a square lattice and a simple cubic lattice at half filling and for different hole-doping

9.2 Dynamical Disorder: DMFT and Beyond

concentrations, i.e., deviations from half filling. The momentum dependent spectral density

$$\rho(\mathbf{k},\omega) = -\frac{1}{\pi}\mathrm{Im}\ G(\mathbf{k},\omega+i\eta) \tag{9.39}$$

has been calculated and an example is shown in Fig. 9.3 for $U=8$ in units of the nearest-neighbor hopping t. There are quasiparticle excitations present, but also appreciable incoherent structure is observed. The results are among the most accurate for the Hubbard model at small hole doping. We draw attention to complementary results in Sect. 10.7 and 10.9.1.

10
Strongly Correlated Electrons

The physics of strongly correlated electrons is a vast field all of its own. But when do we speak of strongly correlated electrons and when of weakly correlated ones? One way of answering this question is by comparing the size of the Coulomb repulsion of electrons on a given site with the hybridization energy, i.e., with the energy gain of an electron by delocalizing. Surely, those two energies work against each other. Delocalization of electrons due to hybridization matrix elements increases the fluctuation of the electron number n_i at a given site i and defined by $\langle (n_i - \langle n_i \rangle)^2 \rangle$, since electrons hop on and off that site (charge fluctuations). This leads to an increase of their Coulomb repulsion, since the latter is optimized by configurations, in which the electrons are distributed over the lattice sites as uniformly as possible. Thus when we characterize the on-site Coulomb repulsion of two electrons by an energy U and the hybridization by a hopping matrix element t to a nearest neighbor site, the ratio $U/|t|$ enables us to distinguish between strong and weak correlations. When $U/|t| \gg t$, the Coulomb repulsion and hence the tendency to minimize charge fluctuations are dominant, while for $U/|t| \ll 1$ the electrons move almost as freely as independent particles. In an alkali metal like Na the overlap of s-wave functions on neighboring sites and hence $|t|$ is large. The functions are relatively extended so that U is not particularly large. Therefore $U/|t|$ is small and the electrons are weakly correlated only. At the other end of the scale are $4f$ electrons which are close to the nuclei. Their wavefunction overlap with neighboring sites is small, while the on-site Coulomb repulsion is large. Therefore $U/|t| \gg 1$ and correlations are strong.

The reductions of charge fluctuations, as compared with the ones one would obtain in a self-consistent field approximation, may be used to define a measure for the strength of electron correlations in a system. Accordingly we may classify, e.g., different bonds in molecules or solids with respect to their correlation strength. Rare-earth ions with incompletely filled $4f$ shells are the strongest correlated valence electrons in solids. Except for Ce^{3+}, which may fluctuate between $4f^1$ and $4f^0$, and Yb^{3+} which fluctuates between $4f^{13}$ and $4f^{14}$, fluctuations in the f-electron number at a site are practically zero, implying that the $4f$ electrons remain localized. Note that we leave aside here the special case of valence fluctuating Sm and Eu ions. An incompletely filled $4f$ shell is characterized by its total angular momentum J, which can be calculated by applying Hund's rules. The $(2J+1)$-fold degeneracy of the ground-state J multiplet is split in the crystalline electric field (CEF) set up by the neighboring ions. Typical splitting energies are of order of a few meV and therefore much smaller than a typical Fermi energy. They define a low-energy scale caused by correlations. This scale prevails even when the $4f$ electrons become slightly delocalized as in the case of Ce^{3+} or Yb^{3+}. While CEF excitations are an almost trivial example of a low-energy scale due to strong correlations, there are other more refined cases. A simple model will be presented. It gives good insight into low-energy excitations of spin degrees of freedom which couple only weakly to charge degrees of freedom.

One goal consists in finding appropriate effective Hamiltonians, which describe the low-energy excitations. They only act on a proper subspace of the full Hilbert space. The high-energy excitations are eliminated and replaced by new interactions within the reduced space. New low-energy scales are an earmark of strong correlations. They lead to a high density of low-energy excitations which show up, e.g., in the low-temperature specific heat. Needless to say, in strongly correlated electron systems nonlocal correlations play an important role. Contrary to local correlations, they involve different lattice sites when they are described in terms of dynamical variables (see Sect. 8.1). In fact, short-range antiferromagnetic correlations, i.e., small antiferromagnetic clusters, which fluctuate in space and time are a well-known example. A discussion of how they may be detected experimentally is found, e.g., in Ref. [255].

One phenomenon caused by correlations has played a particularly large role in condensed matter physics and that is the Kondo effect. It was discovered by *Kondo* when he tried to explain a characteristic minimum in the temperature dependent resistivity, which is observed when certain magnetic impurities are added to a metal. Due to the internal degrees of freedom of the impurity, the scattering of a conduction electron by a magnetic impurity turns out to be a true many-body problem. In metals with strongly correlated electrons, e.g., Kondo lattices the characteristic low-energy scales result in heavy quasiparticles or heavy fermions as they are often called.

The analogue of the *drosophila* fruit fly, used by biologist for genetic studies, is the Hubbard model for the study of correlated electron systems. Trial wavefunctions for the ground state of that model have been used of which the Gutzwiller wavefunction has been an especially popular one. For the excitations spectrum of that Hamiltonian, *Hubbard* has introduced a number of approximations which demonstrate how a single band can split into two in the presence of strong correlations. The simplest one was already met in Sec. 8.2. This splitting serves as an explanation of the metal-insulator transition which may take place at half-filling. Insulators due to correlations are therefore called Mott-Hubbard insulators. When the band filling is very small, we deal with the so-called Kanamori limit. Here, because of the large interelectron spacings, the effect even of strong local correlations remains rather limited. Near half filling strong correlations in the Hubbard model can be treated quite well by transforming it to the so-called $t - J$ model. This model acts on a much smaller Hilbert space than the original Hubbard model, but still contains most of the important features of strong correlations. A subject which can be studied particularly well in the $t - J$ model is the motions of holes doped into a system at half filling.

It is important to know the strong and the weak points of different approximation schemes, in particular of advanced mean-field approximations. They can be tested by applying them to a simple generic model, for which the exact solution is known. One way of avoiding approximations when dealing, e.g., with the Hubbard Hamiltonian is by numerical studies. They can be done only for small systems, but the results give often new insights and serve as guideline for new approximation schemes.

In metallic systems with strong electron correlations a Fermi liquid description is sometimes inapplicable. There is not always a one-to-one correspondence possible between the low-energy excitations of strongly correlated electrons and those of a weakly interacting system of quasiparticles. A special case are systems with a marginal Fermi liquid behavior at low temperatures. The self-energy $\Sigma(\omega)$ in the Green's function has here a different frequency dependence for small ω than is mandatory for a Fermi liquid. This has considerable consequences for measurable quantities, such as the temperature dependent resistivity or other transport coefficients. A survey of different theoretical approaches to strong correlations is found in Ref. [19].

10.1 Measure of Correlation Strengths

As pointed out in the Introduction, the description of a H_2 molecule by a molecular-orbital wavefunction and by a Heitler-London one represent two extremes. The wavefunction $\psi_{MO}^S(\mathbf{r}_1, \mathbf{r}_2)$ in (1.2) describes the electrons as independent or uncorrelated. Consequently, charge fluctuations at a proton site are large, i.e., ionic configurations have 50 % weight. The Heitler-London wavefunction given by (1.1) has no ionic configurations, i.e., interatomic charge fluctuations vanish. Therefore it corresponds to the strong correlation limit. The two limiting cases suggest immediately introducing the reduction of charge fluctuations compared with the uncorrelated case as a measure of the strength of interatomic correlations (see (2.51)). Let $|\psi_0\rangle$ denote the exact ground state of the electrons in the H_2 molecule and $|\Phi_{SCF}\rangle$ that of uncorrelated electrons. The reduction of the normalized mean-square deviation of the electron number n_i at site $i = 1, 2$ can be quantified by computing:

$$\Sigma(i) = \frac{\langle \Phi_{SCF} |\delta n_i^2| \Phi_{SCF}\rangle - \langle \psi_0 |\delta n_i^2| \psi_0\rangle}{\langle \Phi_{SCF} |\delta n_i^2| \Phi_{SCF}\rangle} \quad , \tag{10.1}$$

where $\delta n_i^2 = n_i^2 - \bar{n}_i^2$. Thus $\Sigma(i) = 0$ implies that $|\psi_0\rangle$ equals $|\Phi_{\text{SCF}}\rangle$ while $\Sigma(i) = 1$ describes the Heitler-London limit of strong interatomic correlations.

The above concept for determining the correlation strength can be generalized to different bonds. Then, for a given bond the index i refers to hybrid functions $g_i(\mathbf{r})$ (i = 1,2) of the atoms forming the bond (see Fig. 2.3), e.g., sp^3 hybrids forming a C–C σ bond. Then δn_i^2 describes the mean-square deviation of the electron occupation number of hybrid $g_i(\mathbf{r})$. Often these hybrids are called half bonds. When dealing with heteropolar bonds some charge fluctuations are required because otherwise there would be no charge flow from atom A to atom B forming the bond. Therefore one has in addition to consider a hypothetical wavefunction $|\psi_{pc}\rangle$ which reproduces the charge distribution of $|\psi_0\rangle$ but minimizes the charge fluctuation to the largest possible extent (strong correlation limit). A good way is to determine first

$$\Sigma^b(i) = \frac{\langle \Phi_{\text{SCF}} |\delta n_i^2| \Phi_{\text{SCF}}\rangle - \langle \psi_0 |\delta n_i^2| \psi_0\rangle}{\langle \Phi_{\text{SCF}} |\delta n_i^2| \Phi_{\text{SCF}}\rangle - \langle \psi_{pc} |\delta n_i^2| \psi_{pc}\rangle} \quad (10.2)$$

with $i = 1, 2$ referring to $g_i(\mathbf{r})$ and thereafter to form

$$\Sigma_m = \frac{1}{2}\left(\Sigma_m^b(1) + \Sigma_m^b(2)\right) \quad (10.3)$$

for a characterization of the correlation strength in bond m. Shown in Fig. 10.1 is Σ for a number of different σ and π bonds. The correlation strength may vary between 0 (uncorrelated limit) and 1 (limit of strong correlations). It is noticed that σ bonds are weakly correlated. Ordinary π bonds with $\Sigma \simeq 0.5$ are just in between the uncorrelated and the strong correlations limit, while resonating π bonds with $0.3 \lesssim \Sigma \lesssim 0.35$ are less strongly correlated.

In Fig. 2.4 we showed that the probability distribution of finding ν valence electrons on a carbon site is nearly of Gaussian form when the atom is part of a molecule or a solid. Correlations narrow that distribution. In the strong correlation limit with electrons becoming localized the distribution reduces to a δ-function.

Of particular interest is to know how strong valence electrons are correlated when they are in the Cu-O planes of the high-temperature superconducting cuprates. Let us consider the ground state of La_2CuO_4 and let $P(d^\nu)$ denote the probability of finding ν 3d electrons on a given Cu site. Within the independent electron or Hartree-Fock approximation the average d count is found to be $\bar{n}_d \simeq 9.5$ and the probabilities of different configurations are $P(d^{10}) = 0.56$, $P(d^9) = 0.38$ and $P(d^8) = 0.06$. When correlations are included, i.e., the correlated ground state $|\psi_0\rangle$ is calculated by quantum chemical wavefunction-based methods, the average d electron number changes to $\bar{n}_d \simeq 9.3$ and $P(d^{10}) = 0.29$, $P(d^9) = 0.70$ while $P(d^8) = 0.0$. One notices that the d^8 configurations are almost completely suppressed by correlations, which is in agreement with photoemission experiments. The fluctuations between the d^9 and d^{10} configurations are fixed by the value of $\bar{n}_d$.

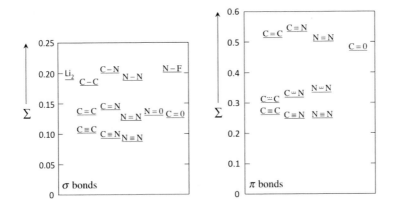

Fig. 10.1. Correlation strength parameter Σ for a number of bonds formed by first-row atoms. Single, double, triple, and aromatic bonds are indicated by single, double, triple and dotted-over solid lines, respectively. (From [352]; actually, the definition of Σ in that paper differs slightly from the present one, but this difference does not affect the figures).

A similar analysis for the oxygen ions reveals that here the $2p^4$ configurations are *not* completely suppressed, because the Coulomb integrals are not as large as for Cu. Indeed, these configurations are important for superexchange to occur, which determines the antiferromagnetic coupling between Cu ions. In accordance with the above considerations one finds $\Sigma(\text{Cu}) \simeq 0.8$ and $\Sigma(\text{O}) \simeq 0.7$. So indeed correlations are quite strong in La_2CuO_4. On the other hand, they are still considerably smaller than those of $4f$ electrons in a system like CeAl_3.

When correlations are so strong that the electrons remain localized like $4f$ electrons in most of the rare-earth systems, we observe a separation of spin and charge degrees of freedom. In that case spin degrees of freedom may lead to excitations in the form of magnons or crystal-field excitations, while charge degrees of freedom are seen in photoemission experiments. In the cuprates the values of Σ are too small in order to expect spin-charge separation.

In addition to interatomic correlations we have to consider intra-atomic correlations. For them a measure of their strength is more difficult to define. One way is to find out to which extent Hund's rule correlations are operative on a given atomic site i. When electrons hop very frequently on and off a site, charge fluctuations at that site are dominated by the kinetic energy and Hund's rules play a minor role only. In the strong correlation limit Hund's rules will be fully operative. A possible measure is the degree of spin alignment at a given atomic site i

$$S_i^2 = \langle \psi_0 | \mathbf{S}^2(i) | \psi_0 \rangle \quad , \tag{10.4}$$

where $\mathbf{S}(i) = \Sigma_m \mathbf{s}_\nu(i)$ and $\mathbf{s}_\nu(i)$ is the spin operator for orbital ν. The quantity S_i^2 should be compared when the SCF ground-state wavefunction $|\Phi_{\text{SCF}}\rangle$ is used and when instead the ground state $|\Phi_{\text{loc}}\rangle$ in the limit of complete suppression of interatomic charge fluctuations is used. Therefore we may define

$$\Delta S_i^2 = \frac{\langle \psi_0 | \mathbf{S}^2(i) | \psi_0 \rangle - \langle \phi_{\text{SCF}} | \mathbf{S}^2(i) | \phi_{\text{SCF}} \rangle}{\langle \phi_{\text{loc}} | \mathbf{S}^2(i) | \phi_{\text{loc}} \rangle - \langle \phi_{\text{SCF}} | \mathbf{S}^2(i) | \phi_{\text{SCF}} \rangle} \qquad (10.5)$$

for a measure of the strength of intra-atomic correlations. Note that $0 \leq \Delta S_i^2 \leq 1$. For example, for the transition metals Fe, Co and Ni the quantity ΔS_i^2 is approximately 0.5.

Those findings show that the much discussed transition metals are just in the middle between the limits of uncorrelated and strongly correlated electrons. Hund's rule correlations are important in them, but relatively large overlaps of atomic wavefunctions on neighboring sites prevent their complete establishment.

10.2 Indicators of Strong Correlations

There are several indicators of strong electron correlations. One is the appearance of low-energy scales. As mentioned before, among all valence electrons, the $4f$ electrons are the most strongly correlated ones due to their nearness to the nuclei. They remain localized in most cases and have low-energy excitations of order meV caused by level splittings by the crystalline electric field of the surrounding. These excitations involve spin and orbital degrees of freedom but no charge degrees of freedom. Therefore we have here a separation between spin (plus orbital) and charge excitations. The latter are of order eV as is known from photoemission experiments. Similar low-energy excitations are found in Ce or Yb intermetallic compounds where the correlations are not quite as strong. Here a small hybridization takes place of the $4f$ electron (hole) with the surroundings. Therefore spin and charge degrees of freedom are no longer independent but become weakly coupled. The associated high density of states gives rise to heavy quasiparticles. Note the special case of one-dimensional electronic systems where spin and charge excitations remain uncoupled, even when correlations are weak (Luttinger liquid).

The energy scale of the quasiparticles is set by the variation of the single-particle energies with wavevector $\mathbf{k}$. A measure of it is the Fermi velocity v_F, which for heavy quasiparticles can be two to three orders of magnitude smaller than in ordinary metals like sodium. Since the conduction electron density and hence p_F is similar to that in other metals, a small v_F implies a large effective mass m^*.

The high density of low-energy excitations holds up only to an energy characterized by a temperature T*. With the total number of excitations being nearly constant, e.g., one per lattice site, the lower T* is, the larger is the density of states of the low-energy excitations. In systems with strong electron correlations T* ranges from a few Kelvin to a few hundred Kelvin. When the correlations are weaker, T* increases until it is no longer useful to speak of a distinct low-energy scale.

The microscopic origin of the low-energy excitations may be quite different. Examples are the Kondo effect, strong intra-atomic correlations, charge ordering, frustrations or nearness to a quantum critical point, to name some of them. One should be aware of this manifold and not associate automatically the appearance of heavy quasiparticles with a Kondo effect. For better insight we study first the simplest possible example for a low-energy scale caused by strong correlations.

10.2.1 Low-Energy Scales: a Simple Model

In the following we want to show the way in which strong correlations may result in new low-energy scales. The simplest example of a system of strongly correlated electrons consists of two electrons distributed over two orbitals. These orbitals are denoted by L (for ligand) and F (for $4f$, for example) and we assume the corresponding orbital energies to be ϵ_l and ϵ_f with $\epsilon_f < \epsilon_l$. Two electrons in the F orbital are expected to repel each other with an energy $U \gg (\epsilon_l - \epsilon_f)$. When they are both in the L orbital, or when one electron is in the L and the other in the F orbital, we neglect their Coulomb interaction. This is justified if the ligand orbital has a large spatial extend. It applies when, for example, the ligand orbital is that of a large molecule. We assume that the hybridization V between the two orbitals is small, i.e., $V \ll (\epsilon_l - \epsilon_f)$. The Hamiltonian of the system depicted in Fig. 10.2 is

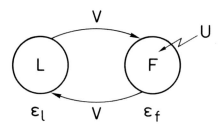

Fig. 10.2. Two orbitals L (for ligand) and F (for f orbital) with orbital energies ϵ_ℓ and ϵ_f which are weakly coupled through a hybridization matrix element V. When two electrons are in orbital F their mutual Coulomb repulsion is U. The L orbital is assumed to be extended and Coulomb interactions between electrons in it are neglected, as are interactions between electrons in an L and an F orbital.

10 Strongly Correlated Electrons

$$H = \epsilon_l \sum_\sigma l_\sigma^+ l_\sigma + \epsilon_f \sum_\sigma f_\sigma^+ f_\sigma + V \sum_\sigma \left(l_\sigma^+ f_\sigma + f_\sigma^+ l_\sigma \right) + U n_\uparrow^f n_\downarrow^f \quad . \tag{10.6}$$

The $l_\sigma^+ (l_\sigma)$, $f_\sigma^+ (f_\sigma)$ create (annihilate) electrons with spin σ in the L and F orbital, respectively; furthermore, $n_\sigma^f = f_\sigma^+ f_\sigma$. When $V = 0$, the ground state of the system has energy $E_0 = \epsilon_l + \epsilon_f$ and is fourfold degenerate. One electron is in the F orbital, while the other is in the L orbital. A state with a doubly occupied F orbital has a high energy, because of the large Coulomb repulsion U. The four states are eigenstates of the total spin S and take the form

$$\begin{aligned}
|\Phi_{S=0}\rangle &= \frac{1}{\sqrt{2}} \left(f_\uparrow^+ l_\downarrow^+ - f_\downarrow^+ l_\uparrow^+ \right) |0\rangle \\
|\Phi_{S=1}^1\rangle &= f_\uparrow^+ l_\uparrow^+ |0\rangle \\
|\Phi_{S=1}^0\rangle &= \frac{1}{\sqrt{2}} \left(f_\uparrow^+ l_\downarrow^+ + f_\downarrow^+ l_\uparrow^+ \right) |0\rangle \\
|\Phi_{S=1}^{-1}\rangle &= f_\downarrow^+ l_\downarrow^+ |0\rangle \quad .
\end{aligned} \tag{10.7}$$

The system has one excited state of the form

$$|\Phi_{\text{ex}}\rangle = l_\uparrow^+ l_\downarrow^+ |0\rangle \quad . \tag{10.8}$$

The energy of the excited state is $E_{\text{ex}} = 2\epsilon_l$. The state $f_\uparrow^+ f_\downarrow^+ |0\rangle$ is excluded from further consideration, since its energy is of order U and we assume that $U \to \infty$.

When the hybridization is turned on, the singlets $|\Phi_{S=0}\rangle$ and $|\Phi_{\text{ex}}\rangle$ are coupled, while the $S = 1$ states $|\Phi_{S=1}\rangle$ remain unchanged. The resulting 2×2 matrix

$$\begin{pmatrix} \epsilon_f + \epsilon_l & V\sqrt{2} \\ V\sqrt{2} & 2\epsilon_l \end{pmatrix} \tag{10.9}$$

is easily diagonalized. For small values of $(V/\Delta\epsilon)$, the eigenvectors are

$$\begin{aligned}
|\psi_0\rangle &= \left[1 - \left(\frac{V}{\Delta\epsilon}\right)^2 \right] |\Phi_{S=0}\rangle - \frac{\sqrt{2}V}{\Delta\epsilon} |\Phi_{\text{ex}}\rangle \\
|\psi_{\text{ex}}\rangle &= \left[1 - \left(\frac{V}{\Delta\epsilon}\right)^2 \right] |\Phi_{\text{ex}}\rangle + \frac{\sqrt{2}V}{\Delta\epsilon} |\Phi_{S=0}\rangle
\end{aligned} \tag{10.10}$$

with $\Delta\epsilon = \epsilon_l - \epsilon_f$. The eigenvalue of $|\psi_0\rangle$ is

$$\tilde{E}_0 = E_0 - \frac{2V^2}{\Delta\epsilon} \tag{10.11}$$

while that of $|\psi_{\text{ex}}\rangle$ is

$$\tilde{E}_{\text{ex}} = E_{\text{ex}} + \frac{2V^2}{\Delta\epsilon} \tag{10.12}$$

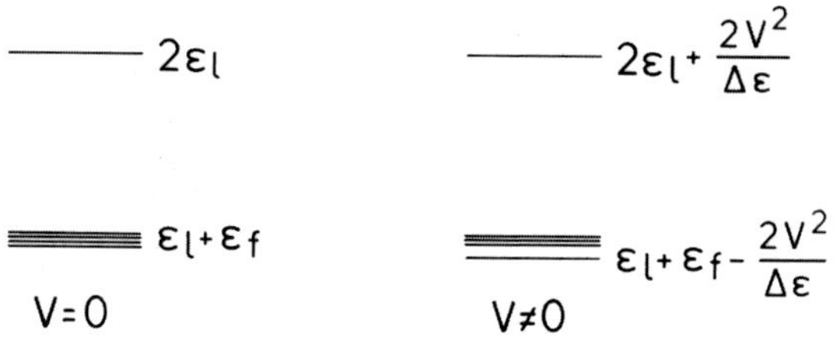

10.2 Indicators of Strong Correlations

```
───── 2ε_l              ───── 2ε_l + 2V²/Δε

═════ ε_l+ε_f           ═════ ε_l+ε_f - 2V²/Δε
   V=0                     V≠0
```

Fig. 10.3. Changes in the two-electron spectrum when $U \to \infty$ and the hybridization V is turned on. A low-lying singlet splits off from the quartet states.

Consider the changes in the spectrum shown in Fig. 10.3. For small values of V there is a low-energy triplet excitation above the singlet ground state. One can attach a characteristic temperature $k_B T^* = 2V^2/\Delta\epsilon$ to the energy gain associated with the singlet formation. The forms of the ground state and of the low-lying excitations are both due to the strong correlations, which forbid double occupancy of the F orbital. In the ground state the occupancy of the F orbital is

$$n^f = 1 - 2\left(\frac{V}{\Delta\epsilon}\right)^2 \quad \text{with} \quad n^f = n^f_\uparrow + n^f_\downarrow \ . \tag{10.13}$$

We conclude that for temperatures $T \ll T^*$ there exist two distinct types of excitations:

a) low-lying excitations with an energy $k_B T^*$ which involve predominantly spin degrees of freedom;

b) an excitation of an f electron into the ligand orbital, with an excitation energy of order $\Delta\epsilon$. This excitation involves charge degrees of freedom.

The separation of excitations into those involving primarily either spin or charge degrees of freedom is what the model enables us to learn. Since the three triplet states have the same energy, spin-rotation symmetry is conserved. Therefore we may re-express the low-energy spin excitation by an effective Hamiltonian

$$H_{\text{eff}} = J \mathbf{S}_\ell \mathbf{S}_f - \frac{J}{4} \ , \quad J = \frac{2V^2}{\Delta\epsilon} \ , \tag{10.14}$$

where $\mathbf{S}_\ell$ and $\mathbf{S}_f$ are the spins of a ligand and an f electron, respectively.

For $T \gg T^*$ the singlet and triplet states are equally populated and the singlet-triplet splitting becomes unimportant. The two electrons in the L and F orbitals act like being effectively coupled via (10.14), with $k_B T \gg J$. The high-energy excitation into the ligand orbital remains possible in all cases.

The above simple model contains key ingredients of the Kondo problem as well as of systems with heavy electrons (heavy fermions). The ground state is a singlet and the magnetic moment of the partially filled F level is zero. As the temperature increases towards T*, the triplet states become thermally populated. Since they have a moment, the magnetic character of the f electron starts to appear; for T≫T* the magnetic moment is fully present. The singlet character of the ground state is noticeable only for T≪T*. The low-lying excitations are intimately connected with the degeneracy of the ground state in the absence of hybridizations. A number of approximation schemes can be tested by applying them to the Hamiltonian (10.7) (see Sect. 10.6.1).

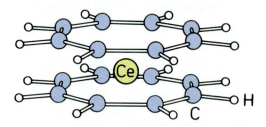

Fig. 10.4. Molecular structure of di-π-cyclo-octatetraene cerium, $(C_8H_8)_2Ce$ (cerocene).

The above simple Hamiltonian describes in essence the ground state of molecules like di-π-cyclo-octatetraene cerium, $(C_8H_8)_2Ce$, abbreviated cerocene. We show its structure in Fig. 10.4. Because the total valence electron number is even, Ce is usually considered to be tetravalent; however, such a statement proves to be misleading. We find instead that the Ce ion belongs almost entirely to a $4f^1$ configuration corresponding to $Ce^{3+}(C_8H_8^{1,5-})_2$. The $4f$ electron forms a singlet with an electron of the highest occupied molecular orbital (HOMO) with e_{2u} symmetry. The $4f^1 e_{2u}^3$ singlet state resembles the state $|\psi_0\rangle$ in (10.10) with two electrons added because the HOMO is fourfold degenerate. This degeneracy results from the C_{8v} symmetry of a C_8H_8 ring. Large-scale MC-SCF calculations with several hundred basis functions confirm this picture [92]. The multiconfigurations on which the SCF calculation is based must include $4f^0 e_{2u}^4$, $4f^1 e_{2u}^3$, and also $4f^2 e_{2u}^2$. One finds that the ground state has in addition to f^1 configurations an admixture of f^0 and f^2 configurations, with weights of 3,8 % and 0,2 %, respectively. The underlying physical picture appears in Fig. 10.5. The calculations provide also the low-lying excitation energies to triplet states; for cerocene, for example, they are on the order of 0.3 eV. Because of this relatively large excitation energy, the Van Vleck paramagnetic contribution is smaller than the diamagnetic one of the ring currents. If we replaced the C_8H_8 rings by more extended molecules, the

HOMO would couple much less to the $4f$ orbital and the excitation energies would correspondingly decrease. The sandwich molecule would then become paramagnetic, since that contribution grows faster than the diamagnetic one with increasing size.

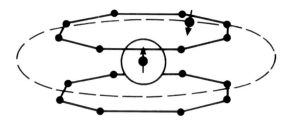

Fig. 10.5. Formation of a singlet in $(C_8H_8)_2Ce$ (cerocene) by an unpaired electron in the HOMO and a $4f^1$ configuration of Ce. The binding energy is approximately 0.3 eV.

10.2.2 Effective Hamiltonians

As discussed in the preceding sections, one indicator of strong correlations are new low-energy scales. The above simple model served us as a specific example for a possible origin of a low-energy scale. It is certainly desirable to describe low-energy excitations of strongly correlated electrons by an appropriate effective Hamiltonian such as (10.14). While the original Hamiltonian contains the large Coulomb interactions acting on the full Hilbert space, the effective Hamiltonian acts on a strongly reduced Hilbert space. It should contain only those degrees of freedom required for describing the low-energy excitations. The remaining degrees of freedom are projected out and enter the effective Hamiltonian in terms of new interactions. For example, the effective Hamiltonian (10.14) has neither the Coulomb repulsion U appearing in it, nor the charge fluctuations in the F orbital. Instead, it contains a spin-spin interaction, a result of the strong correlations, i.e., of $U \gg V$. The reduced Hilbert space consists of all states with one electron each in the L and in the F orbital. The change from (10.6) to (10.14) is a special example of a Schrieffer-Wolff transformation [405] which was originally used to transfer the Anderson impurity Hamiltonian into the Kondo Hamiltonian. First a general route is shown for reducing the Hilbert space. Thereafter we apply it to the above model, i.e., to two electrons described by the Hamiltonian (10.6).

Consider the configurational space of an electron system. This space is divided into two subspaces by using the projection operators P and $Q = 1 - P$. The projection of a wavefunction $|\psi\rangle$ onto the two subspaces is given by

$P|\psi\rangle = |\psi_P\rangle$ and $Q|\psi\rangle = |\psi_Q\rangle$. The Schrödinger equation acting on the full space is written as

$$(H_{PP} - E)|\psi_P\rangle + H_{PQ}|\psi_Q\rangle = 0 ,$$
$$H_{QP}|\psi_P\rangle + (H_{QQ} - E)|\psi_Q\rangle = 0 . \qquad (10.15)$$

The Hamiltonians H_{PP} and H_{QQ} act within the subspaces defined by the projectors P and Q, respectively. The two subspaces are connected through the Hamiltonian $H_{PQ} = PHQ$. By eliminating either $|\psi_Q\rangle$ or $|\psi_P\rangle$ from (10.15) we reduce the problem to one subspace $|\psi_P\rangle$ or $|\psi_Q\rangle$ and obtain

$$\left(\tilde{H}_{PP} - E\right)|\psi_P\rangle = 0 \quad ; \quad \left(\tilde{H}_{QQ} - E\right)|\psi_Q\rangle = 0 . \qquad (10.16)$$

The effective Hamiltonians $\tilde{H}_{PP}$ and $\tilde{H}_{QQ}$ are given by

$$\tilde{H}_{PP} = H_{PP} - H_{PQ}\frac{1}{H_{QQ} - E}H_{QP} , \qquad (10.17a)$$

$$\tilde{H}_{QQ} = H_{QQ} - H_{QP}\frac{1}{H_{PP} - E}H_{PQ} \qquad (10.17b)$$

and act within the subspaces defined by P and Q, respectively. Working with $\tilde{H}_{PP}$, one can limit oneself to the space $|\psi_P\rangle$; the subspace $|\psi_Q\rangle$ is then eliminated from the problem. The Schrieffer-Wolff transformation (10.17) consists of treating the influence of Q on $\tilde{H}_{PP}$ to lowest-order perturbation theory. This is justified as long as H_{PQ} is sufficiently small.

Let us apply this scheme to the Hamiltonian (10.6). In subspace P one electron is in each of the orbitals L and F while in $|\psi_Q\rangle$ both electrons are in the L orbital. The operators H_{PQ} and H_{QP} are

$$H_{QP} + H_{PQ} = V\sum_\sigma \left(\ell_\sigma^+ f_\sigma + f_\sigma^+ \ell_\sigma\right) . \qquad (10.18)$$

The energy denominator in (10.17a) is $\Delta\epsilon$ and, by setting H_{QP} and H_{PQ} into (10.17a), we obtain immediately (10.14) when the ground-state energy of H_{PP} for the two electron system is set equal to zero.

10.3 Kondo Effect

The Kondo effect has been playing a significant role in condensed matter physics. It was discovered in an attempt made by *Kondo* to explain the resistivity $\rho(T)$ of metals with added magnetic impurities. It has a characteristic minimum at low temperatures which had remained unexplained for a long time. Soon it became clear that with Kondo's explanation a road had opened to a whole new class of problems.

Kondo's original work starts from a Hamiltonian of free conduction electrons coupled antiferromagnetically to a local impurity spin $S = 1/2$ via

$$H_{\text{int}} = J\mathbf{s}(0)\mathbf{S}, \quad J > 0 \quad . \tag{10.19}$$

Here $\mathbf{s}(0)$ is the conduction electron spin density at the impurity site which is taken to be at the origin. Therefore this Hamiltonian is often referred to as Kondo Hamiltonian although it had been known long before [474]. Due to the interaction, the impurity spin and the conduction electron spin form a singlet very similar to the one found for the Hamiltonian (10.6) or (10.14). The difference to the simple model considered before is that here we are dealing with a continuum of conduction electron states instead of a single ligand orbital. It turns out that one must go beyond perturbation theory in order to obtain finite results in the low-temperature limit. The scattering of conduction electrons on a magnetic impurity becomes a true many-body problem, since the Fermi distribution function enters explicitly the scattering rate. This is due to the quantum character of the impurity spin and differs fundamentally from the scattering by a nonmagnetic impurity. Here the Fermi function drops out of the scattering rate and the scattering process is a one-electron problem. It is the appearance of the Fermi function in the former case which leads to divergent results for the scattering rate at low temperatures, when the Hamiltonian (10.19) is treated by perturbation theory.

Instead of showing explicitly the failure of perturbation theory, we proceed here differently. We do not start from the Hamiltonian (10.19) but rather from one which is due to *Anderson* and still contains the charge degrees of freedom of the impurity site like (10.6) does. This Hamiltonian will be treated by nonperturbative methods and the formation of a ground-state singlet will be demonstrated.

Assume that the magnetic impurity is Ce^{3+} which implies a $4f^1$ configuration. According to Hund's rule, the lowest j multiplet of an f electron is $j = 5/2$. The z component of $\mathbf{j}$ is denoted by m. The corresponding creation (annihilation) operators are $f_m^+(f_m)$, the number operators are $n_m^f = f_m^+ f_m$. The Coulomb repulsion between f electrons is denoted by U and the limit $U \to \infty$ is later assumed. The conduction electrons in form of Bloch states are created by operators $c_{\mathbf{k}\sigma}^+$. They have a dispersion $\epsilon(\mathbf{k})$ and are considered to be non-interacting. Their weak hybridization with the f electrons is described by the Hamiltonian

$$H = \sum_{\mathbf{k}\sigma} \epsilon(\mathbf{k}) c_{\mathbf{k}\sigma}^+ c_{\mathbf{k}\sigma} + \epsilon_f \sum_m n_m^f + \frac{U}{2} \sum_{m \neq m'} n_m^f n_{m'}^f$$
$$+ \sum_{\mathbf{k}m\sigma} \left[V_{m\sigma}(\mathbf{k}) f_m^+ c_{\mathbf{k}\sigma} + V_{m\sigma}^*(\mathbf{k}) c_{\mathbf{k}\sigma}^+ f_m \right] \quad . \tag{10.20}$$

As all energies are measured from the Fermi energy ϵ_F, the resemblance to the Hamiltonian (10.6) is apparent. The difference is that now the f orbital is ν_f-fold degenerate and the ligand orbital has been replaced by a partially filled band of conduction electrons. Note that the Kondo Hamiltonian (10.19) as well as a generalized form due to *Coqblin* and *Schrieffer* [72] are obtained from the Anderson Hamiltonian by a Schrieffer-Wolff transformation (see preceding section).

Due to crystal-field splitting of the lowest J manifold, the degeneracy ν_f of the ground state of a magnetic rare earth impurity is usually less than $(2J+1)$. We will later analyze the case of Ce^{3+} with one and zero f electrons. Then the total angular momentum j of the $4f^1$ electron equals the total angular momentum J of the incomplete f shell. The most interesting effects stem from the coupling of the impurity to the conduction electrons. Thus we keep only those degrees of freedom of the conduction electrons which couple to the impurity. The matrix elements $V_{m\sigma}(\mathbf{k})$ vary rapidly with the direction of $\mathbf{k}$. The angular average of $V_{m\sigma}^*(\mathbf{k})V_{m'\sigma}(\mathbf{k})$ is small, except when $m' = m$. One may therefore set

$$\sum_\sigma \int \frac{d\hat{k}}{4\pi} V_{m\sigma}^*(\mathbf{k}) V_{m'\sigma}(\mathbf{k}) \simeq V^2(k) \delta_{mm'} \quad , \tag{10.21}$$

where $\hat{k} = \mathbf{k}/|\mathbf{k}|$. This suggests introducing the following orthogonal electronic basis:

$$|k, m\rangle = \frac{1}{V(k)} \sum_\sigma \int \frac{d\hat{k}}{4\pi} V_{m\sigma}^*(\mathbf{k}) |\mathbf{k}\sigma\rangle \quad . \tag{10.22}$$

When expressed in this basis, and provided $\epsilon(\mathbf{k}) = \epsilon(|\mathbf{k}|) = \epsilon(k)$, the Anderson Hamiltonian takes the form

$$H = \sum_{km} \epsilon(k) c_{km}^+ c_{km} + \epsilon_f \sum_m n_m^f + \frac{U}{2} \sum_{m \neq m'} n_m^f n_{m'}^f$$
$$+ \sum_{km} V(k) \left(f_m^+ c_{km} + c_{km}^+ f_m \right) + \tilde{H}_0 \quad . \tag{10.23}$$

The Hamiltonian $\tilde{H}_0$ contains all those degrees of freedom of the conduction electrons which do not couple with the impurity.

In the limit $U \to \infty$, the magnetic ion cannot be in a configuration with more than one f electron. The Hamiltonian (10.23) is then rewritten in the form

$$H = \sum_{km} \epsilon(k) c_{km}^+ c_{km} + \epsilon_f \sum_m \hat{f}_m^+ \hat{f}_m$$
$$+ \sum_{km} V(k) \left(\hat{f}_m^+ c_{km} + c_{km}^+ \hat{f}_m \right) + \tilde{H}_0 \quad , \tag{10.24}$$

where the $\hat{f}_m^+$ create an f electron only if the f site was previously empty, i.e.,

$$\hat{f}_m^+ = f_m^+ |0\rangle\langle 0| \quad . \tag{10.25}$$

The ket $|0\rangle$ denotes the $4f^0$ state and therefore $|0\rangle\langle 0|$ acts like a projection operator. If we denote $f_m^+|0\rangle = |m\rangle$, then we may also write

$$\hat{f}_m^+ = |m\rangle\langle 0|$$
$$= X_{m0} \quad . \tag{10.26}$$

In the literature the operators $X_{m0} = |m\rangle\langle 0|$, $X_{0m} = \hat{f}_m = |0\rangle\langle m|$, $X_{mm} = \hat{f}_m^+\hat{f}_m = |m\rangle\langle m|$ and $X_{00} = |0\rangle\langle 0|$ are frequently used and referred to as standard basis operators or Hubbard operators. The $X_{\nu\mu}$ are no longer fermion operators, because they do not obey simple fermionic anticommutation relations. For example, $[X_{m0}, X_{0m}]_+ = X_{mm} + X_{00} \neq 1$. In addition they must fulfill the subsidiary condition

$$\sum_{m=1}^{\nu_f} X_{mm} + X_{00} = 1 \quad . \tag{10.27}$$

This condition is in compliance with the requirement that one remains in the Hilbert space with either a singly occupied or empty f site. When a Hamiltonian is expressed in terms of the Hubbard operators, double occupancy of the f site is strictly excluded. We return to the Hubbard operators later.

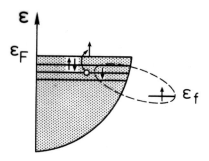

Fig. 10.6. A magnetic impurity with one f orbital placed in a metal. In order to reduce the problem to the one in Fig. 10.2 one must first move one conduction electron to the Fermi surface. The remaining electron of that conduction-electron state and the f electron can form a singlet.

In order to account for the energy gain (10.11), by forming a ground-state singlet, we have to reduce the problem of a magnetic ion in a metal to a two-electron one. We show in Fig. 10.6 how this is done for the case of $\nu_f = 2$. One electron from a given conduction-electron state is moved to the Fermi energy. The remaining electron and the f electron can form a singlet. In contrast to Sect. 10.2.1 where we considered only one ligand orbital, the singlet formation can take place here with many different conduction-electron states. The states close to ϵ_F become particularly important since it takes less energy to move from them one electron up to ϵ_F.

In order to put the above arguments onto a quantitative basis, we assume that
$$|\epsilon_f| \gg \nu_f \Gamma = \nu_f \pi N(0) V^2 \quad , \quad V = V(k_F) \quad , \tag{10.28}$$
where $N(0)$ is the conduction electron density of states per spin direction. The energy Γ is the width of the f level due to the coupling to the conduction electrons. We obtain it when we apply Fermi's golden rule.

For the calculation of the ground-state energy we have to use a nonperturbative approach. We will outline two of them. One is by constructing a proper trial wavefunction and optimizing the parameters contained in it. The second is to use the projection method discussed in Sect. 5.4.1. We begin with the more conventional trial-wavefunction approach of *Varma* and *Yafet* [465]. Let $|\Phi_0\rangle$ represent the filled Fermi sea of the conduction electrons. The occupied spin orbitals contain two electrons each. A proper ansatz for a variational wavefunction is

$$|\psi_{S=0}\rangle = A\left(1 + \frac{1}{\sqrt{\nu_f}} \sum_{km} \alpha(k) f_m^+ c_{km}\right) |\Phi_0\rangle \quad . \tag{10.29}$$

As is seen from Fig. 10.6, the part without an f electron must have small weight as compared with the part when one f electron is present.

The normalization constant A in (10.29) relates to the f electron number by means of
$$|A|^2 = 1 - n_f \quad , \tag{10.30}$$
where
$$n_f = \sum_m n_m^f \quad . \tag{10.31}$$

For $n_f \to 1$ the state $|\Phi_0\rangle$ has indeed little weight in $|\psi_{S=0}\rangle$. By minimizing $\langle \psi_{S=0}|H|\psi_{S=0}\rangle$ one can determine A as well as $\alpha(k)$ and obtain the energy gain due to the formation of a singlet. Instead of doing this here, we want to derive the same result by the projection method, since it is a nice example for the usefulness of that technique.

We start from
$$\mathcal{E}_0 = (H|\Omega) \tag{10.32}$$
and choose for H_0 the Hamiltonian (10.24) with $V(k) = 0$. We are interested in a ground state with total spin $S = 0$ and therefore choose for the ground state of H_0 the singlet state

$$|\tilde{\Phi}_0\rangle = \frac{1}{\sqrt{\nu_f}} \sum_m \hat{f}_m^+ c_{k_F m} |\Phi_0\rangle \quad . \tag{10.33}$$

Note that for a singlet state the total electron number must be even and therefore an electron from $|\Phi_0\rangle$ has to be removed. In constructing $|\Omega)$ we must include the most important microscopic processes caused by H_1. One important feature is that also electrons with $k < k_F$ contribute to the formation of the singlet $|\psi_{S=0}\rangle$ when $V(k) \neq 0$ (Fig. 10.7b). This is achieved by including the operators $A_{km} = c_{km} c^+_{k_F m}$ in the operator set $\{A_\nu\}$ with the help of which $|\Omega\rangle$ is constructed. A second important process is that the f electron can leave the impurity site, which then becomes a $4f^0$ configuration. The latter corresponds to $|\Phi_{\text{ex}}\rangle$ in (10.10) and is described by including $c^+_{k_F m} \hat{f}_m$ in the set $\{A_\nu\}$ (Fig. 10.7c). We discard all processes caused by H_1 (i.e., by $V(k)$) in which a conduction electron is promoted into a state with $k > k_F$. As it turns out, they are higher-order corrections in an expansion in terms of $1/\nu_f$ [162, 163].

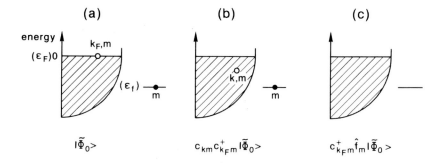

Fig. 10.7. Configurations from which the singlet state $|\psi_{S=0}\rangle$ is constructed.

Led by (5.68) we make the following ansatz for Ω

$$|\Omega\rangle = \left|1 + S + \frac{1}{2} S^2\right\rangle \quad , \tag{10.34}$$

where $S = S_1 + S_2$ and

$$S_1 = \frac{C}{\sqrt{\nu_f}} \sum_{k<k_F, m} \alpha(k) c_{km} c^+_{k_F m} \tag{10.35}$$

$$S_2 = C \sum_m c^+_{k_F m} \hat{f}_m \quad . \tag{10.36}$$

We set

$$\mathcal{E}_0 = E_0 + \epsilon \quad , \tag{10.37}$$

where E_0 is the energy of $|\tilde{\Phi}_0\rangle$ and ϵ is the energy gain due to H_1. From (10.32) we obtain

$$\mathcal{E}_0 = E_0 + (H_1 \mid S_2)$$
$$= E_0 + 2VC \quad . \tag{10.38}$$

Therefore $\epsilon = 2VC$. We obtain a set of equations for C and $\alpha(k)$ from the relations

$$\left(\hat{f}_m^+ c_{k_F m} \mid H\Omega\right) = 0$$
$$\left(c_{km} c_{k_F m}^+ \mid H\Omega\right) = 0 \quad , \tag{10.39}$$

which follow from (5.44b). After replacing C by $\epsilon/(2V)$, we obtain from them the following coupled equations:

$$\epsilon = |\epsilon_f| + \sqrt{\nu_f} V \sum_{k \leq k_F} \alpha(k)$$
$$(\epsilon(k) + \epsilon) \alpha(k) = \sqrt{\nu_f} V \quad . \tag{10.40}$$

Replacing $\alpha(k)$ in the first equation, we find

$$\epsilon = |\epsilon_f| + \nu_f V^2 \sum_{k \leq k_F} \frac{1}{\epsilon + \epsilon(k)}$$
$$= |\epsilon_f| + \nu_f V^2 N(0) \int_{-D}^{0} \frac{d\epsilon(k)}{\epsilon + \epsilon(k)} \quad . \tag{10.41}$$

Here we have assumed a constant density of states. The lower cut-off D is equal to half the conduction-electron bandwidth when the band is half filled. Note that the same equations (10.40,10.41) are obtained from the trial wavefunction (10.29) when its energy is minimized.

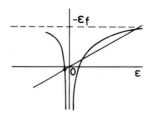

Fig. 10.8. Graphical solution of (10.41). There always exists a solution with $\epsilon < 0$.

The solutions of this equation can be visualized from Fig. 10.8, in which we plot the two sides of (10.41) separately. For small values of V we find three solutions; one of them has $\epsilon < 0$ and is the one we are looking for. It is approximately determined by

$$|\epsilon_f| = \nu_f N(0) V^2 \ln(D/|\epsilon|) \quad . \tag{10.42}$$

The lowering of the energy due to hybridization is therefore

$$\epsilon = -D e^{-|\epsilon_f|/(\nu_f N(0) V^2)} \quad . \tag{10.43}$$

The energy of the singlet $|\psi_{S=0}\rangle$ has to be compared with that of the multiplets $|\psi_{S\neq 0}\rangle$. We start from a corresponding multiplet ground state $|\tilde{\Phi}_{S\neq 0}\rangle$ of H_0. Provided we discard once more processes, in which a conduction electron is promoted by H_1 to a state with $k > k_F$, there is no effect of H_1 on $|\tilde{\Phi}_{S\neq 0}\rangle$ and the corresponding Ω is equal to unity. This can easily be seen, for example, by considering the largest ground-state multiplet and inspecting $|\tilde{\Phi}_{S_{\max}}^m\rangle = \hat{f}_m^+ \hat{c}_{k_F-m}|\Phi_0\rangle$ with $m = (\nu_f - 1)/2$. Processes of the form of (10.36) are not possible, and $\mathcal{E}_0 = E_0$. As expected from Sect. 10.2.1, we find that the energy of the singlet $|\psi_{S=0}\rangle$ is always lower than that of the multiplets $|\psi_{S\neq 0}\rangle$.

It is customary to associate a characteristic temperature T_K (Kondo temperature) with the energy gain due to the formation of the singlet, i.e.,

$$k_B T_K = D \, \exp\left(\frac{-\pi |\epsilon_f|}{\nu_f \Gamma}\right) \quad , \tag{10.44}$$

with Γ as defined in (10.28). The condition $|\epsilon_f| \gg \nu_f \Gamma$ ensures that T_K remains sufficiently small (Kondo regime). In contrast to (10.11), the energy gain is a nonanalytic function of V; its origin may be easily traced back to the fact that the singlet formation involves many different conduction electron k states.

Equation (10.44) does not change when processes involving conduction electrons with $k > k_F$ are included. These contributions are the same for $|\psi_{S=0}\rangle$ and $|\psi_{S\neq 0}\rangle$ and cancel when the energy difference between the singlet and the multiplets is calculated.

The f electron number can be determined by adding a term of the form $\lambda \sum_m \hat{f}_m^+ \hat{f}_m$ to the Hamiltonian and by calculating

$$n_f = \left.\frac{\partial \mathcal{E}_0(\lambda)}{\partial \lambda}\right|_{\lambda=0}$$

$$= 1 + \left.\frac{\partial \epsilon(\lambda)}{\partial \lambda}\right|_{\lambda=0} \quad . \tag{10.45}$$

By means of (10.37,10.38), we obtain

$$n_f = 1 - \frac{k_B T_K \pi}{\nu_f \Gamma} \quad , \tag{10.46}$$

which is a useful relation between the f electron number and the Kondo temperature. As T_K decreases, so does the deviation of n_f from unity.

When the ground-state singlet is formed, the magnetic susceptibility of the impurity χ^{imp} remains finite in the zero-temperature limit. It can become very large, when the energy difference to the excited states is small (Van Vleck susceptibility). The susceptibility can be calculated by including an external field h in the Hamiltonian. It lifts the degeneracy by Zeeman splitting the f levels; we must replace ϵ_f in (10.23) by $\epsilon_f - g_J \mu_B m h$ with $-J \leq m \leq J$.

The factor g_J denotes the Landé factor of the ground-state J multiplet, which for Ce^{3+} is $J = 5/2$ as pointed out before. We write the energy $\mathcal{E}_0(h)$ in analogy to (10.37) in the form

$$\mathcal{E}_0(h) = E_0 + \Delta\epsilon(h) \quad . \tag{10.47}$$

Repeating the above calculation, we find

$$\Delta\epsilon(h) = |\epsilon_f| + V^2 \sum_m \sum_{k \leq k_F} (\Delta\epsilon(h) + g_J \mu_B m h + \epsilon(k))^{-1} \quad . \tag{10.48}$$

This equation generalizes (10.41) to finite magnetic fields. If we take the second derivative with respect to h, we obtain

$$\begin{aligned}
\chi^{\text{imp}} &= - \left(\frac{\partial^2}{\partial h^2} \Delta\epsilon(h) \right)_{h \to 0} \\
&= (g_J \mu_B)^2 \frac{J(J+1)}{3} \frac{1}{\nu_f \Gamma} \frac{n_f}{1 - n_f} \\
&= \frac{(g_J \mu_B)^2}{\pi} \frac{J(J+1)}{3} \frac{1}{k_B T_K} \quad .
\end{aligned} \tag{10.49}$$

This shows that the smaller T_K is, the larger is χ^{imp}. Experiments measuring the magnetic susceptibility, demonstrate that the magnetic impurity loses its moment as the temperature falls below T_K. This is a direct consequence of the singlet formation, which in turn results from the strong electron correlations.

In Sect. 10.2.1 we have seen that a local f orbital, which hybridizes weakly with an extended ligand orbital has a low-lying singlet-triplet excitation. The latter involves predominantly spin degrees of freedom of the system; the same is expected to hold true for a magnetic impurity. Formally, we could determine the excitation energies by computing the poles of the one-particle Green's function of the system, thereby applying the projection technique as before. But, in order to bring out the analogy with Sect. 10.2.1, we prefer instead to generalize the ansatz (10.29) to excited states.

The hole state $|\psi_{pn}^{\text{ex}}\rangle$ with quantum numbers p and n can be written in the form

$$|\psi_{pn}^{\text{ex}}\rangle = A \left(1 + \frac{1}{\sqrt{\nu_f}} \sum_{km} \alpha(k) f_m^+ c_{km} \right) c_{pn} |\Phi_0\rangle \quad , \tag{10.50}$$

which should be compared with (10.29). If we think of that state as a dressed conduction-band state, we notice that the "dress" or "cloud" of the bare state

$$|\Phi_{pn}^{ex}\rangle = c_{pn}|\Phi_0\rangle \tag{10.51}$$

consists of an admixture with the impurity f state. For energies $|\epsilon - \epsilon_F| \ll k_B T_K$ the weight of $|\Phi_{pn}^{ex}\rangle$ in $|\psi_{pn}^{ex}\rangle$ is $|A|^2 = 1 - n_f$ (see (10.30)) and therefore very small. A different way of stating the same is to say that the bare hole must be strongly renormalized in order to become a quasihole.

Consider a photoemission experiment in which the f spectral density is measured. Its weight near ϵ_F is given by the sum over the squared matrix elements

$$\sum_{pn}|\langle \psi_{pn}^{ex}|f_m|\psi_{S=0}\rangle|^2 = |A|^2 n_f/\nu_f \quad, \tag{10.52}$$

and varies as $n_f(1 - n_f)/\nu_f$. We are dealing here with the weighting factor of the well-known Abrikosov-Suhl or Kondo resonance, which appears in the vicinity of the Fermi energy. As we saw in Sect. 10.2.1, the low-energy singlet-triplet excitation has only a small change in f charge associated with it, i.e., $\Delta n_f \propto (1 - n_f)$. The main contributions to the f spectral weight come from an energy regime near ϵ_f, well separated from the Kondo resonance. It corresponds to the high-energy excitation (10.12), in which the f electron is removed from its orbital.

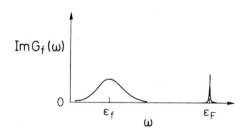

Fig. 10.9. Schematic plot of the spectral function vs. ω. The narrow low-energy peak has weight $(1-n_f)$ and results from spin fluctuations while the broad peak near ϵ_f is due to charge fluctuations. The broadening is obtained from a more advanced theory than presented here.

The f spectral density is schematically shown in Fig. 10.9. Clearly seen are low-energy excitations involving predominantly spin degrees of freedom and high-energy excitations involving mainly charge degrees of freedom.

Before closing, it is important to mention that the Anderson Hamiltonian for one impurity can be solved exactly by Bethe ansatz methods. For more details the reader should consult [32] or the reviews [339, 401, 456]. The exact solution provides a stringent test for any approximation, yet the physics is simpler to grasp from the approach above.

10.4 The Hubbard Model Revisited

The Hubbard Model, which we briefly introduced in Sect. 8.2 has played a major role in the physics of strongly correlated electrons. Therefore we want to discuss it here in more detail. For convenience we rewrite the model Hamiltonian (8.22) in simplified form

$$H = -t \sum_{\langle ij \rangle \sigma} \left(a_{i\sigma}^{+} a_{j\sigma} + h.c. \right) + U \sum_i n_{i\uparrow} n_{i\downarrow} \; , \qquad (10.53)$$

where only nearest neighbor hopping processes are considered. The interesting case is, of course, when $U \gg t$, i.e., when electrons are strongly correlated. Here *Gutzwiller*, *Hubbard* and *Kanamori*, who formulated independently the model Hamiltonian (10.53) took quite different routes in order to extract physical properties from it. We will especially concentrate on the approaches which *Gutzwiller* and *Hubbard* took because they come closest to projection techniques, which are one of the guidelines in this book. But before doing that we apply the simplest possible approximation to (10.53), which is a molecular field approximation. It leads to a spin-density wave state. In a metal with strong electron correlations we have to reduce charge fluctuations. This reduction of charge fluctuations is achieved here by symmetry breaking.

10.4.1 Spin-Density Wave Ground State

It is straightforward to show that the interaction term in the Hubbard Hamiltonian (10.53) can be rewritten as

$$U n_{i\uparrow} n_{i\downarrow} = \frac{U}{2} n_i - \frac{2U}{3} \mathbf{s}_i \cdot \mathbf{s}_i \; , \qquad (10.54)$$

where

$$\mathbf{s}_i = a_{i\alpha}^{+} \boldsymbol{\sigma}_{\alpha\beta} a_{i\beta} \qquad (10.55)$$

is the operator of electron spin at site i (this is seen most easily by noting that both sides of the equation give identical results when acting onto any of the four possible states of site i, i.e., states $|0\rangle$, $|\uparrow\rangle$, $|\downarrow\rangle$ or $|\uparrow\downarrow\rangle$). This shows that the system can lower its energy by forming magnetic moments at each site. These moments may either form a static pattern–so that the expectation values $\langle \mathbf{s}_i \rangle$ have nonvanishing time-independent values–or they may fluctuate so that $\langle \mathbf{s}_i^2 \rangle \neq 0$ but $\langle \mathbf{s}_i \rangle = 0$. The question, of which case is realized is a tricky issue and the answer depends on dimensionality, lattice geometry and temperature.

Spin-density wave theories are based on the mean-field approximation and treat the former case, i.e., one assumes time-independent expectation values

$$\langle \mathbf{s}_i \rangle = Re \left(\mathbf{M} e^{-i \mathbf{Q} \cdot \mathbf{R}_i} \right) \qquad (10.56)$$

where the $\mathbf{R}_i$ are lattice vectors. Choosing, e.g., $\mathbf{M} = (0, 0, m_0)$ yields a state where the ordered moment always points in z-direction but its sign oscillates with wave vector $\mathbf{Q}$. However, choosing $\mathbf{M} = (\sqrt{2}m_0, i\sqrt{2}m_0, 0)$ with a $\mathbf{Q}$ in z-direction gives a state where the ordered moment rotates around the wave vector and so on. In the following we consider a square lattice and for simplicity restrict ourselves to the first case, i.e., an ordered moment in z-direction oscillating with a wave vector $\mathbf{Q} = (\pi, \pi)$.

We decompose H into two terms

$$H = H_{\text{SCF}} + H_{\text{res}}$$
$$H_{\text{SCF}} = -t \sum_{\langle ij \rangle \sigma} \left(a_{i\sigma}^+ a_{j\sigma} + h.c. \right) + U \sum_{i\sigma} \langle n_{i-\sigma} \rangle n_{i\sigma} + E_0$$
$$H_{\text{res}} = U \sum_i \delta n_{i\uparrow} \delta n_{i\downarrow} \quad , \tag{10.57}$$

where we have introduced $\delta n_{i\sigma} = n_{i\sigma} - \langle n_{i\sigma} \rangle$ and

$$E_0 = -U \sum_i \langle n_{i\uparrow} \rangle \langle n_{i\downarrow} \rangle \quad . \tag{10.58}$$

The mean-field approximation is here an unrestricted SCF approximation. It neglects H_{res} and makes an ansatz for the ground state, which breaks spin-rotational and translational symmetry. It has the form of a spin-density wave, which in the case of half filling becomes an antiferromagnetic Néel state. We shall restrict ourselves to the latter; for systems different from half filling one finds solutions in the form of spiral states (compare with Fig. 10.23). We divide the lattice into sublattices A and B, whereby A contains the origin $(0, 0)$, and make the following ansatz

$$\langle n_{i\sigma} \rangle = \frac{1}{2} \left(1 + \sigma m_0 e^{-i\mathbf{Q} \cdot \mathbf{R}_i} \right) \quad . \tag{10.59}$$

The staggered magnetization m_0 is defined as

$$m_0 = \begin{cases} \langle n_{i\uparrow} - n_{i\downarrow} \rangle , & \text{sublattice } A \\ \langle n_{i\downarrow} - n_{i\uparrow} \rangle , & \text{sublattice } B \end{cases} \quad . \tag{10.60}$$

These relations are ensured by the factor $\exp(-i\mathbf{Q} \cdot \mathbf{R}_i)$ in (10.59) because $\exp(-i\mathbf{Q} \cdot \mathbf{R}_i)$ is $+1$ for $\mathbf{R}_i$ on sublattice A and -1 for sublattice B. The vector $\mathbf{Q} = (\pi, \pi)$ is the reciprocal lattice vector in the presence of an AF ground state. The unit cell is doubled in that case and correspondingly the Brillouin zone is reduced by one half. This is indicated in Fig 10.10. When (10.59) is inserted into (10.57) and the Fourier transform is taken, we obtain

$$H_{\text{SCF}} = \sum_{\mathbf{k}\sigma}{}' \left[\left(\epsilon(\mathbf{k}) + \frac{U}{2} \right) a_{\mathbf{k}\sigma}^+ a_{\mathbf{k}\sigma} + \left(-\epsilon(\mathbf{k}) + \frac{U}{2} \right) a_{\mathbf{k}+\mathbf{Q}\sigma}^+ a_{\mathbf{k}+\mathbf{Q}\sigma} \right.$$
$$\left. - \frac{U}{2} \sigma m_0 \left(a_{\mathbf{k}\sigma}^+ a_{\mathbf{k}+\mathbf{Q}\sigma} + a_{\mathbf{k}+\mathbf{Q}\sigma}^+ a_{\mathbf{k}\sigma} \right) \right] + E_0 \quad , \tag{10.61}$$

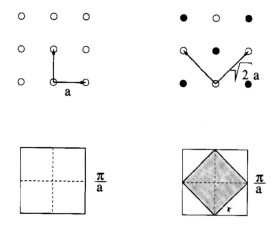

Fig. 10.10. Top: By partitioning the square lattice into two sublattices the primitive translation vectors are rotated by 45° and scaled by a factor of $\sqrt{2}$. Bottom: The reciprocal lattice vectors–which span the Brillouin zone–are thus rotated by 45° as well as shrunk by a factor of $1/\sqrt{2}$. The reduced Brillouin zone is shaded.

where $\epsilon(\mathbf{k}) = -2t(\cos k_x + \cos k_y)$. The dash indicates that the summation is over the reduced Brillouin zone. We have used that $\epsilon(\mathbf{k} + \mathbf{Q}) = -\epsilon(\mathbf{k})$. The Hamiltonian is diagonalized by the transformation

$$\alpha^+_{\mathbf{k}\sigma} = u_\mathbf{k} a^+_{\mathbf{k}\sigma} + \sigma v_\mathbf{k} a^+_{\mathbf{k}+\mathbf{Q}\sigma}$$
$$\beta^+_{\mathbf{k}\sigma} = -\sigma v_\mathbf{k} a^+_{\mathbf{k}\sigma} + u_\mathbf{k} a^+_{\mathbf{k}+\mathbf{Q}\sigma} \qquad (10.62)$$

where $u^2_\mathbf{k} + v^2_\mathbf{k} = 1$. Diagonalization requires that

$$u^2_\mathbf{k} = \frac{1}{2}\left(1 - \frac{\epsilon(\mathbf{k})}{E(\mathbf{k})}\right), \quad v^2_\mathbf{k} = \frac{1}{2}\left(1 + \frac{\epsilon(\mathbf{k})}{E(\mathbf{k})}\right)$$
$$E(\mathbf{k}) = \sqrt{\epsilon(\mathbf{k})^2 + m_0^2 U^2/4} \ . \qquad (10.63)$$

We denote the diagonalized form of H_SCF by H_SDW:

$$H_\text{SDW} = {\sum_{\mathbf{k}\sigma}}' \left[\left(\frac{U}{2} - E(\mathbf{k})\right)\alpha^+_{\mathbf{k}\sigma}\alpha_{\mathbf{k}\sigma} + \left(\frac{U}{2} + E(\mathbf{k})\right)\beta^+_{\mathbf{k}\sigma}\beta_{\mathbf{k}\sigma}\right] + E_0$$
$$= {\sum_{\mathbf{k}\sigma}}' \left[E_1(\mathbf{k})\alpha^+_{\mathbf{k}\sigma}\alpha_{\mathbf{k}\sigma} + E_2(\mathbf{k})\beta^+_{\mathbf{k}\sigma}\beta_{\mathbf{k}\sigma}\right] + E_0 \ . \qquad (10.64)$$

One effect of U is to generate a gap in the excitation spectrum as indicated in Fig. 10.11. The SDW ground state is

$$|\varPhi_{\mathrm{SDW}}\rangle = \prod_{\mathbf{k}\sigma}{}' a_{\mathbf{k}\sigma}^{+}|0\rangle$$
$$= \prod_{\mathbf{k}\sigma}{}' \left[u_{\mathbf{k}}a_{\mathbf{k}\sigma}^{+} + \sigma v_{\mathbf{k}}a_{\mathbf{k}+\mathbf{Q}\sigma}^{+}\right]|0\rangle \quad . \tag{10.65}$$

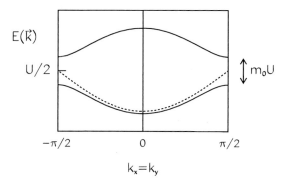

Fig. 10.11. Excitation spectrum $E_{1,2}(\mathbf{k})$ of a SDW state on a square lattice (schematic). Dashed line: the case $m_0 U = 0$.

As is obvious from (10.60) the staggered magnetization is

$$\begin{aligned} m_0 &= \frac{1}{N_0}\sum_i e^{i\mathbf{Q}\cdot\mathbf{R}_i}\langle n_{i\uparrow} - n_{i\downarrow}\rangle \\ &= \frac{1}{N_0}\sum_{\mathbf{k}\sigma} \sigma\langle a_{\mathbf{k}\sigma}^{+} a_{\mathbf{k}+\mathbf{Q}\sigma}\rangle \\ &= \frac{1}{N_0}{\sum_{\mathbf{k}\sigma}}' \sigma\langle a_{\mathbf{k}\sigma}^{+} a_{\mathbf{k}+\mathbf{Q}\sigma} + h.c.\rangle \end{aligned} \tag{10.66}$$

This leads to the following self-consistency relation

$$m_0 = \frac{2}{N_0}\sum_{\mathbf{k}}^{\mathrm{occ}} \frac{m_0 U}{\sqrt{(\epsilon(\mathbf{k}) - \epsilon(\mathbf{k}+\mathbf{Q}))^2 + m_0^2 U^2}} \quad , \tag{10.67}$$

where N_0 is the number of sites. It should be noted that in the present case of a square lattice at half filling with $\epsilon(\mathbf{k}) = -\epsilon(\mathbf{k}+\mathbf{Q})$ we have perfect nesting. This implies that $m_0 \neq 0$ for all values of $U > 0$, i.e., the system becomes unstable with respect to the formation of antiferromagnetic ordering as soon as particles repel each other.

The ground-state $|\varPhi_{\mathrm{SDW}}\rangle$ has fewer configurations with doubly occupied sites than does the ground state without symmetry breaking (spin restricted

SCF ground state). This is caused by the magnetization m_0 and is particularly evident in the limit of large U. In that limit $m_0 \to 1$ and $|\Phi_{\text{SDW}}\rangle$ goes over into a Néel state with no doubly occupied sites.

It is instructive to study the spectral density for the full, i.e., unreduced Brillouin zone. We do this here as an exercise. It is $\bar{A}(\mathbf{k},\omega) = -\pi^{-1} Im\{G^R(\mathbf{k},\omega)\}$. The Green's function is obtained in SDW approximation by replacing in (8.23), (8.1) $|\psi_0\rangle$ by $|\Phi_{\text{SDW}}\rangle$ and by choosing for the set of dynamic variables $\{A_n\}$ the operators $A_1 = a_{\mathbf{k}\sigma}^+$, where $\mathbf{k}$ is inside the reduced Brillouin zone and $A_2 = a_{\mathbf{k}+\mathbf{Q}\sigma}^+$. This leads to a 2×2 Green's function matrix (see (8.25))

$$R_{ij}(\omega) = \left(A_i \mid \frac{1}{\omega - L_{\text{SCF}} + i\eta} A_j \right)_+ \quad , \qquad (10.68)$$

where L_{SCF} refers to H_{SCF} given by (10.57). By using (10.62) and (10.64) equation (8.25) can be solved simply, from which $G^R(\mathbf{k},\omega)$ is obtained as $G^R(\mathbf{k},\omega) = R_{11}(\mathbf{k},\omega)$. For $\mathbf{k}$ in the AF Brillouin zone the result is

$$G^R(\mathbf{k},\omega) = \frac{v_{\mathbf{k}}^2}{\omega - \left(\frac{U}{2} - E(\mathbf{k})\right) + i\eta} + \frac{u_{\mathbf{k}}^2}{\omega - \left(\frac{U}{2} + E(\mathbf{k})\right) + i\eta} \quad , \qquad (10.69)$$

with poles which are obvious from (10.64). Therefore the spectral density is

$$\rho(\mathbf{k},\omega) = v_{\mathbf{k}}^2 \delta\left(\omega - \frac{U}{2} + E(\mathbf{k})\right) + u_{\mathbf{k}}^2 \delta\left(\omega - \frac{U}{2} - E(\mathbf{k})\right) \quad . \qquad (10.70)$$

A schematic plot of the spectral density is shown in Fig. 10.12. When we

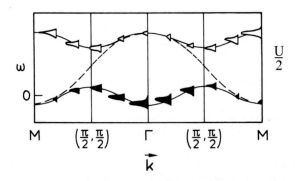

Fig. 10.12. Schematic plot of $\rho(\mathbf{k},\omega)$ in the ω, $\mathbf{k}$ plane within SDW approximation. The momentum vector $\mathbf{k}$ varies between Γ and M. The δ-function peaks have been slightly broadened. The shadow band is obtained by connecting points with the smaller pole strength (dashed line). Compare with Fig. 8.1.

back fold the part from $(\frac{\pi}{2}, \frac{\pi}{2})$ to M of that quantity onto the part from Γ to $(\frac{\pi}{2}, \frac{\pi}{2})$ we obtain excitations with pole strength unity. The behavior of the pole strengths in the full Brillouin zone is of relevance when we deal with paramagnets having pronounced short-range antiferromagnetic correlations. Then the full Brillouin zone must be used, but features of long-range order begin to show up here in the form of shadow bands.

The structure of (10.69) is similar to the one in (8.29) for $n = 1$. In fact, in the limit of large U both expressions agree with each other after a shift of energy by $U/2$. In that case Figs. 8.1 and 10.12 become identical. For a given **k** vector we may consider the pole with the smaller residue as a satellite to the pole with the larger pole strength. Stated differently, for each **k** value we have a band state and a shadow band state. At $(\pm\frac{\pi}{2}, \pm\frac{\pi}{2})$ the quasiparticle picture is breaking down completely because both poles have equal strength.

10.4.2 Gutzwiller's Ground-State Wavefunction

The reduction of doubly occupied sites at half filling due to the on-site repulsion U is also achieved by a trial wavefunction introduced by *Gutzwiller* [164]. The Gutzwiller wavefunction is of the form

$$| \psi_G \rangle = \exp\left(\eta \sum_i n_{i\uparrow} n_{i\downarrow} \right) | \Phi_0 \rangle$$
$$= P_G(\eta) | \Phi_0 \rangle \ . \qquad (10.71)$$

Here $|\Phi_0\rangle$ is the ground state of noninteracting electrons and the exponential prefactor is often referred to as Gutzwiller projector. As demonstrated before (see (5.96)) the Gutzwiller wavefunction is a single-mode approximation to the true ground state of the Hubbard Hamiltonian. The variational parameter η (or $\tilde{\eta}$ in (5.97)) is determined by minimizing the energy

$$E_G = \frac{\langle \psi_G | H | \psi_G \rangle}{\langle \psi_G | \psi_G \rangle} \ . \qquad (10.72)$$

Generally we are not able to compute that expectation value without drastic approximations. Noticeable exceptions are the cases of one dimension [322] and infinite dimensions [323]. In both cases E_G can be computed exactly.

Equation (10.72) can be evaluated my making use of the so-called Gutzwiller approximation [165, 166]. The essence of this approximation is that the hopping matrix element t is replaced by a product $\gamma_\sigma t$. The renormalization factor $\gamma_\sigma < 1$ is caused by the reduced hopping probability when the projector $P_G(\eta)$ is applied to a wavefunction of uncorrelated electrons. The γ_σ factor is obtained from combinatorial considerations when two assumptions are made. One is that all possible electronic configurations, i.e., all possible distributions of electrons over the sites of the system, are assumed to have the same overall phase factor, e.g., $+1$. The other is that the relative weights of the different configurations are determined solely by the number of doubly occupied sites D [351]. No fermionic sign changes are taken into account. This enables us to rederive γ_σ by using single-site bosonic instead of fermionic operators [106]. Thus we introduce four bosonic operators for each site i. They correspond to the four different states of a site, i.e., empty, singly occupied with spin σ and doubly occupied. The operators are e_i^+, $s_{i\sigma}^+$ and d_i^+ [250]. With their help we define a bosonic wavefunction for site i, namely $\nu_i^+|0\rangle$, with

$$\nu_i^+ = \left(1 + \alpha_\uparrow^2 + \alpha_\downarrow^2 + \beta^2\right)^{-1/2} \left(e_i^+ + \alpha_\uparrow s_{i\uparrow}^+ + \alpha_\downarrow s_{i\downarrow}^+ + \beta d_i^+\right) . \quad (10.73)$$

The α_σ, β are related to the densities n_σ with spin σ (i.e., resulting from singly and doubly occupied sites) and d of doubly occupied sites through

$$n_\sigma = \frac{\alpha_\sigma^2 + \beta^2}{1 + \alpha_\uparrow^2 + \alpha_\downarrow^2 + d^2} \quad , \quad d = \frac{\beta^2}{1 + \alpha_\uparrow^2 + \alpha_\downarrow^2 + d^2} . \quad (10.74)$$

Note that for uncorrelated electrons $d = n_\uparrow n_\downarrow$. The equations can be inverted to yield

$$\alpha_\sigma = \sqrt{\frac{n_\sigma - d}{1 - n_\uparrow - n_\downarrow + d}} \quad ; \quad \beta = \sqrt{\frac{d}{1 - n_\uparrow - n_\downarrow + d}} . \quad (10.75)$$

The total wavefunction written in terms of the bosonic operators is

$$|\psi\rangle = P(N_\uparrow)P(N_\downarrow) \prod_i \nu_i^+ |0\rangle \quad (10.76)$$

and the projectors $P(N_\sigma)$ project onto the states with $\sum_i \left(s_{i\sigma}^+ s_{i\sigma} + d_i^+ d_i\right) = N_\sigma$. The state $|\psi\rangle$ has the following properties: The probability that a site is singly or doubly occupied is the same for all sites. All configurations with the same number of doubly occupied sites have the same weight in $|\psi\rangle$ and they all have the same phase.

It is interesting that in the thermodynamic limit the projectors $P(N_\sigma)$ in (10.76) may be omitted. The wavefunction $|\psi\rangle$ is then no longer an eigenfunction of the total particle number operator N_{op}, yet $\langle\psi|\psi\rangle = 1$. When N_{op} is applied to it the distribution of eigenvalues is strongly peaked at N with a mean square deviation $(\Delta N)^2$ of order N. Therefore $\Delta N/N \sim 1/\sqrt{N}$ and vanishes for $N \to \infty$. A similar situation is found in Chapter 15 where the BCS ground state is discussed. Because $|\psi\rangle$ is a product state it is easy to calculate with it expectation values. In order to compute the renormalization factor γ_σ one should notice that the operator $a_{i\sigma}$ is in the bosonic basis equivalent to $s_{i-\sigma}^+ d_i + e_i^+ s_{i\sigma}$. Annihilating an electron with spin σ can either reduce a doubly occupied site to a singly occupied one with spin $-\sigma$ or convert a singly occupied site into an empty one. Note the ignoring of the fermionic sign change. Fermions require a given order of the $a_{i\sigma}^+$ operators when a site is doubly occupied, e.g., $a_{i\uparrow}^+ a_{i\downarrow}^+$ in which case the application of a fermionic operator $s_{i-\sigma}^+$ would give a sign change for one of the two spin directions. This is neglected here. We then find for the hopping term

$$r_\sigma = \langle \psi | \left(s_{i-\sigma}^+ d_i + e_i^+ s_{i\sigma} \right) \left(d_j^+ s_{j-\sigma} + s_{j\sigma}^+ e_j \right) | \psi \rangle$$

$$= \left(\frac{\alpha_\sigma + \beta \alpha_{-\sigma}}{1 + \alpha_\uparrow^2 + \alpha_\downarrow^2 + \beta^2} \right)^2$$

$$= \left(\sqrt{n_\sigma - d}\sqrt{1 - n_\sigma - n_{-\sigma} + d} + \sqrt{d}\sqrt{n_{-\sigma} - d} \right)^2 \quad (10.77)$$

and for the renormalization factor

$$\gamma_\sigma = \frac{r_\sigma(d)}{r_\sigma(d = n_\uparrow n_\downarrow)}$$

$$= \frac{\left[\sqrt{n_\sigma - d}\sqrt{1 - n_\sigma - n_{-\sigma} + d} + \sqrt{d(n_{-\sigma} - d)} \right]^2}{n_\sigma (1 - n_\sigma)} \quad . \quad (10.78)$$

We have to divide by $r_\sigma(d = n_\uparrow n_\downarrow)$, the matrix element for uncorrelated electrons, since in that limit γ_σ has to be equal to unity. We study γ_σ by assuming a paramagnetic ground state with band filling $n \leq 1$, i.e., $n_\sigma = n_{-\sigma} = n/2$. Let us take the limit $U \to \infty$. In that case $d = 0$ and we find that

$$\gamma = \frac{1 - n}{1 - n/2} \quad . \quad (10.79)$$

The numerator is equal to the probability that an electron at site j finds site i to be empty so that it can move to it. Of this probability a fraction $(1 - n_\sigma) = (1 - n/2)$ is already taken into account by Pauli's principle. Therefore, $(1 - n)$ is divided by this factor. For the special case of half filling $n = 1$ we find that $\gamma = 0$. Each site is occupied by one electron and each move would generate a doubly occupied site and cost an infinite amount of energy.

When we keep U finite, we find from (10.78) that for $n = 1$

$$\gamma = 16d(1/2 - d) \quad . \quad (10.80)$$

Therefore the energy per site becomes

$$\frac{E_G(d)}{N_0} = 16d(1/2 - d)\bar{\epsilon} + Ud \quad . \quad (10.81)$$

Here $\bar{\epsilon}$ is the average kinetic energy per electron, when unrenormalized hopping matrix elements t_{ij} are used, i.e., when the electrons are noninteracting. It is negative when the center of the band is set equal to zero. The energy is minimized when

$$d_{\min} = \frac{8\bar{\epsilon} + U}{32\bar{\epsilon}} \quad (10.82)$$

and takes the form
$$\frac{E_{\min}}{N_0} = \bar{\epsilon}(1 - U/U_c)^2 \quad . \tag{10.83}$$

Note that $U_c = 8|\bar{\epsilon}|$ is a critical interaction constant. For $U \geq U_c$ the energy is zero and the number of doubly occupied sites vanishes, i.e., $d_{\min} = 0$. In that case the system is going over into an insulating state with one electron per site (Brinkmann-Rice transition) [45]. Inspecting (10.81) we notice that $q = 16d(\frac{1}{2}-d)$ acts as a renormalization factor of the average kinetic energy caused by correlations. At the metal-insulator transition it vanishes like $(1 - U/U_c)$. This implies that the effective mass m^* diverges, i.e.,

$$\frac{m^*}{m_0} = \frac{1}{1 - U/U_c} \quad , \tag{10.84}$$

where m_0 is the electron mass in the absence of the interaction energy U. The same holds true for the magnetic susceptibility χ_s, which is proportional to the effective density of states and hence to m^*,

$$\chi_s \sim \frac{1}{1 - U/U_c} \quad . \tag{10.85}$$

The Gutzwiller wavefunction suppresses partially charge fluctuations due to the interaction U, yet it fails to properly include spin-spin correlations and density correlations between sites (compare with (5.87)). This shortcoming becomes visible when we try to derive the antiferromagnetic interaction contributions to the energy at half filling in the limit of large ratio U/t. They are of order $J \sim t^2/U$ and cannot be obtained from Gutzwiller's wavefunction, while other approximation schemes like the t–J model produce them easily (see Sect. 10.6). Therefore the Brinkmann-Rice phase transition does not fully describe the metal-insulator transition in finite dimensions. In infinite dimensions though it turns out to be exact.

The momentum distribution $n(\mathbf{p})$ calculated from $|\psi_G\rangle$ shows always a discontinuity at p_F like for a metal, even for large values of U when at half-filling the system is an insulator. The exponential prefactor in (10.71) which describes the correlations is not able to remove the discontinuity, a feature of $|\Phi_0\rangle$. The shortcomings just described can be linked to the single-mode approximation discussed in Sect. (5.4.4).

10.4.3 Hubbard's Approximations and their Extensions

In Sect. 8.1 we discussed the projection method and demonstrated their usefulness by applying it to the Hubbard Hamiltonian (10.53). In particular, the so-called Hubbard approximation takes a very simple form when this technique is applied. The retarded Green's function matrix $R_{ij}(\omega)$ (8.5) reduces to a simple 2 × 2 matrix since the relevant variables are just the operators $a_{i\sigma}^+$ and $a_{i\sigma}^+ \delta n_{i-\sigma}$. With this choice of variables one is able to reduce double occupancies of sites when an on-site Coulomb interaction is operative. For large values of $U/|t|$ the retarded Green's function was shown to be of the form (8.29), which for half filling ($n = 1$) reduces to

$$G(\mathbf{k}, z) = \frac{1/2}{z - \epsilon(\mathbf{k})/2} + \frac{1/2}{z - U - \epsilon(\mathbf{k})/2} \quad , \quad z = \omega + i\eta \quad . \tag{10.86}$$

The poles are given by

$$\omega_1(\mathbf{k}) = \epsilon(\mathbf{k})/2$$
$$\omega_2(\mathbf{k}) = U + \epsilon(\mathbf{k})/2 \quad . \tag{10.87}$$

Since $U/t \gg 1$ we deal with two Hubbard bands separated by a gap. For a square lattice the corresponding spectral density $\rho(\mathbf{k}, \omega)$ was shown in Fig. 8.1. The density of states $\rho(\omega)$ is schematically shown in Fig. 10.13 for half filling and for $n < 1$. From the numerator in (8.29) as well as from Fig. 8.1 it is seen that weight between the two Hubbard bands is shifted when n deviates from $n = 1$. The transfer of spectral density from the upper to the lower band with increasing hole doping is easy to understand:

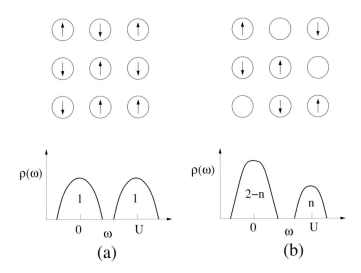

Fig. 10.13. Configurations with different filling factors n in the limit $U \to \infty$ together with the corresponding density of states (DOS) $\rho(\omega)$ (a): $n = 1$ (half filling); (b): $n < 1$. In the lower part we show the integrated DOS of the Hubbard bands, which depends on the filling factor n as indicated.

Consider a system of N_0 sites at half filling in the limit of large U. Then each site is occupied by one electron (Fig. 10.13a). Adding one hole yields N_0 possible ways of ending up with one empty site. Therefore the upper Hubbard band contains N_0 states. Assume now that the system contains M holes corresponding to a filling factor of $n = 1 - M/N_0$. If this is the case, there are only $(N_0 - M)$ ways of creating an empty site when a hole is added. This implies a reduction of the number of states in the upper band by a factor of n. There are $2M$ ways of singly occupying a site, when a electron is added and $(N_0 - M)$ ways of removing an electron from a (singly) occupied site. Therefore, the total number of states in the lower band is $N_0(2 - n)$, of which $N_0 n$ are filled and $2N_0(1 - n)$ are empty. This reasoning, schematically indicated in Fig. 10.13, agrees with the results of the Hubbard I approximation. As we have seen in Sect. 9.2 this picture remains incomplete. In a paramagnetic system a third, e.g., central peak in $\rho(\omega)$ is present, when we deviate from half filling by hole doping (see Fig. 10.27).

Of special interest is the transformation from two Hubbard bands to one band when U decreases, i.e., the Mott-Hubbard transition from an insulator to a metal. It cannot be described by the Hubbard I approximation, which therefore needs improvement. The latter is obtained by including the band shifts $W(\mathbf{k})$ in L_{22} neglected hitherto, but more important by an extension of the relevant variables $\{A_\nu\}$ from which the Green's function is computed. We obtain the Hubbard III approximation by extending the variables to the set $a_{i\sigma}^+, a_{i\sigma}^+ \delta n_{l-\sigma}$ and $a_{i\sigma}^+ \delta(a_{l-\sigma}^+ a_{i-\sigma})$, where $\delta(a_{l-\sigma}^+ a_{i-\sigma}) = a_{l-\sigma}^+ a_{i-\sigma} - \langle a_{l-\sigma}^+ a_{i-\sigma} \rangle$. They stress the importance of *nonlocal* correlations, because they involve different lattice sites.

It is instructive to consider the physical picture behind the Hubbard III approximation, i.e., the particular choice of variables. Consider an electron with spin σ moving through the system. When electrons with spin $-\sigma$ are kept fixed, the moving electron experiences two different potentials. Sites occupied by an electron with spin $-\sigma$ present a potential U, while sites without an electron with spin $-\sigma$ have a vanishing potential. Since we expect the spin $-\sigma$ electrons to be distributed at random, we are dealing with the problem of an electron moving through a random potential. The disorder scattering is treated by the inclusion of the variables $a_{i\sigma}^+ \delta n_{\ell-\sigma} (i \neq \ell)$ in the set $\{A_\nu\}$. The assumption of keeping electrons with spin $-\sigma$ fixed can be relaxed, which allows us to include the effect of their motion on that of the spin σ electron. As a result of this motion, a given site switches back and forth between having a potential 0 and U. This feature is treated by including the variables $a_{i\sigma}^+ \delta(a_{\ell-\sigma}^+ a_{i-\sigma})$ in the set of $\{A_\nu\}$. When the switching rate $1/\tau$ is less than U, we expect a broadening of the two levels 0 and U, while for $1/\tau \gg U$ only a simple resonance is expected. The change from one situation to the other causes a transition from an insulator to a metal. An important feature of the Hubbard III approximation is that for $n \neq 1$ the linewidth of the low-energy excitations does not vanish like ω^2 when $\omega \to 0$, but instead remains finite. This implies a break-down of the quasiparticle picture (see Sect. 7.2). It is related to a remaining static character of the random potential, in which an electron is moving. The variables $a_{i\sigma}^+ \delta(a_{\ell-\sigma}^+ a_{i-\sigma})$ do not completely remedy that shortcoming.

Let us return to the different choices of variables within the two-particle, one-hole operator space, when the Hubbard I and III approximations are made. The most general form of variables within that space is $a_{i\sigma}^+ \delta(a_{j,-\sigma}^+ a_{\ell,-\sigma})$. In the Hubbard I approximation we select those variables for which $i = j = \ell$, while in Hubbard III those variables are kept for which either $j = \ell$ or $i = \ell$. Among alternative approximations for the Hubbard Hamiltonian, the most appropriate one to choose is $a_{i\sigma}^+ \delta(a_{i+\nu_1,-\sigma}^+ a_{i+\nu_2,-\sigma})$ with ν_1 and ν_2 limited to a given number of neighbors (nearest, next-nearest, etc.) of site i. Because of the fixed number of variables associated with each site i, we are dealing here with a matrix problem of relatively small dimension. Therefore the corresponding matrix equations can be solved numerically. But this has not been done yet.

The Hubbard Hamiltonian has also been treated by making use of Hubbard operators. They were briefly introduced before and are defined in site representation by

$$X_i^{\sigma 0} = a_{i\sigma}^+ (1 - n_{i-\sigma}) \quad , \quad X_i^{0\sigma} = \left(X_i^{\sigma 0}\right)^+$$
$$X_i^{2-\sigma} = \sigma a_{i\sigma}^+ n_{i-\sigma} \quad . \tag{10.88}$$

They cause a transition from an empty site to one occupied with a single electron of spin σ, and from a singly occupied site to a doubly occupied site. One notices that the Hubbard operators are composite operators. In addition to the above operators, there are the ones $X_i^{nn} = |n,i\rangle\langle n,i|$ with $n = 0, \sigma, -\sigma, 2$. They specify the four possible configurations of a site. The spin flip operator is $X_i^{\sigma-\sigma} = X_i^{\sigma 0} X_i^{0-\sigma}$. One may define Green's functions for the Hubbard operators and apply various techniques like a diagrammatic one [210, 373] or projection methods [308]. They are not simple since Hubbard operators are not fermion operators. Instead they fulfill the commutation relations

$$\left[X_i^{mn}, X_j^{rs}\right]_\pm = \delta_{ij} \left[\delta_{rn} X_i^{ms} \pm \delta_{ms} X_i^{rn}\right] \quad . \tag{10.89}$$

In addition they have to satisfy the local constraint

$$X_i^{00} + \sum_\sigma X_i^{\sigma\sigma} + X_i^{22} = 1 \tag{10.90}$$

because a site has to be in one of the four different states, i.e., empty, singly occupied with spin $\pm\sigma$ or doubly occupied. In the infinite U limit this condition has to be satisfied without X_i^{22}, i.e., without a possible double occupancy of a site.

When correlations are strong, i.e., for large ratios of U/t the Hubbard Hamiltonian can be expanded in terms of t/U. To leading order this brings us to the so-called $t - J$ Hamiltonian. Since that limit is particularly interesting we devote to it a separate Sect. 10.5.

10.4.4 Kanamori Limit

When we deal with a system with a nearly empty or nearly full band of correlated electrons the Kanamori limit applies [230]. In a dilute gas of Fermi particles or holes the most important processes are those in which two particles scatter repeatedly on each other without affecting the background [141]. A t-matrix approach applies in this case. As a result the Hubbard interaction U goes over into a screened interaction U_{eff} which never exceeds the order of the bandwidth W. This is seen by a simple argument: When U is very large, electrons avoid occupying the same site. This costs them at most kinetic energy of order W. The effective repulsion energy, therefore, cannot be larger than that energy.

The t-matrix approach starts from a Hubbard interaction written in $\mathbf{k}$ space,

$$H_{\text{int}} = U \sum_i n_{i\uparrow} n_{i\downarrow}$$
$$= \frac{U}{N_0} {\sum_{\mathbf{kk'q}}}' a^+_{\mathbf{k+q}\uparrow} a_{\mathbf{k}\uparrow} a^+_{\mathbf{k'-q}\downarrow} a_{\mathbf{k'}\downarrow} \quad . \tag{10.91}$$

The prime on the summation sign indicates that momentum is conserved during the scattering process only up to a reciprocal lattice vector $\mathbf{G}$, i.e., $a^+_{\mathbf{k'-q}\downarrow}$ has to be replaced more generally by $a^+_{\mathbf{k'-q+G}\downarrow}$.

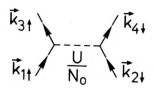

Fig. 10.14. Lowest-order scattering process due to the local Hubbard interaction U.

A single scattering process between two particles of opposite spin in initial states $|\mathbf{k}_1, \mathbf{k}_2\rangle$ is given by the matrix element (see Fig. 10.14)

$$\langle \mathbf{k}_3, \mathbf{k}_4 | H_{\text{int}} | \mathbf{k}_1, \mathbf{k}_2 \rangle = \frac{U}{N_0} \Delta(\mathbf{k}_1 + \mathbf{k}_2; \mathbf{k}_3 + \mathbf{k}_4) \quad . \tag{10.92}$$

The lattice function $\Delta(\mathbf{k}_1+\mathbf{k}_2; \mathbf{k}_3+\mathbf{k}_4)$ is equal to a δ-function with argument $(\mathbf{k}_1+\mathbf{k}_2-\mathbf{k}_3-\mathbf{k}_4+\mathbf{G})$. The t-matrix is the sum of repeated scattering processes with the two-particle propagator $G(\mathbf{k}_1, \mathbf{k}_2)$ between two subsequent scattering events. This is shown in Fig. 10.15. The sum is a geometric series and therefore can be written in the form

$$U_{\text{eff}} = \frac{U}{1 + G(\mathbf{k}_1, \mathbf{k}_2) U} \quad . \tag{10.93}$$

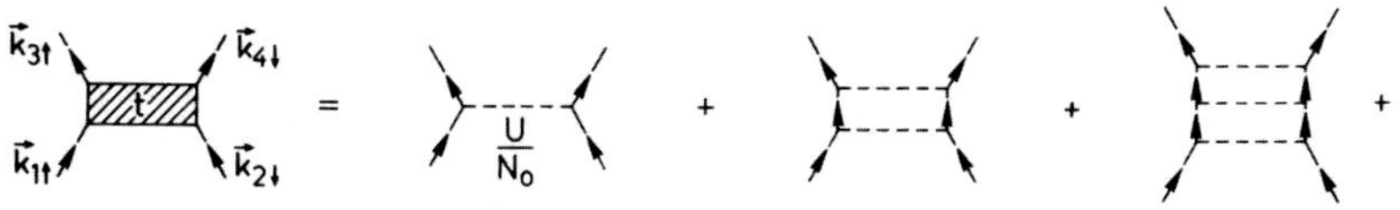

Fig. 10.15. Scattering processes contained in the t-matrix in the low-density or Kanamori limit.

The sign in the denominator must be positive for positive $G(\mathbf{k}_1, \mathbf{k}_2)$ since the interaction is repulsive, i.e., the bare interaction U is screened and not amplified. The two-particle propagator is

$$G(\mathbf{k}_1, \mathbf{k}_2) = \frac{1}{N_0} \sum_{\mathbf{k}_3 \mathbf{k}_4} \frac{\Delta(\mathbf{k}_1 + \mathbf{k}_2; \mathbf{k}_3 + \mathbf{k}_4)}{\epsilon(\mathbf{k}_3) + \epsilon(\mathbf{k}_4) - \epsilon(\mathbf{k}_1) - \epsilon(\mathbf{k}_2) - \Delta\epsilon_{12}} \quad . \tag{10.94}$$

The sum of the unperturbed kinetic energies of two particles is modified by an increment $\Delta\epsilon_{12} = U_{\text{eff}}/N_0$. In order to evaluate $G(\mathbf{k}_1, \mathbf{k}_2)$ we go over from a two-electron problem to a partially filled band. The assumption is that the scattering processes contained in Fig. 10.15 are the dominant ones and the only ones to be taken into account. We neglect the generation of electron-hole excitations out of the Fermi sea. As pointed out before, this approximation is the better, the lower the electron or hole density is. The momenta $\mathbf{k}_1, \mathbf{k}_2$ must be above the Fermi energy for the scattering into the state $|\mathbf{k}_3, \mathbf{k}_4\rangle$ to take place. When the bottom of the band is at a $\mathbf{k}$ point of high symmetry, e.g., at $\mathbf{k} = 0$, we take for $\epsilon(\mathbf{k}_1)$ and $\epsilon(\mathbf{k}_2)$ the energy at the bottom of the band. In this case

$$G(0,0) \simeq \frac{1}{2} \int_{\epsilon_F}^{W} d\epsilon \frac{N(\epsilon)}{\epsilon} \quad , \tag{10.95}$$

where $N(\epsilon)$ is the density of states per spin direction and energies are measured from the bottom of the band. We neglect the energy $\Delta\epsilon_{12}$ and notice that $G(0,0)$ is of order W^{-1}. When $U \to \infty$, the screened interaction is reduced to

$$U_{\text{eff}} \simeq \frac{U}{1 + U/W} \simeq W \quad , \tag{10.96}$$

i.e., to an energy of the order of the bandwidth. This agrees with the intuitive argument presented above.

10.5 The t-J Model

We have shown before that the Hubbard model at half filling describes an insulating state, when the ratio U/t is sufficiently large. In distinction to conventional band insulators the gap in the excitation spectrum of the system is here a consequence of the strong correlations (Mott-Hubbard insulators). Materials which fall into that category are discussed in Chapters 12 and 15. It is especially important to understand how the physics of those systems is affected by doping them with holes or electrons. Here the $t - J$ model is a significant tool. It is obtained by an expansion of the Hubbard Hamiltonian in terms of t/U when only the leading order terms are kept. As it turns out, antiferromagnetic correlations play an important role in that limit. This is immediately seen by starting out from the insulating state at half-filling. Here excitations require an energy of order U. We speak of virtual excitations when an electron hops onto a neighboring site which becomes doubly occupied and then hops back. This is possible only, when electrons on neighboring sites have opposite spins. Those charge fluctuations can be eliminated by a unitary transformation of the form described in Sect. 10.2.2. They result in an effective antiferromagnetic Hamiltonian H_{eff} which acts on a reduced Hilbert space, i.e., one without double occupancies. We introduce a projection operator P which projects here onto a reduced Hilbert space, in which doubly occupied sites are excluded. Obviously it satisfies the relation $P^2 = P$. Then $Q = 1 - P$ projects onto the space of configurations with doubly occupied sites. The hopping term t of the Hubbard-Hamiltonian takes us from the reduced Hilbert space to the full one. To second order in t we may write (10.17a) as

$$\tilde{H} = \text{P}H\text{P} - \frac{1}{U}\text{P}H_0\text{Q}H_0\text{P} \quad , \tag{10.97}$$

where we used that $Q^2 = Q$. The second term reflects the energy increase by U when in the intermediate state a site is doubly occupied. From the definition of P and Q, it follows that

$$\text{Q}H_0\text{P} = \sum_{ij\sigma} t_{ij} n_{i-\sigma} a^+_{i\sigma} a_{j\sigma} (1 - n_{j-\sigma})$$

$$\text{P}H_0\text{Q} = \sum_{ij\sigma} t_{ij} (1 - n_{i-\sigma}) a^+_{i\sigma} a_{j\sigma} n_{j-\sigma} \quad . \tag{10.98}$$

The term QH_0P describes hopping of an electron with spin σ from a singly occupied site j to site i, which is already occupied by an electron with spin $-\sigma$. The second term is the inverse of the first. Only nearest-neighbor hopping is considered ($t_{ij} = -t$). After some rearrangement [170], we obtain

$$\tilde{H} = -t \sum_{\langle ij \rangle \sigma} \left(\hat{a}^+_{i\sigma} \hat{a}_{j\sigma} + \text{h.c.} \right) + \frac{4t^2}{U} \sum_{\langle ij \rangle} \left(\mathbf{S}_i \cdot \mathbf{S}_j - \frac{\hat{n}_i \hat{n}_j}{4} \right)$$

$$- \frac{t^2}{U} \sum_{\langle ijk \rangle \sigma} \left(\hat{a}^+_{k\sigma} \hat{n}_{j-\sigma} \hat{a}_{i\sigma} - \hat{a}^+_{k\sigma} \hat{a}^+_{j-\sigma} \hat{a}_{j\sigma} \hat{a}_{i-\sigma} + \text{h.c.} \right) \quad . \tag{10.99}$$

As before, brackets $\langle ij \rangle$ denote pairs of nearest neighbors, while $\langle ijk \rangle$ stands for three-site terms, $i \neq k$ being nearest neighbors of j. The $\hat{a}^+_{i\sigma}$, $\hat{a}_{i\sigma}$ are electron creation and annihilation operators which act on the reduced Hilbert space. However, they do not satisfy simple anticommutation relations. They are identical with the Hubbard operators $X_i^{\sigma 0}$, $X_i^{0\sigma}$, yet in the present context the above notation is more often used, i.e.,

$$\hat{a}_{i\sigma}^+ = a_{i\sigma}^+ \left(1 - \hat{n}_{i-\sigma}\right)$$
$$\hat{a}_{i\sigma} = a_{i\sigma} \left(1 - \hat{n}_{i-\sigma}\right) \quad . \tag{10.100}$$

The spin operators are $\mathbf{S}_i = (1/2) \sum_{\alpha\beta} \hat{a}_{i\alpha}^+ \sigma_{\alpha\beta} \hat{a}_{i\beta}$ and $\hat{n}_{i\sigma} = \hat{a}_{i\sigma}^+ \hat{a}_{i\sigma}$. The three-site terms in $\tilde{H}$ contribute only when the system deviates from half filling and they describe indirect hopping processes between sites i and k. They are of order t/U smaller when compared with the first, direct hopping term in (10.99). These terms are often discarded for low doping concentrations, although not always justifiably. If we do so, the Hamiltonian (10.53) transforms into the $t - J$ model Hamiltonian

$$H_{t-J} = -t \sum_{\langle ij \rangle \sigma} \left(\hat{a}_{i\sigma}^+ \hat{a}_{j\sigma} + \text{h.c.}\right) + J \sum_{\langle ij \rangle} \left(\mathbf{S}_i \cdot \mathbf{S}_j - \hat{n}_i \hat{n}_j / 4\right)$$
$$= H_t + H_J \quad , \tag{10.101}$$

The $t - J$ Hamiltonian can be considered as the leading term of an expansion of the original Hubbard Hamiltonian in powers of t/U. Higher order terms generate ring exchange processes, spin-dependent three-sites hopping etc.

We want to decompose the spin-spin interaction part H_J into an Ising part

$$H_{\text{Ising}} = J \sum_{\langle ij \rangle} \left(S_i^z S_j^z - \hat{n}_i \hat{n}_j / 4\right) \tag{10.102}$$

and a remaining part

$$H_1 = \frac{J}{2} \sum_{\langle ij \rangle} \left(S_i^+ S_j^- + S_i^- S_j^+\right) \quad . \tag{10.103}$$

In the following we investigate the Hamiltonian on a square lattice. In the limit of $n = 1$, i.e., at half-filling H_{t-J} is the Hamiltonian of a 2D Heisenberg antiferromagnet (AF). Its ground-state energy can be calculated by deploying either the projection technique, Monte Carlo methods, exact diagonalizations or variational methods.

When holes are doped into the system, they move according to the first term in the Hamiltonian (10.99). In an antiferromagnetic background with its two sublattices A and B a hole hopping between nearest neighbors must necessarily create magnetic frustration, because each hop of the hole shifts one spin to the opposite sublattice. To make this more quantitative we note that a bond connecting two antiparallel spins contributes an energy of $-J/2$ to the expectation value $\langle H_{\text{Ising}} \rangle$. A bond which connects either two parallel

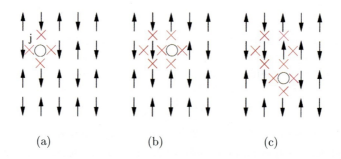

Fig. 10.16. Motion of a hole in a 2D Néel state $|\Phi_N\rangle$. (a) The hole is generated at site j, (b) after one hop seven bond defects have been created. They are marked by crosses, (c) after two loops nine bond defects are present.

spins or a spin and a hole contributes zero. We call the latter type of bond 'defects' and note that there are no defect bonds in the Néel state.

Removing a spin from – say – a site j belonging to the A-sublattice of the Néel state creates $z_L = 4$ defect bonds (see Fig. 10.16a) and thus increases the expectation value $\langle H_{\text{Ising}}\rangle$ by $z_L \cdot J/2$ (z_L is the coordination number of the square lattice). The first hop away from j creates another $z_L - 1$ defects and increases the energy by $(z_L - 1)J/2$, see Fig. 10.16b. For most of the paths any further hop creates two additional defects, see Fig. 10.16c. The last statement is in fact exact for all paths with length $\nu \leq 3$. The lattice constant is set equal to unity here.

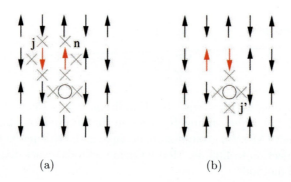

Fig. 10.17. The truncation of a string of defects by the term H_1. Acting with $\frac{J}{2}S_j^+ S_n^-$ onto the string of length $\nu = 2$ starting at j shown in (a) produces the 'string of length 0' starting at j' in (b).

It thus might seem that the hole is self-trapped in an effective potential due to the string of spin defects it generates and which increases roughly linearly with the number of hops. At this point, however, the transverse part of the Heisenberg exchange, H_1, becomes of crucial importance. Namely H_1 can flip a pair of 'wrong' spins and thus shorten the string of defects by 2. Simultaneously, the starting point of the string is shifted to a second ((1,1)-like) or third ((2,0)-like) neighbor, see Fig. 10.17. We thus arrive at the following picture of hole motion in an antiferromagnet: the hole executes a rapid zig-zag motion around a site j on a time scale $\propto t^{-1}$. Thereby it remains tied to j by a string of spin defects. Occasionally – i.e. on a timescale given by the inverse coupling constant J – the term H_1 acts and shifts the starting point of the string to a second nearest neighbor j'. Then, the hole oscillates around j' and so on.

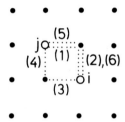

Fig. 10.18. Motion of a hole along a Trugman path. The hole is generated at site j and moves in six steps to position i, indicated by dotted lines and labeled by numbers in parentheses. No bonds are frustrated. By the spiraling motion, the hole eliminates the disordered bonds it generated before.

Thus a hole created at site j with spin σ hops with matrix element $-t$ around that site and generates a cloud of disordered spins around it, often referred to as spin bag. The hole with its bag moves through the system giving rise to a quasiparticle dispersion $E(\mathbf{k})$ of bandwidth J rather than $t \gg J$.

It should be noted here that it is in fact possible for a hole to delocalize even without the help of H_1, namely by executing a complicated spiral motion where it passes each site twice. This has been discussed by *Trugman* and is shown in Fig. 10.18. The process is of little importance though.

In order to describe the hole motion and to calculate $E(\mathbf{k})$ we apply the projection method and use the following set of relevant variables:

$$A_{j,0} = \hat{a}_{j\uparrow}$$

$$A_{j,1} = \frac{1}{\sqrt{z_L}} \sum_m S_j^- \Delta_{j,m} \hat{a}_{m\downarrow}$$

$$A_{j,2} = \frac{1}{\sqrt{z_L}} \sum_{m \neq n} S_j^- \Delta_{j,m} S_m^+ \Delta_{m,n} \hat{a}_{n\uparrow} \quad , \text{ etc.} \quad (10.104)$$

While $A_{j,0}$ removes an electron with spin ↑ at site j, the $A_{j,\nu\neq 0}$ operators describe to subsequent motion of the hole to sites m, n etc followed by spin flips. As discussed above, the latter annihilate defects generated by the motion of the hole. Here $\Delta_{m,n} = 1$ if m and n are nearest neighbors and zero otherwise. Let us briefly discuss the normalization factors: there are z_L paths of length 1 and since each hop couples a path of length $\nu > 0$ to $z_L - 1$ longer paths and one shorter path, the total number of paths of length n is $N_\nu \leq z_L(z_L-1)^{\nu-1}$. This right-hand site is an upper bound because for $\nu > 3$ the paths may intersect themselves and these paths should be excluded. Since the number of self-intersecting paths is only a small fraction of the total number of paths, we will henceforth ignore this complication and set $N_\nu = z_L(z_L - 1)^{\nu-1}$. This is called the Bethe-lattice approximation.

For a description of the coherent motion of the hole together with its spin bag we introduce the Fourier transform

$$A_\nu(\mathbf{k}) = \sqrt{\frac{2}{N_0}} \sum_{j \epsilon A} e^{-i\mathbf{k}\cdot\mathbf{R}_j} A_{j,\nu} \quad . \tag{10.105}$$

We have divided the square lattice into sublattices A and B and denoted with A the sublattice of the Néel state with spin ↑ sites.

With the above choice of dynamic variables, we define a projector P onto the set of variables $\{A_\nu(\mathbf{k})\}$:

$$P = \sum_{\nu=0}^{\nu_{\max}} |A_\nu(\mathbf{k})) \chi_{\nu\mu}^{-1} (A_\mu(\mathbf{k})| \quad . \tag{10.106}$$

The dispersion $E(\mathbf{k})$ of the hole is obtained from

$$R_\uparrow(\mathbf{k},\omega) = \left(\hat{a}_{\mathbf{k}\uparrow} \left| P \frac{1}{z-L} P \right| \hat{a}_{\mathbf{k}\uparrow}\right) \tag{10.107}$$

and we determine this function with the help of (8.18). We need to calculate the susceptibility matrix $\chi_{\mu\nu}$ as well as the frequency matrix $\omega_{\mu\nu}$ which enter that equation. Within the present approximation, the matrix $\chi_{\mu\nu}(\mathbf{k}) = (A_\mu(\mathbf{k})|A_\nu(\mathbf{k}))$ is equal to the unity matrix. In analogy to (8.19) the frequency matrix is determined from $\omega_{\mu\nu}(\mathbf{k}) = (A_\mu(\mathbf{k})|LA_\nu(\mathbf{k}))$. For its computation, we decompose the Liouvillean L into

$$L = L_t + L_{\text{Ising}} + L_1 \quad , \tag{10.108}$$

where L_{Ising} corresponds to the Ising part H_{Ising} of H_J, i.e., $L_{\text{Ising}} A = [H_{\text{Ising}}, A]_-$ while L_t and L_1 correspond to the hopping Hamiltonian H_t and to H_1, respectively.

The part L_{Ising} describes the increase of magnetic energy due to the strings of defects generated by the moving hole. We have seen that the number of frustrated bonds is 4, 7, 9, ... for paths of length 0, 1, 2, ... Therefore, we have

$$(A_\mu(\mathbf{k}) \mid L_{\text{Ising}} \mid A_\nu(\mathbf{k})) = \delta_{\mu,\nu} \cdot \frac{J}{2} [(z_L - 2)\nu + 5 - \delta_{\nu,0}] \ . \quad (10.109)$$

The matrix element of L_t is calculated as follows: according to the Bethe-lattice approximation there are $N_\nu = z_L(z_L - 1)^{\nu-1}$ paths of length $\nu > 0$. Each of these has a prefactor of $N_\nu^{-1/2}$ and is coupled to $z_L - 1$ paths of length $\nu + 1$ by H_t, all of which have a prefactor of $N_{\nu+1}^{-1/2}$. The matrix element of L_t contains therefore the effective hopping

$$t(z_L - 1)\sqrt{\frac{N_\nu}{N_{\nu+1}}} = t\sqrt{z_L - 1} \ . \quad (10.110)$$

Analogous considerations apply for $\nu = 0$ and therefore we arrive at:

$$(A_\mu(\mathbf{k}) \mid L_t \mid A_\nu(\mathbf{k})) = -\delta_{\mu,\nu+1}\tilde{t}_\nu - \delta_{\mu+1,\nu}\tilde{t}_\mu \ , \quad (10.111)$$

where $\tilde{t}_\nu = t\sqrt{z_L}$ for $\nu = 0$ and $\tilde{t}_\nu = t\sqrt{z_L - 1}$ otherwise.

Finally, we need the matrix elements of L_1. As mentioned before, this term truncates or extends the string of defects by 2 and shifts the starting point j to a second- or third-nearest neighbor. Therefore it is the only term which is **k**-dependent. It is straightforward to see that there are two possibilities to reach a (1,1)-type neighbor by the string truncation process shown in Fig. 10.17, yet only one possibility to reach a (2,0)-type neighbor. The **k**-dependence is therefore given by:

$$g(\mathbf{k}) = 2 \cdot 4\cos(k_x)\cos(k_y) + 2(\cos(2k_x) + \cos(2k_y))$$
$$= 4\left((\cos(k_x) + \cos(k_y))^2 - 1\right) \quad (10.112)$$

in units of the lattice vector. To obtain the remaining prefactor of the matrix element we note that $N_\nu/(z_L(z_L - 1))$ of the paths of length $\nu > 2$ which start at j pass through a (2,0)-type neighbor j' (it would be twice this number for a (1,1)-type neighbor!). Each of them has a prefactor of $N_\nu^{-1/2}$ and is transformed by H_1 into a path of length $\nu - 2$ starting at j'. In turn, each of these latter paths has a prefactor of $N_{\nu-2}^{-1/2}$. We thus obtain for the matrix element an overall factor of

$$\frac{J}{2}\frac{1}{z_L(z_L - 1)}\sqrt{\frac{N_\nu}{N_{\nu-2}}} = \frac{J}{2} \cdot \frac{1}{z_L} \ . \quad (10.113)$$

The case $\nu = 2$ can be treated in an analogous way and we find

$$(A_\mu(\mathbf{k}) \mid L_1 \mid A_\nu(\mathbf{k})) = \frac{J}{2}g(\mathbf{k})\left(\delta_{\mu,\nu+2}f_\nu + \delta_{\mu+2,\nu}f_\mu\right) \ , \quad (10.114)$$

where $f_\nu = 1/\sqrt{z_L(z_L-1)}$ for $\nu = 0$ and $1/z_L$ otherwise.

For the frequency matrix a dimension of $\nu_{\max} \lesssim 20$ is sufficient. The matrix can be diagonalized numerically. Let ϵ_i and $\{\alpha_\mu^{(i)}(\mathbf{k})\}$ denote the eigenvalues and eigenstates of this matrix. The spectral function $\rho(\mathbf{k}, \omega)$ is defined in terms of them by

$$\rho(\mathbf{k}, \omega) = \sum_i^{\nu_{\max}} \left|\alpha_0^{(i)}(\mathbf{k})\right|^2 \delta(\omega - \epsilon_i) \quad . \tag{10.115}$$

The index 0 refers to the variables $A_{j,0}$ (see (10.104)) but here with the site index j replaced by the momentum vector $\mathbf{k}$. We show the result for $\nu_{\max} = 20$ in Fig. 10.19 and compare it with the one obtained from the diagonalization of a 4×4 cluster with periodic boundary conditions using the Lanczos method (Appendix F). The agreement between the two types of calculations is very satisfactory despite the dramatic reduction in the number of all possible dynamical variables. It should be pointed out that the exact treatment of a 4×4 cluster requires the diagonalization of a matrix of order 5×10^4, while with the projection method we need to diagonalize for each $\mathbf{k}$-point a matrix of dimension 20 only.

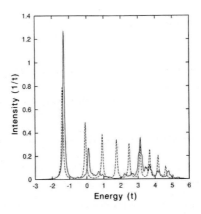

Fig. 10.19. Spectral function $\rho(\mathbf{k}, \omega)$ for $\mathbf{k} = (\pi/2, \pi/2)$ of a hole moving on an antiferromagnetic square lattice according to (10.115) (dashed line). Also shown are the results of the diagonalization of a 4×4 cluster (solid line). (From [103])

The sequence of smaller peaks accompanying the main peak in Fig. 10.19 is due to the internal degrees of freedom of the spin bag. In a simplified picture we may think of them as eigenmodes of a droplet. When we plot the energy of the main peak, i.e., the coherent quasiparticle peak as function of $\mathbf{k}$ we obtain $E(\mathbf{k})$. We find that $E(\mathbf{k})$ has a minimum along the line $|k_x| + |k_y| = \pi$ in the Brillouin zone (compare with Fig. 10.10). This degeneracy is lifted, when the small contributions of Trugman paths are taken into account. Then only four minima at the **k** points $\mathbf{k} = (\pm\pi/2, \pm\pi/2)$ remain. We will return to this point later.

It is interesting that we obtain a result of comparable quality when a self-consistent Born approximation is applied to the $t - J$ model [314]. Starting point is again (10.101 - 10.103) but for later purposes with a general anisotropic spin interaction, i.e.,

$$H_{\text{int}} = H_{\text{Ising}} + \alpha H_1 \quad . \tag{10.116}$$

For $\alpha = 1$, H_{int} reduces to H_J. Note that for $\alpha = 0$ the spin interaction reduces to the Ising limit and the $t - J$ model goes over into a $t - J^z$ model. The ultimate goal is to calculate the Green's function

$$G(\mathbf{p}, \omega) = \left\langle \psi_0 \left| \hat{a}^+_{\mathbf{p}\sigma} \frac{1}{\omega - H_{t-J} + E_0} \hat{a}_{\mathbf{p}\sigma} \right| \psi_0 \right\rangle \tag{10.117}$$

for a hole. The ground state $|\psi_0\rangle$ is that of the AF Heisenberg Hamiltonian with energy E_0. We divide the lattice into sublattices A and B and rotate the spins on sublattice B by π about the x axis. Then $S_i^\pm \to S_i^\mp$ and $S_i^z \to -S_i^z$ for sites i on B. The Hamiltonian H_{int} becomes

$$H_J = J \sum_{\langle ij \rangle} \left[-S_i^z S_j^z + \frac{1}{2} \left(S_i^+ S_j^+ + S_i^- S_j^- \right) - \hat{n}_i \hat{n}_j / 4 \right] \tag{10.118}$$

and the ground state is ferromagnetic. We assume that spins are up. We note that in distinction to a conventional ferromagnetic ground state the one we are dealing with here contains quantum fluctuations caused by the operators $S_i^+ S_j^+$ and $S_i^- S_j^-$.

In the following we proceed in several steps. First we reformulate the $t - J$ Hamiltonian so that the interaction of a hole with spin waves becomes apparent. In a second step we compute instead of (10.117) a Green's functions for a holon which we shall define appropriately. Finally, we show that the holon Green's function agrees approximately with the one in (10.117).

We begin by introducing new annihilation operators with respect to the ground state. They consist of spinless fermion operators f_i^+, which create what is called a holon at site i. Thus a separation is being made between charges and spins. Therefore we define

$$\hat{a}_{i\uparrow} = f_i^+ \quad , \tag{10.119}$$

and composite operators

$$\hat{a}_{i\downarrow} = f_i^+ S_i^+ \quad . \tag{10.120}$$

The latter cause a spin flip before a hole on a spin $\downarrow$ site is generated and allow for a description of spin-wave emission processes induced by the motion of the hole. Note that neither the right nor the left side of the above two equations do satisfy simple fermionic commutation relations since double occupancies of sites are excluded. Yet matrix elements calculated with this reduced Hilbert space are the same when the right or left hand side of the equations are used. Instead of having three states per site, i.e., empty or singly occupied with spin σ, we deal here with four states, namely products of $|\text{holon}\rangle \otimes |\text{spin}\rangle$ states. The holon number is 0 or 1. Each of the holon states has a spin attached to a site. The state $|1,\downarrow\rangle$ is unphysical and must be excluded. Because of (10.120) it is not possible to create a holon at site i with a remaining spin $\downarrow$ left at that site. A holon together with an up spin is, of course, possible.

Next a Holstein-Primakoff transformation for $S = 1/2$ is applied. The $\mathbf{S}_i$ operators are expressed in terms of boson operators b_i as

$$S_i^+ = \left(1 - b_i^+ b_i\right)^{1/2} b_i$$
$$S_i^- = b_i^+ \left(1 - b_i^+ b_i\right)^{1/2}$$
$$S_i^z = \frac{1}{2} - b_i^+ b_i \quad . \tag{10.121}$$

When a b_i^+ operator is applied to the vacuum, which here is a conventional ferromagnetic lattice, it creates a spin $\downarrow$ at site i. In linear spin wave theory $(1 - b_i^+ b_i)^{1/2}$ is set equal to unity. This might seem questionable in the present case since for the ground state of a Heisenberg AF on a square lattice $\langle b_i^+ b_i \rangle = \frac{1}{2} - \langle S_i^z \rangle \simeq 0.2$, i.e., $\langle b_i^+ b_i \rangle$ is not small. However, numerical studies show that the approximation is better than expected. The constraint on the Hilbert space, i.e., the exclusion of $|1,\downarrow\rangle$ can now be taken into account by adding a term

$$H_\lambda = \lambda \sum_i f_i^+ f_i b_i^+ b_i \tag{10.122}$$

with $\lambda \to \infty$ to the Hamiltonian H_{t-J}. Then those states obtain an infinite energy and drop out. When we make the above replacements in (10.101) it becomes

$$H_{t-J} = -t \sum_{i,\tau} \left(f_i f_{i+\tau}^+ b_{i+\tau} + h.c.\right)$$
$$+ \frac{J}{4} \sum_{i,\tau} \left(1 - n_i^f\right) \left[b_i b_{i+\tau} + b_i^+ b_{i+\tau}^+ + n_i^b + n_{i+\tau}^b\right] \left(1 - n_{i+\tau}^f\right)$$
$$- \frac{J}{2} N_0 (1-\delta)^2 \quad . \tag{10.123}$$

Here we have used the notation $n_i^b = b_i^+ b_i$ and $n_i^f = f_i^+ f_i$. Furthermore, the sum over τ runs over the four nearest neighbors of the different sites. The factors $(1 - n_i^f)$ and $(1 - n_{i+\tau}^f)$ account for the loss in magnetic energy in the presence of holes. The Hamiltonian H_{t-J} is bilinear in the b-operators. Therefore we can diagonalize it with respect to these operators. After a Fourier transformation this is done with the help of a Bogoliubov transformation which is similar, but not identical to (10.62), i.e.,

$$\alpha_{\mathbf{q}} = u_{\mathbf{q}} b_{\mathbf{q}} - v_{\mathbf{q}} b^{+}_{-\mathbf{q}} \quad . \tag{10.124}$$

Note that $u_{\mathbf{q}}^2 - v_{\mathbf{q}}^2 = 1$. Introducing the structure factor for a square lattice $\gamma_{\mathbf{q}} = (\cos q_x + \cos q_y)/2$, we obtain

$$H_{t-J} = \frac{4t}{\sqrt{N_0}} \sum_{\mathbf{p}\cdot\mathbf{q}} [f_{\mathbf{p}}^+ f_{\mathbf{p}-\mathbf{q}} \alpha_{\mathbf{q}} (u_{\mathbf{q}} \gamma_{\mathbf{p}-\mathbf{q}} + v_{\mathbf{q}} \gamma_{\mathbf{p}}) + h.c.]$$
$$+ \sum_{\mathbf{q}} \omega_{\mathbf{q}} \alpha_{\mathbf{q}}^+ \alpha_{\mathbf{q}} + E_0 \quad . \tag{10.125}$$

The $\alpha_{\mathbf{q}}^+$ operators generate spin waves of energy $\omega_{\mathbf{q}}$. One finds that $\omega_{\mathbf{q}} = 2J(1-\delta)^2 \nu_{\mathbf{q}}$ where $\nu_{\mathbf{q}} = (1 - \gamma_{\mathbf{q}}^2)^{1/2}$. The hole concentration δ is defined through $\delta = \langle n_i^f \rangle$. It is seen that H_{t-J} describes fermions, i.e., holons coupled to spin waves of energy $\omega_{\mathbf{q}}$. The functions $u_{\mathbf{q}}$ and $v_{\mathbf{q}}$ are determined by diagonalization of (10.125) when $t=0$. In the isotropic limit it is

$$u_{\mathbf{q}} = \sqrt{\frac{1+\nu_{\mathbf{q}}}{2\nu_{\mathbf{q}}}} \quad , \quad v_{\mathbf{q}} = -\mathrm{sgn}\gamma_{\mathbf{q}} \cdot \sqrt{\frac{1-\nu_{\mathbf{q}}}{2\nu_{\mathbf{q}}}} \quad . \tag{10.126}$$

Apart from a trivial factor $(1-\delta)^2$ in $\omega_{\mathbf{q}}$ the effect of holes back on $\omega_{\mathbf{q}}$ has not been taken into account (see Sect. 15.5). The present case resembles that of spinless electrons coupled to phonons. For that reason the above Hamiltonian is often referred to as that of spin polarons. However, there is no *free* Fermion term $\sum_{\mathbf{p}} \epsilon_{\mathbf{p}} n_{\mathbf{p}}^f$ like in the electron-phonon problem. Instead the motion of holons is directly coupled to a generation of spin waves. Without the $S_i^+ S_j^+$ and $S_i^- S_j^-$ terms the hole is tied to the site where it was created. We notice that the form factor

$$M(\mathbf{p}, \mathbf{q}) = (u_{\mathbf{q}} \gamma_{\mathbf{p}-\mathbf{q}} + v_{\mathbf{q}} \gamma_{\mathbf{p}}) \tag{10.127}$$

in (10.125) is zero when $\mathbf{q} = 0$ or (π, π). It is largest for $\mathbf{q}$ vectors in between these two points, showing that the coupling of holons to *short-ranged* spin fluctuations is most important.

In a next step we calculate the retarded Green's function for holons, i.e.,

$$G_f(\mathbf{p}, \omega) = \left\langle 0 \left| f_{\mathbf{p}} \frac{1}{\omega - H_{t-J} + E_0 + i\eta} f_{\mathbf{q}}^+ \right| 0 \right\rangle \quad . \tag{10.128}$$

The vacuum state $|0\rangle$ is here the *quantum* Néel state. It includes fluctuations. In the spirit of Sect. 5.4.4 we write it in the form

$$|0\rangle = \exp\left(\sum_{\mathbf{q}} \frac{v_{\mathbf{q}}}{u_{\mathbf{q}}} \alpha_{\mathbf{q}}^+ \alpha_{-\mathbf{q}}^+\right) |F\rangle \quad . \tag{10.129}$$

In this case $|F\rangle$ is the ferromagnetic state, or classical Néel state before the spin rotation in sublattice B. The Green's function is determined by applying the self-consistent Born approximation. In that scheme the self-energy $\Sigma(\mathbf{p},\omega)$ is computed in lowest order perturbation theory using for the hole propagator the full Green's function $G_f(\mathbf{p},\omega)$. The corresponding diagrams are shown in Fig. 10.20. With four nearest neighbors per site we obtain for the self-energy due to emission and reabsorption of a spin wave

$$\Sigma(\mathbf{p},\omega) = \frac{16t^2}{N_0} \sum_\mathbf{q} \frac{M^2(\mathbf{p},\mathbf{q})}{\omega - \omega_\mathbf{q} - \Sigma(\mathbf{p}-\mathbf{q},\omega-\omega_\mathbf{q})} \quad . \tag{10.130}$$

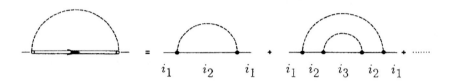

Fig. 10.20. Diagrams in self-consistent Born approximation. Solid and dashed lines represent holon and spin wave propagators in the Ising limit of H_{int}. The double solid line denotes the full holon Green's function. The real space indices refer to different sites i_n. (From [314])

This self-consistent equation for $\Sigma(\mathbf{p},\omega)$ can only be solved numerically, in particular when $t > J$ (strong coupling case).

It is quite instructive to study the Ising case, i.e., when the anisotropy coefficient in (10.116) is $\alpha = 0$. In that case $M(\mathbf{p},\mathbf{q})$ simplifies to $M(\mathbf{p},\mathbf{q}) = \gamma_{\mathbf{p}-\mathbf{q}}$, $\omega_\mathbf{q} = \omega_0 = 2J^z$ and $(N_0)^{-1}\sum_\mathbf{q} \gamma_\mathbf{q}^2 = \frac{1}{4}$. This leads to a $\mathbf{p}$-independent, i.e., local self-energy

$$\Sigma(\omega) = \frac{4t^2}{\omega - \omega_0 - \Sigma(\omega-\omega_0)} \quad . \tag{10.131}$$

A numerical solution of that equation shows that the spectral function $\rho(\mathbf{k},\omega)$ derived from (10.128) has the form of a ladder of spin excitations. The hole is confined, i.e., it is tied by a string to its origin. Trugman paths are not included in the Born approximation. Therefore the string of disordered spins cannot be healed in the $t-J^z$ model as spin-flip processes are not possible. Figure 10.21 should be compared with Fig. 10.19 where holes are deconfined.

Of course, this situation differs when the isotropic case, i.e., H_{t-J} is considered. Here $\rho(\mathbf{k},\omega)$ has always for fixed momentum $\mathbf{k}$ a distinct quasiparticle peak separated from a quasicontinuum. The coherent motion of the holes gives rise to a doubly degenerate quasiparticle band, because of the two sublattices with minima at momenta $(\pm\pi/2,\pm\pi/2)$. The dispersion $E_\mathbf{k}$ is shown in Fig. 10.22. For a small hole concentration we obtain hole pockets in the Brillouin zone at those $\mathbf{k}$ points. Finally, we notice that the spin-wave velocity $\omega_\mathbf{q}/q$ is proportional to $(1-\delta)^2$ and vanishes only for a hole concentration of $\delta = 1$.

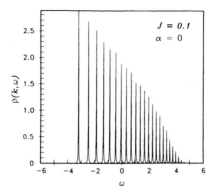

Fig. 10.21. Spectral function $\rho(\mathbf{k},\omega)$ for the $t-J^z$ model in units of t with $J = 0.1$. The result was obtained by solving (10.129) for a 16×16 square lattice. (From [314])

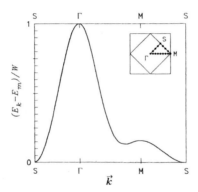

Fig. 10.22. Dispersion relation E_k of the quasiparticle band relative to $E_m = E(\pi/2, \pi/2)$ normalized by the bandwidth $W = E(0,0) - E(\pi/2, \pi/2)$ and along symmetry lines in the AF Brillouin zone: $J = 0.8$. With increasing hole doping, states near S are filled (hole pockets). Inset: allowed k points for a 16×16 lattice along the symmetry lines. (From [314])

What remains to be discussed is the relation between the original electron Green's function (10.117) and the one for holons (10.128). By expressing $\hat{a}_{\mathbf{k}\sigma}$ in terms of holons, thereby taking for the spins on the two sublattices their average value into account, one can show that for small hole concentrations $G(\mathbf{p}, \omega) \simeq G_f(\mathbf{p}, \omega)$ [314].

An important effect which has been left out so far is that of holes acting back on the antiferromagnetic properties. An exception is the trivial reduction of the spin-wave energy due to the number of empty sites or holes. *Shraiman* and *Siggia* have pointed out that a Néel state is unstable under doping and that instead a spiral magnetic phase forms at low hole concentrations. The wavelength of the spiral grows with diminishing hole concentration. Although the arguments for a spiral phase were originally given for a classical spin background, one can prove that the instability is also present when quantum fluctuations are taken into account. The two-fold degeneracy of the dispersive hole bands is then split and it is found that a spiral state with twisting vector $\mathbf{q}_s = (1,1)$ has a lower energy than a state with $\mathbf{q}_s = (1,0)$. For illustration we show in Fig. 10.23 the two twisted magnetic states. For further details we refer to the original literature[1].

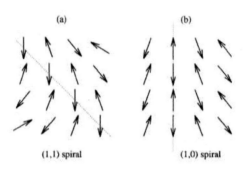

Fig. 10.23. Schematic representation of two spiral spin states with $\mathbf{q}_s$ along the (1,1) direction (a), and along the (1,0) direction (b). The dotted line connects spins with untwisted directions. (From [229])

The previous considerations have started from an ordered AF ground state. It is known that a Heisenberg antiferromagnet on a square lattice has long-range order only at $T = 0$. The Mermin-Wagner theorem excludes that order for any finite $T \neq 0$. It is instructive to know which features of the calculated quasiparticle spectrum require long-range AF order. In order to study the effect of short-range magnetic order on the spectrum we consider the motion of a hole at finite temperatures, where long-range order is absent.

For that purpose we have to go over to still another representation of electron creation and annihilation operators $\hat{a}_{i\sigma}^+, \hat{a}_{i\sigma}$. We apply a slave-fermion Schwinger-boson representation by setting

$$\hat{a}_{i\sigma}^+ = f_i b_{i\sigma}^+ \quad , \quad \hat{a}_{i\sigma} = f_i^+ b_{i\sigma} \quad . \tag{10.132}$$

As before, the spinless fermion operators f_i^+ create a holon at sites i while the bosons $b_{i\sigma}^+, b_{i\sigma}$ keep track of the spins [14]. The subsidiary condition is here

$$f_i^+ f_i + \sum_\sigma b_{i\sigma}^+ b_{i\sigma} = 1 \tag{10.133}$$

---

[1] see, e.g., [52, 104, 202, 217, 314, 413]

since a site i is either empty $(f_i^+ f_i = 1)$ or occupied by a spin σ. Thus the previously independent existence of spins and charges is given up here. Yet we allow for possible spin-charge separation in the sense that there may be excitations which solely involve spin degrees of freedom and others which involve solely charge degrees of freedom. Within that representation the $t-J$ Hamiltonian has the form

$$H_{t-J} = -t \sum_{\langle ij \rangle \sigma} \left(f_i f_j^+ b_{i\sigma}^+ b_{j\sigma} + h.c. \right) + J \sum_{\langle ij \rangle} f_i f_i^+ f_j f_j^+ \left[-\frac{1}{2} A_{ij}^+ A_{ij} \right] , \quad (10.134)$$

where

$$A_{ij} = b_{i\uparrow} b_{j\downarrow} - b_{i\downarrow} b_{j\uparrow} . \quad (10.135)$$

First we want to determine the spin-wave dispersion. We assume that there are no holes or holons, in which case we may replace $f_i f_i^+ f_j f_j^+$ by unity. Next we apply a spin rotation to the b operators on sites of sublattice B by replacing $b_{j\uparrow} \to -b_{j\downarrow}$, $b_{j\downarrow} \to b_{j\uparrow}$. Then a Lagrange multiplier of the form $\lambda \sum_{i\sigma} b_{i\sigma}^+ b_{i\sigma}$ is added to (10.134). It ensures that the subsidiary condition $\sum_\sigma \langle b_{i\sigma}^+ b_{i\sigma} \rangle = 1$ can be satisfied. The average taken is here a *thermal* one. We apply a mean-field approximation to the product $A_{ij}^+ A_{ij}$, in which case

$$A_{ij}^+ A_{ij} = \langle A_{ij} \rangle A_{ij}^+ + \langle A_{ij}^+ \rangle A_{ij} - |\langle A_{ij} \rangle|^2 \quad (10.136)$$

with $\langle A_{i,i+\tau} \rangle = \langle b_{i\uparrow} b_{i+\tau\downarrow} - b_{i\downarrow} b_{i+\tau\uparrow} \rangle \equiv \Delta$. Here $i + \tau$ denotes the nearest neighbors of site i. Because of the mean-field approximation, the J dependent part of H_{t-J} can be diagonalized.

With these approximations the Hamiltonian takes the form:

$$\begin{aligned} H &= \lambda \sum_{i,\sigma} b_{i,\sigma}^+ b_{i,\sigma} - \frac{J\Delta}{2} \sum_{\langle i,j \rangle} \sum_\sigma \left(b_{i,\sigma}^+ b_{j,\sigma}^+ + h.c. \right) \\ &= \lambda \sum_{\mathbf{k},\sigma} b_{\mathbf{k},\sigma}^+ b_{\mathbf{k},\sigma} - \frac{J\Delta}{2} \sum_{\mathbf{k},\sigma} \left(b_{\mathbf{k},\sigma}^+ b_{\mathbf{k},\sigma}^+ + h.c. \right) . \end{aligned} \quad (10.137)$$

The Hamiltonian can be diagonalized by a bosonic Bogoliubov transformation similar to (10.124) and we obtain for the spin excitation energy

$$E(\mathbf{k}) = \sqrt{\lambda^2 - (2J\gamma_{\mathbf{k}} \Delta)^2} . \quad (10.138)$$

The excitation energy has a minimum at $\mathbf{k} = 0$ and is gapped. The order parameter Δ and λ are determined from self-consistency conditions [17]. With the spin-excitation spectrum determined, we can treat the kinetic energy part of H_{t-J} by applying again the self-consistent Born approximation. But this time temperature Green's functions have to be used.

A detailed description of these calculations is beyond the scope of this book. It is worthwhile nonetheless to discuss briefly the outcome. The spectral

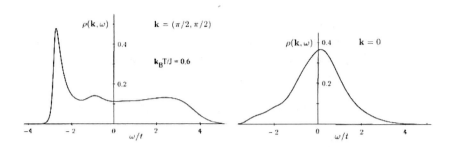

Fig. 10.24. Spectral density for holons when $J/t = 0.3$ and $k_B T/J = 0.6$ for $\mathbf{k} = (\pi/2, \pi/2)$ and $(0,0)$. Results are for a 12×12 site cluster. (From [202])

function for the slave fermions, i.e., holons computed for a cluster of 12×12 sites shows a pronounced quasiparticle peak. At low T, e.g., $k_B T/J = 0.05$ it is particularly strong at $\mathbf{k} = (\pi/2, \pi/2)$. At a temperature of $k_B T/J = 0.6$ the quasiparticle peak is still pronounced at $(\pi/2, \pi/2)$ (see Fig. 10.24) while at $\mathbf{k} = (0,0)$ it has been replaced by a broad incoherent structure. The dispersion of the holons at low temperatures is found to be rather similar to the one in Fig. 10.22, although it is not identical with it, because of the mean-field approximation (10.136) which has been made. The absence of long-range antiferromagnetic order has therefore little effect on the holon spectrum as long as there is strong short-range antiferromagnetic order. In particular, minima are found again at $(\pm\pi/2, \pm\pi/2)$ as in Fig. 10.22.

An interesting result is obtained when the Green's function of the *physical* hole is calculated, instead of the one for a holon. In the previous case, where only spin-wave emission was considered, both Green's functions were nearly equal. Here it is found that there is a bound state between a holon and a Schwinger boson, as the energy $\epsilon_B(\mathbf{k})$ of the physical hole is lower than the quasiparticle energy $\epsilon(\mathbf{k})$ of a holon and a boson. This demonstrates that within the $t - J$ model there is no spin-charge separation taking place when holes are doped into a square lattice. This finding is in agreement with the string picture of a moving hole discussed earlier. There we found that a hole has a spin bag attached to it. The energy of the bound state has again minima at $(\pm\pi/2; \pm\pi/2)$ suggesting that at low doping small pocket-like Fermi surfaces are formed in the Brillouin zone. This result, which holds in the absence of long-range magnetic order, agrees again with the previous result for $T = 0$. These findings are expected to break down when the magnetic correlation length becomes shorter than the diameter of the spin bag. This is the case at large hole dopings and at high temperatures. The small pocket-like Fermi surface at $(\pm\pi/2, \pm\pi/2)$ transforms then into a large Fermi surface to which all electrons contribute. This subject will come up again in Sect. 12.1 and especially in 15.5.5.

Summarizing the above discussion, we note that a Mott insulator doped with holes and described by a $t-J$ model does not fulfill Landau's criteria for a Fermi liquid. Nevertheless, there still exist coherent quasiparticle excitations. Yet when hole doping becomes large, correlations become less important and Fermi liquid behavior is reestablished. Finally, before concluding we want to mention an extensive review of the $t-J$ model by *Ogata* and *Fukuyama* [350].

10.6 Mean-Field Approximations

By studying the $t-J$ Hamiltonian we have seen that mean-field approximations are very helpful. We were able to gain insight into new physical phenomena. Nevertheless, they require care. Sometimes they predict symmetry breakings which actually do not occur. This was pointed out in Sect. 2.4 where a H_2 molecule was used in order to discuss this unwanted phenomenon. The origin of symmetry breaking as an artifact is an incomplete treatment of important correlation effects. By breaking a symmetry some of the correlation effects such as a reduction of charge fluctuations are simulated. On the other hand, there are certainly solids existing with broken symmetries. In that case a symmetry broken solution has a lower energy than the phase with unbroken symmetry, independent of how well correlations are treated.

The challenge consists in finding out in which cases a symmetry broken solution is an artifact and when it is real. These comments should be kept in mind in the following discussion. Mean-field approximations were not made with respect to electrons or holes but rather with respect to auxiliary or slave fields. This was possible because electron creation and annihilation operators were written in the form of composite operators which relate to these auxiliary fields. In the following, different forms of composite operators are considered. They allow for different kinds of mean-field approximations. Depending on the physical situation one or the other form is preferable.

In the presence of an AF ground state the forms (10.119,10.120) have been used, i.e., $\hat{a}_{i\uparrow} = f_i^+$ and $\hat{a}_{i\downarrow} = f_i^+ S_i^+$. Another form, which does not require an AF ground state was introduced in (10.132), i.e, by setting $\hat{a}_{i\sigma}^+ = f_i b_{i\sigma}^+$ and $\hat{a}_{i\sigma} = f_i^+ b_{i\sigma}$. The f_i^+'s are spinless fermion operators while the bosons $b_{i\sigma}$ take care for the spin degrees of freedom. The subsidiary condition eliminating unphysical states is here given by (10.132) and the Hilbert space is again the one without double occupancies of sites.

A third form which plays a prominent role in the so-called resonating valence bond (RVB) theory is

$$\hat{a}_{i\sigma} = f_{i\sigma} b_i^+ \quad ; \quad \hat{a}_{i\sigma}^+ = f_{i\sigma}^+ b_i \quad . \tag{10.139}$$

The spin degree of freedom is here attached to the fermionic field and the corresponding particle is called a spinon while the bosonic operator b_i^+ creates a charge defect which is named a holon. Contrary to the previous fermionic holon here it is a boson. There is again a subsidiary condition required which here is of the form

$$\sum_\sigma f_{i\sigma}^+ f_{i\sigma} + b_i^+ b_i = 1 \qquad (10.140)$$

and ensures that a site is either occupied by a holon or a spinon. Note that the reduced Hilbert space excludes doubly occupied sites. Specifically we denote with

$$|0(i)\rangle = b_i^+ |0\rangle$$
$$|\sigma(i)\rangle = f_{i\sigma}^+ |0\rangle \qquad (10.141)$$

a holon and a spinon at site i. We want to express the $t-J$ Hamiltonian solely in terms of spinons. This can be done by using (10.139) and the condition (10.140). We obtain for the kinetic energy

$$H_0 = -t \sum_{\langle ij \rangle, \sigma} \left(b_i f_{i\sigma}^+ f_{j\sigma} b_j^+ + h.c. \right) \ . \qquad (10.142)$$

In order to rewrite the exchange term we introduce the following two combinations of f-operators:

$$s_{ij}^+ = f_{i\uparrow}^+ f_{j\downarrow}^+ - f_{i\downarrow}^+ f_{j\uparrow}^+$$
$$\tau_{ij} = \sum_\sigma f_{i\sigma}^+ f_{j\sigma} = \tau_{ji}^+ \ . \qquad (10.143)$$

While s_{ij}^+ denotes a spin singlet of two spionons, τ_{ij} does the same for a particle-hole pair of spinons. It is then easy to see that

$$s_{ij}^+ s_{ij} = -\left(S_i^+ S_j^- + S_i^- S_j^+ \right) + \left(n_{i\uparrow} n_{j\downarrow} + n_{i\downarrow} n_{j\uparrow} \right) \ ,$$
$$\tau_{ij} \tau_{ji} = -\left(S_i^+ S_j^- + S_i^- S_j^+ \right) - \left(n_{i\uparrow} n_{j\uparrow} + n_{i\downarrow} n_{j\downarrow} \right) + \sum_\sigma n_{i\sigma} \qquad (10.144)$$

It follows that the Heisenberg exchange along a given bond can be written as

$$J \sum_{\langle ij \rangle} \mathbf{S}_i \mathbf{S}_j = -\frac{J}{4} \sum_{\langle ij \rangle} \Bigl[s_{ij}^+ s_{ij} +$$
$$+ \frac{1}{2} \left(\tau_{ij} \tau_{ji} + \tau_{ji} \tau_{ij} \right) - \frac{1}{2} \sum_\sigma \left(n_{i\sigma} + n_{j\sigma} \right) \Bigr] \ . \qquad (10.145)$$

The quartic terms in this expression are now in a form which immediately allows for a mean-field decomposition. We introduce the order parameters

$$\chi_{ij} = \langle \tau_{ij} \rangle$$
$$\Delta_{ij} = \langle s_{ij} \rangle \qquad (10.146)$$

and absorb terms proportional to the spinon number into a renormalization of the chemical potential. This leaves us with following mean-field Hamiltonian

$$H_J = -\frac{J}{4} \sum_{\langle i,j \rangle, \sigma} \left(\chi_{ji} f_{i,\sigma}^+ f_{j,\sigma} + h.c. \right)$$
$$-\frac{J}{4} \sum_{\langle i,j \rangle} \left(\Delta_{ij} s_{ij}^+ + h.c. \right)$$
$$-\mu \sum_i f_{i,\sigma}^+ f_{i,\sigma} \quad . \tag{10.147}$$

In order to understand better what these order parameters imply let us consider a system at half-filling. In this case there are no holons, i.e., $H_0 = 0$ and we deal with a Heisenberg Hamiltonian. As pointed out before the ground state of this Hamiltonian on a square lattice is antiferromagnetic. A conventional mean-field approximation for the Heisenberg Hamiltonian with a staggered field $m_i = \langle \mathbf{S}_i \rangle = \frac{1}{2} \sum_{\sigma\sigma'} \langle f_{i\sigma}^+ \boldsymbol{\sigma}_{\sigma\sigma'} f_{i\sigma'} \rangle$ describes this feature correctly, yet with a relatively poor ground-state energy E_0 per site, i.e., $E_0/N_0 = -2J/4$. This has to be compared with the true energy $E_0/N_0 = -2J \cdot 0.335$ which includes fluctuations.

With (10.147) we have applied a more sophisticated mean-field approximation in terms of spinons. The first term in (10.147) describes a band of spinon excitations with a bandwidth of order J. Obviously, these must be spin-wave like excitations since charge degrees of freedom are ruled out by a Heisenberg Hamiltonian. When $\chi_{ij} = \chi_0$ for nearest neighbors i and j and zero otherwise, this term becomes

$$H_{\text{RVB}} = -\frac{J}{2} \chi_0 \sum_{\mathbf{k}\sigma} \left(\cos k_x + \cos k_y \right) f_{\mathbf{k}\sigma}^+ f_{\mathbf{k}\sigma} \quad . \tag{10.148}$$

We have attached to this Hamiltonian a RVB label, since it is the one used in the uniform RVB theory. In that approach the order parameter $\Delta_{ij} = 0$. Because of the mean-field approximation manifested by (10.148), the subsidiary condition (10.140) is considered to be fulfilled on average only, i.e., $\sum_\sigma \langle f_{i\sigma}^+ f_{i\sigma} \rangle = 1$. This implies that doubly occupied sites are no longer ruled out. They are present like in a SCF theory. The dispersion of the spinons

$$\epsilon(\mathbf{k}) = -\frac{J\chi_0}{2} \left(\cos k_x + \cos k_y \right) \quad . \tag{10.149}$$

vanishes for $k_x \pm k_y = \pm \pi$ like in Fig. 10.10. The dispersion is that of uncorrelated spinons with an effective hopping matrix element $t_{\text{eff}} = J\chi_0/4$. To interpret $\mathbf{k}$ points with $\epsilon(\mathbf{k}) = 0$ as a Fermi surface (of spinons) would be somewhat misleading. We are dealing with an insulating state and therefore those $\mathbf{k}$ points do not represent a Fermi surface in the usual sense. Note that (10.149) must not be interpreted as describing antiferromagnetic magnons.

The latter vanish at points (0,0) and $(\pm\pi, \pm\pi)$ and not on lines. Within a spinon description magnons are particle-hole excitations. Transitions from, e.g., $(\pi/2, \pi/2)$ to $(-\pi/2, -\pi/2)$ in Fig. 10.10 give rise to a momentum transfer of (π, π). Thus the phase space occupied with spinons has a surface given by $\epsilon(\mathbf{k}) = 0$ which looks like a Fermi surface. The ground-state energy based on H_{RVB} is improved as compared with the one of the Néel state. In fact one finds a value of $E_0/N_0 = -2J \cdot 0.27$, which is lower than the one of the Néel state [499]. From that perspective the method serves its purposes.

The study of the mean-field Hamiltonian (10.147) has mainly concentrated on the case that $\Delta_{ij} \neq 0$. Obviously, the order parameter Δ_{ij} and with it the Hamiltonian (10.147) do not conserve the spinon number. The model Hamiltonian looks formally very similar to the BCS Hamiltonian for superconductors (see Chapter 15). The symmetry, which is broken in superconductivity and by (10.147) when $\Delta_{ij} \neq 0$ is gauge symmetry. The Hamiltonian does not commute with the particle number operator $\hat{N}$, i.e., $[H, \hat{N}] \neq 0$ and therefore a gauge transformation, which replaces $f_j \to e^{i\phi} f_j$ and $f_j^+ \to e^{-i\phi} f_j$, does not leave H invariant.

A Heisenberg Hamiltonian does not only have a global gauge invariance, but also a local one, i.e., a $U(1)$ symmetry. Since electrons are localized, we may perform local gauge transformation $f_j \to e^{i\phi_j} f_j$ without changing H. In order to ensure that local gauge invariance holds for (10.147) we have to require the following transformation rules to hold:

$$f_i \to e^{i\phi_i} f_i \quad ; \quad \chi_{ij} = e^{-i\phi_i} \chi_{ij} e^{i\phi_j} \quad ; \quad \Delta_{ij} \to e^{-i\phi_i} \Delta_{ij} e^{i\phi_j} \quad . \quad (10.150)$$

This $U(1)$ invariance allows for the development of a gauge theory for the above Hamiltonian and also for an extension of it in the presence of holes. For a detailed discussion we refer to the comprehensive review of Lee et al. [274]. One particular self-consistent solution of the Hamiltonian (10.147) has played a special role. It is of the form

$$\Delta_{ij} = \begin{cases} \Delta_0 & , \quad \text{if } \mathbf{j} = \mathbf{i} + \hat{\mathbf{x}} \\ -\Delta_0 & , \quad \text{if } \mathbf{j} = \mathbf{i} + \hat{\mathbf{y}} \end{cases}, \quad (10.151)$$

where $\hat{x}, \hat{y}$ are the unit lattice vectors in x and y direction. The order parameter has here a d-wave symmetry. The excitation spectrum is the same as that of a superconductor with d-wave pairing, i.e.,

$$E_k = \sqrt{(\epsilon(\mathbf{k}))^2 + \Delta_{\mathbf{k}}^2} \quad (10.152)$$

with $\epsilon(\mathbf{k})$ given by (10.149) and

$$\Delta_{\mathbf{k}} = \frac{\Delta_0 J}{2} (\cos k_x - \cos k_y) \quad . \quad (10.153)$$

We notice that along the energy contour $\epsilon(\mathbf{k}) = 0$ the excitation spectrum is gapped, except at the points $(\pm\frac{\pi}{2}, \pm\frac{\pi}{2})$. These points are often referred to as nodal points. The ground-state wavefunction has the form of a BCS ground-state wavefunction (see (15.39)). We want to point out that these findings have little to do with superconductivity. We deal here with an unrestricted SCF solution to a Heisenberg Hamiltonian which, as we want to show, gives a much improved ground-state energy as compared to the classical Néel state. The spinon excitations resemble again those of a SDW system (see (10.63)), but with a gap of d-wave symmetry (unconventional or d-wave SDW). Note that here again the occupation of a lattice site by spinons is equal to unity only *on average*, implying that double occupancies of sites do occur. The gap in (10.152) vanishes at $(\pm\pi/2, \pm\pi/2)$ like for a d-wave SDW.

Double occupancies of sites can be excluded by applying a Gutzwiller projector $P_G(\eta)$ with $\eta \to \infty$ (see (10.71)) to the BCS-like mean-field ground-state wavefunction $|\Phi_{\mathrm{MF}}\rangle$ of (10.147), i.e.,

$$|\psi_0\rangle = P_G(\eta \to \infty) | \Phi_{\mathrm{MF}}\rangle$$

$$|\Phi_{\mathrm{MF}}\rangle = \prod_{\mathbf{k}} \left(u_{\mathbf{k}} + v_{\mathbf{k}} f^+_{\mathbf{k}\uparrow} f^+_{-\mathbf{k}\downarrow} \right) | 0 \rangle \quad . \tag{10.154}$$

The $f^+_{\mathbf{k}\sigma}$ are the Fourier transforms of the $f^+_{i\sigma}$ and $|0\rangle$ is the vacuum state. The form of the coefficients $u_{\mathbf{k}}$ and $v_{\mathbf{k}}$ is discussed in Sect. 15.1.2.

The ground-state energy

$$E_0 = \frac{\langle \Phi_{\mathrm{MF}} | P_G \, H \, P_G | \Phi_{\mathrm{MF}}\rangle}{\langle \Phi_{\mathrm{MF}} | P_G | \Phi_{\mathrm{MF}}\rangle} \tag{10.155}$$

can be determined only numerically by using Monte Carlo methods. One finds for a d-wave order parameter [2] a value of $E_0/N_0 = -2J \cdot 0.319$ [3] which is very close to the exact value. The goal to calculate the ground-state energy of a Heisenberg AF on a square lattice by a sophisticated mean-field approximation has therefore been achieved.

However, there are much simpler ways based on conventional spin-wave theory or on the projection method to obtain results of comparable quality. We mention here the one using the projection method, since results can be obtained analytically, i.e., without resorting to numerical calculations. Starting point is the Ising part H_{Ising} of the Hamiltonian (10.102) while the remaining part H_1 (see (10.103)) is treated by projection methods. Already lowest-order perturbation theory in H_1 gives an energy change δE_0 of the form

$$\frac{\delta E_0}{N_0} = -2J \left(\frac{1}{12} + \frac{J}{16 \cdot 9} \lim_{z \to 0} \left(\frac{8}{z - 4J} + \frac{28}{z - 5J} - \frac{46}{z - 6J} \right) \right)$$

$$= -2J \cdot 0.083 \tag{10.156}$$

and therefore a ground-state energy $E_0/N_0 = -2J \cdot 0.333$. Details as well as further improvements when the projection method is used instead, are found, e.g., in [26].

---
[2] see [16]
[3] see [157]

Another simple method is the application of spin-wave theory. Starting from the classical Néel state a Holstein-Primakoff transformation (10.121) is applied. The zero-point energies of the spin waves correct the energy of the Néel state and yield a value of $E_0/N = -2J \cdot 0.332$ [4]. So the question might be asked why we were looking for a mean-field theory in terms of composite operators when the goal, i.e., an improved ground-state energy can be obtained much simpler by calculating analytically the lowest order perturbation corrections. The answer is that the main motivation for searching for a mean-field theory is the hope to gain better insight into the underlying physics when holes are added to the system. This is the situation we encounter in the high-T_c superconducting cuprates. Therefore this subject will be taken up again in Sect. 15.5.

10.6.1 Test of Different Approximation Schemes

In Sect. 10.2.1 we have considered the simplest case of a system with strong electron correlations, i.e., two electrons distributed over two non-equivalent orbitals. Here we want to apply several different computational methods in order to test how well they approximate that particular system. Since the exact solution of it is known, the model serves as a testing ground for the ability of those methods to describe strongly correlated electrons. The results are most interesting.

Starting point is the Hamiltonian (10.6) with U being large. First we solve the Hamiltonian in the independent-electron or *SCF approximation*. We express $U n_\uparrow^f n_\downarrow^f$ in terms of $\delta n_\alpha^f = n_\alpha^f - \langle n_\alpha^f \rangle$ and neglect the contribution $U \delta n_\uparrow^f \delta n_\downarrow^f$. This leads to the following replacement:

$$U n_\uparrow^f n_\downarrow^f \to U \left(\langle n_\downarrow^f \rangle n_\uparrow^f + \langle n_\uparrow^f \rangle n_\downarrow^f - \langle n_\uparrow^f \rangle \langle n_\downarrow^f \rangle \right) \qquad (10.157)$$

and the resulting SCF Hamiltonian is trivially diagonalized. The resulting eigenstates are bonding and antibonding states. The corresponding creation operators (B_σ^+) and (A_σ^+) can be expressed in terms of the operators l_σ^+ and f_σ^+ as

$$\begin{aligned} B_\sigma^+ &= l_\sigma^+ \cos\theta + f_\sigma^+ \sin\theta \\ A_\sigma^+ &= -l_\sigma^+ \sin\theta + f_\sigma^+ \cos\theta \end{aligned} \qquad (10.158)$$

For $U \gg V$ we find for the angle

$$\theta = -(V/U)^{1/3} \quad . \qquad (10.159)$$

The energies of the two single-particle states are

---

[4] see [16]

$$\epsilon_\sigma^B = \epsilon_l - V\left(\frac{V}{U}\right)^{1/3}$$
$$\epsilon_\sigma^A = \epsilon_f + (V^2 U)^{1/3} \quad . \tag{10.160}$$

We notice that in the SCF ground state

$$|\Phi_{\text{SCF}}\rangle = \prod_\sigma B_\sigma^+ |0\rangle \tag{10.161}$$

the electrons are predominantly in the ligand orbital. Because there is a contribution of the F orbital to the molecular bonding orbital, the probability is $\sin^2\theta$ that the F orbital is doubly occupied in $|\Phi_{\text{SCF}}\rangle$. The ground-state energy is

$$\tilde{E}_{\text{SCF}} = \sum_\sigma \epsilon_\sigma^B - U\langle n_\uparrow^f \rangle \langle n_\downarrow^f \rangle$$
$$= 2\epsilon_l - 3V(V/U)^{1/3} \tag{10.162}$$

and therefore considerable above the true ground-state energy (10.11). Note that for $U \to \infty$ one finds $\tilde{E}_{\text{SCF}} = 2\epsilon_l$, while according to (10.11), the ground state energy is $\tilde{E}_0 \simeq \epsilon_l + \epsilon_f - 2V^2/\Delta\epsilon$. Due to the missing correlations, the SCF approximation gives unphysical results.

The situation improves if we use an *unrestricted SCF approximation* (Sect. 2.4), in which case we attribute different orbitals to the two electrons of opposite spin. We find that one electron is localized on the ligand orbital L while the other electron is predominantly in the F orbital, yet lowers its energy by hybridization. The ground-state wavefunction takes the form

$$|\Phi_{\text{USCF}}\rangle = \left\{\left[1 - \frac{1}{2}\left(\frac{V}{\Delta\epsilon}\right)^2\right]f_\uparrow^+ - \frac{V}{\Delta\epsilon}l_\uparrow^+\right\} l_\downarrow^+ |0\rangle \quad , \tag{10.163}$$

and the corresponding energy is

$$\tilde{E}_{\text{USCF}} = E_0 - \frac{V^2}{\Delta\epsilon} \quad . \tag{10.164}$$

We notice that this energy is much better than the previous $\tilde{E}_{\text{SCF}}$, but the wavefunction certainly does not describe the real situation correctly. The exact wavefunction of the two electron system does not break a symmetry. Nevertheless, we ought here to reconfirm the conclusions drawn in Sect. 2.4. There we stated that, within the independent-electron approximation, symmetry-broken solutions strongly suppress charge fluctuations. Thus they simulate correlation effects as far as the energy is concerned, but the wavefunction remains in error. This point was reemphasized before in this Section.

We continue with *density-functional theory*. This is an excellent exercise for applying the basic idea of that theory and various approximations to it.

Thereby we follow rather closely [395]. By setting $\epsilon_l = 0$ we find from (10.11) that in the limit $U \to \infty$ the ground-state energy (4.1) is

$$E[\rho, V_{\text{ex}}(\mathbf{r})] = \epsilon_f - \frac{2V^2}{|\epsilon_f|} \quad . \tag{10.165}$$

Note that ρ is identified here with the f electron number n_f since the total electron number is fixed. The functional $F[\rho]$ is according to (4.1)

$$\begin{aligned} F[\rho] &= E[\rho, V_{\text{ex}}(\mathbf{r})] - \epsilon_f \rho \\ &= \epsilon_f(1 - \rho) + \frac{2V^2}{\epsilon_f} \quad , \end{aligned} \tag{10.166}$$

where we have used that the orbital energy ϵ_f acts on the electrons like an external potential $V_{\text{ex}}(\mathbf{r})$. Since $F[\rho]$ does not depend on $V_{\text{ex}}(\mathbf{r})$ (see (4.1)) we have to express ϵ_f in terms of ρ. This relation is given by (10.13), i.e.,

$$\rho = 1 - \frac{2V^2}{\epsilon_f^2} \tag{10.167}$$

from which $\epsilon_f = -V(2/(1-\rho))^{1/2}$ is obtained. Therefore $F[\rho]$ is of the form

$$F[\rho] = -2\sqrt{2\rho(1-\rho)}V \quad . \tag{10.168}$$

We want to point out that this density functional is *exact* for the model Hamiltonian. In order to derive an exact expression for $E_{\text{xc}}[\rho]$ we identify the last equation with (4.10) and evaluate the Hartree term and $T_0[\rho]$.

The Hartree term is given by

$$E_H = \frac{1}{2}\rho^2 U \quad . \tag{10.169}$$

The term $T_0[\rho]$ is the kinetic energy of a fictitious noninteracting two-electron system in a fictitious external potential η. The latter must be chosen so that the density ρ is the one of the true ground state. This is necessary because the correct density ρ must enter the expression for the kinetic energy. The ground-state energy $2\epsilon_0$ of the noninteracting system is obtained from

$$\begin{vmatrix} -\epsilon_0 & V \\ V & \eta - \epsilon_0 \end{vmatrix} = 0 \tag{10.170}$$

and therefore given by $\epsilon_0 = \frac{\eta}{2} - \sqrt{\frac{\eta^2}{4} + V^2}$. From the corresponding eigenvector we obtain $\rho = \frac{2\epsilon_0^2}{(\epsilon_0^2 + V^2)}$ or

$$\epsilon_0 = -V\sqrt{\frac{\rho}{1-\rho}} \quad . \tag{10.171}$$

10.6 Mean-Field Approximations

By equating this expression with the previous one we find for the external potential

$$\eta = \frac{2V(1-\rho)}{\sqrt{\rho(2-\rho)}} \quad . \tag{10.172}$$

The kinetic energy is derived from

$$2\epsilon_0 = T_0[\rho] + \eta\rho \tag{10.173}$$

and given by

$$T_0[\rho] = -2V\sqrt{\rho(2-\rho)} \quad . \tag{10.174}$$

The exchange-correlation contribution to the true ground-state energy is therefore

$$E_{\text{xc}}[\rho] = F[\rho] - E_H - T_0[\rho]$$
$$= -2\sqrt{2(1-\rho)}V - \frac{1}{2}\rho U^2 + 2V\sqrt{\rho(2-\rho)} \quad . \tag{10.175}$$

We see that $E_{\text{xc}}[\rho]$ must compensate the large Hartree term and is therefore not small! The exchange-correlation potential follows from (4.14)

$$v_{\text{xc}}[\rho] = -\rho U + \sqrt{\frac{2}{1-\rho}}V + O(V) \quad , \tag{10.176}$$

where $O(V)$ denotes additional contributions of order V, which remain regular when $\rho \to 1$. Therefore, if V is small it may be neglected in comparison with the second term. The effective potential V_{eff} which enters the Kohn-Sham equation (4.15) is therefore

$$V_{\text{eff}} = V_{\text{ex}} + v_H + v_{\text{xc}}$$
$$= \epsilon_f + \sqrt{\frac{2}{1-\rho}}V + \ldots \ldots \quad , \tag{10.177}$$

with the external potential and Hartree contribution given by $V_{\text{ex}} = \epsilon_f$ and $v_H = \rho\, U$, respectively.

We want to see whether or not the eigenvalues of the Kohn-Sham equation may be interpreted as excitation energies. They are obtained from

$$\begin{vmatrix} -E^{\text{DF}} & V \\ V & V_{\text{eff}} - E^{\text{DF}} \end{vmatrix} = 0 \quad , \tag{10.178}$$

i.e.,

$$E_{1,2}^{\text{DF}} = \frac{V_{\text{eff}}}{2} \pm \sqrt{\frac{V_{\text{eff}}^2}{4} + V^2} \quad . \tag{10.179}$$

We notice that $E_{1,2}^{\text{DF}}$ are of order V. This shows that we should not interpret $\delta E^{\text{DF}} = E_1^{\text{DF}} - E_2^{\text{DF}}$ as an excitation energy as is very often done when density functional theory is applied. In the limit $\rho \to 1$ or $V \to 0$ we find that $\delta E^{\mathrm{DF}} = 2V$ whereas from Fig. 10.3 one sees that the low-energy excitation of the two-electron system is given by $\delta E = \frac{2V^2}{\Delta \epsilon} = \sqrt{2} V \sqrt{1-\rho}$. This implies

$$\delta E = \delta E^{\mathrm{DF}} \sqrt{\frac{1-\rho}{2}} \;, \tag{10.180}$$

i.e., for small values of the hybridization V the excitation energy is much less than the difference of the Kohn-Sham eigenvalues would suggest. We want to stress that the above findings hold irrespective of whether or not any approximation to the density functional is made. This reemphasizes the notion that density functional theory is a ground-state theory. The eigenvalues of the Kohn-Sham equation *should not* be interpreted as excitation energies in particular when electron correlations are strong. This holds true irrespective of any approximation to the density functional.

In the following we want to study the modifications that occur when a *local density approximation* (LDA) or *local spin-density approximation* (LSDA) is made. We write

$$V_{\mathrm{eff}}(\sigma) = \epsilon_f + v_H[\rho_\uparrow, \rho_\downarrow] + v_{\mathrm{xc}}[\rho_\uparrow, \rho_\downarrow, \sigma] \tag{10.181}$$

and use a phenomenological local form for the exchange-correlation potential suggested in [443] for a single site of a Hubbard chain. Thereby the coefficients which appear in v_{xc} are adjusted so that the energy of the ground state and first excited state of a hydrogen atom is reproduced

$$v_{\mathrm{xc}}[\rho, \rho_\sigma] = -U \left[0.15 \rho^{\frac{1}{3}} + 0.46 \rho_\sigma^{\frac{1}{3}} \right] \;. \tag{10.182}$$

Note that at the ligand site L it is $v_{\mathrm{xc}} \equiv 0$. It is seen that the potential favors spin polarization because it makes it more negative, i.e., attractive. When the corresponding 2×2 Hamiltonian matrix is diagonalized and the eigenvalues and eigenfunctions are evaluated for $U \to \infty$, we find unphysical solutions, i.e., the energy increases linear with U. This holds true for the polarized (LSDA) as well as the unpolarized (LDA) case. Although the Hartree and exchange-correlation potentials cancel for large values of U when the exact density functional is used, this does not hold true for the LSDA or LDA ground-state energies. They diverge as U increases.

The situation is considerably improved when we include self-interaction corrections (SIC) (see Section 4.2). In accordance with (4.31) we set

$$E_{\mathrm{xc}}^{\mathrm{SIC}}[\rho_\uparrow, \rho_\downarrow] = E_{\mathrm{xc}}^{\mathrm{LSD}}[\rho_\uparrow, \rho_\downarrow] - \frac{U}{2}\left(\rho_\uparrow^2 + \rho_\downarrow^2\right) - E_{\mathrm{xc}}^{\mathrm{LSD}}[\rho_\uparrow, 0] - E_{\mathrm{xc}}^{\mathrm{LSD}}[0, \rho_\downarrow] \;. \tag{10.183}$$

This leads to a modified exchange-correlation potential (10.182)

$$v_{\mathrm{xc}}^{\mathrm{SIC}}[\rho, \rho_\sigma] = -0.15 U (\rho^{\frac{1}{3}} - \rho_\sigma^{\frac{1}{3}}) - U \rho_\sigma \;. \tag{10.184}$$

When the eigenvalues and eigenfunctions of the corresponding 2×2 matrix are evaluated we find that they agree with the ones of the unrestricted SCF solutions (10.163,10.164).

Next we consider the case that H is replaced by an *effective single-particle Hamiltonian*

$$\tilde{H} = \tilde{\epsilon}_f \sum_\sigma f_\sigma^+ f_\sigma + \tilde{V} \sum_\sigma (f_\sigma^+ l_\sigma + l_\sigma^+ f_\sigma) \quad , \tag{10.185}$$

in which the effect of the interaction term $U n_\uparrow^f n_\downarrow^f$ is replaced by renormalized single-particle quantities $\tilde{\epsilon}_f$ and $\tilde{V}$. For simplicity we have again set $\epsilon_l = 0$. We determine $\tilde{\epsilon}_f$ and $\tilde{V}$ by requiring that the charge distribution of the ground state $|\tilde{\psi}_0\rangle$ of $\tilde{H}$ is the same as the one of the ground state of (10.6), and furthermore that the expectation value of the hybridization term is the same in both cases. We make the following ansatz for the ground state of (10.185)

$$|\tilde{\psi}_0\rangle = \prod_\sigma (\sin \tilde{\vartheta} f_\sigma^+ + \cos \tilde{\vartheta} l_\sigma^+) |0\rangle \tag{10.186}$$

with

$$\tanh \tilde{\vartheta} = \frac{-2\tilde{V}}{\left[\left(\tilde{\epsilon}_f^2 + 4\tilde{V}^2\right)^{\frac{1}{2}} + \tilde{\epsilon}_f\right]} \quad . \tag{10.187}$$

The requirement $\langle \tilde{\psi}_0 | n_\sigma^\ell | \tilde{\psi}_0 \rangle = \langle \psi_0 | n_\sigma^\ell | \psi_0 \rangle$ with $n_\sigma^\ell = l_\sigma^+ l_\sigma$ leads to

$$\langle \tilde{\psi}_0 | n_{\ell\sigma} | \tilde{\psi}_0 \rangle = \cos^2 \tilde{\vartheta}$$
$$= \frac{1}{2} + \frac{V^2}{\epsilon_f^2} \tag{10.188}$$

and similarly for n_σ^f

$$\langle \tilde{\psi}_0 | n_\sigma^f | \tilde{\psi}_0 \rangle = \sin^2 \tilde{\vartheta}$$
$$= \frac{1}{2} - \frac{V^2}{\epsilon_f^2} \quad . \tag{10.189}$$

Furthermore, the condition $\langle \tilde{\psi}_0 | \tilde{V} f_\sigma^+ l_\sigma | \tilde{\psi}_0 \rangle = \langle \psi_0 | V f_\sigma^+ l_\sigma | \psi_0 \rangle$ leads to

$$\langle \tilde{\psi}_0 | \tilde{V} f_\sigma^+ l_\sigma | \tilde{\psi}_0 \rangle = \tilde{V} \cos \tilde{\vartheta} \sin \tilde{\vartheta}$$
$$= -\frac{V^2}{|\epsilon_f|} \quad . \tag{10.190}$$

This implies that to leading order

$$\tilde{V} = \frac{2V^2}{|\epsilon_f|} + \ldots \tag{10.191}$$

$$\tilde{\epsilon}_f = \frac{8V^3}{\epsilon_f^2} + \ldots\ldots \quad . \tag{10.192}$$

It is noticed that the energy $\tilde{\epsilon}_f$ of the renormalized f level is larger than zero, which is the energy of the ℓ level. In addition V is reduced to $\tilde{V} = V\sqrt{1-n_f}$. The ground state of $\tilde{H}$ is a singlet and there exist four excited states with excitation energy $\tilde{\epsilon}_f$. This should be compared with the excited triplet state of the exact solution and its energy. Within the present approximation the two-electron system has also an excited singlet state of energy $2\tilde{\epsilon}_f$.

A similar result is obtained when a *slave boson mean-field approximation* is made. This approximation is often used for the description of heavy quasiparticles (see Chapter 13). The basic idea is to account for strong correlations by a renormalized hybridization $\tilde{V}$ and a renormalized f orbital energy $\tilde{\epsilon}_f$. Both renormalizations are achieved by auxiliary bosonic degrees of freedom. The Hamiltonian is rewritten in that case as

$$H_{\text{SB}} = \epsilon_\ell \sum_\sigma \ell_\sigma^+ \ell_\sigma + (\epsilon_f + \lambda) \sum_\sigma f_\sigma^+ f_\sigma + V \sum_\sigma \left(b^+ \ell_\sigma^+ f_\sigma + b f_\sigma^+ \ell_\sigma\right) + \lambda(b^+ b - 1) \quad . \tag{10.193}$$

The boson operator b^+ creates an empty f orbital and $b f_\sigma^+$ ensures that an f electron is created only when the orbital was previously empty. This eliminates the interaction term in (10.6) in the limit $U \to \infty$. The Lagrange parameter λ refers to the subsidiary condition

$$\sum_\sigma f_\sigma^+ f_\sigma + b^+ b = 1 \quad , \tag{10.194}$$

i.e., the F orbital is either occupied with spin σ or it is empty. In mean-field approximation we set $\langle b^+ \rangle = \langle b \rangle = r$ so that $\langle b^+ b \rangle = r^2$ and $\sum_\sigma \langle f_\sigma^+ f_\sigma \rangle = 1 - r^2$. The ground-state energy E_{SB} is obtained by requiring that $\partial E_{\text{SB}}/\partial r = 0$ and $\partial E_{\text{SB}}/\partial \lambda = 0$. We find that

$$E_{\text{SB}} = \epsilon_f - \frac{V^2}{|\epsilon_f|} \qquad n_f = 1 - \frac{V^2}{\epsilon_f^2} \quad . \tag{10.195}$$

One notices a missing factor of two in the V-dependent term as compared with the exact results (10.11). The ground state is a singlet and the excited states form a quartet (singlet + triplet) but with nearly the correct excitation energy. The high energy state at $|\epsilon_f|$ of the exact solution has been shifted to low energies, i.e., it takes part in the quartet (see Fig. 10.25). This shows that the slave boson mean-field approximation is only limited suitable for the interpretation of spectroscopic data.

Altogether the slave boson mean-field approximation yields very reasonable results, the most serious deficiency being the wrong multiplicity of the excited states.

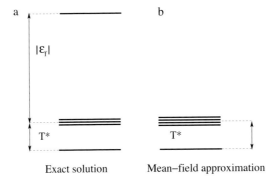

Fig. 10.25. (a) Exact solution from Fig. 10.3 with $\epsilon_\ell = 0$ and (b) solution in slave-boson mean-field approximation where the triplet is replaced by a quartet. (From [395])

It is instructive to investigate a *Gutzwiller-type ansatz* (see(10.71)) for the ground-state wavefunction, which we write here in the form of (5.97), i.e.,

$$|\psi_G\rangle = (1 - \tilde{\eta}\, n_{f\uparrow}n_{f\downarrow})\,|\Phi_{\mathrm{SCF}}\rangle \quad (10.196)$$

with $|\Phi_{\mathrm{SCF}}\rangle$ given by (10.161). With this ansatz the configuration with a doubly occupied F orbital is reduced, but the relative weight of the f^0 and f^1 configurations in $|\Phi_{SCF}\rangle$ remains unchanged. Therefore the ground-state energy remains grossly in error.

Considerable improvement is attained by starting from a modified Gutzwiller-type ansatz

$$|\psi_{\mathrm{MG}}\rangle = (1 - \tilde{\eta} n_{f\uparrow}n_{f\downarrow}) \prod_\sigma \left(\sin\vartheta f_\sigma^+ + \cos\vartheta l_\sigma^+\right)|0\rangle \quad . \quad (10.197)$$

The parameters $\tilde{\eta}$ and ϑ are determined by minimizing the ground-state energy. For small values of $\frac{V}{|\epsilon_f|}$ we find

$$\cos\vartheta = -\frac{2V}{|\epsilon_f|} + \ldots \quad . \quad (10.198)$$

In distinction to $|\Phi_{\mathrm{SCF}}\rangle$ the occupied single-particle state $B_\sigma^+ = \sin\vartheta\, f_\sigma^+ + \cos\vartheta\, l_\sigma^+$ is here mainly localized on the F site. Reducing the doubly occupied f state with the help of the prefactor in (10.196) leads to the correct ground-state energy and density distribution.

The local-ansatz wavefunction

$$|\psi_{\mathrm{LA}}\rangle = e^{\eta\delta n_{f\uparrow}\delta n_{f\downarrow}}\,|\Phi_{\mathrm{SCF}}\rangle \quad (10.199)$$

with $\delta n_{f\sigma} = n_{f\sigma} - \langle\Phi_{\mathrm{SCF}}|n_{f\sigma}|\Phi_{\mathrm{SCF}}\rangle$ (see Sect. 5.4) yields the exact ground-state energy and f electron number to leading order in $\frac{V}{|\epsilon_f|}$ when η is determined by minimization of the energy.

10.7 Metal-Insulator Transitions

There are several ways in which a system, which is metallic at high temperatures may become an insulator at low temperatures. The simplest and most trivial one is that of a band insulator with a small band gap. Such a system has completely filled bands at zero temperature, while at finite temperatures a number of electrons are excited from the valence bands into the conduction bands. They may result in a semiconducting or insulating like behavior depending on the size of the gap. Electron interactions play a secondary role here, except that they may contribute to the small size of the gap.

Another prototype are systems with a half-filled conduction band, for which the unit cell doubles at low temperatures due to structural changes. In this case the Brillouin zone is reduced to half its size as the temperature approaches zero and the previously half-filled bands become completely filled. Thus we are back to a band insulator. An example is trans-polyacetelene $(CH)_n$ which is discussed in Sect. 14.1 in a different context. Here the unit cell doubles because of a dimerization of π bonds (Peierls distortion). Lattice degrees of freedom are crucial, while electron-electron interactions remain unimportant for an understanding of the basic physical process.

A generalization of the Peierls distortion is a metal-insulator transition based on nesting. When the Fermi surface of a metal has parallel sections in **k**-space, the charge-density correlation function as well as the spin-density correlation function diverge at low temperature at a nesting vector **Q**, i.e., a vector which connects two parallel sections of the Fermi surface (see Fig. 10.26). Then along the **Q** direction the system resembles a one-dimensional one over a finite region in momentum space. Remember that the Fermi surface of a one-dimensional system has the shape of a slab with $\epsilon_{\mathbf{k+Q}} = \epsilon_{\mathbf{k}}$. The divergence of the spin- and charge susceptibility is seen from Sect. 10.4.1 or (11.99). When $\chi_0(\mathbf{q}, \omega \to 0)$ is calculated, the energy denominator vanishes for **q** vectors on the nesting portion of the Fermi surface. This causes a singularity in the response function $\chi_0(\mathbf{q})$ and hence an instability of the system. Thus a metal with nesting properties of the Fermi surface forms a charge-density (CDW) or spin-density wave (SDW) ground state. In that case the Fermi surface is partially or totally reduced due to the formation of a gap. We show in Fig. 10.26 a Fermi surface with nesting portions and characteristic nesting vectors $\mathbf{Q}_1$ and $\mathbf{Q}_2$. Here only a partial gapping of the Fermi surface is expected. By the opening of a gap kinetic energy is gained (see Sect. 14.1). When the Fermi surface is fully gapped the system becomes an insulator. A simple example considered earlier is that of a square lattice with one orbital per site at half filling. When electron hopping is limited to nearest neighbor sites we obtain a Fermi surface, which is enclosing the dashed area in Fig. 10.10. The parallel parts of the Fermi surface are connected by two orthogonal nesting vectors $\mathbf{Q}_1$ and $\mathbf{Q}_2$, which are reciprocal lattice vectors of the reduced Brillouin zone. The divergent static spin susceptibility leads to a SDW, but as discussed in Sect. 10.4.1 a gap will open only when there is a finite on-site

Coulomb repulsion U present. The gap extends here over the full Fermi surface and a metal-insulator transition takes place at sufficiently low temperatures.

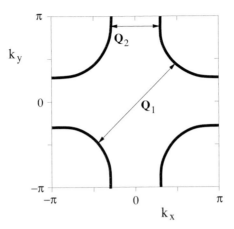

Fig. 10.26. Example of a Fermi surface with partial nesting. Parts of the Fermi surface in the vicinity of the nesting vectors $\mathbf{Q}_1$, or $\mathbf{Q}_2$ are nearly parallel and give rise to large response functions at those wavevectors.

Another example of a doubling of the unit cell is the onset of antiferromagnetic (AF) order within the framework of the $t - J$ model. Assume electrons on a bipartite lattice with a half-filled conduction band. When electrons on neighboring sites interact via an antiferromagnetic Heisenberg spin-spin interaction, the system will become an antiferromagnet at sufficiently low temperatures with a corresponding reduction of the Brillouin zone. Again, the band in the reduced zone is now completely filled and the system becomes an insulator. Consider La_2CuO_4, the parent compound of the high-temperature superconducting cuprates. With La^{3+} and O^{2-} we have Cu^{2+} ions with one hole in the $3d$ shell. This results in a half-filled $3d_{x^2-y^2}$ conduction band (see Sect. 12.1). The system is an antiferromagnetic insulator at low temperatures. But in distinction to a SDW it remains an insulator above the Néel temperature $T_N \simeq 80$ K. Obviously, electron-electron interactions play here a more important role than simply causing antiferromagnetic spin-spin interactions.

This leads over to a discussion of the role of electron correlations in metal-insulator transitions. We want to consider the consequences when the Coulomb repulsion between electrons is much larger than their energy gain due to delocalization. In trying to minimize their mutual repulsions, electrons will localize as much as possible. For a homogeneous electron gas this is achieved by the formation of a Wigner crystal, a topic described in Sect. 3.3. When the electronic crystal is pinned to the lattice so that it cannot move as a whole, the system is insulating. It does not have a gap in the excitation spectrum though, since the Wigner lattice can support phonon-like electronic excitations. Formation of a

Wigner crystal requires quite large values of r_s or low densities. This is rather different when the electrons are closely tied to ions of an underlying atomic lattice. The overlap of atomic-like wavefunctions between neighboring sites and hence the kinetic energy gain due to delocalization may be quite small in that case. An extreme case are $4f$ electrons which are known to be close to the nuclei. The mutual overlap of $4f$ orbitals on neighboring sites can be really small and so is the associated energy gain due to delocalization. Therefore charge order of $4f$ electrons commensurate with the underlying lattice structure can take place and connected with it a metal-insulator transition. The involved densities can be of the order of one electron per atom. A prototype system for charge ordering is Yb_4As_3, a system discussed extensively in Sect. 13.2.

Most studies for capturing the essence of metal-insulator transitions are done by using the Hubbard Hamiltonian (10.53). In this model a metal-insulator or Mott-Hubbard transition, as it is commonly called, is obtained at half filling and sufficiently large ratio of $U/|t|$. The Hubbard III approximation discussed in Sect. 10.4.3 describes this transition in terms of a retarded Green's function. The essential assumption is a coherent potential approximation (CPA). With its help one can treat the effect of the disordered potential an electron experiences when the other electrons are kept frozen.

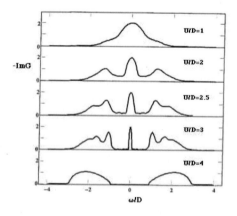

Fig. 10.27. Local spectral density (or equivalently $-$ Im G $(\mathbf{r} = 0, \omega)$) for a half-filled Hubbard system with a semi-circular density of states. Results for different ratios U/D (D is the half-bandwidth) are obtained from DMFT and apply to infinite dimension. (From [147])

An improved approximation takes into account, that the disorder potential is a dynamic one, since the positions of the other electrons fluctuate. This is accounted for in the DMFT as well as in the dynamical CPA or equivalents. They were discussed in Chapter 9. The fluctuating disorder potential results in a new feature of the metal-insulator (M-I) transition. Between the lower and upper Hubbard band a narrow band appears which is symmetric with respect to the Fermi energy. At a critical ratio of $U/|t|$ this band disappears and a M-I transition takes place. This aspect of the Mott-Hubbard transition was first found for infinite dimensions by DMFT [147] and is shown in Fig. 10.27.

However, care has to be taken when interpreting that finding. The narrow band is found only for a *paramagnetic* ground state. The DMFT, being a single-site approximation, has problems in stating under which conditions a system is a paramagnet or an antiferromagnet. It can compare the energies of the two ground-states. Yet because of missing intersite correlations that comparison contains considerable uncertainties. The situation is improved when instead a cluster DMFT is used (see Sect. 9.2), in which case the ground state of a square- or cubic lattice is found to be an antiferromagnetic insulator and the narrow central band is absent. Assuming a paramagnetic ground state, the details of the density of states change somewhat when the momentum dependence of $\Sigma(\mathbf{p}, \omega)$ is included by a procedure outlined in Sect. 9.2 (see Fig. 10.28).

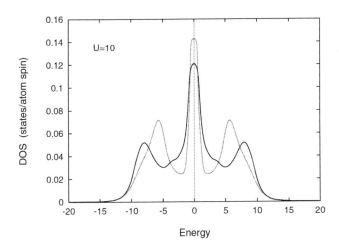

Fig. 10.28. Density of states for a paramagnetic Hubbard model on a cubic lattice at half-filling when $\Sigma(\mathbf{p}, \omega)$ is calculated within the FSCP method (solid line) and when a single-site approximation is made implying a momentum independent $\Sigma(\omega)$ (dotted line). U is in units of t. (From [227])

One would like to know which low-energy excitations give rise to the narrow peak in the density of states around $\omega \simeq 0$. As in Fig. 9.3 the peak results from a narrow band of quasiparticle like excitations. They must involve predominantly spin degrees of freedom. The latter couple increasingly weakly to charge degrees of freedom as U (in units of t) approaches the critical value U_c at which the peak disappears. Therefore the width of the peak must be related to the exchange interaction J. It is however not simply proportional to it. In the insulating phase we are left with the J dependent part of H_{t-J} only, i.e., with spin-wave excitations. They are uncoupled from charge excitations, do not contribute to the electronic DOS and have a bandwidth of order J. It should be noticed that there is considerable uncertainty what the critical value U_c is at which a M-I transition takes place, even for the simple case of a square lattice.

The description of a metal-insulator transition in a real system by a one-band Hubbard model is, of course, a highly idealized one. An important generalization is the inclusion of more than one orbital per site. Due to different overlap of those orbitals with the ones of neighboring sites the criteria for a metal-insulator transition, or more precisely for localization, may first be satisfied by *one* of the different orbitals. Electrons in that orbital become localized while electrons in other orbitals are still delocalized. It may also happen that the criterion is fulfilled first by *two* of the orbitals. This is the case for some of the intermetallic uranium compounds to which the *dual model* applies. The latter is discussed in Sect. 13.3. Here $5f$ electrons localize in some of the j_z orbitals while they remain delocalized in others. Hund's rule correlations play an important role in establishing such an incomplete Mott-Hubbard transition.

Another feature one should be aware of is the following. Approaching a Mott-Hubbard transition may cause redistributions of orbital occupancies at a given site. In order to demonstrate this point by means of a gedanken experiment, take Li metal. If we were able to increase the lattice constant arbitrarily, moving this way towards the atomic limit, we would observe a redistribution of electrons in $2s$ and $2p$ orbitals. A metal-insulator transition would take place at some stage of the increase of the lattice constant.

Up to here we have always presumed a Hubbard model at half filling and varied the size of the hopping matrix element as compared with the on-site Coulomb repulsion U. One might also consider an approach to the Mott-Hubbard transition, when $U/|t|$ is kept fixed while the filling or chemical potential is slightly changed. Some aspects of this topic are discussed in Sect. 12.1, where we deal with doped cuprates.

10.8 Numerical Studies

Due to its importance for understanding strongly correlated electrons as function of temperature, the Hubbard model has been studied by many different numerical methods in particular near or at half filling. It cannot be our aim here to discuss all these approaches. Instead we limit ourselves to a few computational schemes. They give an impression of what can be presently achieved.

Thereby we will consider often a Hubbard Hamiltonian on a square lattice because of its relevance for high-temperature superconductivity.

The numerical method which comes immediately to mind is, of course, exact diagonalization of a cluster with, e.g., open boundary conditions. However, the Hubbard model has four states per site: empty, singly occupied with spin σ and doubly occupied. Therefore, the dimension of the involved Hilbert space grows like $N_H = 4^N$ where N is the number of sites. For a 16 sites cluster this implies a total of $43 \cdot 10^9$ states and puts enormous requirements on the computer memory. By exploiting various symmetries together with the electron number conservation this number of states can be considerably reduced. Those symmetries, respective conservation laws include the conservation of magnetization ($S^2_{\text{tot}} = $ const.), the translational symmetry (i.e., momentum conservation), and the point group symmetry (i.e., parity and angular momentum conservation). The full SU(2) symmetry (i.e., conservation of $\mathbf{S}^2_{\text{tot}}$) cannot be used, since it is practically not possible to set up a basis of eigenstates of S^2_{tot}.

In order to give an explicit example we consider the strong correlation limit of the Hubbard model at half filling, i.e., the t-J model. The Hilbert space is here the same as for the Heisenberg Hamiltonian. On a square lattice with 40 sites there are 2^{40} configurations. Thus the dimension of the Hilbert space is $N_H \simeq 10^{12}$. Restricting ourselves to the $S^2_{\text{tot}} = 0$ sector reduces the dimension to $N_H \simeq 140 \cdot 10^9$ (i.e., $40!/(20!\,20!)$). The space group has 160 symmetry elements (i.e., 40 translations times a 4-fold symmetry axis). Together with spin inversion symmetry we end up with a basis of approximately $N_H = 4.3 \cdot 10^6$ states to be handled. The matrices which have to be diagonalized are of dimension $N_H \times N_H$. They are sparse and have only a relatively small number of nonvanishing matrix elements. Matrices up to $N_H \simeq 10^7$ have been treated by applying the Lanczos algorithm or variations of it. In Appendix F the Lanczos method is described and it is shown how the ground state and low-energy excited states of the system can be computed. Yet we would like to be able to calculate various physical properties even at finite temperatures, i.e., for $T > 0$.

Although a generalization of the Lanczos method to finite temperatures is possible, the Quantum Monte Carlo technique has proven in this case an invaluable tool. Starting point is the partition function

$$Z = Tr e^{-\beta H} \quad . \tag{10.200}$$

Since H_0 and H_{int} given by (8.22) do not commute, we decompose $\exp[-\beta(H_0 + H_{\text{int}})]$ according to *Suzuki* and *Trotter* by dividing β into L different *imaginary time* slices, i.e., $\beta = L\Delta\tau$. This enables us to write

$$\begin{aligned} e^{-\beta H} &= \left[e^{-\Delta\tau(H_o + H_{\text{int}})} \right]^L \\ &\simeq \left[e^{-\Delta\tau H_0} e^{-\Delta\tau H_{\text{int}}} \right]^L \quad . \end{aligned} \tag{10.201}$$

The error introduced by the finite size of the time steps $\Delta\tau$ is of order $O(\Delta\tau^2 tU)$. By reducing $\Delta\tau$ it can be made as small as necessary. With (10.201) we can factorize the right-hand side of (10.200) into

$$Z = Tr\left[e^{-\Delta\tau H_0} e^{-\Delta\tau\left[U\sum_i n_{i\uparrow}n_{i\downarrow} - \mu\sum_i n_i\right]}\right]^L \quad . \tag{10.202}$$

By introducing the chemical potential μ we have gone over to a grand canonical description. The electron-electron interaction can be rewritten as a one particle term by making for each time slice use of the identity

$$e^{-\Delta\tau U(n_{i\uparrow}-1/2)(n_{i\downarrow}-1/2)} = \frac{e^{-\frac{U}{4}\Delta\tau}}{2}\sum_{\xi_i=\pm 1} e^{-\Delta\tau\xi_i\lambda(n_{i\uparrow}-n_{i\downarrow})} \quad . \tag{10.203}$$

Here λ must satisfy the equation $\cosh(\Delta\tau\lambda) = \exp(\Delta\tau U/2)$. We do not prove that equation but refer to [184] instead. Equation (10.203) can be checked immediately by applying the right-hand side to the four possible configurations of site i. Note that $\xi_i(\tau)$ is a discrete Hubbard-Stratonovich field which for each time slice can take the two values ± 1. The on-site Coulomb repulsion U between electrons has been replaced here by a fluctuating discrete Ising field ξ_i acting on an electron. It should be compared with the continuous field $z_i(\tau)$ used instead in Sect. 11.3.1.

Equation (10.203) enables us to rewrite (10.202) in a form which factorizes the spin up and down contributions

$$Z = Tr_\xi Tr \left[\prod_{\ell=1}^L e^{-\Delta\tau H_0} e^{\Delta\tau\sum_i(\lambda\xi_i(\ell)-\mu')n_{i\uparrow}}\right] \times$$
$$\left[\prod_{\ell=1}^L e^{-\Delta\tau H_0} e^{\Delta\tau\sum_i(-\lambda\xi_i(\ell)-\mu')n_{i\downarrow}}\right] \quad . \tag{10.204}$$

Here we have set $\mu' = \mu - U/2$ and the index ℓ refers to the ℓ-th time slice. We introduce an effective Hamiltonian for electrons of spin σ on time slice ℓ be defining

$$H_\sigma(\xi(\ell)) = \sum_{ij} c_{i\sigma}^+ h_\sigma(\xi_i(\ell)) c_{j\sigma}$$
$$h_\sigma(\xi_i(\ell)) = -t\delta_{\langle ij\rangle} + \delta_{ij}\left(-\mu' + \lambda\sigma\xi_i(\ell)\right) \quad , \tag{10.205}$$

where $\delta_{\langle ij\rangle} = 1$ when i and j are nearest neighbors and zero otherwise. Thus $h_\sigma(\xi_i(\ell))$ contains an effective, spin-dependent random local field for time slice ℓ. Then the partition function can be expressed in terms of the effective Hamiltonian as

$$Z = Tr_\xi Tr \prod_{\sigma=\pm 1}\prod_{\ell=1}^L e^{H_\sigma(\xi(\ell))\Delta\tau} \quad . \tag{10.206}$$

Since $H_\sigma(\ell)$ is bilinear in the c^+, c operators, the trace over the fermions can be taken explicitly. The result is

$$Z = \sum_{\{\xi\}} \det L_\uparrow(\xi) \det L_\downarrow(\xi) \quad , \tag{10.207}$$

where the sum is over all configurations of the Ising field $\xi_i(\ell)$. The matrix $L_\sigma(\xi)$ has dimensions $N^2 \times N^2$ where N is the number of sites and is of the form

$$L_\sigma(\xi) = \mathbb{1} + B_L^\sigma B_{L-1}^\sigma \ldots B_1^\sigma \tag{10.208}$$

with B_ℓ^σ given by

$$B_\ell^\sigma = e^{-\Delta\tau h_\sigma(\xi(\ell))} \quad . \tag{10.209}$$

The step going from (10.206) to (10.207) is proven in Ref. [184] to which we refer. It makes use of the identity

$$Tr e^{-c_i^+ A_{ij} c_j} e^{-c_i^+ B_{ij} c_j} = \det\left[1 + e^{-A} e^{-B}\right] \tag{10.210}$$

which is proven there too. The sum over the Ising variables in (10.208) is done with the help of the Monte Carlo Method where the role of the Boltzmann weights (see Appendix (H)) is taken by the product $\det L_\uparrow(\xi) \det L_\downarrow(\xi)$ [39]. Indeed, for the Hubbard model at half filling and with nearest neighbor hopping only, this product is always positive. We conclude this by first noticing that under these circumstances the Hubbard Hamiltonian is invariant under a particle-hole transformation $c_{i\sigma} = (-1)^i c_{i\sigma}^+$. The prefactor $(-1)^i$ ensures that the sign differs for nearest neighbor sites. This is needed for the kinetic energy term to remain invariant under this transformation. The term $\lambda\sigma \sum_i \xi_i(\ell) n_{i\sigma}$ in (10.205) goes over into $-\lambda\sigma \sum_i \xi_i(\ell)(1 - n_{i\sigma})$. This implies that

$$\det L_\downarrow(\xi) = \prod_{i\ell} e^{\lambda \Delta\tau \xi_i(\ell)} \det L_\uparrow(\xi) \tag{10.211}$$

and therefore that the product $\det L_\uparrow(\xi) \det L_\downarrow(\xi)$ is non-negative for all configuration ξ. The Monte Carlo importance sampling is done with probabilities

$$P(\xi) = \frac{1}{Z} \det L_\uparrow(\xi) \det L_\downarrow(\xi) \quad . \tag{10.212}$$

By selecting C independent configurations with a probability distribution given by (10.212) the thermal average of an arbitrary operator A is then given by

$$\langle A \rangle = \frac{1}{C} \sum_{\{\xi\}} A(\xi) \quad , \tag{10.213}$$

where $A(\xi)$ is the matrix element of A calculated for a given ξ configuration. For hole doped systems the product of the two determinants may become negative, an example of the famous sign problem for fermions. Then we must replace (10.212) by

$$P(\xi) = \frac{|\det L_\uparrow(\xi)\det L_\downarrow(\xi)|}{\sum_{\{\alpha\}} |\det L_\uparrow(\xi)\det L_\downarrow(\xi)|} \quad . \tag{10.214}$$

In order to find the correct thermal average $\langle A \rangle$ we have to include the sign changes by defining the quantity

$$s(\xi) = \text{sgn}\left(\det L_\uparrow(\xi)\det L_\downarrow(\xi)\right) \tag{10.215}$$

and using it in

$$\langle A \rangle = \frac{\langle As \rangle}{\langle s \rangle} \quad . \tag{10.216}$$

The average is calculated by using $P(\xi)$ as given by (10.214). When $\langle s \rangle$ becomes small implying that there are nearly as many positive and negative products $\det L_\uparrow(\xi)\det L_\downarrow(\xi)$ the method looses its advantages. For a given statistical error the number of required independent configurations in the sampling procedure increases like $\langle s \rangle^{-2}$. One can show that with lowering of the temperature $\langle s \rangle$ decreases exponentially, thus limiting the method to higher temperatures. The average $\langle s \rangle$ decreases also when U is of the order or larger than the bandwidth. Figure 10.29 shows how $\langle s \rangle$ changes with temperature for a system with moderate hole doping and strong correlations.

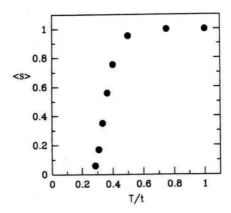

Fig. 10.29. Averaged sign $\langle s \rangle$ of the product $\det L_\uparrow(\xi)\det L_\downarrow(\xi)$ for a 8×8 square lattice with $\langle n \rangle = 0.87$ and $U = 8t$ as function of temperature. The plot shows the restrictions of Monte Carlo calculations to high temperatures. (From [399])

Monte Carlo calculations for the Hubbard model on a square lattice with nearest neighbor hopping have led to a number of interesting results. Despite

10.8 Numerical Studies

the limitations to temperatures $T/t > 0.3$ results for an antiferromagnetic ground state could be extracted [184]. The order parameter is much smaller though than a mean-field treatment would suggest. The findings are made possible by considering a set of N × N lattices of increasing size. By extrapolating to the $T = 0$ limit for each lattice and by providing for a finite-size scaling analysis to the bulk limit, various ground-state properties have been obtained. They reconfirm that the half-filled system is an insulator with a gap in the excitation spectrum for all values of $U \neq 0$. As previously discussed, this is due to nesting properties of the Fermi surface.

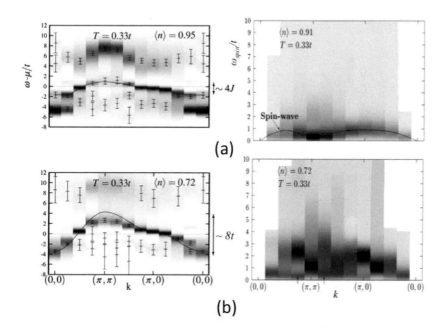

Fig. 10.30. Spectral weight $\rho(\mathbf{k},\omega)$ shown by the intensity of dark colour and spin-excitation energies ω_{spin}/t as obtained from a QMC calculation for the Hubbard model on a square lattice with different deviations from half filling. (a): $\langle n \rangle = 0.95$, i.e., 5 % hole doping. (b): $\langle n \rangle = 0.72$, i.e., 28 % hole doping. The lattice is 8 × 8 and $U/t = 8$. The Fermi energy is at $\omega - \mu/t = 0$. (From [379])

Another interesting finding obtained by the same method concerns the single-particle spectral weight $\rho(\mathbf{k},\omega)$ when the system is hole doped. Results are shown in Fig. 10.30 for small doping, i.e., $\langle n \rangle = 0.95$ and large doping, i.e., $\langle n \rangle = 0.72$. In the former case spectral weight changes dramatically as the Fermi energy is crossed. We notice the presence of an upper and a lower Hubbard band. At the same time spin-wave like excitations are found. They are obtained from the computed magnetic susceptibility and explain why the spectral weight of the quasiparticle-like low-energy band is essentially limited to the reduced antiferromagnetic Brillouin zone. This is quite different when the doping is large. Here spin waves can no longer be identified and consequently the spectral weight of the low-energy excitations is appreciable in the whole, i.e., paramagnetic Brillouin zone [379]. Another interesting result obtained from QMC calculations is the behavior of the equal time spin-spin correlation function

$$S(\mathbf{q}) = \frac{1}{3} \sum_i \langle \mathbf{s}(\mathbf{R}_i) \cdot \mathbf{s}(\mathbf{0}) \rangle e^{-i\mathbf{q}\mathbf{R}_i} \quad (10.217)$$

where the sum extends over the sites of the square lattice.

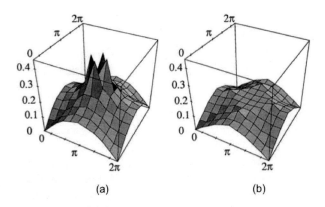

Fig. 10.31. Magnetic structure factor $S(\mathbf{q})$ as obtained by QMC calculations for a 10×10 square lattice with $U/t = 4$. (a) for hole doping $\delta = 1 - \langle n \rangle = 0.18$ and (b) $\delta = 0.5$. (From [140])

At half filling $S(\mathbf{q})$ has a divergent peak at $\mathbf{Q} = (\pi, \pi)$ because of AF order. Yet, when the system is hole doped the peak splits into four. We denote the corresponding $\mathbf{q}$ values by $\mathbf{q}_{\text{inc}}$. The incommensurate peaks $S(\mathbf{q}_{\text{inc}})$ have finite height with a value which scales like $1/\delta$ where $\delta = 1 - \langle n \rangle$ is the doping concentration [140]. Furthermore, $\mathbf{q}_{\text{inc}}$ seems to be of the form $\mathbf{q}_{\text{inc}} = (\pi, \pi \pm c\delta), (\pi \pm c\delta, \pi)$ with c being constant. This indicates short-range antiferromagnetic order with an antiferromagnetic correlation length ξ_m which scales like $\xi_m \sim 1/\delta^{1/2}$. The function $S(\mathbf{q})$, shown in Fig. 10.31 for different hole doping concentrations, is obtained for a 10×10 square lattice and $U = 4t$.

Another approach used for finite temperatures is the Finite Temperature Lanczos Method. The main idea is to combine the Lanczos method described in Appendix E with a stochastic sampling of the full Hilbert space spanned by the different configurations. Since only operations of the type $|\psi\rangle = H|\phi\rangle$ are needed and the matrices are very sparse, one can treat larger systems by full diagonalization. A detailed description of the method is found in Ref. [213].

10.9 Break-down of Fermi Liquid Description

In Sect. 7.2 we have discussed the concept of quasiparticles and of Fermi liquids. The basic assumption of Fermi liquid theory is a one to one correspondence between the low-energy excitations of an interacting electron system and those of a noninteracting system with renormalized parameters. It is surprising that even in systems with strong electron correlations the Fermi liquid concept is applicable in many cases. An extreme case are systems with heavy quasiparticles. Here the renormalization of the effective quasiparticle mass may exceed a factor of hundred when compared with the free electron mass (see Chapter 13). Despite of this Fermi liquid theory is still applicable. But there are also a number of situations where a Fermi liquid description is inapplicable. In the following we want to discuss a few of them.

At first we consider the Hubbard model (10.53) away from half filling when $U/|t| \gg 1$. In that case two Hubbard bands form. The number of states which are available, e.g., for intraband excitations depends on the filling factor n. This was pointed out before and is illustrated in Fig. 10.13. A one to one correspondence between the excitations in a Hubbard band with partial filling and those of a nearly free electron gas is here not possible. It was also pointed out in Sect. 10.4.3 that in the Hubbard III approximation the imaginary part of $\Sigma(\mathbf{p}, \omega)$ does not vanish like ω^2 when the limit $\omega \to 0$ is taken but remains instead constant. This is due to static disorder scattering. The self-energy resembles therefore that of free electrons scattered by impurities with a scattering rate τ^{-1}, i.e.,

$$Im\Sigma(\mathbf{p}, \omega) = i/\tau \quad . \tag{10.218}$$

An important difference between the Hubbard model in Hubbard III approximation and impurity scattering is that the former system is translationally invariant while the latter is not. Therefore, in the presence of elastic impurity scattering, one can in principle always replace the momentum $\mathbf{p}$ of an electron, which is no longer conserved, by three other electronic quantum numbers. They are denoted by the compact index n and define the single-particle eigenstates in the presence of a given distribution of scattering centers. When the retarded Green's function is expressed in term of them, the self-energy vanishes, i.e.,

$$G_{nn'}(\omega) = \frac{\delta_{nn'}}{\omega - \epsilon_n + i\delta} \tag{10.219}$$

and the system is a Fermi liquid because the distribution of the ϵ_n is similar to that of the $\epsilon_\mathbf{p}$. For a translationally invariant system with momentum conservation like the Hubbard model, a similar argument cannot be used.

Another interesting example of a breakdown of Fermi liquid behavior is obtained when a Mott-Hubbard system with strong electron correlations is slightly doped with holes. Here a square lattice deserves special attention because it is considered to be the minimum model for the Cu-O planes of the high-temperature superconducting cuprates. In Sect. 10.5 we have discussed the $t - J$ model which is the effective low-energy version of the Hubbard Hamiltonian. There it was shown that near half filling there is no one to one correspondence between the low-energy excitations of the system with a few holes and those of independent electrons with a Fermi surface which encloses half of the Brillouin zone. A different question is whether such a correspondence exists but with charge carriers of a density which equals the hole concentration. This seems to be the case, since we have seen in Fig. 10.24 that for $\mathbf{k}$ vectors near $(\pi/2, \pi/2)$ a quasiparticle-like peak shows up in the spectral density $\rho(\mathbf{k}, \omega)$. Its dispersion can be interpreted as that of a quasiparticle. Also, the self-energy $\Sigma(\mathbf{p}, \omega)$ calculated with the FSCP operator method is in the low-frequency limit of the form (7.21), which is characteristic for a Fermi liquid.

10.9.1 Marginal Fermi Liquid Behavior

A number of cuprates have properties like the optical conductivity, Raman scattering intensity or nuclear relaxation rate which cannot be explained within the frame of Fermi liquid theory. Best known is the anomalous behavior of the resistivity. In the normal state of several superconducting cuprates a temperature dependent resistivity of the form $\rho(T) = AT$ rather than $\rho \sim T^2$ is observed experimentally. It came as a surprise that many of the anomalous properties can be explained by making a single, yet rather bold assumption, namely that the imaginary part of the density susceptibility $\chi_e(\mathbf{q}, \omega)$ as well as of the spin susceptibility $\chi(\mathbf{q}, \omega)$ is proportional to

$$Im\chi_{(e)}(\mathbf{q}, \omega) \sim \begin{cases} -N(0)\,(\omega/T) & , \quad \text{for } |\omega| < T \\ -N(0)\,\text{sgn}\,\omega & , \quad \text{for } T < |\omega| \end{cases} \quad (10.220)$$

Both susceptibilities are response functions. They describe the response of the system to a space and time dependent perturbation acting on the density and spin density, respectively. Up to here we have not yet discussed the effects of correlations on these two functions, in particular what their form is when a Fermi liquid description applies. For the spin susceptibility this is done in Chapter 11, in particular in Sect. 11.3[5]. From the form of $Im\chi(\mathbf{q}, \omega)$ one can determine $Re\chi(\mathbf{q}, \omega)$ via Kramers-Kronig relations. Associated with the

---

[5] see, e.g., Fig. 10.15

response function is a corresponding contribution to the electron self-energy $\Sigma(\mathbf{p}, \omega)$. This is plausible, since an electron (or hole) moving through the system acts as a time- and space-dependent perturbation of the medium. A pictorial way how $\Sigma(\mathbf{p}, \omega)$ is affected is seen in Fig. 11.17, where the role of $r(\mathbf{k}, \omega)$ is played here by $\chi(\mathbf{q}, \omega)$. As a result of (10.220) the following form of $\Sigma(\mathbf{p}, \omega)$ is obtained without proof

$$\Sigma(\mathbf{p}, \omega) = g^2 N^2(0) \left[\omega \ln(x/\omega_c) - i\pi x/2\right] \quad (10.221)$$

where $x = \text{Max}(|\omega|, T)$ and g is a coupling constant. Furthermore, ω_c is a high-frequency cut off. We notice that the self-energy is quite different from the one in a Fermi liquid theory. As discussed in Sect. 7.2 we find that for a Fermi liquid $Re\Sigma(\mathbf{p}, \omega) \sim \omega$ and $Im\Sigma(\mathbf{p}, \omega) \sim \omega^2$ in the zero temperature limit. A metallic system with a self-energy given by (10.221) is called a marginal Fermi liquid [464]. The renormalization constant $Z(\mathbf{p})$ in the single-particle Green's function (7.22) can be calculated from (7.20). In the present case it reduces to $Z(\mathbf{p}) \sim \ln^{-1}(\omega_c/\epsilon_{\mathbf{p}})$ and vanishes at the Fermi energy where $\epsilon_{\mathbf{p}}$ goes to zero. Thus the momentum distribution $n(\mathbf{p})$ has is no longer a discontinuity at p_F which in a Fermi liquid theory is a signature of the Fermi surface (see Fig. 7.1). Likewise, there are no quasiparticles existing in the low-frequency limit, in which case the Green's function becomes completely incoherent.

This raises the question whether a one band Hubbard model, considered to be the minimal model for the Cu-O planes in the high-T_c superconducting materials can show under special circumstances a marginal Fermi liquid behavior.

Very accurate calculations for the self-energy $\Sigma(\mathbf{p}, \omega)$ of the zero temperature retarded Green's function have been done by applying the FSCP operator technique to a Hubbard model on a square lattice. It is found that the form (7.21) for the self-energy $\Sigma(\mathbf{p}, \omega)$ and therefore Fermi liquid-like behavior is prevailing in all cases, i.e., independent of the doping concentration. In passing we note that it is necessary to apply the fully-consistent projection operator method. A simplified treatment in which only the diagonal self-energy matrix elements $\tilde{\Sigma}_{ii}(z)$ are calculated self-consistently yields erroneously a marginal Fermi liquid behavior for very small hole doping concentrations. We mention that feature in order to make the reader aware of the accuracies which treatments of this kind of problem require. However, quantum Monte Carlo as well as FSCP operator calculation show an interesting feature which we want to discuss briefly.

It is well known that the density of states of a square lattice with nearest-neighbor interactions has a logarithmic van Hove singularity in the center of the band. At half filling the Fermi energy ϵ_F coincides with this singularity. It turns out that the self-energy $\Sigma(\omega)$ has consequently a form like (10.221). Thus, at half filling, the system is expected to behave like a marginal Fermi liquid. However, that changes when the system is doped, e.g., with holes. In that case the Fermi energy moves away from the van Hove singularity and $\Sigma(\mathbf{p}, \omega)$ reduces again to (7.21) in the low-frequency limit. It came as a

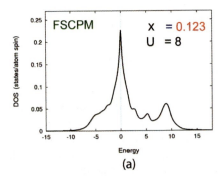
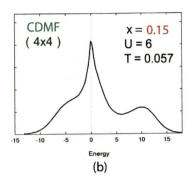

Fig. 10.32. (a) Density of states for a doped Hubbard system on a square lattice for $x = 0.123$ when the FSCPM is applied [226]. (b) The same function when a (4×4) cluster with similar doping concentration is used in a CDMF treatment. The temperature is in units of $|t|$ [469].

surprise that at a hole concentration of $x = 0.123$, i.e., of nearly 1/8 the Fermi energy returns to the peak in the DOS [226]. This is due to a doping-dependent transfer of spectral density between the two Hubbard bands. A similar result is found in CDMF theory for a 4×4 cluster [469]. A comparism of the two results is shown in Fig. 10.32. Since the singularity has been replaced by a high peak the self-energy $\Sigma(\omega)$ shows again Fermi liquid behavior, i.e., $Re\Sigma(\omega) \sim \omega$ and $Im\Sigma(\omega) \sim \omega^2$ for $\omega \to 0$. However, due to the narrow peak of the DOS, at higher frequencies the self-energy resembles more the one of a marginal Fermi liquid. Also we expect that at this doping concentration the system is particularly sensitive to instabilities which reduce the DOS. This is of interest in view of the stripe formation of holes, which takes place at $x = 1/8$ in some of the cuprates (see Sect. 15.5.6).

10.9.2 Charged and Neutral Quasiparticles

A quite different reason for non-Fermi liquid behavior is found in the low-temperature phase of Yb_4As_3, a system with heavy quasiparticles. This system is discussed at length in Sect. 13.2. It has a first-order phase transition near $T_0 = 292$ K and is metallic in the high temperature phase and semi-metallic in the low-temperature phase. The change is connected with partial electronic charge order in form of well separated Yb^{3+} chains with an effective spin 1/2 per site. In such a chain the holes are expected to be immobile (Mott-Hubbard insulator). It is well known that a Heisenberg spin chain has a low-temperature specific heat of the form $C = \gamma T$ like a metal. It is due to spin excitations (spinons) which obey Fermi statistics[6]. The coefficient γ is large here because of a weak coupling of the spins in a chain and therefore

---

[6] Note that in 1D bosons can be transformed into fermions and vice versa.

the specific heat resembles that of heavy quasiparticles. But the charge carriers, which are mainly $4f$ holes in the high-temperature phase consist in the low-temperature phase of a small number of As $4p$ holes. Therefore one may speak of spin-charge separation and a breakdown of the conventional Fermi liquid picture. The spin excitations take place in the Yb^{3+} chains while the charge excitations are associated with As p holes. A one-to-one correspondence between the excitations in the low temperature phase and those of an independent electron system is no longer given. In fact, we are dealing here with *two* Fermi liquids, i.e., one with charge-neutral heavy quasiparticles (spin excitations) and the other with charged light quasiparticles (As $4p$ holes). The light quasiparticles observed in cyclotron resonance experiments are scattered off by the neutral heavy quasiparticles. This results in a large A coefficient in $\rho(T) = AT^2$. While only the light quasiparticles contribute to the resistivity, the thermal conductivity is dominated by the neutral, i.e., heavy ones. This explains why the system has many properties of an ordinary metal with heavy quasiparticles.

10.9.3 Hubbard Chains

A breakdown of the Fermi-liquid approach is also taking place in one-dimensional (1D) systems. Consider again the Hubbard model. For 1D it was solved exactly by *Lieb* and *Wu* [283]. For half filling, an analytic form can be derived for the ground-state energy per electron E_0/N. It is found that

$$\frac{E_0}{N} = -4|t| \int_0^\infty \frac{dx\, J_0(x) J_1(x)}{x\left[1 + \exp\left(xU/2(2|t|)\right)\right]} \;, \tag{10.222}$$

where the $J_\nu(x)$ are Bessel functions. From this expression, the following asymptotic form can be derived for small ratios $U/|t|$:

$$\begin{aligned}
\frac{E_0}{N} &= -\frac{4|t|}{\pi} + \frac{U}{4} + \frac{7\zeta(3)}{8\pi^3} \frac{U^2}{|t|} \\
&= -4\frac{|t|}{\pi} + \frac{U}{4} - 0.017\frac{U^2}{|t|} \;,
\end{aligned} \tag{10.223}$$

where $\zeta(x)$ is the zeta function. In the opposite limit, i.e., for $|t|/U \ll 1$, we obtain the result

$$\frac{E_0}{N} = -\frac{4t^2}{U} \ln 2 \;, \tag{10.224}$$

precisely the energy of a one-dimensional Heisenberg antiferromagnet with $J = 4t^2/U$.

Another interesting result is that for the half-filled case the ground state is insulating for all values of $U > 0$. It is proven by showing that the chemical potential for adding an electron, μ_+, and that for removing an electron, μ_-, differ, i.e., $\mu_+ > \mu_-$. We encounter the same situation in a semiconductor with a gap.

The Bethe ansatz technique [32] makes the solution of the 1D Hubbard model possible. With its help one can derive the explicit form of the ground-state wavefunction $\psi(x_1, \ldots, x_N)$, where $x_1, \ldots, x_M$ are the coordinates of the spin-down electrons and $x_{M+1}, \ldots, x_N$ are those of the spin-up electrons. One finds especially that in the limit $U \to \infty$, where double occupancies of sites are excluded, the wavefunction factorizes into the form [409]

$$\psi(x_1, \ldots, x_N) = \det \left| e^{ik_j x_j} \right| \Phi(y_1, \ldots, y_M) \quad . \tag{10.225}$$

We have introduced here "pseudo coordinates" $y_1, \ldots, y_M$ for the spin-down electrons, thereby omitting all empty sites. The first part is a Slater determinant of noninteracting spinless fermions with momenta $k_1, \ldots, k_N$ describing the charge degrees of freedom of the electron system. In contrast to (2.15), the spin functions are here excluded, i.e., one is dealing with spinless fermions. The second part $\Phi(y_1, \ldots, y_M)$ is the exact wavefunction of a 1D Heisenberg spin chain. In fact, the form (10.225) for the ground-state wavefunction can be guessed from elementary considerations, instead of using the Bethe ansatz solutions. For this purpose consider the effective Hamiltonian H_{t-J} of (10.101), which holds for large values of $U/|t|$. When $J = 4t^2/U = 0$, i.e., for $U \to \infty$, the electrons cannot exchange their positions within the chain. The eigenstates of H_{t-J} prove to be degenerate with respect to the electron spins. Only the hopping term of H_{t-J} is left in this limit, the ground state of which is the Slater determinant of noninteracting spinless fermions. As the interaction J is turned on, the 2^N-fold spin degeneracy is lifted. Taking the expectation value of H_{t-J} given by (10.101) with respect to the Slater determinant of the spinless fermions, we arrive at an effective Hamiltonian $H_{\text{eff}} = \langle H_{t-J} \rangle$ describing the spin degrees of freedom. The following result is obtained [409]

$$H_{\text{eff}} = -\frac{2t}{\pi} N_0 \sin(\pi n) + J_{\text{eff}} \sum_i (\mathbf{S}_i \cdot \mathbf{S}_{i+1} - 1/4) \quad . \tag{10.226}$$

As pointed out before, empty sites between the spins are omitted in the term proportional to J_{eff}. The effective coupling constant J_{eff} is given by

$$J_{\text{eff}} = n^2 J \left(1 - \frac{\sin(2\pi n)}{2\pi n} \right) \quad , \tag{10.227}$$

where $n = N/N_0$ is the electron density (as before N_0 is the number of sites). J_{eff} decreases rapidly with departure from half filling. In $\psi(x_1, \ldots, x_N)$ spins can only interact when they are nearest neighbors. The density dependence of J_{eff} is related to the probability that this interaction does occur, and is determined by the different configurations of the Slater determinant. The ground state of H_{eff} is that of a 1D Heisenberg chain, i.e., $\Phi(y_1, \ldots, y_M)$, with only the positions of the down spins specified.

10.9 Break-down of Fermi Liquid Description

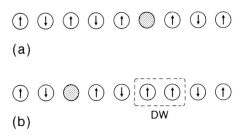

Fig. 10.33. Motion of a hole in a Hubbard chain in the limit of large U: (a) a hole has been generated (shaded site), (b) after three hops the hole has separated from a domain wall (DW) which it has created. Hole and DW can move independently of each other.

The factorization (10.225) of the ground-state wavefunction into one part involving charge degrees of freedom and another involving spin degrees of freedom only constitutes a remarkable result. It is then not surprising that also the excitations separate into a class involving spin degrees of freedom and another class involving only charge degrees of freedom. A near separation of these two classes of excitations showed up first when only two electrons were considered (Sect.10.2.1). It is a general feature of strongly correlated electronic systems.

An intuitive interpretation of spin-charge separation has frequently been made by considering a Hubbard chain at half filling in the limit of large U. In that case double occupancies of sites are strongly reduced and one can simulate the antiferromagnetic correlations by a local Néel-like order (Fig. 10.33a). When an electron is removed from the Hubbard chain, the hole is surrounded by two parallel spins. This changes after the first hop of the hole to a neighboring site has taken place which requires an energy $J_{\text{eff}}/2$. Thereby a domain wall has been created with two neighboring parallel spins. This is a spin excitation (spinon). As the hole continues hopping, it is surrounded by anti-parallel spins. Under these conditions, it can move freely, i.e., without cost of magnetic energy (compare with Fig. 10.33b). Also the domain wall may move without loss of magnetic energy. However, this argument disregards the attractive δ-function like potential between the hole and the domain wall, which in one dimension should always result in a bound state between the two. Thus the problem is more subtle than the above argument suggests. Nevertheless, we gain an intuitive picture of how dimensionality enters spin-charge separation. As this separation persists to very low energies, we find two types of elementary excitations namely of spins and of charges and the system is no longer a Fermi liquid. The statement that a 1D Hubbard chain does not represent a Fermi liquid is provable if we study the momentum distribution $n(p)$. In the limit $U \to \infty$, we may use the wavefunction (10.225) to calculate

$\langle c_{p\sigma}^+ c_{p\sigma}\rangle$. We find that $n(p)$ does *not* have a discontinuity at p_F, but rather a singularity of the form [419]

$$n(p) = -\frac{1}{2}C\,|p-p_F|^\alpha \operatorname{sgn}(p-p_F) \quad. \tag{10.228}$$

The exponent α is found to be $\alpha = 1/8$ in the limit considered here [421]. On account of the missing discontinuity, the system does not have a Fermi surface in the usual sense.

Correlation functions prove to be more difficult to compute for the 1D Hubbard model because they cannot be obtained directly from Bethe-ansatz solutions for the ground- and excited states. We can solve this problem by deploying conformal field theory [123, 406]. Here the behavior of the spin-spin and density-density correlation functions for large distances is related to physical quantities (i.e., compressibility κ and velocity v_c of charge excitations). They can be derived from the Bethe-ansatz solutions. For example, it is found that the spin-spin correlation function $\langle S_i^z S_j^z\rangle$ for large distances $|R_j - R_i|$ between lattice sites j and i decays like

$$\langle S_i^z S_j^z\rangle \propto \frac{\cos[2p_F(R_i-R_j)]}{|R_i-R_j|^\eta}\ln^{1/2}|R_i-R_j| \quad. \tag{10.229}$$

The exponent $\eta(=1+\pi n^2\kappa v_c/2)$ is calculated to be $3/2$ for all values of band filling, provided $U\to\infty$. The same value is found for arbitrary values of U, provided $n=1$ (half-filled case).

In contrast, the density-density correlation function $\langle n_i n_j\rangle$ decays at large distances like

$$\langle n_i n_j\rangle \propto \frac{\cos[4p_F(R_i-R_j)]}{|R_i-R_j|^{4(\eta-1)}} \quad. \tag{10.230}$$

One notices a different power of the denominators in (10.229 - 10.230) as well as a different oscillatory length of the cosine functions. This is again an indication of spin-charge separation. The doubling of the argument of the cosine function in (10.230) is due to spinless fermions. They require twice as much volume in momentum space as the same number of spin 1/2 fermions, of which two can occupy a given k state; consequently, the Fermi momentum becomes twice as large. For free electrons the corresponding expressions for $\langle n_i n_j\rangle$ and $\langle S_i^z S_j^z\rangle$ are both proportional to $\cos[2p_F(R_i-R_j)]$ and describe Friedel oscillations and Ruderman-Kittel-Kasuya-Yosida (RKKY) oscillations, respectively. It should be pointed out that the behavior of $n(p)$ and $\langle S_i^z S_j^z\rangle$ corresponds to that found for other model Hamiltonians for 1D electronic systems as proposed by *Tomonaga*[7] and *Luttinger*[8]. For this reason the 1D Hubbard system has also been called a Tomonaga-Luttinger liquid[9]. For a number of other properties of the 1D Hubbard model derivable from the Bethe-ansatz solution see, e.g., Ref. [232].

---

[7] see [453]
[8] see [296]
[9] see [167]

10.9.4 Quantum Critical Point

Another example of non-Fermi liquid behavior is found in systems with a quantum critical point (QCP). We speak of a QCP when, as a function of an external control parameter, a phase transition ends with a transition temperature $T_c = 0$. The external parameter can be pressure, magnetic field, change in chemical composition etc. Near such a point the behavior of the system is dominated by quantum fluctuations rather than thermal fluctuations. Yet in an electronic system with a QCP quantum fluctuations are those processes which go beyond a SCF theory. They are just the correlation contributions to, e.g., the ground-state energy or wavefunction of a system. In Chapter 6 we have discussed them extensively. When the functional integral formalism is used like in Sect. 11.3, quantum fluctuations show up in form of the τ dependence of the Hubbard-Stratonovich fields $\xi(\tau)$ and $\eta(\tau)$. It is obvious that quantum fluctuations, i.e., correlation effects influence the critical interaction strength at which, e.g., a metal becomes an antiferromagnet or a ferromagnet at $T = 0$.

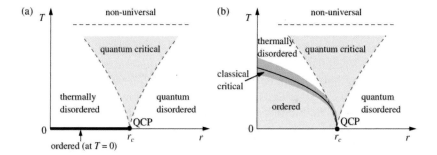

Fig. 10.34. Two types of phase transitions ending in a quantum critical point (QCP) as function of an external parameter r. (a) Absence of long range order at $T \neq 0$ in low dimensional systems like a 2D Heisenberg AF due to strong fluctuations. (b) Long-range order which terminates in a QCP at r_c. Classical, i.e., temperature dependent fluctuations of the order parameter are indicated in dark. In the quantum disordered regime a Fermi liquid description prevails. (From [470])

The simultaneous effect of thermal- and quantum fluctuations determines the phase diagram of a system. Here we distinguish between two different cases. In one case a phase transition is possible only at $T = 0$ like in a 2D Heisenberg antiferromagnet. The thermal fluctuations destroy any long range order at finite temperature because of their high density in two dimensions (Mermin-Wagner theorem). In the second case long-range order is diminished by a variable external parameter r. At a critical value r_c of that parameter $T_c \to 0$ and a quantum critical point is approached. Both cases are illustrated in Fig. 10.34. Above the QCP we deal with a quantum critical regime. The boundaries of that regime are approximately given by the condition that $k_B T \simeq \omega_c$ where ω_c is a typical energy of long-distance fluctuations of the order parameter. It depends on $t = (T - T_c)/T_c$ like $\omega_c \sim t^{\nu z}$. The exponent ν relates the correlation length ξ to t, i.e., $\xi \sim t^\nu$ and the exponent z relates the correlation time τ_c to the correlation length, i.e., $\tau_c \sim \xi^z$. The correlation time is the time it takes for a fluctuation of the spatial extend ξ to decay. From $\omega_c \tau_c \simeq 1$ the above relation follows.

Within the quantum critical regime a Fermi liquid description does not apply. The order parameter fluctuations couple here to electronic low-energy modes. That results in a self-energy, which does not have the form required for a Fermi liquid, i.e., $Im \Sigma(\mathbf{k}, \omega) \sim \omega^2$ for small ω and $T \ll \omega$. A detailed description of $\Sigma(\mathbf{k}, \omega)$ is still missing though, in particular since details of the Fermi surface become important. On some magnetic transitions in electron systems light has been shed by the work of *Hertz* [179] and *Millis* [325].

Considerable experimental work has been devoted to CeCu$_{6-x}$Au$_x$. Here a phase transition takes place between an AF, which exists for $x < 0.1$ and a system with heavy quasiparticles we deal with when $x > 0.1$ [287]. It was found that for $x = 0.1$ and $T < 1K$ the specific heat behaves like $C(T)/T \sim -\ln T/T_0$ while the resistivity is of the form $\rho(T) = \rho_0 + AT$ over more than one order of magnitude in temperature. The material is considered a prototype for non-Fermi liquid behavior near a QCP. The unusual temperature dependence of the resistivity can be qualitatively explained by a strong interplay between quantum fluctuations such as spin fluctuations and isotropic impurity scattering. For examples, scattering of electrons by antiferromagnetic spin fluctuations is strongest near lines on the Fermi surface separated by the AF wavevector $\mathbf{Q}$, i.e., the so-called *hot* lines. Without impurity scattering those contributions are of little importance because they are short circuited by the contributions from the remaining *cold* parts of the Fermi surface. Remember that for the average scattering rate τ^{-1} the relation $\tau^{-1} = \tau_{\text{hot}}^{-1} + \tau_{\text{cold}}^{-1}$ holds, where both contributions enter with their respective weight. Without impurity scattering $\tau_{\text{hot}}^{-1} \ll \tau_{\text{cold}}^{-1}$ leading to the above mentioned short circuitry. As a result the resistivity is $\rho \sim \tau_{\text{cold}}^{-1} \sim T^2$. Yet, when impurity scattering is present, the two contributions are no longer so different, i.e., the hot lines become important. Therefore strong deviations from the T^2 behavior are expected [388]. Detailed discussions of quantum phase transitions are found in Refs. [70, 396].

Last not least we want to point out that systems with fractionalized excitations discussed in Chapter 14 are clearly outside a Fermi liquid description. This concludes the brief survey of non-Fermi liquid systems.

11
Transition Metals

In order to understand transition metals thoroughly, a proper description of their magnetic properties proves important. They have been fascinating and intriguing physicists for a long time. It is the interplay of delocalized versus localized features of electrons that challenges theorists. This takes place in the presence of fairly strong correlations, in particular in $3d$ systems. The question posed is whether the correlations are so strong that important atomic properties caused by Hund's rules are significant in solids too. To give a specific example: we would like to know to what extent the spin of an Fe site in a solid resembles the one of a single Fe atom. In order to work out criteria for magnetic order, correlations among the electrons must be properly treated. Otherwise, a symmetry broken ground state might be found in calculations merely because it has reduced charge fluctuations at an atomic site which simulate incompletely treated correlations and diminish Coulomb repulsions. This reduction is most clearly seen by considering "strong ferromagnets". They have a filled majority-spin band and a partially filled minority-spin band. Therefore, only electrons with minority spin may move and charge fluctuations are strongly reduced as compared with the nonmagnetic case. If we fail to sufficiently treat electron correlations in the nonmagnetic state, we may overestimate the energy gain due to magnetic order and, therefore, favor too much a magnetic ground state. Since correlations are quite strong in d electron systems, there are important incoherent excitations appearing in some of them. They show up in photoemission experiments as satellite or shake-up peaks.

In studying transition metals we wish to begin with a discussion of the ground-state wavefunction. It will give us an insight into where those systems are with respect to the two extremes, i.e., uncorrelated electrons and the strongly correlated or Hund's-rule-dominated limit. Finding this out requires the use of a model Hamiltonian. Density-functional-based ab initio calculations are of no help here since they avoid calculating many-body wavefunctions. Calculations using quantum chemical methods to determine the ground state have not yet been performed, yet are expected to become feasible in the near future. The model Hamiltonian used here is the five-band Hubbard

Hamiltonian. The discussion of the ground state is followed by a discussion of excited states.

Correlations have three different effects on the energy dispersion of excitations, i.e.,

(a) the dispersion curves obtain a finite lifetime and consequently they broaden, while without correlations they would remain infinitely sharp;

(b) there are **k**-dependent energy shifts of the dispersion curves which generally result in a reduction of the d band widths;

(c) shake-up or satellite peaks appear when the ratio of the one-site Coulomb repulsion to the d-band width becomes sufficiently large.

Finally we discuss the role of correlations when finite temperatures are considered. They have profound effects on the magnetic properties. With increasing temperature, transition metals show more and more local moment features, while itineracy and delocalization of d electrons seem to become less important. An example are the magnetization curves of Fe, Co and Ni. When plotted as a function of temperature they resemble Brillouin curves that one expects to obtain when a localized electron picture applies. In order to describe this cross-over, the simplest possible model, i.e., that of a one-band Hubbard Hamiltonian given by (8.22) we will use. Technically the treatment of finite temperatures is done by means of the functional integral method (see Sect. 11.3.1) to which a single-site approximation is applied. It is supposed to take care of the most important fluctuations when the ferromagnets Fe, Ni or Co are considered, although it is not clear whether at high temperatures there are no small ferromagnetic clusters remaining. The approximation is certainly inappropriate when Pd or transition metal alloys like Ni_3Ga are considered. They are paramagnetic yet close to a ferromagnetic phase transition. Here the amplitudes of short-range magnetic fluctuations are relatively small, and long wavelength ($q \simeq 0$) fluctuations are the most important ones. This leads to the (self-consistently renormalized) spin-fluctuation theory.

11.1 Ground-State Wavefunction

For a discussion of correlations in the ground state of transition metals we use a model Hamiltonian for the five d orbitals. We assume that the electrons interact with each other only when they are at the same site. The s- and p-electrons are neglected. Their effect is understood to be incorporated in the effective interaction parameters of the d-electron system.

Operators $a_{i\sigma}^+(\ell)$ $(a_{i\sigma}(\ell))$ which create (destroy) electrons at site ℓ in the atomic orbital i are introduced. Orbitals at different sites are assumed to be orthogonal with respect to each other. We choose the model Hamiltonian to be of the form

11.1 Ground-State Wavefunction

$$H = H_0 + \sum_\ell H_1(\ell)$$

$$H_0 = \sum_{\nu\sigma\mathbf{k}} \epsilon_\nu(\mathbf{k}) n_{\nu\sigma}(\mathbf{k}) \quad,$$

$$H_1(\ell) = U \sum_i n_{i\uparrow}(\ell) n_{i\downarrow}(\ell) + \left(U' - \frac{J}{2}\right) \sum_{j>i} n_i(\ell) n_j(\ell)$$

$$- 2J \sum_{j>1} \mathbf{s}_i(\ell) \mathbf{s}_j(\ell) \quad. \tag{11.1}$$

The notation $n_{i\sigma}(\ell) = a_{i\sigma}^+(\ell) a_{i\sigma}(\ell)$, $n_i(\ell) = \sum_\sigma n_{i\sigma}(\ell)$ and $\mathbf{s}_i(\ell) = (1/2) \sum_{\alpha\beta} a_{i\alpha}^+(\ell) \boldsymbol{\sigma}_{\alpha\beta} a_{i\beta}(\ell)$ has been used here. The Hamiltonian H_0 describes the canonical d bands with energy dispersions $\epsilon_\nu(\mathbf{k})$, where ν is a band index; the latter are known from LDA calculations. The creation operators for the corresponding Bloch eigenstates are

$$c_{\nu\sigma}^+(\mathbf{k}) = \frac{1}{\sqrt{N_0}} \sum_{i\ell} \alpha_i(\nu,\mathbf{k}) a_{i\sigma}^+(\ell) e^{i\mathbf{k}\cdot\mathbf{R}_\ell} \quad. \tag{11.2}$$

The vectors $\mathbf{R}_\ell$ denote the positions of the N_0 different sites and the $\alpha_i(\nu,\mathbf{k})$ stand for the projections of the Bloch states onto the atomic orbitals. The occupation-number operators $n_{\nu\sigma}(\mathbf{k})$ are

$$n_{\nu\sigma}(\mathbf{k}) = c_{\nu\sigma}^+(\mathbf{k}) c_{\nu\sigma}(\mathbf{k}) \quad. \tag{11.3}$$

The operator $H_1(\ell)$ describes the interactions at site ℓ. It contains three parameters, i.e., U, U' and J. The Coulomb repulsion of two electrons in the same d orbital is denoted by U while the one of electrons in different orbitals is U' when their spins are opposite and $(U' - J)$ when they are parallel. For the definition of the exchange integral J compare with (2.27) and (2.2). A rotational invariant environment implies that $U = U' + 2J$. The nonmagnetic SCF ground state of the Hamiltonian is of the form

$$|\Phi_0\rangle = \prod_{\substack{\nu,\sigma \\ |\mathbf{k}| \leq k_F}} c_{\nu\sigma}^+(\mathbf{k}) |0\rangle \quad. \tag{11.4}$$

We are interested here in the residual interactions only $H_{\text{res}}(\ell) = H_1(\ell) - \langle H_1(\ell) \rangle$, i.e., we assume that the SCF part of $H_1(\ell)$ is included in H_0. In this case $n_{i\sigma}(\ell)$ in $H_{\text{res}}(\ell)$ has to be replaced by $\delta n_{i\sigma}(\ell) = n_{i\sigma}(\ell) - \langle n_{i\sigma}(\ell) \rangle$ and there is also a term of the form $J a_{i\uparrow}^+(\ell) a_{i\downarrow}^+(\ell) a_{j\uparrow}(\ell) a_{j\downarrow}(\ell)$ appearing in $H_{\text{res}}(\ell)$. Imagine that $|\Phi_0\rangle$ is decomposed with the help of (11.2) into a sum of products of operators $a_{i\sigma}^+(\ell)$. Each term in the sum represents a different configuration, two of which appear in Fig. 11.1. They differ considerably in their respective interaction energy: configuration (a) has a comparatively small repulsion energy, while that of configuration (b) is large due to a significant deviation

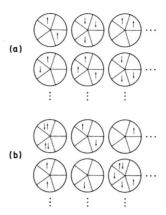

Fig. 11.1. Schematic representation of a favorable configuration (a) and an unfavorable configuration (b) contained in a nonmagnetic SCF ground-state wavefunction $|\Phi_0\rangle$. The circles symbolize atoms and the five segments represent the different d orbitals. The average d electron occupancy per atom is 2.5. One notices that in (a) electrons obey Hund's-rule correlations. We also note that in (b) charge fluctuations between different sites are large.

of the atomic charges from their mean values. One notices also that in configuration (a) the electrons are predominantly aligned according to Hund's rule. Thus it is expected that correlations increase the relative weight of that configuration considerably, whereas they strongly suppress configurations of the type depicted in Fig. 11.1b.

For a more quantitative discussion we choose the following variables $A_{ij}(\ell)$ for the wave operator Ω

$$A^{(1)}_{ij}(\ell) = \begin{cases} \delta n_{i\uparrow}(\ell)\delta n_{i\downarrow}(\ell) & i = j \\ \delta n_i(\ell)\delta n_j(\ell) & i \neq j \end{cases}$$
$$A^{(2)}_{ij}(\ell) = \mathbf{s}_i(\ell)\mathbf{s}_j(\ell) \quad , \tag{11.5}$$

so that (5.60) reads

$$|\Omega\rangle = \left|1 + \sum_{\ell,\nu}\sum_{i,j}\eta^{(\nu)}_{ij}A^{(\nu)}_{ij}(\ell)\right) \quad . \tag{11.6}$$

The site-independent coefficients $\eta^{(\nu)}_{ij}$ are obtained from (5.62) and (5.63). While the $A^{(1)}_{ij}(\ell)$ reduce density or charge fluctuations between orbitals at a given site ℓ, the variables $A^{(2)}_{ij}(\ell)$ generate Hund's-rule correlations. One should notice that the variables (11.5) resemble the ones of the Local Ansatz (5.86). When the expressions (5.62), (5.65) are evaluated, we make the additional approximation that correlations at different sites are treated as being

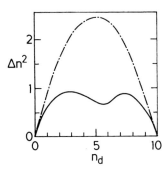

Fig. 11.2. Mean-square deviations Δn^2 of the charges from their average value (charge fluctuations) for a bcc structure as a function of d band filling. Upper curve without and lower curve with inclusion of electron correlations. The parameters are $\frac{U'}{W} = 0.5$ and $\frac{J}{W} = 0.1$ and $U = U' + 2J$ has been assumed (Kanamori parameterization). (From [135])

independent of each other. Thus we keep only those matrix elements in which $A_{ij}^{(\nu)}(\ell)$, $H_1(\ell')$ and $A_{mn}^{(\mu)}(\ell'')$ refer to the same site, i.e., for which $\ell = \ell' = \ell''$, an approximation proposed by *Friedel* and coworkers and called the "$R = 0$ approximation" [222, 454]. Note that a similar local approximation is done in the DMFT. In what follows we present a number of results of such a model calculation.

We consider first a *nonmagnetic* ground state. Of particular interest are the degree of suppression of charge fluctuations, the importance of Hund's-rule correlations, and the various correlation contributions to the ground-state energy.

a) Partial Suppression of Charge Fluctuations
A measure of the degree of suppression of charge fluctuations is the mean square deviation

$$\Delta n^2 = \left(\Omega|n^2(\ell)\Omega\right) - \left(\Omega|n(\ell)\Omega\right)^2 \quad,$$
$$\text{where} \quad n(\ell) = \sum_i n_i(\ell) \tag{11.7}$$

is the total d electron number operator for site ℓ. Results for various degrees of d band filling n_d are shown in Fig. 11.2 for a parameter choice of $U/W = 0.7$ and $J/W = 0.1$. Here W denotes the d bandwidth. One notices that $\Delta n^2 \lesssim 1$, which implies that, e.g., for $n_d = 3.5$ only configurations with three of four d electrons at a site have appreciable weight. All configurations with larger charge fluctuations are strongly suppressed.

b) Build-up of Hund's-Rule Correlations
A measure of the degree of spin alignment at an atomic site is the quantity

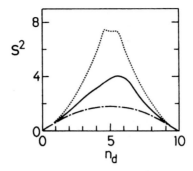

Fig. 11.3. On-site spin correlations S^2 (11.8) as a function of d band filling for a bcc structure (Hund's-rule correlations). *Upper curve:* atomic limit. *Lower curve:* independent-electron approximation. The *solid curve* corresponds to the correlated ground state, i.e., $|\Omega\rangle$ given by (11.6). The parameters are $\frac{U'}{W} = 0.5$ and $\frac{J}{W} = 0.1$ (Kanamori parameterization). (From [135])

$$S^2 = \langle\Omega|S^2(\ell)\Omega\rangle \quad , \tag{11.8}$$

where $\mathbf{S}(\ell) = \sum_i \mathbf{s}_i(\ell)$. We show these results in Fig. 11.3, where S^2 is compared with S^2_{SCF} and S^2_{loc}. These are the corresponding expectation values of $S^2(\ell)$ when the nonmagnetic SCF ground state is used (i.e., $\Omega = 1$), and when the ground state is in the localized limit $|\Phi_{\text{loc}}\rangle$, i.e., for large atomic distances. We can see from the figure that the relative spin alignment

$$\Delta S^2 = \frac{S^2 - S^2_{\text{SCF}}}{S^2_{\text{loc}} - S^2_{\text{SCF}}} \tag{11.9}$$

is rather constant and approximately $1/2$, which indicates that, for the above choice of parameters, one is - with respect to Hund's rule correlations - half way between the cases of uncorrelated and fully localized electrons. With increasing ratio U/W, the value of ΔS^2 increases continuously towards $\Delta S^2 = 1$. There exists no particular threshold value of the ratio U/W at which a local moment sets in. Instead, according to Hund's rule, the alignment of d electrons at an atomic site increases steadily as the electron interactions increase in units of W.

c) Ground-State Energy

We consider of particular interest the correlation energy's contribution to the cohesive energy. The prevailing opinion states that the s electrons contribute an amount to the cohesive energy which is approximately independent of the d-band filling. We can thus attribute the *variation* in the cohesive energy to the d electrons [125, 126]. For an analysis, it suffices to restrict oneself to a bcc structure. Furthermore, it is convenient to subtract from all interaction energies the energy of n_d electrons equally distributed among the different d spin orbitals, i.e.,

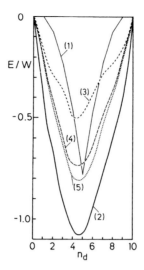

Fig. 11.4. Various energies per site as function of d-electron number for a bcc structure when $U'/W = 0.5$ and $J/W = 0.1$. (1): ΔE_{atom}; (2): E_{kin}; (3): $\langle H \rangle - \tilde{E}_0$, where $\langle H \rangle$ is the energy of uncorrelated electrons; (4): with the inclusion of density correlations; (5): $(\mathcal{E}_0 - \tilde{E}_0)$, where $\mathcal{E}_0$ is the energy of the correlated ground state. (From [135])

$$\tilde{E}_0 = \frac{n_d}{2}(n_d - 1)\frac{1}{9}(U + 8U' - 4J)$$
$$= \frac{n_d}{2}(n_d - 1)\bar{U} \quad , \tag{11.10}$$

where $\bar{U}$ is the average repulsive energy. We obtain the prefactor by calculating the interaction energy of one electron with a second one. On an isolated atom, the d electrons are *not* equally distributed among the spin orbitals; instead, they prefer parallel spin alignment (first Hund's rule). The gain in interaction energy compared with $\tilde{E}_0$ is therefore

$$\Delta E_{\text{atom}} = -\frac{1}{2}\tilde{n}_d(\tilde{n}_d - 1)\frac{1}{9}(U - U' + 5J) \quad , \tag{11.11}$$

where

$$\tilde{n}_d = \begin{cases} n_d & \text{if } n_d \leq 5 \\ 10 - n_d & \text{if } n_d > 5 \end{cases} \quad , \tag{11.12}$$

ΔE_{atom} has a minimum for $n_d = 5$, which corresponds to a maximal spin alignment.

In a solid, d electrons gain kinetic energy by delocalization. This energy gain E_{kin} is shown in Fig. 11.4 for a bcc structure. Delocalization results in fluctuations of the d electron number of an atom, and therefore in an increase in interaction energy. In the SCF or independent-electron approximation, the interaction energy of electrons in equally populated spin orbitals is

$$E_{\text{SCF}}^{\text{int}} = \frac{n_d^2}{2} \frac{9}{10} \bar{U} \quad . \tag{11.13}$$

The increment in energy as compared with $\tilde{E}_0$ is then

$$\Delta E_{\text{SCF}}^{\text{int}} = \frac{n_d}{2} \left(1 - \frac{n_d}{10}\right) \bar{U} \quad . \tag{11.14}$$

The sum $\left(E_{\text{kin}} + \Delta E_{\text{SCF}}^{\text{int}}\right)$ is shown in Fig. 11.4 as a function of d-band filling. Again a bcc structure has been considered and allowance has been made for different partial occupancies of e_g and t_{2g} orbitals. We can see from Fig. 11.4 that, for d-band filling in the range $4 \leq n_d \leq 6$, the energy of the state with localized d electrons (curve (1)) has a lower energy than the SCF state (curve (3)). Hence, in the independent-electron approximation, any symmetry-breaking solution which reduces the charge fluctuations and allows for partial spin alignment will be favored in that range of n_d values.

As pointed out before, charge fluctuations are also reduced by electron correlations rather than symmetry breaking. When taking into account density as well as spin correlations, we find that the energy is always lower for delocalized than for localized electrons. Most of the energy reduction is due to density correlations.

The cohesive energy is the difference between the energy of localized electrons and that of the correlated ground state, i.e., between curves (1) and (5) in Fig. 11.4. It shows the well-known double-peak structure as a function of n_d with a maximum value of $0.3W$ or 1.5 eV when $W = 5$ eV is assumed, a point in qualitative agreement with the known d electron contributions to experimental binding energies. A more quantitative discussion would require the incorporation of s electrons and, in particular, of $s - d$ charge transfers.

11.2 Satellite Structures

One important effect of electron correlations are satellite structures or shake-up peaks which appear in spectral densities and are seen in photoemission experiment.

Let us now consider photoemission from band states of Ni or Fe. In a photoemission experiment, a hole is generated in a d band. Correlation effects cause additional electron-hole excitations to accompany such a process. The essential features for Ni are a broad structure (main line) due to the d bands with a width of order of 4 eV and a satellite peak positioned about 6 eV below the Fermi energy ϵ_F. Calculations of d bandwidths within the LSD approximation yield results which are too large by approximately 10 % for Fe and 25 % for Ni. Also the satellite structure remains unexplained within that approximation.

One method of improvement is the application of the GW approximation [15]. For Ni it results in a band narrowing of the right size and also explains the broadening of the quasiparticle peaks. Yet the satellite structure at 6 eV remains unexplained, which is hardly surprising since explaining the satellite peak requires a good treatment of short-ranged correlations, and that is outside the scope of the GW approximation. There have also been perturbation calculations which yield shifts, lifetime effects and also shake-up processes. But in order to explain the latter quantitatively one must determine the scattering matrix. Since the hole number in Ni is small, i.e., 0.6 holes per atom, one may use the Kanamori limit (see Sect. 10.4.4) and improvements of it[1]. Here we want to show how the projection technique may be applied to that problem.

Starting point is the retarded Green's function matrix (8.5). We select the variables $\{A_\mu(\mathbf{k})\}$ and include, as was previously explained,

$$A_\nu^{(0)}(\mathbf{k}) = c_{\nu\uparrow}^+(\mathbf{k}) \tag{11.15}$$

because we want to calculate eventually the Green's function for this operator. Without loss of generality, we have set $\sigma = \uparrow$. But the set of $A_\mu(\mathbf{k})$ should also include those local two-particle–one-hole operators which are obtained when $[H_1(\ell), a_{i\uparrow}^+(\ell)]_-$ is calculated. They are of the form

$$A_{ij}^{(1)}(\ell) = \begin{cases} a_{i\uparrow}^+(\ell)\delta n_{i\downarrow}(\ell) &, \quad i = j \\ a_{i\uparrow}^+(\ell)\delta n_j(\ell) &, \quad i \neq j \end{cases}$$

$$A_{ij}^{(2)}(\ell) = \frac{1}{2}\left(a_{i\uparrow}^+(\ell)s_j^z(\ell) + a_{i\downarrow}^+(\ell)s_j^+(\ell)\right)$$

$$A_{ij}^{(3)}(\ell) = \frac{1}{2}a_{j\downarrow}^+(\ell)a_{j\uparrow}^+(\ell)a_{i\downarrow}(\ell) \quad . \tag{11.16}$$

The density- and spin operators are the same as previously defined (see Sect. 11.1). The selected relevant operators $A_\mu(\mathbf{k})$ are therefore of the form (11.15) and

$$A_{ij}^{(\tau)}(\mathbf{k}) = \frac{1}{\sqrt{N_0}}\sum_\ell A_{ij}^{(\tau)}(\ell)e^{i\mathbf{k}\mathbf{R}_\ell} \quad , \qquad \tau = 1,2,3 \quad . \tag{11.17}$$

For a given value of $\mathbf{k}$ their total number is $1 + 25 + 20 + 20 = 66$. When the 66×66 matrices $\mathbf{L}$ and $\boldsymbol{\chi}$ are computed (see (8.15, 8.10)), the $R = 0$ approximation is made again. Those matrix elements which involve products of more than three of the operators H_1 and $A_{ij}^{(\tau)}(\ell)$ are neglected. For more details we refer to [462].

The single-particle excitation spectrum of paramagnetic Ni which results from the above calculations is shown in Fig. 11.5. The following set of parameters has been used: $U'/W = 0.56$, $J/W = 0.22$, and an anisotropy parameter, $\Delta J/J = 0.14$. They are obtained from fits of experiments which measure the multiplet structure of Ni ions embedded in a simple metal such as Ag [313]. One notices a strong satellite structure around $-1.2W$. With a bandwidth of

---

[1] see [201, 203, 284, 360]

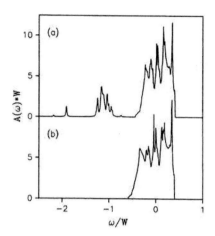

Fig. 11.5. Single-particle excitation spectrum of Ni with a d electron number of $n_d = 9.4$ and the following choice of parameters: $U'/W = 0.56$, $J/W = 0.22$, $\Delta J/W = 0.031$; (a) by applying projection techniques (b) in SCF approximation. (From [462])

$W = 4.3$ eV as obtained by spin averaging the LSDA bandwidths [476], this maximum is approximately 6.8 eV below the top of the d bands and should be compared with an experimental value of 6.3 eV [101, 196]. The structure of the satellite peak reflects the form of the atomic d^8 multiplet. Isotropic exchange splits the latter into three peaks corresponding to a ^1S singlet, degenerate singlets ^1G and ^1D and degenerate triplet states ^3P and ^3F. The energy difference between ^1S and ^1G is $5J$ and the one between ^1G and ^3F is $2J$. The three structures at $-1.9W$, $-1.1W$, and $-0.7W$ show a comparable energy splitting. An exchange anisotropy splits the main peak at $-1.1W$ into smaller peaks.

11.3 Temperature-Dependent Magnetism

As pointed out above, a description of magnetism at finite temperatures requires distinguishing between two different situations. One concerns the magnetism of the transition metals Fe, Co and Ni. Here spin fluctuations involve large amplitudes and seem to be nearly local. How local remains an open question, but in the lowest approximation one may assume a single-site approximation. Quite another situation prevails when nearly ferromagnetic materials like Pd and Ni$_3$Ga or weakly ferromagnetic systems like ZnSn$_2$ and Sc$_3$In are considered. The most important magnetic fluctuations in these materials have small amplitudes and long wavelengths. They require, therefore, a different description than the ones in Fe, Co or Ni. The RPA-like spin-fluctuation theory is an appropriate tool here. We shall start with local spin fluctuations

11.3.1 Local Spin Fluctuations

A very useful tool for describing local spin fluctuations is the functional integral method. It replaces the interactions of an electron with the other electrons by fictitious space- and time-dependent external fields acting on that electron. Stated differently, the fictitious fields have the same effect on an electron as would the interactions with the other electrons. A Gaussian average over these external fields must be taken. In the classical limit, these fields depend only on space and not on time and play a similar role as the fluctuating forces acting on a Brownian particle. In this (static) approximation, only a uniform field - the Stoner field - is left when we take the limit $T = 0$. Therefore, no correlations remain in the zero temperature limit. This shows that ground-state correlations result from the time dependence of the fictitious fields. There are a number of ways to go beyond the static approximation, the most advanced ones working with a dynamical coherent potential. They were discussed in Sect. 9.2 to which we refer here. However, at $T \neq 0$, we find correlations even within the static approximation. They result from the *spatial* fluctuations of the fictitious fields; they are studied in order to explain, among other phenomena, the large entropy changes near the magnetic ordering temperature T_c in Fe and Ni and the Curie-Weiss susceptibility above T_c. We shall start out by introducing first the functional integral method before we apply it to the description of local spin fluctuations.

The functional integral method provides a scheme for calculating approximately the partition function Z of a system of correlated electrons. This problem is reduced to finding appropriate approximations for the spatial and temporal behavior of fluctuating external fields. For a demonstration we consider here the simplest possible example, namely that of an atom with one orbital which is coupled to a heat bath and an electron reservoir. The Hamiltonian of the system is assumed to be

$$H - \mu N_{\text{el}} = (\epsilon_0 - \mu)(n_\uparrow + n_\downarrow) + U n_\uparrow n_\downarrow \quad . \tag{11.18}$$

Here ϵ_0 is the orbital energy and U is the Coulomb repulsion of two electrons in the atomic orbital. Furthermore, n_σ is the number operator for an electron with spin σ in that orbital. The partition function is easily evaluated, i.e.,

$$\begin{aligned} Z &= Tr\left\{e^{-\beta(H-\mu N_{\text{el}})}\right\} \\ &= 1 + 2e^{-\beta(\epsilon_0-\mu)} + e^{-\beta[2(\epsilon_0-\mu)+U]} \quad . \end{aligned} \tag{11.19}$$

The first term is the contribution from the state with the empty orbital, while the second and third terms result from the states in which the orbital is singly and doubly occupied, respectively. In order to express the partition function as that of an electron in a fluctuating external field, the following operator identity is used (Hubbard-Stratonovich transformation[2]):

$$e^{-\beta A^2} = \int_{-\infty}^{+\infty} dx \, e^{-\pi x^2 - 2i(\beta\pi)^{1/2}xA} \quad . \tag{11.20}$$

The identity becomes immediately obvious by completing the square in the exponent.

Equation (11.20) is easily generalized to the product of two commuting operators AB. The electron interaction $Un_\uparrow n_\downarrow$ is of this form, i.e., $A = \sqrt{U}n_\uparrow$, $B = \sqrt{U}n_\downarrow$ and $[A,B]_- = 0$. We start with the identity

$$AB = \frac{1}{4}(A+B)^2 - \frac{1}{4}(B-A)^2 \tag{11.21}$$

and, by applying (11.20), obtain

$$e^{-\beta AB} = \int_{-\infty}^{+\infty} dxdy \, e^{-\pi(x^2+y^2)} \, e^{\sqrt{\pi\beta}(B-A)x - i\sqrt{\pi\beta}(A+B)y} \quad . \tag{11.22}$$

If we introduce $z = x + iy$ and $d^2z = dxdy = (i/2)dzdz^*$, where $*$ denotes the complex conjugate), we can write this expression in the form

$$\begin{aligned} e^{-\beta AB} &= \int_{-\infty}^{+\infty} d^2z \, e^{-\pi|z|^2} e^{-\sqrt{\pi\beta}(Az - Bz^*)} \\ &= \left\langle e^{-\sqrt{\pi\beta}(Az - Bz^*)} \right\rangle_{GA} \quad . \end{aligned} \tag{11.23}$$

Here $\langle \ldots \rangle_{GA}$ stands for taking a Gaussian average over the complex fictitious field z.

For the special model under consideration, the noninteracting part of the Hamiltonian $H_0 = (\epsilon_0 - \mu)n$ with $n = n_\uparrow + n_\downarrow$ commutes with the interaction part $H_1 = Un_\uparrow n_\downarrow$. Thus,

$$Z = Tr\left\{ e^{-\beta(\epsilon_0 - \mu)n} e^{-\beta U n_\uparrow n_\downarrow} \right\} \tag{11.24}$$

with

$$e^{-\beta U n_\uparrow n_\downarrow} = \left\langle e^{-\sqrt{\pi\beta}(n_\uparrow z - n_\downarrow z^*)} \right\rangle_{GA} \quad . \tag{11.25}$$

Often a different composition of $n_\uparrow n_\downarrow$ is of advantage. If we set

$$n_\uparrow n_\downarrow = \frac{1}{4}n^2 - s_z^2 \quad , \tag{11.26}$$

with $s_z = (n_\uparrow - n_\downarrow)/2$ and change variables, it follows from (11.22) that

---

[2] see [193, 438]

$$e^{-\beta U n_\uparrow n_\downarrow} = \frac{\beta U}{4\pi} \int\limits_{-\infty}^{+\infty} d\xi d\eta\, e^{-(\beta U/4)(\xi^2+\eta^2)} e^{\beta U(s_z \xi - i n \eta/2)} \quad . \tag{11.27}$$

When this expression is substituted into (11.24), the partition function becomes

$$Z = Tr \left\{ \frac{\beta U}{4\pi} \int\limits_{-\infty}^{+\infty} d\xi d\eta\, e^{-(\beta U/4)(\xi^2+\eta^2)} e^{-\beta H_{\text{eff}}(\xi,\eta)} \right\} \quad , \tag{11.28}$$

where the effective Hamiltonian $H_{\text{eff}}(\xi, \eta)$ is given by

$$H_{\text{eff}} = (\epsilon_0 + iU\eta/2 - \mu)n - U s_z \xi \quad . \tag{11.29}$$

The partition function is thus reduced to that of a one-particle Hamiltonian in the presence of two fields ξ and η, over which a Gaussian average is taken. The two fields act on the spin and on the density of the electrons, respectively.

For interacting electron systems H_0 and $H_1 = AB$ do usually *not* commute. In spite of this we would like to split off a factor $\exp(-\beta H_0)$ in order to treat the remaining part by appropriate approximations. This can be done by introducing time-dependent operators $A(t)$ according to (7.1), i.e., $A(t) = \exp(iH_0 t) A \exp(-iH_0 t)$. In terms of them we may write

$$e^{-i(H_0+H_1)t} = e^{-iH_0 t} T e^{-\int_0^t d\tau H_1(\tau)} \quad . \tag{11.30}$$

This equation was proven in Sect. 7.1.1.

In a next step we change the product $i \cdot t$ into β so that (11.30) goes over into

$$e^{-\beta(H_0+AB)} = e^{-\beta H_0} T_\tau e^{-\int_0^\beta d\tau A(\tau) B(\tau)} \quad . \tag{11.31}$$

The "time" ordering operator T_τ is defined in analogy to (7.64). The decomposition (11.30) has to be made now for each imaginary "time" τ. For this purpose the interval $[0, \beta]$ is divided into M segments $\Delta \tau = \beta/M$. Finally the limit $M \to \infty$ is taken (compare with Sect. 10.8). With the functional differential

$$D^2 z(\tau) = \lim_{M \to \infty} \prod_{i=1}^M \left(\frac{d^2 z(\tau_i)}{M} \right) \quad , \tag{11.32}$$

we obtain from (11.31)

$$e^{-\beta(H_0+AB)} = e^{-\beta H_0} \int D^2 z(\tau) \exp\left(-\frac{\pi}{\beta} \int_0^\beta d\tau\, |z(\tau)|^2 \right)$$

$$\times T_\tau \exp\left(-\sqrt{\frac{\pi}{\beta}} \int_0^\beta d\tau\, [A(\tau) z(\tau) - B(\tau) z^*(\tau)] \right) \quad . \tag{11.33}$$

When dealing with a lattice of sites with one orbital each, the field $z(\tau)$ contains an additional site index i, and so do the operators $A(\tau)$ and $B(\tau)$. For example, when

$$H_1 = U \sum_i n_{i\uparrow} n_{i\downarrow} , \qquad (11.34)$$

$AB \to \sum_i A_i B_i$ with $A_i = \sqrt{U} n_{i\uparrow}$ and $B_i = \sqrt{U} n_{i\downarrow}$.

As we have already pointed out, the computation of the partition function Z of an interacting electron system becomes equivalent to that of noninteracting electrons moving in complex auxiliary fields $z_i(\tau)$. These fields fluctuate in space indicated by the site label i and time τ, thus connecting the theory of interacting electrons to that of disordered systems. There electrons are moving in randomly distributed external potentials.

In the following discussion, the static approximation will play an important role. It neglects the τ dependence of $z_i(\tau)$. Stated differently, when we use the Fourier decomposition

$$z_i(\tau) = \sum_{\nu=-\infty}^{+\infty} z_i(\nu) e^{-i\omega_\nu \tau}, \qquad \omega_\nu = \frac{2\pi\nu}{\beta} , \qquad (11.35)$$

only the term $z_i(\nu = 0) = z_i$ is kept. The partition function Z becomes

$$Z_{\text{stat}} = \int d^2 z \, e^{-\beta \tilde{\Omega}(z)} ,$$

$$\beta \tilde{\Omega}(z) = \pi |z|^2 - \ln Y(z) \qquad (11.36)$$

with

$$Y(z) = Tr \left\{ e^{-\beta(H_0 - \mu N_{el})} e^{-\sqrt{\pi \beta}(Az - Bz^*)} \right\} . \qquad (11.37)$$

Neglecting the τ dependence of $z(\tau)$ is equivalent to assuming that H_0 and H_1 commute. This is the case if we treat the system in the classical high-temperature limit instead of in the quantum mechanical one.

Even within the static approximation, the evaluation of the partition function usually proves impossible without further approximations. Here we shall discuss the saddle-point approximation, i.e., the method of steepest descent as it is often called. It operates on the principle that for large values of β, i.e., low temperatures, the largest contribution to the integral (11.36) comes from the region near the minimum of $\tilde{\Omega}(z)$. The stationary conditions are thereby

$$\frac{\partial \tilde{\Omega}(z)}{\partial z} = 0 \quad \text{and} \quad \frac{\partial \tilde{\Omega}(z)}{\partial z^*} = 0 .$$

We should consider first that in the case of only one real variable x an expansion around the stationary point x_s yields

$$\exp\left[-\beta \tilde{\Omega}(x)\right] = \exp\left[-\beta \tilde{\Omega}(x_s)\right] \exp\left[\left(-\frac{\beta}{2} \frac{\partial^2 \tilde{\Omega}}{\partial x^2}\right)\bigg|_{x=x_s} (x - x_s)^2\right] \qquad (11.38)$$

and, after performing the integral in (11.36), the partition function goes over into that of the saddle-point approximation

$$Z_{sp} = \sqrt{\frac{2\pi}{\beta\left(\partial^2\tilde{\Omega}/\partial x^2\right)_{x=x_s}}} e^{-\beta\tilde{\Omega}(x_s)} \quad . \tag{11.39}$$

When $\tilde{\Omega}$ depends on two variables $x_1 = x, x_2 = y$ (or z and z^*), the argument under the square root is replaced by $(2\pi)^2/\left[\beta \det\left(\partial^2\tilde{\Omega}/\partial x_\nu \partial x_\mu\right)\right]$. In the zero-temperature limit, i.e., $\beta \to \infty$, the partition function reduces to that of a mean-field theory. Indeed, in that limit only the term $\pi|z|^2$ of $\beta\tilde{\Omega}(z)$ (11.36) contributes to the second derivative and the square root reduces to 1. The electron then moves in a constant field z_s. Thus

$$\lim_{\beta\to\infty} Z_{sp} = Z_{\text{MF}} e^{-\beta\tilde{\Omega}(x_s)} \quad . \tag{11.40}$$

Now we shall apply the saddle-point approximation to the example of a single orbital coupled to an electron reservoir (11.18). We return to (11.24, 11.25) and write the partition function as

$$Z = Tr\left\{\left\langle e^{-\beta(E_\uparrow n_\uparrow + E_\downarrow n_\downarrow)}\right\rangle_{\text{GA}}\right\} \tag{11.41}$$

with

$$\begin{aligned}E_\uparrow &= \epsilon_0 - \mu + \sqrt{\pi U/\beta}\ z \quad , \\ E_\downarrow &= \epsilon_0 - \mu - \sqrt{\pi U/\beta}\ z^* \quad .\end{aligned} \tag{11.42}$$

From (11.36) it follows that

$$\beta\tilde{\Omega}(z) = \pi|z|^2 - \ln\left[\left(1 + e^{-\beta E_\uparrow}\right)\left(1 + e^{-\beta E_\downarrow}\right)\right] \quad . \tag{11.43}$$

The stationary point is

$$\begin{aligned}z_s &= \sqrt{\frac{\beta U}{\pi}}\frac{1}{1 + e^{\beta E_\downarrow}} \quad , \\ z_s^* &= -\sqrt{\frac{\beta U}{\pi}}\frac{1}{1 + e^{\beta E_\uparrow}} \quad .\end{aligned} \tag{11.44}$$

If these values are substituted into (11.42) we obtain

$$\begin{aligned}E_\uparrow &= \epsilon_0 - \mu + U\left\langle n_\downarrow\right\rangle_{\text{MF}} \quad , \\ E_\downarrow &= \epsilon_0 - \mu + U\left\langle n_\uparrow\right\rangle_{\text{MF}} \quad .\end{aligned} \tag{11.45}$$

The averages are

$$\left\langle n_\sigma\right\rangle_{\text{MF}} = \frac{1}{1 + e^{\beta E_\sigma}} \quad . \tag{11.46}$$

11 Transition Metals

This demonstrates the mean-field character of the saddle-point approximation.

The above expansion started from a form of $\tilde{\Omega}$ in the static approximation. If we also include in the expansion the time variable τ and the fluctuations around the static path, we can show that the saddle-point approximation corresponds to the random-phase approximation.

After having explained the main features of the functional integral method we return to our primary problem, i.e., a proper description of the magnetism of Fe, Co and Ni. As pointed out before, when the magnetization of these itinerant d-electron systems is plotted as a function of temperature it resembles that of localized electrons. Similarly, the observed changes in the specific heat of Fe at the magnetic ordering temperature T_c correspond to changes in the entropy of order $k_B \ln 3$, as for localized electrons. In a band picture, this value is much smaller because only electrons near the Fermi energy are involved in the magnetic ordering. Other experimental findings are in agreement with this. For example, the susceptibility above the magnetic ordering temperature $\chi(T)$ shows Curie-Weiss-like behavior, i.e., $\chi(T) \propto (T-T_c)^{-1}$. If the independent-electron theory applied, the susceptibility would be Pauli-like, i.e., independent of T as long as $T \ll T_F$, where T_F is the Fermi temperature. Also the observed Curie temperature T_c itself is much smaller than the one obtained from a band theory. For example, in a band calculation based on the LSDA, a finite temperature enters only through the Fermi distribution function which replaces the step function when the occupied single-electron states are determined [compare (4.22)]. The transition from a magnetic to a nonmagnetic state is described in band theory as shown in Fig. 11.6a.

In order to compute the temperature-dependent magnetization in the simplest way we use a one-band Hubbard Hamiltonian (8.22) instead of the multi-band one (11.1). It is a highly simplified model Hamiltonian for a transition metal which, for example, does not allow for intra-atomic correlations like Hund's-rule coupling. On the other hand, it has the virtue of relative simplicity, sufficient to attempt generalizations to a five-band Hamiltonian. In order to apply the Hubbard-Stratonovitch transformation, the interaction part must be written in a quadratic form, three examples of which are listed below:

$$U \sum_i n_{i\uparrow} n_{i\downarrow} = \frac{U}{4} \sum_i n_i^2 - U \sum_i (s_i^z)^2 , \quad (11.47a)$$

$$= \frac{U}{4} \sum_i n_i^2 - \frac{U}{3} \sum_i \mathbf{s}_i^2 , \quad (11.47b)$$

$$= \frac{U}{4} \sum_i n_i^2 - U \sum_i (\mathbf{e}_i \cdot \mathbf{s}_i)^2 . \quad (11.47c)$$

Here $n_i = n_{i\uparrow} + n_{i\downarrow}$ and $s_i^z = \frac{1}{2}(n_{i\uparrow} - n_{i\downarrow}) = \sigma_i^z/2$. Furthermore, $\mathbf{e}_i$ is an arbitrary unit vector at site i. The last two forms have the advantage that they are rotationally invariant. If no approximations were made in evaluating the functional integrals, it would not matter which of the three forms (11.47)

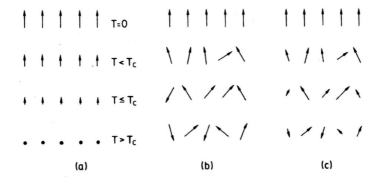

Fig. 11.6. Effect of finite temperatures on ferromagnetic order. (a) Conventional band theory: the difference in population of Bloch states with spin up and down decreases continuously with increasing temperature. Above T_c, the two populations are equal to each other and the magnetization vanishes at each site. (b) Localized description: the spins at different sites fluctuate more and more with increasing T. At $T \geq T_c$ the net magnetization is zero. (c) Correlated delocalized electrons: the magnetic moment at different sites fluctuate in magnitude and direction due to finite temperatures.

is chosen. In practice, rather drastic approximations become necessary when performing these calculations. In particular, the static approximation is often made. In that case, the final result depends rather sensitively on the particular quadratic form chosen. This shows one of the intrinsic difficulties of the functional integral approach. We will be using here the form (11.47a), known to reduce to the Hartree-Fock result when a saddle-point approximation is made.

For describing the magnetic properties the partition function Z must be known. It is expressed in terms of the thermodynamic potential Ω as $Z = \exp(-\beta\Omega)$. We calculate Ω by applying the functional integral technique and make the static approximation. According to (11.36, 11.37)

$$Z = \int d^2z \, e^{-\beta\tilde{\Omega}(z)} \quad , \tag{11.48}$$

with

$$\tilde{\Omega}(z) = \frac{U}{4}\sum_i |z_i|^2 - \frac{1}{\beta}\ln Tr\left\{e^{-\beta[\tilde{H}_0(z)-\mu N_{\text{el}}]}\right\} \tag{11.49}$$

and

$$\tilde{H}_0(z) = H_0 - \frac{U}{2}\sum_i (n_{i\uparrow}z_i - n_{i\downarrow}z_i^*) \quad . \tag{11.50}$$

The complex space-dependent external field z_i (two-field case) acts on the electron density n_i and on the magnetization s_i^z. This is seen if we introduce

$$x_i = \frac{1}{2}(z_i + z_i^*) \quad , \quad iy_i = \frac{1}{2}(z_i - z_i^*) \tag{11.51}$$

and rewrite

$$n_{j\uparrow}z_j - n_{j\downarrow}z_j^* = 2s_j^z x_j + in_j y_j \quad . \tag{11.52}$$

When set into (11.50) we obtain

$$\tilde{H}_0(x,y) = H_0 - \frac{U}{2}\sum_{j\sigma}(\sigma x_j + iy_j)n_{j\sigma} \quad . \tag{11.53}$$

The prefactor $\sigma = \pm 1$ is depending on the spin direction. One notices that $\tilde{H}_0(x,y)$ has the form of a one-particle Hamiltonian in the presence of disorder. The latter is caused by the fields x_i and y_i, which lead to different energies at different sites. This finding proves important because it links the problem of correlations at finite temperatures to another important branch of solid-state theory, namely disordered systems [75]. Originally enunciated by *Hubbard*, this link has furthered progress in both fields.

It is usually assumed that the disorder caused by the fields y_i is less important than that caused by the field x_i. In view of the discussion presented in Sect. 11.1, where we showed that density correlations play an important role, this distinction does not seem justified. However, for the sake of simplicity, and because we are particularly interested in understanding magnetic quantities like the magnetization or susceptibility at finite T, we shall also eliminate the fields y_j by simply replacing them by $y_j = \bar{n}$, i.e., the average site occupancy. In doing so the two-field problem is reduced to a one-field problem. Because of the elimination of the y_i, the Hamiltonian $\tilde{H}_0$ is reduced to a one-particle Hamiltonian with a random field x_i acting at sites i on the spins of the electrons. Equation 11.48 reduces to

$$Z_{\text{st}} = \int \prod_i \left(\frac{\beta U}{4\pi}\right)^{1/2} dx_i e^{-\beta \tilde{\Omega}(x,T)} \quad . \tag{11.54}$$

The potential $\tilde{\Omega}(x,T)$ is obtained from $\tilde{\Omega}(x,y,T)$ given by (11.49) by replacing y_i^2 by $\bar{n}^2$ and $\tilde{H}_0(x,y)$ by $\tilde{H}_0(x)$ with

$$\tilde{H}_0(x) = H_0 + U\bar{n}\sum_i n_i - \frac{U}{2}\sum_{i\sigma}\sigma x_i n_{i\sigma} \quad . \tag{11.55}$$

One notices that the field y_i has been treated in a SCF or Hartree-Fock approximation. If we also treat the field x_i with a saddle-point approximation, we find

$$x_i = \langle n_{i\uparrow} - n_{i\downarrow}\rangle \tag{11.56}$$

and the Hamiltonian $H_0(x)$ goes over into the SCF Hamiltonian

$$H_{\text{SCF}} = \sum_{i\neq j} t_{ij} a_{i\sigma}^+ a_{j\sigma} + U\sum_{i\sigma} n_{i\sigma} + \frac{U}{2}\sum_i \langle n_{i\uparrow} - n_{i\downarrow}\rangle_{\text{SCF}}(n_{i\uparrow} - n_{i\downarrow}) \quad . \tag{11.57}$$

11.3 Temperature-Dependent Magnetism

The expectation value is over a thermodynamic ensemble with respect to H_{SCF}. The Hamiltonian (11.57) may be used in order to calculate a phase diagram for ferromagnetic or antiferromagnetic phases within the independent-electron approximation.

Returning to the alloy (or disorder) problem as defined by (11.55), it is interesting to rewrite $\tilde{\Omega}(x,T)$ in the form of increments

$$\tilde{\Omega}(x) = \sum_i \tilde{\Omega}_1(x_i) + \sum_{i \neq j} \tilde{\Omega}_2(x_i, x_j) + \ldots \quad . \tag{11.58}$$

The first term corresponds to a *single-site approximation*. When one limits oneself to it, the expression (11.54) for Z_{st} factorizes into N_0 independent integrals. Neglecting correlations between fields at different sites is justified when the important fluctuations in the electronic system are predominantly *local*. A different point of view would be to assume that fields at neighbouring sites are strongly correlated and change only slightly from site to site. This assumption would put emphasis on the long-wavelength fluctuations and suggest using only the terms $\tilde{\Omega}_2(x_i, x_j)$ in (11.58).

We know from the theory of random alloys that the best single-site approximation is the CPA (see Chapter 9). Therefore, we proceed according to it. The Hamiltonian $\tilde{H}_0(x)$ given by (11.55) describes a system of noninteracting electrons with site diagonal disorder. In order to determine the coherent potential, we have to know the concentration $c(x_i)$ of sites with a given value of the external field x_i. It is reasonable to assume that this concentration is given by

$$c(x_i) = \frac{e^{-\beta \tilde{\Omega}_1(x_i)}}{\int dx_i e^{-\beta \tilde{\Omega}_1(x_i)}} \quad . \tag{11.59}$$

This ensures that the concentration matches the weight with which each value x_i enters the partition function Z_{st}, see (11.54). The effective medium is characterized by the retarded one-particle Green's function

$$\tilde{G}_{ij}^\sigma(\omega) = \frac{1}{N_0} \sum_{\mathbf{k}} \frac{e^{-i\mathbf{k} \cdot (\mathbf{R}_i - \mathbf{R}_j)}}{\omega - \epsilon_{\mathbf{k}} - \langle E_\sigma \rangle - \Sigma_\sigma(\omega) + \mu} \quad , \tag{11.60}$$

where $\mathbf{R}_i, \mathbf{R}_j$ are site positions. The kinetic energy is simply

$$\epsilon_{\mathbf{k}} = \frac{1}{N_0} \sum_{i \neq j} t_{ij} e^{i\mathbf{k} \cdot (\mathbf{R}_i - \mathbf{R}_j)} \tag{11.61}$$

and the average energy $\langle E_\sigma \rangle$ is defined by

$$\langle E_\sigma \rangle = \int dx_i c(x_i) E_{i\sigma} \tag{11.62}$$

with

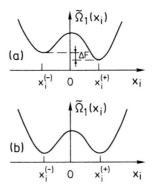

Fig. 11.7. Functional form of $\tilde{\Omega}_1(x_i, T)$ in the local moment case. (a) For $T < T_0$ the two minima have different depth. (b) Form of the function $\tilde{\Omega}_1$ for $T > T_0$.

$$E_{i\sigma} = \frac{U}{2}(\bar{n} - \sigma x_i) \quad . \tag{11.63}$$

The frequency-dependent self-energy $\Sigma_\sigma(\omega)$ is determined by requiring that the site average of the single-site t matrix vanishes, i.e., $\langle \tilde{t}_i^\sigma \rangle = 0$ (see (9.12)). This leads to the CPA equation for $\Sigma_\sigma(\omega)$ from which this quantity can be determined. Within the effective medium, a given site i of the system acts like an impurity. Its scattering potential is

$$v_{i\sigma}(\omega) = E_{i\sigma} - \langle E_\sigma \rangle - \Sigma_\sigma(\omega) \quad . \tag{11.64}$$

The corresponding scattering Hamiltonian is here complex and of the form

$$H_{\text{scatt}}(i) = \sum_\sigma v_{i\sigma}(\omega) n_{i\sigma} \quad . \tag{11.65}$$

After the self-energy $\Sigma_\sigma(\omega)$ has been determined one may proceed by computing the thermodynamic potential $\tilde{\Omega}_1(x_i)$. It follows from (11.49) when the scattering Hamiltonian $H_{\text{scatt}}(i)$ is included. It is beyond the scope of this introductory book to elaborate further on the computation of $\tilde{\Omega}_1(x_i)$. Instead we refer to the literature, (see, e.g., Ref. [225]). Here we want to discuss the different forms it may have.

Consider a ferromagnetic system in which case $\tilde{\Omega}_1(x_i)$ may have the form shown in Fig. 11.7. At $T = 0$ the function has two minima of different depth. Which one is lower depends on the sign of the magnetization M_0. The site i, when considered as an impurity embedded in a ferromagnetic effective medium, lines up ferromagnetically with its surroundings. For temperatures larger than a characteristic temperature T_0, the two minima are symmetric with respect to $x_i = 0$. Obviously, T_0 is the Curie temperature. This situation is called the local-moment case. For better physical insight, we apply the two-saddle-point approximation. The two saddle points $x_i^{(+)}$ and $x_i^{(-)}$ are obtained from $\partial \tilde{\Omega}_1(x_i)/\partial x_i = 0$. Here the average external field

is of the form
$$\langle x \rangle = \int dx_i c(x_i) x_i \tag{11.66}$$

$$\langle x \rangle = \frac{x^{(+)} + x^{(-)} e^{-\beta \Delta F}}{1 + e^{-\beta \Delta F}} \quad . \tag{11.67}$$

The subscript i is omitted from now on, since there is nothing special about site i. ΔF denotes the difference between the two minima, i.e., $\Delta F = \tilde{\Omega}_1(x^{(-)}) - \tilde{\Omega}_1(x^{(+)})$. It can be shown and, in fact, is plausible that for large values of U the positions of the minima are related by $x^{(+)} = -x^{(-)} = x^*$. From (11.67) we find $\langle x \rangle$ to be

$$\langle x \rangle = x^* \tanh(\beta \Delta F / 2) \quad . \tag{11.68}$$

Note that ΔF is a function of $\langle x \rangle$. The last relation resembles the mean-field approximation of a Heisenberg ferromagnet, provided one replaces ΔF by

$$\Delta F = \nu \langle x \rangle J_{\text{ex}} \quad , \tag{11.69}$$

where J_{ex} is the exchange interaction of a localized spin with its ν nearest neighbors. Therefore, the single-site approximation provides a link between the itinerant Hubbard Hamiltonian and localized spins as they appear in a Heisenberg Hamiltonian. Clearly, for $T > T_0$ the susceptibility shows Curie-Weiss behavior.

It is worth noticing that the ordinary Stoner theory of ferromagnetism discussed subsequently is obtained when we evaluate the integrals at *one* saddle point $x_i^{(+)}$ only. Moreover, the present theory will fail when the temperature T is of order $k_{\text{B}} T \gtrsim \Delta F$.

Of particular interest is also the entropy obtained for the local-moment case shown in Fig. 11.7 since it reflects the localized-spin picture. The entropy S of a system of noninteracting electrons moving in a random alloy consists of two parts

$$S = S_1 + S_2 \quad . \tag{11.70}$$

The first part S_1 is the entropy of independent electrons. It is (per lattice site)

$$S_1 = -k_{\text{B}} \sum_\sigma \int_{-\infty}^{-\infty} d\omega N_\sigma(\omega) \{f(\omega) \ln f(\omega) + [1 - f(\omega)] \ln [1 - f(\omega)]\} \quad , \tag{11.71}$$

where $N_\sigma(\omega)$ is the spin-dependent density of states. The second part S_2 is a configurational entropy determined by the number of different ways in which one can distribute different sites in external fields x_i when their respective probabilities are $c(x_i)$. This part is given by

$$S_2 = -k_{\text{B}} \langle \ln c(x_i) \rangle \quad , \tag{11.72}$$

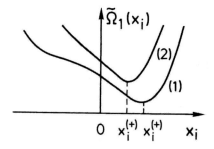

Fig. 11.8. Functional form of $\tilde{\Omega}_1(x_i, T)$ for $T = 0$ when only one minimum is present. The case (1) with a pronounced shoulder seems to apply to Fe, while Ni corresponds more to case (2). When $\tilde{\Omega}_1$ has one deep minimum only, local moments are absent.

and the average is defined as in (11.66). When the two-saddle-points approximation is made, S_2 reduces to the simple form

$$S_2 = -k_B \left[c^{(+)} \ln c^{(+)} + c^{(-)} \ln c^{(-)} \right] , \qquad (11.73)$$

where $c^{(\pm)} = c(x^{(\pm)})$. When the system is nonmagnetic, we have $c^{(+)} = c^{(-)} = 1/2$ and $S_2 = k_B \ln 2$, which is the entropy per site of a spin 1/2 system. We notice that the alloy analogy again provides for a description of the local features of correlated itinerant electrons.

Before continuing we want to point out that it may happen that the relative heights of the two minima in Fig. 11.7a are inverted. In that case, a ferromagnetic state is not stable. Since the spin of the impurity i is aligned antiferromagnetically to the medium, we expect an antiferromagnetic ground state here. A second type of behavior of $\tilde{\Omega}_1(x)$ is shown in Fig. 11.8, in which only one minimum exists at $T = 0$ with or without an additional shoulder at negative x. If there is only one deep minimum, only fluctuations around it will be of importance. This case - known as the case of no local moment - applies primarily to metals with weak ferromagnetism and also to systems in which the interactions are relatively weak. One can show that here $\langle x_i^2 \rangle \propto T$ at low temperatures. We obtain again a Curie-Weiss behavior of the magnetic susceptibility, but as discussed below for a different physical reason than in the local-moment case. As pointed out before, weak ferromagnetism requires a proper treatment of long-wavelength fluctuations. Although a single-site approximation is inappropriate in that case, we have mentioned weak ferromagnetism here in order to point out the different forms which $\tilde{\Omega}_1(x_i)$ can take.

The above theory has been applied to Fe and Ni [173,174]. We begin with ferromagnetic iron, using for its description a single band. The d-electron number is $n_d = 7.2$ or $n = 1.44$ per orbital. We choose for the d bandwidth a value of $W = 6$ eV. The model density of states for the bcc structure is shown

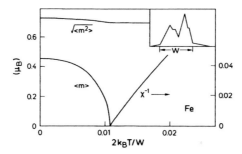

Fig. 11.9. Magnetization $\langle m \rangle$, inverse susceptibility χ^{-1} and amplitude of the local moment $\sqrt{\langle m^2 \rangle}$ for Fe as a function of temperature T in units of the bandwidth W. The inset shows the model density of states. (In analogy to [173])

in the inset of Fig. 11.9. The parameter U is chosen so as to obtain a zero-temperature magnetization per orbital of $\langle m(T=0)\rangle = 0.44\mu_B$ ($= 2.2:5$). The value should *not* be compared with the one used in Figs. 11.3, 11.4 because we use here a one-band model. After calculating $\tilde{\Omega}_1(x_i)$, we find that at $T=0$ the function has one minimum and a pronounced shoulder as indicated in Fig. 11.8. The results depend greatly on the choice of parameters and it is therefore also possible that $\tilde{\Omega}_1(x_i; T=0)$ has two minima instead. In any case, Fe must be considered as belonging to the local-moment case. Figure 11.9 shows the temperature dependence of $\langle m(T)\rangle$, which is close to a Brillouin function. The inverse susceptibility is almost linear in T, indicating a Curie-Weiss behavior.

Fcc Ni has $n = 1.8$ ($= 9.0:5$) d electrons per orbital. The reproduction of a moment $\langle m(T=0)\rangle = 0.12\mu_B$ ($= 0.62:5$) requires a value of $U = 6.7$ eV when a bandwidth of $W = 4.8$ eV is chosen. When $\tilde{\Omega}_1(x_i)$ is calculated, we find that it has one minimum and a light shoulder only. Nevertheless, we cannot speak of being in the regime of no moment since $\langle m^2(T)\rangle^{1/2}$ remains practically unchanged when the Curie temperature is crossed. The Curie temperatures estimated within the single-site approximation are $T_c \simeq 2000$ K for Fe and $T_c \simeq 700$ K for Ni. They are much smaller than the values which would follow from a Stoner theory, but still somewhat larger than the experimental values of $T_c = 1044$ K and 630 K, respectively.

The static approximation can be improved by employing a dynamical CPA approximation, like DCPA or DMFT (see Sect. 9.2). Without going into more detail it is instructive to look at the outcome of this generalization. We present results for Ni here [224] because a comparison can be made with those obtained in the preceding section. There the projection technique was applied. In Fig. 11.10 we show the density of states of Ni in the DCPA approximation. It is obtained from the Green's function of the effective medium when using (9.35). A model density of states for noninteracting electrons on a Ni fcc lattice is assumed and parameter values $U/W = 1.7$, $k_B T/W = 0.018$ and $W \simeq 4.76$ eV are chosen. Furthermore, $n = 1.8$ which corresponds to nine d-

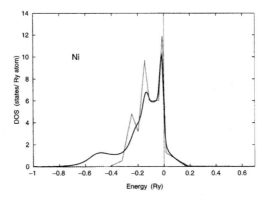

Fig. 11.10. Calculated density of states in the dynamical CPA (solid curve) for paramagnetic Ni. The density of states of noninteracting electrons is shown by the dotted curve. (From [224])

electrons. One notices that the DCPA leads to a narrowing of the free-electron density of states and a satellite structure. This agrees with the findings in Sect. 11.2 which are based on the projection method and shown in Fig. 11.5. The ratio U/W differs considerably in the two cases because in one case it is a single-band model which is treated while in the other it is a multiband model Hamiltonian. It is interesting that the density of states obtained in the static CPA is *broader* than the one of noninteracting electrons, a consequence of the strong spin fluctuations. The latter are considerably suppressed in the DCPA which contrary to the static CPA reduces to the correlated ground state in the zero temperature limit. Finally we want to mention that the DCPA reduces the Curie temperature by roughly a factor of $1/2$ as compared with the static CPA.

11.3.2 Long-Wavelength Spin Fluctuations

We continue by discussing spin fluctuations in nearly ferromagnet systems like Pd and Ni_3Ga and weakly ferromagnetic compounds like $ZrSn_2$, Sc_3In, Ni_3Al or $Ni_{0.43}Pt_{0.57}$. In the former case the magnetic susceptibility χ_S is strongly enhanced due to a large enhancement factor $S = (1 + F_0^a)^{-1}$ named after *Stoner*. It is due to the interactions of quasiparticles as discussed previously in Sect. 7.2 (compared with (7.98)). In Pd as well as Ni_3Ga the Landau parameter $F_0^a \gtrsim -1$ which implies closeness to a divergence in χ_S and hence to a ferromagnetic instability. This has as a consequence that overdamped magnetic excitations (paramagnons) exist. Weak ferromagnets have $F_0^a \lesssim -1$ and are identified by a low Curie temperature T_c, e.g., $T_c = 28$ K for $ZrZn_2$ and $T_c = 6$ K for Sc_3In.

In order to study magnetic fluctuations in almost or weakly ferromagnetic systems we assume that the electrons interact through a hard-core interaction.

11.3 Temperature-Dependent Magnetism

The model Hamiltonian is therefore of the form

$$H = \sum_{\mathbf{k}\sigma} \epsilon_{\mathbf{k}} c^+_{\mathbf{k}\sigma} c_{\mathbf{k}\sigma} + H_{\text{int}} \quad ,$$

$$H_{\text{int}} = \frac{U}{\Omega} \sum_{\mathbf{k p q}} c^+_{\mathbf{k}\uparrow} c_{\mathbf{k}+\mathbf{q}\uparrow} c^+_{\mathbf{p}\downarrow} c_{\mathbf{p}-\mathbf{q}\downarrow} \quad . \tag{11.74}$$

Alternatively, one may assume a system of lattice sites with an on-site Hubbard interaction

$$H_{\text{int}} = U \sum_i n_{i\uparrow} n_{i\downarrow} \quad . \tag{11.75}$$

Depending on convenience, either form of the interaction will be used in the following analysis.

The above repulsive short-range interaction results in an enhancement of the magnetic susceptibility and eventually in ferromagnetic order, depending on the size of U. In order to demonstrate this, we decompose H_{int} according to (11.47a) into a density-dependent and a spin-dependent part by writing

$$n_{i\uparrow} n_{i\downarrow} = \frac{1}{4} n_i^2 - (s_i^z)^2 \tag{11.76}$$

and consider just the spin-dependent part, i.e.,

$$\tilde{H}_{\text{int}} = -U \sum_i (s_i^z)^2 \quad . \tag{11.77}$$

We remark that we could have also chosen the decomposition (11.47b) which is rotationally invariant in spin space; for the present purpose however (11.77) is sufficient. With this interaction we want to calculate the susceptibility, but first a few definitions and general relations need to be listed.

We relate the magnetization, i.e., the thermodynamic expectation value of the magnetization operator $\mathbf{M}(\mathbf{r})$, to an external magnetic field $\mathbf{h}(\mathbf{r}, t)$ through

$$\langle \mathbf{M}(\mathbf{r}, t) \rangle = \int d^3 r' dt' \chi(\mathbf{r} - \mathbf{r}', t - t') \mathbf{h}(\mathbf{r}', t') \quad . \tag{11.78}$$

This defines the magnetic susceptibility tensor $\chi(\mathbf{r}, t)$. Causality requires that $\chi(\mathbf{r} - \mathbf{r}', t - t') = 0$ for $t < t'$, which implies that the Fourier transform is

$$\chi(\mathbf{q}, \omega) = \int d^3 r \int_{-\infty}^{+\infty} dt \chi(\mathbf{r}, t) e^{-i(\mathbf{q}\cdot\mathbf{r} - \omega t)}$$

$$= \int_0^{\infty} dt \chi(\mathbf{q}, t) e^{i(\omega + i\eta)t} \quad . \tag{11.79}$$

In order to ensure convergency, an infinitesimal imaginary part has been added to the frequency. Another consequence of causality is that the real and imaginary parts of $\chi(\mathbf{q},\omega) = \text{Re}\{\chi(\mathbf{q},\omega)\} + i\,\text{Im}\{\chi(\mathbf{q},\omega)\}$ are connected with each other through Kramers-Kronig relations. We have

$$\text{Re}\{\chi(\mathbf{q},\omega)\} = \frac{1}{\pi}\mathscr{P}\int_{-\infty}^{+\infty} d\omega' \frac{\text{Im}\{\chi(\mathbf{q},\omega')\}}{\omega' - \omega} \quad , \tag{11.80}$$

where $\mathscr{P}$ implies the principal value of the integral. Another general relation to be used later is given by the fluctuation-dissipation theorem[3]. It relates the fluctuations of the system described by a correlation function to the dissipations described by the imaginary part of a susceptibility.

We continue with the discussion of the above Hamiltonian by calculating the static spin susceptibility. In mean-field approximation the interaction (11.77) contributes a molecular field

$$h_{\text{mf}} = -\frac{U}{\mu_B}\langle s_i^z \rangle \tag{11.81}$$

to the effective field h_{eff} acting on an electron spin. The spin susceptibility in the presence of $\tilde{H}_{\text{int}}$ is therefore related to the one in the absence of the interactions, χ_0, through

$$\Omega^{-1}\langle M \rangle = \chi h = \chi_0\,(h + h_{\text{mf}}) \quad , \tag{11.82}$$

where Ω is the volume of the probe. The external field $\mathbf{h(r)}$ is assumed to be constant and for an isotropic system the tensor χ is proportional to the unit matrix. With $\Omega^{-1}\langle M \rangle = 2\mu_B\langle s_z \rangle$ this results in

$$\langle M \rangle = \frac{\chi_0 \Omega h}{1 - U\chi_0/2\mu_B^2} \quad . \tag{11.83}$$

The susceptibility of a system of free electrons is $\chi_0 = 2\mu_B^2 N(0)$, where $N(0)$ is the density of states (per spin) at the Fermi energy and therefore

$$\chi = \frac{2\mu_B^2 N(0)}{1 - N(0)U} \quad . \tag{11.84}$$

Thus within this mean-field approximation, the Landau parameter F_0^a is

$$F_0^a = -N(0)U \quad . \tag{11.85}$$

Almost ferromagnetic materials like Ni_3Ga or Pd are characterized by a positive value

$$[1 - N(0)U] \ll 1 \quad , \tag{11.86}$$

---

[3] see [51, 254]

i.e., U is close to the critical interaction strength $U_c = 1/N(0)$ at which a ferromagnetic instability occurs. The above mean-field theory goes back to *Stoner* and *Slater*, who applied it to the magnetic phase of transition metals and their alloys. Despite great successes in explaining a number of important properties (for example, the noninteger Bohr magneton number of the spontaneous magnetization, low-temperature specific heat, large cohesive energy), the mean-field theory also has severe shortcomings. They were discussed extensively before. Therefore, we have to go beyond the mean-field theory by including spin fluctuations. Hereby one has to realize that a treatment of fluctuations around a mean-field equilibrium configuration is not sufficient since the equilibrium state itself changes when fluctuation are considered. It is no longer that of the mean-field theory. When this change is accounted for, we arrive at the self-consistent renormalization theory[4].

We describe now the changes in the equilibrium state due to fluctuations. We use a simple, yet sufficiently accurate approach, based on the concept of *Onsager*'s reaction field [353]. The reaction field allows a step to be taken beyond the mean-field approximation. It describes the changes in the molecular field acting on a spin when the latter takes a different direction. These changes occur because a spin contributes to its own molecular field. According to *Onsager*, this contribution, the reaction field, must be subtracted when determining the effective field orienting the spin.

For a discussion of the reaction field, consider the general spin Hamiltonian

$$H = \sum_{ij} \lambda_{ij} \mathbf{S}_i \cdot \mathbf{S}_j \quad . \tag{11.87}$$

The molecular field at site i is given by

$$\mathbf{h}_{\mathrm{mf}}(i) = \frac{2}{g\mu_B} \sum_j \lambda_{ij} \langle \mathbf{S}_j \rangle \quad , \tag{11.88}$$

where g is the Landé factor. The reaction field is then given by

$$\mathbf{h}_{\mathbf{r}}(i) = \frac{2}{g\mu_B} \langle \mathbf{S}_i \rangle \sum_j \lambda_{ij} \langle \mathbf{S}_i \cdot \mathbf{S}_j \rangle \quad , \tag{11.89}$$

i.e., the alignment of a spin at site i influences that of a spin at site j through the spin-spin correlation function. The effective molecular field $\mathbf{h}_{\mathrm{eff}}(i)$ which must be added to the external field $\mathbf{h}(i)$ when calculating the spin alignment is then given by

$$\begin{aligned}\mathbf{h}_{\mathrm{eff}}(i) &= \mathbf{h}_{\mathrm{mf}}(i) - \mathbf{h}_{\mathbf{r}}(i) \\ &= \frac{2}{g\mu_B} \sum_j \left(\langle \mathbf{S}_j \rangle - \langle \mathbf{S}_i \rangle \langle \mathbf{S}_i \cdot \mathbf{S}_j \rangle \right) \quad . \end{aligned} \tag{11.90}$$

---

[4] see [331] and also [180]

Using the fluctuation-dissipation theorem we can relate the correlation function $\langle \mathbf{S}_i \cdot \mathbf{S}_j \rangle$ in the paramagnetic phase to the imaginary part of the space- and frequency-dependent susceptibility $\chi_{ij}(\omega)$ by

$$g^2 \mu_B^2 \langle \mathbf{S}_i \cdot \mathbf{S}_j \rangle = \frac{1}{\pi} \int d\omega \frac{1}{1 - e^{-\beta\omega}} \text{Im}\{\chi_{ij}(\omega + i\eta)\} \quad . \tag{11.91}$$

From the last two equations and (11.84), we can derive the static susceptibility. We apply those relations to the Hamiltonian $\tilde{H}_{\text{int}}$ of (11.77) in which case $\lambda_{ij} = U\delta_{ij}$ and the effective molecular field becomes

$$H_{\text{eff}}(i) = \frac{U}{\mu_B} \langle s_i^z \rangle \left[1 - \langle (s_i^z)^2 \rangle\right] \quad . \tag{11.92}$$

Note that the spin is treated here classically. In analogy to (11.84), we obtain

$$\chi(T) = \frac{2\mu_B^2 N(0)}{1 - N(0)U\left[1 - \langle (s_i^z)^2 \rangle\right]} \quad , \tag{11.93}$$

a substantial improvement on the previously presented mean-field theory. For example, the temperature dependence of $\chi(T)$ now occurs on a temperature scale much lower than the Fermi temperature T_F. We can see this change if we use (11.91) with $g = 2$, which in the present case reads

$$\langle (s_i^z)^2 \rangle = \frac{1}{4\pi\mu_B^2} \int_{-\infty}^{+\infty} d\omega \frac{1}{1 - e^{-\beta\omega}} \text{Im}\{\chi_{ii}(\omega + i\eta)\} \quad . \tag{11.94}$$

For temperatures which are high compared with the frequency spectrum of $\chi_{ij}(\omega)$, this equation reduces to

$$\langle (s_i^z)^2 \rangle = \frac{k_B T}{4\pi\mu_B^2} \mathcal{P} \int_{-\infty}^{+\infty} \frac{d\omega}{\omega} \text{Im}\{\chi_{ii}(\omega + i\eta)\}$$

$$= \frac{k_B T}{4\mu_B^2} \chi_{ii}(\omega = 0) \quad . \tag{11.95}$$

An important point to be added is that in almost ferromagnetic materials and weakly ferromagnetic systems $\chi_{ii}(0)$ is T independent. This occurs because the Fourier transform $\chi(\mathbf{q}, T)$ shows a Curie-Weiss behavior only for small values of $\mathbf{q}$ when the present theory is generalized to arbitrary $\mathbf{q}$ values. Therefore $\Omega^{-1} \sum_q \chi(\mathbf{q}, T) = \chi_{ii}$ is practically T independent. We observe a different behavior in systems with local moments discussed before. There the susceptibility has a Curie-Weiss behavior also for large q components and, in contrast to (11.95), the expectation value $\langle (s_i^z)^2 \rangle$ is almost T independent.

When (11.95) is inserted in (11.93), we obtain for almost ferromagnetic metals a temperature-dependent susceptibility of the form

$$\chi(T) = \frac{2\mu_B^2 N(0)}{[1 - N(0)U] + AT} \quad , \tag{11.96}$$

where $A = N(0)Uk_B\chi_{ii}/4\mu_B^2$. Note that in deriving this relation we have assumed that $k_B T$ is larger than the characteristic frequency spectrum of the spin excitations (classical limit). As shown below, this spectrum is low in almost ferromagnetic metals.

For weak ferromagnets the denominator vanishes at a Curie temperature T_c, i.e.,

$$1 - N(0)U + AT_c = 0 \quad . \tag{11.97}$$

For temperatures $T > T_c$, we obtain from (11.96)

$$\chi(T) = \frac{2\mu_B^2 N(0)}{A} \frac{1}{T - T_c} \quad . \tag{11.98}$$

As pointed out before, the origin of the Curie-Weiss-type susceptibility differs here from that in systems like Fe or Ni.

We consider next spin fluctuations around the equilibrium state at $T = 0$ described by the poles of the full frequency- and wavevector-dependent spin susceptibility $\chi(\mathbf{q}, \omega)$ [208]. The inclusion of spin fluctuations changes the mean-field equilibrium state and that effect has to be included in the theory. We will include here only paramagnons, that is, spin fluctuations in almost ferromagnetic metals [28]. After a discussion of $\chi(\mathbf{q}, \omega)$, we determine their influence on the conduction-electron effective mass.

We obtain the susceptibility $\chi(\mathbf{q}, \omega)$ in mean-field or RPA approximation if we replace χ_0 in (11.83) by the frequency and wave-number dependent expression

$$\chi_0(\mathbf{q}, \omega) = \frac{2\mu_B^2}{\Omega} \sum_{\mathbf{p}} \frac{f(\mathbf{p}) - f(\mathbf{p}+\mathbf{q})}{\omega - \epsilon_{\mathbf{p}+\mathbf{q}} + \epsilon_{\mathbf{p}} + i\eta} \quad . \tag{11.99}$$

The function corresponds to the creation and subsequent annihilation of an electron-hole pair of momentum $\mathbf{q}$ and energy ω in response to an external perturbing field $\mathbf{h}(\mathbf{q}, \omega)$ (electron-hole bubble in the language of diagrams).

For a derivation of $\chi_0(\mathbf{q}, \omega)$ and $\chi(\mathbf{q}, \omega)$ consider an external field $\mathbf{h}(\mathbf{r}, t) = \mathbf{h}(\mathbf{q})e^{i(\mathbf{q}\cdot\mathbf{r} - \omega t)}$. The Zeeman term to be added here to the Hamiltonian (11.74) is

$$H_{Ze} = -\int d^3r \mathbf{M}(\mathbf{r}) \cdot \mathbf{h}(\mathbf{r}, t)$$
$$= -\mathbf{M}(-\mathbf{q}) \cdot \mathbf{h}(\mathbf{q})e^{i\omega t} \quad , \tag{11.100}$$

where $\mathbf{M}(\mathbf{q})$ is the Fourier transform of $\mathbf{M}(\mathbf{r})$. The last equation can be written as

$$H_{\text{Ze}} = \mu_B \left[s_+(-\mathbf{q})h_-(\mathbf{q}) + s_-(-\mathbf{q})h_+(\mathbf{q}) + 2s_z(-\mathbf{q})h_z(\mathbf{q}) \right] e^{i\omega t} \quad , \tag{11.101}$$

where $h_\pm = (h_x \pm i h_y)$ and similarly $s_\pm = (s_x \pm i s_y)$. Next the equation of motion for

$$M_+(\mathbf{q}) = -2\mu_B \sum_{\mathbf{k}} c^+_{\mathbf{k}-\mathbf{q}\uparrow} c_{\mathbf{k}\downarrow} \tag{11.102}$$

is set up. It is

$$\dot{M}_+(\mathbf{q}) = i \left[(H + H_{\text{Ze}}), M_+(\mathbf{q}) \right]_- \quad . \tag{11.103}$$

The commutator is easily evaluated. Products of four operators are factorized in a mean-field-like approximation so that we obtain

$$\begin{aligned} i\frac{d}{dt} c^+_{\mathbf{k}-\mathbf{q}\uparrow} c_{\mathbf{k}\downarrow} &= \left(\epsilon_{\mathbf{k}} - \epsilon_{\mathbf{k}-\mathbf{q}} + \frac{U}{\Omega} \sum_{\mathbf{p}} \langle n_{\mathbf{p}\uparrow} - n_{\mathbf{p}\downarrow} \rangle \right) c^+_{\mathbf{k}-\mathbf{q}\uparrow} c_{\mathbf{k}\downarrow} \\ &\quad - \frac{U}{\Omega} \langle n_{\mathbf{k}-\mathbf{q}\uparrow} - n_{\mathbf{k}\downarrow} \rangle \sum_{\mathbf{q}'} c^+_{\mathbf{k}-\mathbf{q}'-\mathbf{q}\uparrow} c_{\mathbf{k}-\mathbf{q}'\downarrow} \\ &\quad + \mu_B h_+(\mathbf{q}) \langle n_{\mathbf{k}+\mathbf{q}\uparrow} - n_{\mathbf{k}\downarrow} \rangle e^{i\omega t} \quad . \end{aligned} \tag{11.104}$$

The expectation value is taken with respect to the unperturbed system. One notices that the time evolution of $M_+(\mathbf{q})$ involves $h_+(\mathbf{q})$ only. For a nonmagnetic system $\langle n_{\mathbf{p}\uparrow} \rangle = \langle n_{\mathbf{p}\downarrow} \rangle = f(\mathbf{p})$, which is the Fermi function. Obviously, the time dependence of the operator $c^+_{\mathbf{k}-\mathbf{q}\uparrow} c_{\mathbf{k}\downarrow}$ is of the form $e^{i\omega t}$. Using this and taking the sum over $\mathbf{k}$ we obtain

$$-\frac{1}{2\mu_B} \langle M_+(\mathbf{q},\omega) \rangle = - \left(\frac{1}{2\mu_B} \langle M_+(\mathbf{q},\omega) \rangle \frac{U}{\Omega} + \mu_B h_+(\mathbf{q}) \right) \\ \times \sum_{\mathbf{k}} \frac{f(\mathbf{k}-\mathbf{q}) - f(\mathbf{k})}{\omega - \epsilon_{\mathbf{k}-\mathbf{q}} + \epsilon_{\mathbf{k}} + i\eta} \quad . \tag{11.105}$$

The added term $i\eta$ is in accordance with (11.79). With the notation

$$u(\mathbf{q},\omega) = \frac{1}{\Omega} \sum_{\mathbf{k}} \frac{f(\mathbf{k}-\mathbf{q}) - f(\mathbf{k})}{\omega - \epsilon_{\mathbf{k}-\mathbf{q}} + \epsilon_{\mathbf{k}} + i\eta} \tag{11.106}$$

(11.105) reduces to

$$\frac{1}{\Omega} \langle M_+(\mathbf{q},\omega) \rangle = \frac{2\mu_B^2 u(\mathbf{q},\omega)}{1 - U u(\mathbf{q},\omega)} h_+(\mathbf{q}) \quad . \tag{11.107}$$

The susceptibility $\chi_{+-}(\mathbf{q},\omega)$ is defined through

$$\frac{1}{\Omega} \langle M_+(\mathbf{q},\omega) \rangle = \chi_{+-}(\mathbf{q},\omega) h_+(\mathbf{q}) \quad , \tag{11.108}$$

with the result that

$$\chi_{+-}(\mathbf{q},\omega) = \frac{2\mu_B^2 u(\mathbf{q},\omega)}{1 - U u(\mathbf{q},\omega)} \quad . \tag{11.109}$$

This form of $\chi_{+-}(\mathbf{q},\omega)$ is referred to as RPA. Setting $U = 0$ one finds that $\chi_0(\mathbf{q},\omega) = 2\mu_B^2 u(\mathbf{q},\omega)$, which proves (11.99). The function $u(\mathbf{q},\omega)$ is known as Lindhard's function. The sum over $\mathbf{k}$ in (11.106) can be explicitly performed, but the result is a rather lengthy expression, which we do not want to write down explicitly. In the limit $\omega \to 0$ it simplifies, in which case we obtain

$$u(x) = N(0)\left[\frac{1}{2} + \frac{1-x^2}{4x}\ln\left|\frac{1+x}{1-x}\right|\right] \quad ; \quad x = \frac{q}{2k_F} \quad . \tag{11.110}$$

A special feature of $u(x)$ is a singulary in the derivative at $x = 1$ (or $q = 2k_F$). Related with it are characteristic oscillations in $u(x)$ (Ruderman-Kittel-Kasuya-Yosida oscillations in case of spin response and Friedel oscillations in case of charge response). Lindhard's function describes the excitation of an electron out of the Fermi sea so that a hole is left behind. It is the response of a noninteracting electron system to an external perturbation. Furthermore, it is plausible that the generated electron-hole pair can be represented by a product of two Green's functions, one of which describes the electron and the other the hole,

$$u(\mathbf{q},\omega) = \frac{i}{2\pi}\int d\omega' \int \frac{d^3p}{(2\pi)^3} G_0(\mathbf{p},\omega') G_0(\mathbf{p}+\mathbf{q},\omega'+\omega) \quad . \tag{11.111}$$

$G_0(\mathbf{p},\omega')$ is the Green's function for free electrons and holes. In terms of diagrams the product of the Green's functions is an electron-hole bubble (see Fig. 11.11). Expressing $u(\mathbf{q},\omega)$ in terms of Green's functions has the advantage that we can generalize the susceptibility to interacting electron systems. In the simplest mean-field approximation this was already done in (11.109) where U enters. However, by using Green's functions and Feynman diagrams we can go beyond the mean-field approximation in a systematic way. We describe this generalization here although we do not need it for the paramagnons we want to describe. However, we will need it in Chapter 15 when we discuss possible Cooper pair formations due to spin fluctuations.

When the electron is interacting with the hole, their mutual scattering is described by a t-matrix. It depends on the momenta and spins of the in- and outgoing particles (see Fig. 11.12) and plays the role of an effective, retarded interaction $V_{\text{eff}}(\mathbf{q},\omega)$. This makes it differ from the instantaneous bare interaction $v_\mathbf{q}$ in (3.4, 3.5). In case of rotational invariance in spin space the t-matrix, which is a spinor, can always be expressed in terms of two scalar functions. We saw that already in Sect. 7.2 (see (7.95))

$$t_{\alpha\beta\gamma\delta} = \delta_{\alpha\gamma}\delta_{\beta\delta}\, t^s + \boldsymbol{\sigma}_{\alpha\gamma}\boldsymbol{\sigma}_{\beta\delta}\, t^a \quad . \tag{11.112}$$

Note that both terms contain singlet and triplet parts. That is discussed in more detail in Sect. 15.5 when dealing with singlet and triplet pairing.

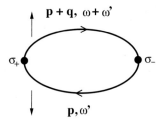

Fig. 11.11. Electron-hole bubble which describes Lindhard's function in form of a diagram. Integration is with respect to the internal variables $\mathbf{p}, \omega'$.

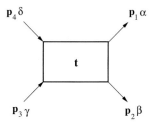

Fig. 11.12. t-matrix $t_{\alpha\beta\gamma\delta}(\mathbf{p}_1, \mathbf{p}_2, \mathbf{p}_3, \mathbf{p}_4)$. When particles with the center of mass at rest scatter, it depends on momentum transfer $\mathbf{q}$ and frequency ω only, i.e., $t(\mathbf{q}, \omega)$.

When approximations to the t-matrix are made, we must ensure that they do not violate conservation laws. Approximations usually consist in selecting among all possible diagrams only a very limited number with the property that they can be summed up to infinite. This corresponds to selecting and summing up special terms in perturbation theory. Conservation laws require relations between approximations to the t-matrix and those for the self-energy $\Sigma(\mathbf{p}, \omega)$ of the Green's function. Rules have been derived by *Baym* [23] which ensure conserving approximations. For the Hubbard Hamiltonian they are in short the following:

When we choose a set of diagrams for the t-matrix, and related with it for the ground-state energy, then a generating function $\Phi(G; U)$ is associated with this selection. It depends on the Green's function $G(\mathbf{p}, \omega)$ and on the two-particle interaction U. In order that this approximation be a conserving one, the self-energy $\Sigma(\mathbf{p}, \omega)$ contained in G (see (7.16)) must be identified with the functional derivative

$$\Sigma(G; U) = \frac{\delta \Phi(G; U)}{\delta G} \quad . \tag{11.113}$$

When temperature Green's functions are used as discussed in Sect. 7.1.2, the function Φ is closely related to the free energy F. Its diagrammatic representation is formally the same as for the ground-state energy. More precisely,

$$\beta F(G; U) = \Phi(G; U) - Tr(\Sigma(G; U) \cdot G) \quad . \tag{11.114}$$

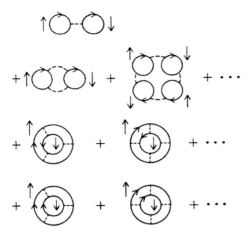

Fig. 11.13. Diagrams in FLEX approximation for the computation of the ground-state energy in the Hubbard model. Dashed lines symbolize the bare interaction U, solid lines represent electron Green's functions. (From [36])

For the ground-state energy of the Hubbard model and also for Φ the diagrams shown in Fig. 11.13 are chosen. They have the advantage that they can be summed up to infinite and consist of bubbles and ladders. Note that the number of bubbles always increases by two, since only electrons of opposite spins are interacting in the Hubbard model. We also want to point out that the interaction denoted by dashed lines in that figure is the bare repulsion U. The self-energy is obtained according to (11.113) by cutting an electron line in each of the diagrams of Fig. 11.13. The resulting diagrams are shown in Fig. 11.14. Note that solid lines denote the full Green's function, i.e., the one with the self-energy included. The bubble corresponds therefore to $\bar{u}(\mathbf{q},\omega)$ where in (11.111) G_0 is replaced by G. We see that three types of scattering processes contribute to $\Sigma(\mathbf{p},\omega)$. The first row contains a molecular field or Hartree contribution (first diagram) plus a sequence of terms. We can interpret the latter as a replacement of one of the dotted lines in the second diagram by

$$U_{\text{eff}}(\mathbf{q},\omega) = \frac{U}{1 - (U\bar{u}(\mathbf{q},\omega))^2} \quad . \tag{11.115}$$

The factor $(U\bar{u}(\mathbf{q},\omega))^2$ results from the geometric series with the number of bubbles increasing by two in each term. The second and third row of diagrams in Fig. 11.14 describe repeated interactions in the particle-hole and particle-particle channel, respectively. One notices that in both cases the t-matrix has the form of ladders. Concerning the case we are interested in here, i.e., close to a ferromagnetic instability, the particle-hole channel is crucial, while in superconductivity the particle-particle channel is the important one. This subject will be taken up again in Sect. 15.5.

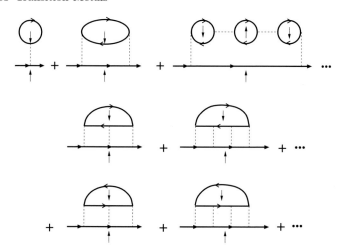

Fig. 11.14. Self-energy diagrams which ensure conservation laws when choosing for the ground-state energy the diagrams in Fig. 11.13.

As discussed before, the self-energy $\Sigma(\mathbf{p}, \omega)$ has to be determined self-consistently since $G(\mathbf{p}, \omega)$, containing $\Sigma(\mathbf{p}, \omega)$ enters the diagrams for it. This can be done by numerical work only. The above approximation scheme is often called fluctuation-exchange or FLEX approximation [36]. It describes spin- as well as charge fluctuations. While the FLEX is limited to an onside interaction, the rules of conserving approximations apply also to more general forms, i.e., momentum dependent two-particle interactions. In that case also electrons with parallel spins can interact when they are situated on different sites. The limitation to bubbles and ladders has the advantage of allowing for infinite summations. Yet, there is no reason why the diagrams which are left out, e.g., those with overlapping interaction lines, should be unimportant. In fact, for a good description of the short-range part of the correlation hole they have to be taken into account. However, in that case, as has been repeatedly pointed out, it is much better to work in r-space rather than in k-space and to use projection techniques rather than diagrams. The FLEX and its limitations will be further discussed in Sect. 15.5 where strong correlations become very important.

In order that correlation functions or susceptibilities are treated by conserving approximations additional rules to those for the one-particle Green's function are required. We shall not discuss them here, since in practice they are not applied. One would not be able to evaluate the resulting diagrams. Instead, when one determines the susceptibility by starting from $\bar{u}(\mathbf{q}, \omega)$, this function is calculated with Green's functions in FLEX approximation. The electron interactions are included by using the t-matrix in the particle-hole channel as it appears in the second row of Fig. 11.14. Explicitly we find

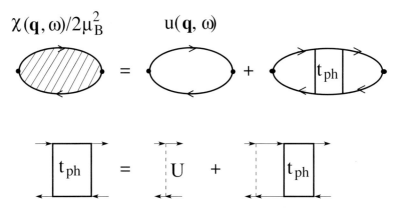

Fig. 11.15. Diagrams which correspond to the mean-field or RPA form of the spin susceptibility. The particle-hole t-matrix is denoted by t_{ph}.

$$\chi_{+-}(\mathbf{q},\omega) = \frac{2\mu_B^2 \bar{u}(\mathbf{q},\omega)}{1 - U\bar{u}(\mathbf{q},\omega)} \quad . \tag{11.116}$$

The corresponding diagrams are those of the RPA and shown in Fig. 11.15. However, here Green's functions in FLEX approximation are used. Therefore, Eq. (11.116) is a renormalized RPA susceptibility.

The renormalization of the susceptibility can also be done self-consistently. As we have seen above, a selection of diagrams for the ground-state energy or free energy like in Fig. 11.13 implies a selected set of diagrams for the self-energy $\Sigma(\mathbf{p},\omega)$ like in Fig. 11.14. The latter enters the Green's function in the diagrams for the susceptibility. But the susceptibility is related back to the free energy in an applied magnetic field h via $-\partial^2 F/\partial h^2 = \chi$. Alternatively, we may relate the susceptibility to the second functional derivative of the generalting function $\Phi(G;U)$. This implies that the diagrams in Fig. 11.13 are cut each at two different places instead of one as for $\Sigma(G;U)$.

When these different relations are solved self-consistently by choosing the RPA diagrams shown in Figs. 11.13, 11.14, we speak of the self-consistent renormalized RPA (SCR) of *Moriya* and *Kawabata* [231, 332]. When the calculations are done for finite temperatures by using temperature Green's functions one finds like in (11.96) a term linear in T in the denominator of the susceptibility. Thus a Curie type of susceptibility results here from a diagrammatic expansion approach. This way an interpolation between itinerant magnetism and that of local moments is achieved.

For a discussion of paramagnons, which is our next goal, it suffices to work with the simple RPA form of (11.109). Paramagnons are overdamped long-wavelength spin fluctuations. They are present if the interaction U is slightly less than the critical value U_c, at which the system becomes unstable against ferromagnetic order. Therefore we are interested in the behavior of $u(\mathbf{q},\omega)$ in the limit of long wavelengths and low frequencies. For this purpose it is useful to introduce the dimensionless quantities $\bar{q} = q/2k_F$ and $\bar{\omega} = \omega/v_F k_F$. For small values of $\bar{\omega}$ and $\bar{q}$ the function $u(\mathbf{q}, \omega)$ reduces to

$$u(\mathbf{q}, \omega) = 1 - \frac{1}{3}\bar{q}^2 + \frac{i\pi}{4}\frac{\bar{\omega}}{\bar{q}} \quad . \tag{11.117}$$

When this expression is substituted into (11.109) we find

$$\chi_{+-}(\mathbf{q}, \omega) = \frac{2\mu_B^2 N(0)}{1 - I} \frac{1 - \bar{q}^2/3 + i\pi\bar{\omega}/4\bar{q}}{1 + \frac{I}{1-I}\frac{\bar{q}^2}{3} - \frac{iI}{1-I}\frac{\pi}{4}\frac{\bar{\omega}}{\bar{q}}} \quad , \tag{11.118}$$

where for almost ferromagnetic systems $(1 - I) = [1 - N(0)U] \ll 1$.

The pole of $\chi(\mathbf{q}, \omega)$ describes the dispersion of the spin fluctuations (paramagnons). The latter is given by

$$\omega = -i\frac{2}{\pi}v_F \frac{(1-I)}{I} q$$
$$= -isq \quad , \tag{11.119}$$

where

$$s = \frac{2}{\pi}\frac{(1-I)}{I} v_F \tag{11.120}$$

is the paramagnon velocity. One notices that the spin fluctuations are overdamped. Because $(1 - I) \ll 1$, the paramagnon velocity is small compared with the Fermi velocity v_F of the electrons.

The inelastic part of the differential neutron scattering cross-section is proportional to $\text{Im}\{\chi_{+-}(\mathbf{q}, \omega)\}$, i.e.,

$$\frac{d^2\sigma}{d\hat{\Omega}d\omega} = \sigma_0 \frac{k'}{k} \text{Im}\{\chi_{+-}(\mathbf{q}, |\omega|)\} \begin{cases} [1 + n(-\omega)] & , \quad \omega < 0 \\ n(\omega) & , \quad \omega > 0 \end{cases} , \tag{11.121}$$

where $d\hat{\Omega}$ is an angular segment and $\mathbf{q}$ ($= \mathbf{k}' - \mathbf{k}$) and ω ($= E' - E$) are the momentum and energy transfer respectively, from the neutron to the probe. The quantities of the outgoing neutron are labeled by a prime. The parameter σ_0 is usually independent of $\mathbf{q}$ and ω. The function $n(\omega) = [\exp(\omega/k_B T) - 1]^{-1}$ represents a Bose factor.

From (11.118) we obtain

$$\frac{1}{2\mu_B^2 N(0)} \text{Im}\{\chi_{+-}(\mathbf{q}, \omega)\} = \frac{\pi}{4}\frac{\bar{\omega}}{\bar{q}} \frac{1}{((1-I) + (I/3)\bar{q}^2)^2 + (I^2\pi^2/16)(\bar{\omega}/\bar{q})^2} \quad . \tag{11.122}$$

This expression can be rewritten in the form

$$\frac{1}{2\mu_B^2 N(0)} \text{Im}\{\chi_{+-}(\mathbf{q}, \omega)\} = \frac{1}{(1-I) + I\bar{q}^2/3} \cdot \frac{\bar{\omega}\Gamma(\bar{q})}{\bar{\omega}^2 + \Gamma(\bar{q})^2} \quad , \tag{11.123}$$

where

11.3 Temperature-Dependent Magnetism

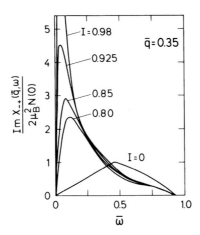

Fig. 11.16. Plot of Im$\{\chi_{+-}(\mathbf{q},\omega)\}$ for fixed value of $\bar{q} = q/2k_F$ as a function of $\bar{\omega} = \omega/v_F k_F$. For small values of $(1-I)$ and of q the peaks becomes more and more pronounced so that one may speak of a quasiparticle-like excitation (paramagnon). (From [95])

$$\Gamma(\bar{q}) = \frac{4\bar{q}}{\pi I}\left((1-I) + I\bar{q}^2/3\right) \tag{11.124}$$

denotes the Lorentzian linewidth. We show the form of Im$\{\chi_{+-}(\mathbf{q},\omega)\}$ in Fig. 11.16. It is peaked at an ω value given by (11.119) and the peak becomes more pronounced as q and $(1-I)$ decrease. When comparing theory with experiments one should take into account that (11.123,11.124) are based on an RPA rather than on a renormalized RPA. In order to incorporate approximately the effects of the latter, one replaces I by $\bar{I}(\mathbf{q})$. In this case $\Gamma(\bar{q})$ is of the general form

$$\Gamma(\bar{q}) = \gamma\bar{q}\left(1/\chi_{+-} + c\bar{q}^2\right) \tag{11.125}$$

[remember that $\chi_{+-} = \chi_{+-}(0,0) \propto (1-I)^{-1}$]. The experimental data for systems like Ni$_3$Ga, Pd, TiBe$_2$ and Ni$_3$Al follow well the form predicted by (11.123) and we can determine experimentally the parameters γ and c in (11.125).

Another effect of spin fluctuations is that they enlarge the effective mass m^* of conduction electrons in the vicinity of the Fermi surface. The mass enhancement can be determined according to (4.35) if we compute $\Sigma(\mathbf{p},\omega)$ as defined in (7.16). Although not done here, this computation can be carried out by starting from (7.15) and making appropriate approximations when evaluating the left-hand side of that equation. Instead, we treat the susceptibility $\chi_{+-}(\mathbf{q},\omega)$ as the propagator of a bosonic excitation, i.e., a paramagnon, with which the conduction electrons interact. The result is a self-energy $\Sigma(\mathbf{p},\omega)$ and a mass renormalization m^*/m, as obtained from other boson excitations with which the electrons interact, e.g., phonons. Let $r(\mathbf{k},\omega)$ denote a general boson propagator and g the coupling constant of the interaction with the conduction electrons. Then $\Sigma(\mathbf{p}, \omega)$ is of the form

$$\Sigma(\mathbf{p}, \omega) = g^2 \int \frac{d^3k}{(2\pi)^3} \int \frac{d\omega'}{2\pi} r(\mathbf{k}, \omega') G(\mathbf{p} - \mathbf{k}, \omega - \omega') \quad . \tag{11.126}$$

A proof of this equation is found, for example, in [117] or [4]. One may associate a diagram with it of the form shown in Fig. 11.17. Here the boson is given by $r(\mathbf{k}, \omega) = (2\mu_B^2)^{-1}\chi_{\pm}(\mathbf{k}, \omega)$ and $g = U$. We have seen before that the interaction U becomes in general renormalized. Since it is an adjustable parameter anyway we consider it here as containing already possible renormalization effects. Starting from (11.126) one can show that m^*/m as given by (4.35) reduces to the form [134]

$$\frac{m^*}{m} = 1 + \frac{g^2 N(0)}{2k_F^2} \int_0^\infty dk\, kr(k, 0) \quad . \tag{11.127}$$

For the bosonic propagator due to paramagnons we find from (11.118)

$$r(\mathbf{k}, \omega) = \frac{N(0)}{1 - I - \frac{i\pi}{2}\frac{\omega}{kv_F} + \frac{1}{12}\frac{k^2}{k_F^2}} \quad . \tag{11.128}$$

When this expression is inserted into (11.127) we obtain for $(1 - I) \ll 1$

$$\frac{m^*}{m} = 1 + 3I^2 \ln\left(1 + \frac{k_c^2}{12k_F^2}\frac{1}{1 - I}\right) \quad . \tag{11.129}$$

The momentum k_c is a cutoff which has been introduced for the k integration. For Pd, $k_c \simeq 1.6 k_F$. Without it one would have to use the original form of the Lindhard function instead of the expansion (11.117). We notice that m^*/m diverges logarithmically as $I = N(0)U$ approaches the value 1. Finally, there is a correction factor $3/2$ to $(m^*/m - 1)$ when the overall mode counting is done better than here.

Up to here we have assumed that the conduction electrons have a spherical Fermi surface so that $u(\mathbf{q}, \omega)$ defined by (11.106) is given by Lindhard's function. That function, which we have explicitly written down only in the long wavelength and low frequency limit (see (11.117)) has its maximum in the static limit at $\mathbf{q} = 0$ where $u(0, 0) = 1$. Therefore, with increasing values of U the denominator in (11.109) vanishes first at $q = 0$. However, that may change when, e.g., the Fermi surface has nesting properties like the one of the Hubbard model on a square lattice near half filling. For a nesting vector $\mathbf{Q}$ defined by the requirement that $\epsilon_{\mathbf{p}} \simeq \epsilon_{\mathbf{p}+\mathbf{Q}}$ over a finite part of the Brillouin zone, the function $u(\mathbf{q} = \mathbf{Q}, \omega)$ may become very large or even diverge. Then a magnetic instability will occur at $\mathbf{q} = \mathbf{Q}$. When $\mathbf{Q}$ is a reciprocal lattice vector we may have an antiferromagnetic instability. When U is slightly less than the critical value U_c defined by

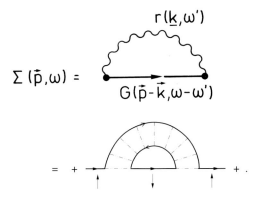

Fig. 11.17. Feynman diagram for the self-energy $\Sigma(\mathbf{p},\omega)$ when conduction electrons of momentum $\mathbf{p}$ and energy ω (solid line) interact with bosonic excitations (wavy line). The electron propagator is $G(\mathbf{p}-\mathbf{k},\omega-\omega')$ while the propagator of the boson is $r(\mathbf{k},\omega')$. We show also one of the ladder diagrams when the boson is an overdamped spin-wave excitation in which case $r(\mathbf{k},\omega)$ is given by (11.128). Dash-dotted lines represent the on-site interaction U.

$$N(0)U_c u(\mathbf{Q},0) = 1 \tag{11.130}$$

the system will contain overdamped antiferromagnetic spin excitations.

In analogy to (11.118) we want to expand $\chi(\mathbf{q},\omega)$ for small $(\mathbf{q}-\mathbf{Q})$ and ω. First we expand the bare susceptibility, i.e., $u_\mathbf{Q}(\mathbf{q},\omega)$. The subscript should indicate that this is not the usual Lindhard function but the modified form in the presence of a nesting vector $\mathbf{Q}$. In analogy to (11.122) we assume that the real and imaginary parts of $u_\mathbf{Q}(\mathbf{q},\omega)$ are related to each other by

$$\lim_{\omega \to 0} \operatorname{Im} u_\mathbf{Q}(\mathbf{q},\omega) = \pi \frac{\omega}{\Gamma(\mathbf{q})} u_\mathbf{Q}(\mathbf{q},0) \quad, \tag{11.131}$$

where $\Gamma(\mathbf{q})$ is a characteristic energy of spin fluctuations with momentum $\mathbf{q}$. We expand $u_\mathbf{Q}(\mathbf{q},0)$ with respect to $(\mathbf{q}-\mathbf{Q})$ and write

$$u_\mathbf{Q}(\mathbf{q},0) = u_\mathbf{Q}(\mathbf{Q},0) - q^2/\xi_0^2 \quad. \tag{11.132}$$

Furthermore, we set near the AF instability, i.e., for $U \lesssim U_c$

$$1 - N(0)U u_\mathbf{Q}(\mathbf{Q},0) = (\xi_0/\xi)^2 \quad. \tag{11.133}$$

The correlation length ξ is a measure how close we are to an AF instability. When these expansions are set into (11.109) we obtain for overdamped antiferromagnetic spin fluctuations

$$\frac{1}{2\mu_B^2 N(0)} \operatorname{Im}\chi_{-+}(\mathbf{q},\omega) = \pi \frac{u_\mathbf{Q}(\mathbf{Q},0)}{\Gamma(\mathbf{q})} \frac{\omega(\xi/\xi_0)^4}{(1+q^2\xi^2)^2 + (\pi^2\omega^2/\Gamma(\mathbf{q}))^2(\xi/\xi_0)^4} \quad. \tag{11.134}$$

These spin fluctuations play a role in the interpretation of experiments on high-T_c superconducting cuprates when they are in the normal and in the superconducting state. They also contribute to the binding of electrons to Cooper pairs [325].

12
Transition-Metal Oxides

Several correlation effects found in transition metals reappear in amplified form in some of the transition-metal oxides. In fact, correlations can be so strong that Mott-Hubbard transitions from a metallic to an insulating state may occur. When these systems are doped with holes, often rich phase diagrams result, among them a variety of magnetic phases. Not only spin-, but also orbital order plays a significant role. Both types of order may influence each other. Manganites are much studied systems in this respect. They have in addition a significant technical potential. The cuprate perovskites are another class of very extensively investigated materials, interesting because of their extraordinary superconducting properties, discussed in Chapter 15. Only in a few cases can one perform true ab initio calculations based on wavefunction methods for strongly correlated systems. They are of interest despite that calculations within the LDA to density-functional theory have been able to describe well or even predict numerous experimental findings. Yet it is also known that calculations of that kind become unreliable when correlations are strong. An extension to LDA+U described in Sect. 4.4 leads to significant improvements. However, those calculations are no longer free of parameters and the involved approximations are uncontrolled. Therefore, it is important to design simplified models with the help of which we may better understand qualitatively and often semiquantitatively the basic physical properties of the oxides.

In this chapter we start out with a discussion of cuprates. They are prototypes of doped correlation-induced insulators. After that we continue with the phenomenon of orbital ordering. Although this type of order was first observed in rare-earth systems such as CeB_6, we shall limit ourselves here to its appearance in transition metal oxides. Finally the discussion turns to examples of the interplay between charge-, structural-, spin- and orbital orderings.

12.1 Doped Charge-Transfer Systems: the Cuprates

Most of the high-temperature superconductors, in particular those with the highest transition temperatures are cuprates with a perovskite structure. This explains the exceptional interest this class of materials has generated. However, the doped cuprates have also very interesting physical properties in the metallic normal state. Many of them can be linked to strong-electron correlations which prevail in those systems. In the most studied cuprates the important structural element are copper-oxide planes with a unit cell CuO_2. Planes are formed from octahedra, pyramids or squares. In each case the Cu atom is surrounded by O atoms as shown in Fig. 12.1.

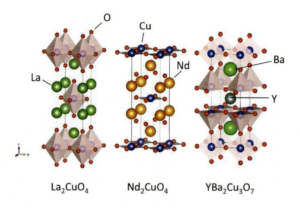

Fig. 12.1. Different Cu based perovskites which play an important role in high-T_c superconductivity. An important element are layers and chains ($YBa_2Cu_3O_4$) of oxygen shared CuO_4 plaquettes. (Courtesy of T. Takimoto)

Examples are La_2CuO_4, where the planes are formed from octahedra; $YBa_2Cu_3O_7$, where they are formed from pyramids (actually, the Cu atoms are slightly buckling here); and Nd_2CuO_4, where they are built from squares. La_2CuO_4 and Nd_2CuO_4 are insulators while $YBa_2Cu_3O_7$ is a metal, in fact a superconductor when temperatures are low enough.

In La_2CuO_4, the CuO_6 octahedra are elongated due to a Jahn-Teller distortion. The Cu–O distances are 190 pm within the plane and 240 pm perpendicular to it. The distortion lifts the degeneracy of the Cu d orbitals, which in octahedral symmetry is twofold ($e_g = \{d_{x^2-y^2}, d_{3z^2-r^2}\}$) and threefold ($t_{2g} = \{d_{xy}, d_{xz}, d_{yz}\}$).

There is direct experimental evidence for strong-electron correlations in the Cu–O planes. We consider La_2CuO_4 as an example and start by simply counting electrons. The valency of La is 3+, which implies that two La atoms

12.1 Doped Charge-Transfer Systems: the Cuprates

donate six electrons. Oxygen has a valency of 2- and therefore O_4 accommodates a total of eight electrons. This leaves for Cu a valency of 2+ implying a $3d^9$ configuration (remember that a Cu atom has a $[Ar]3d^{10}4s^1$ electron configuration). The hole in the 3d-shell is placed into the highest antibonding Cu–O state, which has predominantly $3d_{x^2-y^2}$ character (Fig. 12.2). With one hole per formula unit of La_2CuO_4, one would expect that system to be metallic with a half-filled conduction band, provided the picture of independent-or weakly correlated electrons holds. In reality the material is an antiferromagnetic semiconductor with a Néel temperature of $T_N \simeq 280$ K. Semiconducting behavior is also found at temperatures $T > T_N$ and is therefore *not* related to a doubling of the unit cell when antiferromagnetic order is present. Instead, it results from strong correlations of the electrons in the Cu–O planes. We are dealing here with what is usually called a Mott-Hubbard insulator (see Sect. 10.7). As will be discussed later, a more appropriate description is that of a charge-transfer insulator, for which the Hubbard band splitting is a prerequisite. Correlations are also responsible for the observed antiferromagnetic ground state. Similar arguments apply to Nd_2CuO_4.

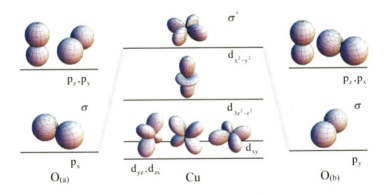

Fig. 12.2. Bonding between a Cu^{2+} and two O^{2-} ions. Only the $3d^9$ electrons of Cu and the $2p^2_{x(y)}$ electrons of O are assumed to hybridize. The antibonding σ^* orbital has predominantly $d_{x^2-y^2}$ character and contains the hole of the unit cell. The bonding σ orbital is predominantly of $p_{x(y)}$ character. The splitting of the d_ν-orbitals is caused by a CEF of tetragonal symmetry.

La_2CuO_4 can be doped with holes by partially replacing La^{3+} by Sr^{2+} or Ba^{2+}. Similarly, Nd_2CuO_4 can be doped with electrons by a replacement of Nd^{3+} by $Ce^{3.5+}$. The Cu–O planes of $YBa_2Cu_3O_7$ contain holes without modification of the material, because charge is moved from the planes into the chains (self-doping). Often Cu–O planes or pairs of planes are separated by blocks of insulating layers. One important family are the bismuth strontium calcium copper oxides, named BSCCO. Members of that family are labeled according to the sequence of the numbers of the metallic ions. For example BSCCO-2212 is $Bi_2Sr_2Ca_1Cu_2O_8$. For illustration the unit cell of three different compounds is shown in Fig. 12.3.

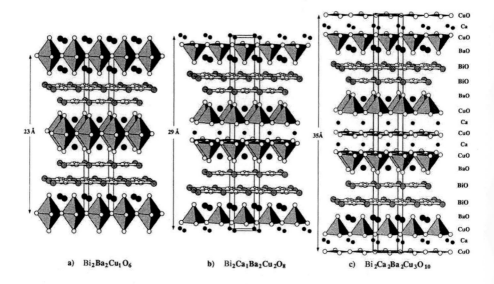

Fig. 12.3. Schematic representation of the crystal structure of some Bi-based cooper oxide superconductors: Bi-2201, Bi-2212 and Bi-2223 (From [181])

12.1.1 Quasiparticle–like Excitations

We want to understand and describe the dispersion of a small number of doped holes in a Cu-O plane. First we note that the electronic structure of the cuprates has been investigated with the help of LDA calculations. They reproduce neither the semiconducting behavior of La_2CuO_4 nor the magnetic properties of the ground state. However, they reproduce well other properties of the systems, such as the electronic charge distributions, which depend only slightly on the strong correlations. This can be explicitly checked by comparing the calculated charge densities with those obtained from quantum chemical calculations on negatively charged planar clusters of $(CuO_3)_nO(n = 1, \ldots, 4)$.

An insulating ground state is obtained when the LDA is replaced by LDA+U. Yet, in order to include properly the strong electron correlations and at the same time to describe the CuO_2 planes as detailed as possible, one has to apply wavefunction based methods, using a CASSCF calculation as a starting point. Before demonstrating this, we want to show which features of the excitation spectrum relate solely to the lattice structure.

12.1 Doped Charge-Transfer Systems: the Cuprates

We consider first the following simple Hamiltonian of noninteracting particles in hole-, instead of electron representation

$$H_0 = \epsilon_d \sum_{i\sigma} n_{d\sigma}(i) + \epsilon_p \sum_{j\sigma} n_{p\sigma}(j) + \sum_{\langle ij \rangle \sigma} V_{ij} \left(d_{i\sigma}^+ p_{j\sigma} + p_{j\sigma}^+ d_{i\sigma} \right) \quad . \quad (12.1)$$

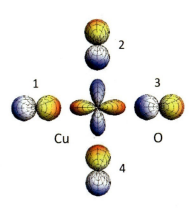

Fig. 12.4. Cu $3d_{x^2-y^2}$ and O $2p_{x(y)}$ orbitals which hybridize most strongly and form the lowest bonding and highest antibonding states. Blue: positive and red: negative sign of atomic wavefunction. For the orbitals at positions 1 and 2, $\alpha_{ij} = 0$, while for those at positions 3 and 4, $\alpha_{ij} = 1$.

The notation $\langle ij \rangle$ refers as usual to pairs of nearest neighbors i and j. The operator $d_{i\sigma}^+$ creates a hole with spin σ in the Cu $3d_{x^2-y^3}$ orbital at site i and the operator $p_{j\sigma}^+$ does the same with respect to the O $2p_x$ (or $2p_y$) orbital at site j (Fig. 12.4). The corresponding hole number operators are $n_{d\sigma}(i)$ and $n_{p\sigma}(j)$. The hybridization between the two types of orbitals is given by the matrix element V_{ij}. From Fig. 12.4 we observe that $V_{ij} = (-1)^{\alpha_{ij}} t_{pd}$ with $\alpha_{ij} = 0$ or 1 depending on the position of the O atom relative to the Cu atom. There are three orbitals per unit cell CuO_2 which are connected by the hybridization term. After diagonalizing the 3 × 3 Hamilton matrix the following three bands are obtained.

$$\epsilon_{1,2}(\mathbf{k}) = \frac{\epsilon_p + \epsilon_d}{2} \pm \sqrt{\left(\frac{\epsilon_p - \epsilon_d}{2}\right)^2 + 4t_{pd}^2 \left(\sin^2 \frac{k_x a}{2} + \sin^2 \frac{k_y a}{2}\right)} ,$$

$$\epsilon_3(\mathbf{k}) = \epsilon_p \quad . \quad (12.2)$$

We show a plot of the bonding and antibonding bands $\epsilon_{1,2}(\mathbf{k})$ in Fig. 12.5a. Note that in electron representation the bonding band is predominantly of oxygen $2p$ character, while the antibonding band is primarily of $3d$-character (compare with Fig. 12.2). In hole representation the situation is reverse.

286   12 Transition-Metal Oxides

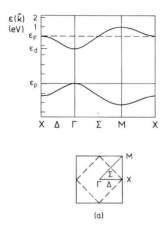
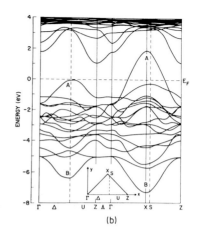

Fig. 12.5. (a) Bonding, antibonding and nonbonding solutions $\epsilon(\mathbf{k})$ in electron representation as given by (12.2) for a square lattice. The parameter values are $\epsilon_p - \epsilon_d = 3.6$eV and $t_{pd} = 1.3$eV. Also shown is the Brillouin zone for a square lattice. (b) Results of LDA band-structure calculations for La_2CuO_4 in electron representation. The Brillouin zone is for a body-centered tetragonal phase and is shown in the inset. Therefore the notation differs from that in (a). The X point corresponds to M while the Z point is in the k_z direction. Notice the similarity of the bonding (B) and antibonding (A) bands to those in (a). The 15 intermediate bands correspond to the ones formed from the remaining four d-orbitals and to nonbonding bands. (From [316])

One notices that the bonding and antibonding bands agree reasonably well with those of a full LDA calculation shown in Fig. 12.5b. In both cases La_2CuO_4 is found to be metallic. The reason is different in the two cases. While in (12.1) Coulomb repulsions are excluded, they are included in the LDA but treated insufficiently.

We want to return to the following question: what is the dispersion of a propagating hole doped into a CuO_2 plane? For such a purpose we first have to understand the most important changes in the density of states caused by strong-electron correlations. As pointed out previously, La_2CuO_4 would be metallic and not an antiferromagnetic semiconductor were it not for those correlations. The density of states corresponding to the bands in Fig. 12.5a is shown in Fig. 12.6a and illustrates that point. From the discussions of the Hubbard model in Sect. 8.2 (consult also Sect. 10.4), we would expect a density of states of the form shown in Fig. 12.6b, i.e., with the antibonding band of predominantly d-character split into a lower and an upper Hubbard band.

In drawing Fig. 12.6 it has been assumed that the splitting of the antibonding band into two Hubbard bands is larger than the one between the

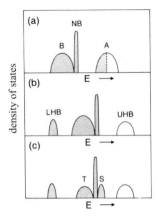

Fig. 12.6. Schematic plot of the density of states for the CuO_2 planes at half filling in electron representation: (a) within the independent-electron approximation; B and A denote contributions from the bonding and antibonding band, respectively, NB labels the one from the nonbonding band; (b) in the presence of a Hubbard-band splitting, LHB and UHB denote the lower and upper Hubbard bands. The latter is separated from the NB peak by a charge transfer gap; (c) when the singlet-triplet (S,T) splitting of a hole on a Cu and an O site is taken into account (Zhang-Rice singlet).

bonding and antibonding one, i.e., the Hubbard interaction U is larger than the difference in the orbital energies $(\epsilon_d - \epsilon_p)$. Therefore we are dealing here with a charge-transfer insulator[1] where the excitation gap is of order $(\epsilon_d - \epsilon_p)$ rather than U. The opposite case, i.e., when $(\epsilon_d - \epsilon_p) \gg U$ gives raise to a Mott-Hubbard insulator for which the excitation gap is given by U.

However, the situation is more complex than that. When a hole is doped into La_2CuO_4, it goes predominantly into a $2p$ orbital of an oxygen ion because two holes on a Cu site would strongly repel each other. Together with the hole on a Cu site, it forms a spin singlet state commonly named after *Zhang* and *Rice* and schematically shown in Fig. 12.7. It is the same type of singlet state discussed in Sect. 10.2.1 with the difference that the ligand orbital denoted there by L is replaced here by a superposition of four equivalent oxygen p orbitals and that two holes instead of two electrons are considered. When the formation of a singlet is taken into account, the density of states is of the form shown in Fig. 12.6c. This reasoning will be confirmed in the discussion that follows.

---

[1] see [128, 129, 506]

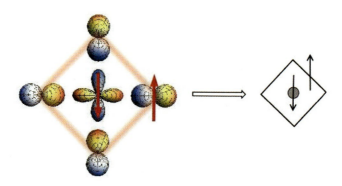

Fig. 12.7. Schematic presentation of a Zhang-Rice singlet formed by two holes on a CuO$_4$ plaquette. Because of a large repulsion of two holes on a Cu site, the second hole is primarily situated on O 2p-orbitals.

After these introductory considerations we want to compute the dispersion of a doped hole by a wavefunction based calculation using quantum chemical methods. It is parameter free and therefore free of arbitrariness. Because the correlations are strong, a simple SCF calculation is not a good starting point. The corrections required would be too large. However, a CASSCF calculation (see Sect. 5.1) is suitable for our purpose. It should include the most important correlations so that the remaining ones are weak and can be treated by perturbation theory or CEPA. For a CASSCF treatment the active space must be defined. In order to find it consider a system to which one hole has been added. Then one plaquette in the system will accommodate two holes as symbolized by the square in Fig. 12.7. This plaquette together with its surrounding is identified with the correlated local state $|\mathbf{R}n\sigma\}$ in (7.114) from which the quasiparticle Bloch state $|\psi_{\mathbf{k}\nu}^{N-1}\rangle$ is constructed. The state $|\mathbf{R}\sigma\}$ is schematically shown in Fig. 12.8. A quantum number n is not required here. If the Coulomb repulsion of two holes on a Cu site were infinitely strong, the singlet state formed on the special plaquette would have the form

$$|\psi\rangle = \frac{1}{\sqrt{2}} \left(d_\uparrow^+ p_\downarrow^+ - d_\downarrow^+ p_\uparrow^+ \right) |0\rangle \quad . \tag{12.3}$$

When applied to the state $|0\rangle$ with no hole on the plaquette, the operator d_σ^+ creates a hole at the Cu site, while p_σ^+ creates a hole on the four surrounding O sites (see Fig. 12.7), i.e.,

$$p_\sigma^+ = \frac{1}{2} \left(p_{1\sigma}^+ + p_{2\sigma}^+ - p_{3\sigma}^+ - p_{4\sigma}^+ \right) \quad . \tag{12.4}$$

This is the Zhang-Rice singlet, i.e., the analogue of the Heitler-London singlet of a H$_2$ molecule. By now the choice of the active space has become clear.

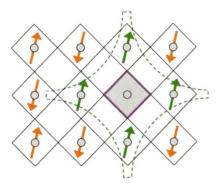

Fig. 12.8. Plaquette with an added hole of spin ↑ forming a Zhang-Rice singlet. The nearest neighbor plaquettes are ferromagnetically aligned with the added hole and form a spin bag or spin polaron.

For the plaquette with the two holes it must comprise the $3d_{x^2-y^2}$-orbital of the Cu sites and in addition the combination (12.4) of $2p_{x(y)}$-orbitals. As explained in Sect. 7.3.3 calculations of $|\mathbf{R}\sigma\}$ require only the treatment of an embedded cluster of sufficient size. We include here the distinct plaquette together with its four nearest-neighbor plaquettes. For these five plaquettes also the apical oxygens are taken into account. The employed basis set used has triple-zeta quality. The cluster is embedded into a much larger system by a proper embedding scheme. For an initial guess of the spin orbitals the ones obtained from a SCF calculation for a hypothetical Cu $3d^{10}$, O $2p^6$ closed-shell configurations are used [192]. Note that we base all calculations on the software MOLCAS [329].

We recall that in a CASSCF calculation the wavefunction $|\psi\rangle$ is written as a linear combination of configurations $|m\rangle$, i.e., $|\psi\rangle = \sum_m \alpha_m |m\rangle$. This is a spin and symmetry adapted superposition of Slater determinants. The Slater determinants in turn are constructed from orthonormal spin orbitals of the active space. Those are supplemented by orbitals which remain occupied in all determinants. The orbitals are variationally optimized simultaneously with the coefficients α_m. If it helps understanding, one may consider a CASSCF calculation for the special plaquette and its surrounding as a determination of the single-site scattering matrix S_I in (5.53).

We find the following results. The Zhang-Rice singlet is essentially a superposition of three configurations, i.e., $|d^1_{x^2-y^2}, p^1\rangle$ given by (12.3) as well as $|d^2_{x^2-y^2}, p^0\rangle$ and $|d^0_{x^2-y^2} p^2\rangle$ where p is the combination (12.4) of $2p_{x(y)}$-orbitals. The respective weights of the three configurations are found to be 0.70, 0.14 and 0.11[2]. The deviations from a Heitler-London-like state are appreciable, since the latter requires that $|d^1_{x^2-y^2}, p^1\rangle$ has weight 1. Therefore

---

[2] see [84, 191]

the correlation strength in the cuprates as defined (10.1-10.2) should not be overrated. The missing 5% in the sum of the three weights is due to the inclusion of the apical oxygens. The latter play a role in generating a ferromagnetic surrounding of the plaquette with an added spin σ hole (see Fig. 12.8). Because of that particular feature one often speaks of a spin-bag or ferromagnetic polaron [314, 404]. Experiments prove that doped holes go predominantly to O sites as found by the quantum chemical calculations. They are based on electron energy loss spectroscopy (EELS) [347]. In those experiments, a 1s electron of an O atom is excited into an empty valence state at the same site. Changes with doping of the p hole number on the O sites are therefore directly measurable.

Of special interest is the dispersion of the (generalized) Zhang-Rice singlet when it moves in form of a Bloch wave through the system. To a large extent it is determined by the geometry of the lattice. For a square lattice it is of the general form

$$\epsilon(\mathbf{k}) = -2t_{\text{eff}}\left(\cos k_x a + \cos k_y a\right) + 4t'_{\text{eff}} \cos k_x a \cos k_y a$$
$$-2t''_{\text{eff}}\left(\cos 2k_x a + \cos 2k_y a\right) \quad (12.5)$$

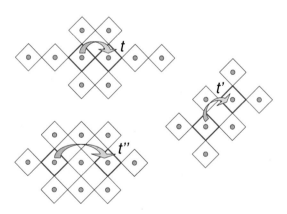

Fig. 12.9. Hopping matrix elements between different plaquettes. The shown clusters, properly embedded, were used for the determination of the effective matrix elements t_{eff}, t'_{eff} and t''_{eff} of Zhang-Rice singlets.

where the hopping matrix elements t, t' and t'' refer to processes shown in Fig. 12.9. Also shown in that figure are the embedded clusters for which t_{eff}, t'_{eff} and t''_{eff} were calculated. Hopping processes between 4th nearest-neighbor plaquettes and beyond are neglected. For holes moving in an antiferromagnetic background one has $t_{\text{eff}} = 0$, i.e., the hole has to remain on the sublattice on which it was generated. In processes characterized by t'_{eff} and t''_{eff}, hopping takes place within the same sublattice and therefore both matrix elements differ from zero. We obtain the corresponding unrenormalized or bare hopping matrix elements t, t' and t'' by fully spin polarizing the background. In that case the hole does not drag a bag along and we find $t = 0.54\text{eV}$, $t' = 0.31\text{eV}$ and $t'' = 0.12\text{eV}$. The renormalized parameters are determined in close analogy to Sect. 7.3.3. A finite cluster like the one shown in Fig. 12.9 will always give a finite value of t_{eff} instead of zero, since the mismatch of the spin order is limited to the size of the cluster when the matrix element H_{ij} (with i,j being nearest neighbor plaquettes) is calculated. However, finite clusters are sufficient when matrix elements are calculated for hopping processes which take place on the same sublattice. They are also suitable for the paramagnetic state when the antiferromagnetic correlation length is of the order of the cluster size. From those calculations we obtain $t'_{\text{eff}} = 0.13\text{eV}$ and $t''_{\text{eff}} = 0.05\text{eV}$. These values are surprisingly close to the one of Ref. [441], where a $t - J$ model was applied with the needed value of J adjusted to experiments. They lead to a dispersion shown in Fig. 12.10 and hence to the formation of hole pockets at $(\pi/2, \pi/2)$ when the doping level increases. The implicit assumption is hereby that a change in doping concentration leads merely to a shift in the chemical potential. It is also found that the energy difference $\Delta E = E(N+1) + E(N-1) - 2E(N) > 0$ is a few eV. Here $E(N\pm1)$ are the energies of the electron addition and electron removal states. That implies that CASSCF calculations yield an insulating state for the undoped N electron system.

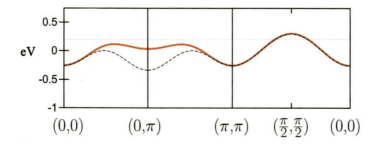

Fig. 12.10. Dispersion of the Zhang-Rice singlet labeled S in Fig. 12.6 without and with the d_{z^2} orbital excluded (dashed line). Thin solid line: lowered chemical potential due to a small hole doping. (From [192])

High resolution ARPES experiments find lense-like hole pockets at small doping (see Fig. 12.11). In the quasiparticle approximation used here, we cannot reproduce the different intensities of the two sides of the hole pockets which are observed experimentally. However, the asymmetry can be under-

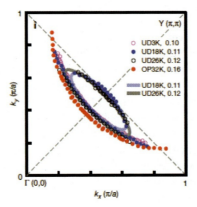

Fig. 12.11. Fermi surface of La-Bi 2201 at different doping concentrations $n_c = 0.10 - 0.16$. UD 18 K (OP 32 K) means underdoped (optimally doped) sample with superconducting $T_c = 18$ K (32 K). The presence of two Fermi surfaces, i.e., pocket as well as arc might be due to sample inhomogeneities. (From [318])

stood by realizing that for a Hubbard system at half filling and $U/|t| \gg 1$ the momentum distribution $n_\sigma(\mathbf{k})$ is *not* constant but varies as $n_\sigma(\mathbf{k}) = \frac{1}{2} + (2gt/U)^2(\cos k_x a + \cos k_y a)$, where $g = \frac{1}{4} - \frac{1}{3}\langle \mathbf{s}_i \mathbf{s}_j \rangle$ and i and j are nearest neighbors. For small hole doping we find small hole pockets, as shown schematically in Fig. 12.12.

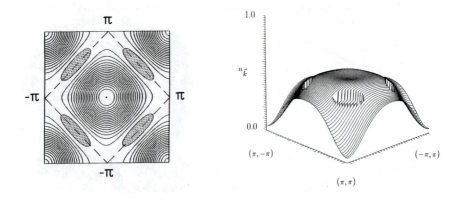

Fig. 12.12. Momentum distribution for a half-filled Hubbard model on a square lattice in the strong correlation limit when the system is doped with holes. Note the asymmetric rim of the lenses in $n_\mathbf{k}$ near $(\pm\pi/2, \pm\pi/2)$ which are actually seen in ARPES experiments (see Fig. 12.11). (Courtesy of R. Eder)

12.1 Doped Charge-Transfer Systems: the Cuprates

When the hole doping is large, long-range antiferromagnetic order is destroyed and only short-range antiferromagnetic correlations remain. Assume that the latter extend over an area of the size of the cluster shown in Fig. 12.8. In that case $t_{\text{eff}} \neq 0$ and a value of $t_{\text{eff}} = 0.1$eV obtained from CASSCF for a cluster of eight plaquettes is reasonable. When set into (12.5) we obtain a dispersion for the quasiparticles, which in distinction to Fig. 12.10 has maxima at $(\pm\pi, \pm\pi)$ instead of $(\pm\frac{\pi}{2}, \pm\frac{\pi}{2})$. The effect of $t_{\text{eff}} > 0$ is to move up the excitation energy at (π, π) seen in that figure. When the chemical potential is adjusted to the hole concentration we obtain a large electron Fermi surface. It is centered at the Γ point and the volume enclosed by the Fermi surface varies according to changes in the electron, not hole number. The intermediate regime is difficult to cover within the present computational schemes. It would require a Green's function approach based on CASSCF. Nevertheless, we attempt here a phenomenological description for increasing hole concentration.

Assume an AF ground state and a hole concentration which is sufficiently small, so that we have hole pockets in the Brillouin zone. At the edges of the pockets the excitation energy is zero, but outside the pockets all electronic excitations are gapped. This implies that charge as well as spin response to external perturbations remain small at low energies. When the hole concentration increases the hole pockets grow, the asymmetry between the edges increases (see Fig. 12.12) until eventually a Fermi surface of the form shown in Fig. 10.26 evolves. Thus density and spin response increase in parallel, since a growing number of low-energy excitations becomes possible. Eventually, at a certain hole concentration n_h^c a full (large) Fermi surface has evolved and the system has become a conventional Fermi liquid. For $n_h < n_h^c$ and at low temperatures the electron system behaves like having a pseudogap because of the reduced low-energy response. The pseudogap refers to both charge and spin response. This subject will be taken up again in Chapter 15.

The above calculations aimed at determining the dispersion of the Zhang-Rice singlet. In order to reproduce the spectral density sketched in Fig. 12.6 we have to proceed differently, i.e., by using Green's functions and a model Hamiltonian. Thus we have to compute the spectral function from

$$\rho(\mathbf{k}, \omega) = -\frac{1}{\pi} Im G^R(\mathbf{k}, \omega)$$

$$= \frac{1}{\pi} \frac{Im\Sigma(\mathbf{k}, \omega)}{[\omega - \epsilon_{\mathbf{k}} - Re\Sigma(\mathbf{k}, \omega)]^2 + [Im\Sigma(\mathbf{k}, \omega)]^2} \quad . \quad (12.6)$$

The retarded Green's function cannot be determined with the same accuracy as the dispersion shown in Fig. 12.10, in which quantum chemical accuracy was achieved. Instead, we have to compromise by using a simplifying model Hamiltonian. We choose the following one, again in hole representation

$$H = H_0 + \sum_{\langle jj' \rangle \sigma} t_{pp}(j,j') \left(p_{j\sigma}^+ p_{j'\sigma} + h.c.\right) + U_d \sum_i n_{d\uparrow}(i) n_{d\downarrow}(i)$$

$$+ U_p \sum_j n_{p\uparrow}(j) n_{p\downarrow}(j) + U_{pd} \sum_{\langle ij \rangle} n_d(i) n_p(j) + U_{pp} \sum_{\langle jj' \rangle} n_p(j) n_p(j') \quad .$$
(12.7)

Like (12.1), the Hamiltonian includes Cu $3d_{x^2-y^2}$ and O $2p_{x(y)}$ orbitals only (Emery model). The noninteracting part H_0 is given by (12.1). The remaining part contains the Coulomb repulsion U_d between two holes on a Cu site and $U_p < U_d$ between two holes on an O site. Repulsions between a d-hole on a Cu site and a p-hole on a neighboring O site are denoted by U_{pd}. Also included is a direct O–O hybridization term $t_{pp'}$. Doped holes go predominantly to O sites since $\epsilon_p < \epsilon_d$. This was demonstrated before by the quantum chemical type of calculations and is in agreement with experiments.

The parameters of the Hamiltonian are usually fitted to a constrained LDA calculation, where energy changes are calculated away from the ground-state density distribution. For example, $n_d(i)$, the d-electron number at a Cu site i, is constrained to a given value $n_d^c(i)$ and the ground-state energy functional $E[n]$ is minimized under this subsidiary condition, i.e.,

$$E[n_d^c(i)] = \min \left\{ E[n] + \lambda \int_{MT} d^3r \left[n_d(\mathbf{r}) - n_d^c(\mathbf{r})\right] \right\} \quad .$$
(12.8)

Here a large value of λ ensures that the subsidiary condition is met. The integral is taken over a muffin-tin sphere, i.e., over the volume of the Cu atom i. The energies $E[n_d^c(i)]$ describe, within the LDA, the response of the system due to a local change of the d-electron number (charge fluctuation). In order to determine the parameters of the Hamiltonian (12.7), equivalent constrained calculations are done for its self-consistent field part H_{SCF}. By bringing the result of the two calculations into agreement, one may determine the parameters of H. The results for La$_2$CuO$_4$ are listed in Table 12.1.

The spectral function $\rho(\mathbf{k}, \omega)$ can be calculated within the above model in two ways. One is by numerical studies of small clusters, e.g., by diagonalization of four units of CuO$_2$ with periodic boundary conditions. The second and preferable one is to determine $\rho(\mathbf{k}, \omega)$ by application of the projection method outlined in Sects. 5.4.1 and 8.1.

We define spectral functions $\rho_{mn}(\mathbf{k}, \omega)$ for a set of operators $\{A_n(\mathbf{k})\}$ by relating them to the (retarded) Green's functions

Table 12.1. Parameter values in eV as obtained from a constrained LDA [200].

$\epsilon_p - \epsilon_d$	t_{pd}	t_{pp}	U_d	U_p	U_{pd}	U_{pp}
3.6	1.3	0.65	10.5	4	1.2	0

12.1 Doped Charge-Transfer Systems: the Cuprates

$$G_{mn}(\mathbf{k},t) = -i\theta(t)\left\langle\psi_0\left|[A_m^+(\mathbf{k},t), A_n(\mathbf{k},0)]_+\right|\psi_0\right\rangle, \qquad (12.9)$$

or more precisely, to the imaginary part of their Laplace transforms

$$G_{mn}(\mathbf{k},z) = \left(A_m(\mathbf{k})\left|\frac{1}{z-L}\right.A_n(\mathbf{k})\right)_+, \qquad (12.10)$$

[for notation see 8.1]. According to (7.9) the relationship is

$$\rho_{mn}(\mathbf{k},\omega) = -\frac{1}{\pi}\lim_{\eta\to 0}\mathrm{Im}\{G_{mn}(\mathbf{k},\omega+i\eta)\}. \qquad (12.11)$$

The Liouvillean L refers to the Hamiltonian (12.7), but with $U_{pd}=0$, for simplicity. The operators $\{A_n(\mathbf{k})\}$ specify the relevant variables (or microscopic processes) to which we limit the calculations. Their choice is discussed below.

We use (8.18) in order to rewrite (12.10) in matrix notation as

$$\mathbf{G}(\mathbf{k},z) = \boldsymbol{\chi}(\mathbf{k})\left[z\boldsymbol{\chi}(\mathbf{k}) - \boldsymbol{\omega}(\mathbf{k})\right]^{-1}\boldsymbol{\chi}(\mathbf{k}), \qquad (12.12)$$

with the susceptibility and frequency matrix defined by

$$\chi_{mn}(\mathbf{k}) = (A_m(\mathbf{k}) \mid A_n(\mathbf{k}))_+$$

$$\omega_{mn}(\mathbf{k}) = (A_m(\mathbf{k}) \mid LA_n(\mathbf{k}))_+, \qquad (12.13)$$

respectively. The memory matrix is zero here since calculations are restricted to the Liouville space spanned by the set $\{A_m\}$. For a given value of $\mathbf{k}$, the dimension of the matrices is given by the number of selected variables. The static expectation values (12.13) can be computed by the methods described in Chapt. 8, but the details are quite involved [461].

The proper choice of the relevant operators $\{A_n(\mathbf{k})\}$ is essential for a high-quality, yet simple, calculation of the spectral density of the electronic excitations. In the present case they must first of all include the hole operators

$$A_p(\alpha,\mathbf{k}) = p^+_{\alpha\mathbf{k}\uparrow}, \qquad A_d(\mathbf{k}) = d^+_{\mathbf{k}\uparrow}, \qquad (12.14)$$

where $\alpha=1,2$ is a band index required because there are two O atoms per unit cell. But the Fourier transforms $\bar{p}^+_{\alpha\mathbf{k}\uparrow}, \bar{d}^+_{\mathbf{k}\downarrow}$ of the local operators $\bar{p}^+_{i\uparrow} = p^+_{i\uparrow}n_{p\downarrow}(i)$ and $\bar{d}^+_{j\uparrow} = d^+_{j\uparrow}n_{d\downarrow}(j)$ must also be included in the set $\{A_n(\mathbf{k})\}$, i.e.,

$$A_{\bar{p}}(\alpha,\mathbf{k}) = \bar{p}^+_{\alpha\mathbf{k}\uparrow}, \qquad A_{\bar{d}}(\mathbf{k}) = \bar{d}^+_{\mathbf{k}\uparrow}. \qquad (12.15)$$

They ensure a properly reduced weight of configurations with doubly occupied Cu and O orbitals, which is an effect of the on-site repulsions U_d and U_p. If we were to limit the $\{A_n\}$ to (12.14-12.15), the calculations would correspond to a Hubbard I approximation (Sect. 8.2). However, there are other microscopic processes which also need to be included.

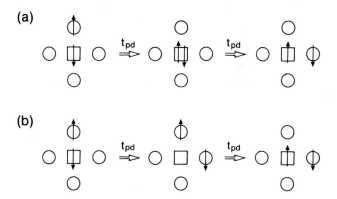

Fig. 12.13. Two processes which cause spin flips of a hole on a Cu site. Process (a) becomes ineffective in the limit $U_d \to \infty$.

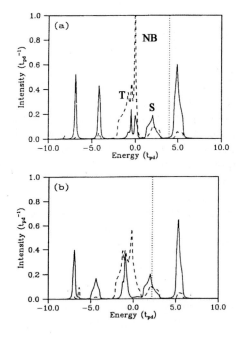

Fig. 12.14. Oxygen (dashed lines) and Cu (solid lines) spectral weight in electron (not hole) representation calculated by projection techniques. The parameter values are (in units of t_{pd}): $U_d = 8$, $U_p = 3$, $t_{pp} = 0.5$, and $\epsilon_p - \epsilon_d = 4$. (a) Half filling and (b) 25 % hole doping. The dotted line indicates the position of the Fermi energy. S, T and NB label singlet, triplet and nonbonding contributions. Fig. (a) should be compared with the schematic plot in Fig. 12.6(c). (From [461])

12.1 Doped Charge-Transfer Systems: the Cuprates

As discussed before, the ground state of two 3d- and 2p-holes consists of a linear superposition of the states $p^+_\uparrow p^+_\downarrow |0\rangle$, $d^+_\uparrow d^+_\downarrow |0\rangle$ and the singlet $2^{-1/2}\left(p^+_\uparrow d^+_\downarrow - p^+_\downarrow d^+_\uparrow\right)|0\rangle$, where $|0\rangle$ denotes a plaquette consisting of a Cu $d_{x^2-y^2}$ orbital and the four nearest neighbor O $2p_{x(y)}$ orbitals, without any holes. We have to ensure that the most important microscopic processes leading to the correlated ground state are also included in the treatment of the Cu–O plane. Therefore we must also include among the set of operators $\{A_n(\mathbf{k})\}$ the following ones

$$A_f(\mathbf{k}) = \frac{1}{\sqrt{N}} \sum_I e^{-i\mathbf{k}\cdot\mathbf{R}_I} p^+_{I\downarrow} S^+_I$$

$$A_a(\mathbf{k}) = \frac{1}{\sqrt{N}} \sum_I e^{-i\mathbf{k}\cdot\mathbf{R}_I} p^+_{I\uparrow} n_{d\downarrow}(I)$$

$$A_c(\mathbf{k}) = \frac{1}{\sqrt{N}} \sum_I e^{-i\mathbf{k}\cdot\mathbf{R}_I} p^+_{I\uparrow} p^+_{I\downarrow} d_{I\downarrow} \quad . \tag{12.16}$$

There are N unit cells labeled by I with lattice vectors $\mathbf{R}_I$. The operator $S^+_I = d^+_{I\uparrow} d_{I\downarrow}$ causes a spin flip of a hole at the Cu site I, while $p^+_{I\sigma}$ denotes the combination (12.4) for a unit with the Cu site I in the center. Spin flips of holes on Cu and O sites as described by $A_f(\mathbf{k})$ are microscopic processes required in forming singlet and triplet states of two holes. Two different processes leading to a spin flip are shown in Fig. 12.13. The other operator needed for a description of singlet and triplet states is $A_a(\mathbf{k})$, which describes antiferromagnetic correlations between holes on Cu and O sites. A Cu-orbital with two holes, which appears as an intermediate step in Fig. 12.13a is described by the variable $A_{\bar{d}}(\mathbf{k})$. The intermediate configuration in Fig. 12.13b with no hole on the Cu site is described by the charge transfer operator $A_c(\mathbf{k})$. The operators (12.16) follow directly by applying L to (12.15). With a total of nine operators, the 9×9 matrix (12.12) has to be diagonalized for each $\mathbf{k}$ point. The static expectation values $\langle A^+_m(\mathbf{k}) A_n(\mathbf{k}) \rangle$ are evaluated self-consistently by using the fluctuation-dissipation theorem

$$\langle A^+_m(\mathbf{k}) A_n(\mathbf{k}) \rangle = \int_{-\infty}^{+\infty} d\omega \, \rho_{mn}(\mathbf{k}, \omega) f(\omega) \quad , \tag{12.17}$$

where $f(\omega)$ is the Fermi function. A proof of this theorem is found in text books like [343]. The calculations remain unchanged when in (12.14 - 12.16) the spin indices are interchanged. The resulting spectral densities for p- and d-holes and electrons are shown in Fig. 12.14 for half filling and for 25 % hole doping. They show good agreement with the results of numerical diagonalization of the Hamiltonian for four CuO_2 units[3]. The basic structure sketched

---
[3] see [425, 450]

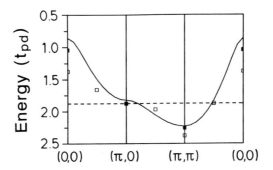

Fig. 12.15. Dispersion of the Zhang-Rice singlet (in hole representation) when the ground state is paramagnetic. Parameters (in units of t_{pd}): $U_d = 6$, $U_p = 0$, and $\epsilon_p - \epsilon_d = 4$. The hole concentration is $n_h = 1.25$ (25 % hole doping). (From [461]). Also shown are the results of quantum Monte Carlo calculations for a 4×4 CuO$_2$ cluster and $(k_B T)^{-1} = 10 t_{pd}^{-1}$ (open squares) and of an exact diagonalization of a 2×2 cluster of CuO$_2$ units (solid squares). (From [96])

in Fig. 12.6c is reproduced. In the above calculations a nonmagnetic ground state has been assumed, but we could have equally well used an antiferromagnetic one. The difference is not important when general features of the spectrum are considered because the energy scale for magnetic order is small as compared with the bandwidths.

Consider first the case of half filling. One notices in Fig. 12.14a a structure labeled S which, when analyzed, can be shown to have predominantly singlet character. The peak from the corresponding triplet configuration is labeled T in the same figure. The high density of states near zero energy corresponds to the nonbonding oxygen band and is denoted by NB in that figure. The singlet structure is separated by a gap of order $(\epsilon_p - \epsilon_d)$ from the upper Hubbard band. The latter corresponds to a $d^9 \to d^{10}$ transition, i.e., to filling the Cu d-shell. As discussed before, we are dealing here with a charge transfer insulator. Other examples of insulators of that kind are CuO, CuCl$_2$, CuBr$_2$ as well as NiCl$_2$ and NiBr$_2$. When we dope the system with 25 % holes, we also find a transfer of spectral weight from the upper Hubbard band to the singlet states (see Fig. 12.14b). A diagonalization of the 9×9 Green's function matrix yields a dispersion curve $E_\nu(\mathbf{k})$ ($\nu = 1, \ldots, 9$) for each eigenmode. The one of lowest energy, i.e., that of the singlet is found to have a dispersion as shown in Fig. 12.15. It resembles very much the one obtained by quantum chemical calculations and shown in Fig. 12.10 if we used in the corresponding equation (12.5) a value of $t_{\text{eff}} \simeq 0.14$eV. Such a value corresponds to a small antiferromagnetic correlation length, i.e., of order of the Cu–Cu distance. This seems reasonable in view of the large hole doping concentration. The above findings demonstrate that the projection method provides a convenient tool for the treatment of strong electron correlations. These findings complete the description of holes in the CuO_2 planes.

12.2 Orbital Ordering

Orbital ordering requires orbital degeneracy. This prerequisite is fulfilled in $4f$- and $5f$- as well as $3d$- and $4d$-electronic systems. In rare-earth compounds spin-orbit interactions are strong. Therefore the states are classified according to the total angular momentum J. Hund's rules determine the J value of the ground state of the incomplete $4f$ shell. The first rule states that electrons occupy the f-shell (or similarly the s-, p- or d-shell) so that they maximize their total spin S. Should this prescription not uniquely specify the electronic configuration the second rule applies. It states that the degeneracies are to be removed in favor of the particular configuration which maximizes the total orbital angular momentum L. The third rule applies to incomplete $4f$-shells only and requires that $J = L - S$ for a less than half-filled shell and $J = L + S$ if the shell is more than half filled. The $(2J+1)$-fold degenerate ground-state multiplet is split by the crystalline electric field (CEF) set up by the neighboring ions. When the $4f$ electron number is odd, it follows from Kramers' theorem that each CEF energy level is at least two-fold degenerate. This Kramers' degeneracy is a consequence of time-reversal symmetry. If the $4f$ electron number is even, the CEF levels can be nondegenerate. Beyond the required Kramers' degeneracy, additional degeneracies may be present depending on the point symmetry of the rare-earth site. Interactions between neighboring sites may lead to orbital ordering. It is favoured when the CEF levels are degenerate or nearly degenerate. Because of the strong spin-orbit interactions it involves necessarily not only charge-, but also spin-degrees of freedom, as manifested by J. A well known example of multipolar, here quadrupolar ordering is CeB_6[4].

The situation is different for d electrons in transition metal oxides. Here CEF splittings are much larger than the spin-orbit interactions. Therefore the latter may be neglected in first approximation and the states may be classified separately according to their orbits and spins. In a cubic environment the 5-fold orbital degeneracy is split into a t_{2g} triplet and a e_g doublet (see Fig. 12.16) with the former being lower in energy in most cases. Orbital ordering can occur if the degenerate energy eigenstates are only partially filled. It is caused by anisotropic hybridizations together with Hund's rule coupling. The Jahn-Teller effect plays also an important role here.

We recall that in the cuprates the hole in the $3d$ shell of Cu^{2+} is in the e_g doublet. Because of a slight elongation of the surrounding oxygen octahedron in z direction however this e_g doublet is split, with the $d_{x^2-y^2}$ being higher in

---

[4] see, e.g., [260]

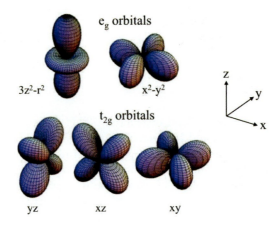

Fig. 12.16. d orbitals split by a crystalline electric field of cubic symmetry into a e_g doublet and t_{2g} triplet.

energy (see Fig. 12.2). This implies that the hole is in that orbital. We have also seen that the hybridization with the oxygen ions results in an effective antiferromagnet interaction $J\mathbf{S}_i\mathbf{S}_j$ between neighboring Cu sites like in the one-band Hubbard model (see Fig. 12.8). However, the situation is quite different in oxides in which neighboring oxygen octahedra, unlike in the cuprates, share edges instead of corners. Here two orthogonal oxygen p orbitals are hybridizing with neighboring Cu sites (see Fig. 12.17). This implies a ferromagnetic interaction between the Cu spins. The interaction is caused by configurations with two holes in an intermediate state of the hybridizing oxygens. Hund's rule correlations on that oxygen site favor ferromagnetic alignment of these two holes in different $2p_{x(y)}$ orbitals. This proves that the indirect exchange between two Cu^{2+} spins, or more generally, two transition-metal ions depends on the path which the exchange takes. It is over a 90°-angle in edge-sharing octahedra and over a 180°-angle in site-sharing octahedra. The latter situation was previously found in the Cu–O planes. This structural dependence is part of the Goodenough-Kanamori rules [149]. Those rules are useful for the determination of interatomic spin-spin interactions between two transition-metal ions. The latter are mediated either by direct hopping processes between the two ions or by a shared anion (often oxygen) in between. The rules state that the interactions, called superexchange, are antiferromagnetic when the effective interaction is between d orbitals which are half-filled and when it is mediated by *one* orbital of an anion. The latter is in most cases oxygen (see Fig. 12.17b). The interaction of two Cu^{2+} ions in the Cu–O planes serves as an example here. We recall that the $d_{x^2-y^2}$ orbital is singly occupied and therefore half filled. When *two* orbitals of the anions are participating, the coupling is ferromagnetic (see Fig. 12.17a). In both cases the effect is proportional to

t^2, where t is the hopping matrix element to the anion. The interactions are also ferromagnetic when hopping takes place from a half-filled to an empty orbital or from a filled to a half-filled one. This is called double exchange.

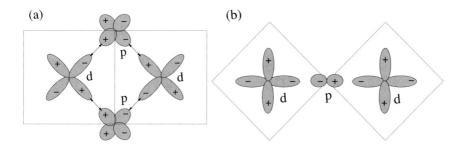

Fig. 12.17. Hybridization between a Cu 3d- and oxygen 2p-orbital: (a) when neighboring oxygen octahedra share edges, and (b) when they share corners (compare with Fig. 12.4). Only orbitals involved in a d–d bonding are shown. In (a) the coupling is ferromagnetic while in (b) it is antiferromagnetic.

An example is here hole doped LaMnO$_3$, in which hopping of holes converts Mn^{3+} sites into Mn^{4+} sites and couples the Mn ions ferromagnetically. In manganites which have perovskite structure, the Mn^{3+} ions are in a $3d^4$ configuration with strong Hund's-rule correlations. Thus a high-spin state forms with $S = 2$. Three of the four d electrons occupy t_{2g} orbitals. The fourth electron is in the e_g doublet and we deal here with a twofold orbital degeneracy. The coupling is ferromagnetic (double exchange) since an electron in an e_g orbital hops into an empty e_g orbital at a neighboring site. This is shown in Fig. 12.18. An enormous amount of research went into the study of manganites because of the colossal magnetoresistance found in some of them. There is a high technological potential associated with this physical effect.

There are a number of different forms of orbital degeneracies observed in the transition-metal oxides. In the vanadates the V^{3+} ions are in a $4d^2$ configuration with the electrons being in t_{2g} states with strong Hund's-rule correlations. Therefore they form a $S = 1$ high-spin state. In the titanates Ti^{3+} is in a $4d^1$-configuration and hence again in a t_{2g} orbital. The ruthenates have Ru^{4+} in a $4d^4$ configuration. The d electrons are again in t_{2g} orbitals since Hund's-rule correlations are not strong enough to overcome the $t_{2g} - e_g$ CEF splitting. Therefore one of the t_{2g} orbitals is doubly occupied while the remaining two electrons form a $S = 1$ state.

As pointed out before, orbital degeneracy may be lifted by orbital ordering. In perovskite structures the various types of cooperative ordering can be divided into four different classes which are shown in Fig. 12.19. They differ in symmetry properties of the ordered state. The following considerations intend a microscopic description of the different forms of orbital ordering. Starting

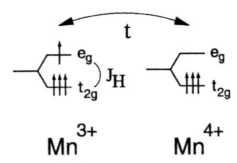

Fig. 12.18. Double exchange due to hopping of an e_g electron into an empty e_g orbital of a neighboring site.

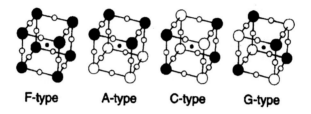

Fig. 12.19. Different types of cooperative orbital order in perovskite structures (A_2BO_4). Centers of the MO_6 octahedra (B-sites) are shown by large filled and empty spheres which represent two different highest occupied d orbitals. Oxygen ions are marked by small open spheres. Small spheres at the body center cubes denote A-sites, e.g., La, Ca or else. (From [169])

Hamiltonian is one of the form of (12.7) but here generalized to five d bands and three p bands. From this Hamiltonian we want to eliminate the oxygen p orbital degrees of freedom and to replace them by effective interactions between neighboring transition metal ions. This can be done by a Schrieffer-Wolff transformation of the form described in Sect. 10.2.2. Consider a d orbital with index $\nu = 1, 2$ at site i and an orbital ν' at site j which couple via an oxygen ion as shown in Fig. 12.17. Then we obtain an effective hopping matrix element $t_{\nu\nu'}$ between sites i and j of the form

$$t_{\nu\nu'} = \frac{V_{i\nu}^{(O)} V_{j\nu'}^{(O')}}{(\epsilon_p - \epsilon_d)} \tag{12.18}$$

in hole notation. Here $V_{i\nu}^{(O)}$ is the hopping matrix element between the d orbital ν at site i and the oxygen orbital O and similar for $V_{j\nu'}^{(O')}$. Therefore

12.2 Orbital Ordering

we may start from the multiband Hubbard Hamiltonian

$$H = -\sum_{\langle ij \rangle} \sum_{\nu\nu'\sigma} \left(t_{\nu\nu'} a^+_{\nu\sigma}(i) a_{\nu'\sigma}(j) + h.c. \right) + H_1 \quad, \tag{12.19}$$

with H_1 given by (11.1). The *direct* hopping matrix element between orbital ν at site i and ν' at site j is assumed to be included in $t_{\nu\nu'}$ too.

From (12.19) we want to extract an effective Hamiltonian for the low-energy excitations in the spirit of *Kugel* and *Khomskii* [256], thereby taking into account that correlations are strong. It should involve spin- and orbital-degrees of freedom. The latter act like an isospin τ which fulfills the same commutation relations as the spin operators, i.e., $[\tau_n, \tau_m] = i\epsilon_{nm\ell}\tau_\ell$. But before we derive the effective Hamiltonian we want to point out the important role of the Jahn-Teller effect. For that purpose we consider a single electron in a degenerate orbital, e.g., in an e_g orbital. The ion Mn^{3+} in $LaMnO_3$ serves as a specific example. The neutral atomic configuration of Mn is $[Ar]\, 3d^5 4s^2$ and therefore Mn^{3+} is in a $3d^4$ configuration. As pointed out before, the t_{2g} orbitals are singly occupied in a high spin $S = 3/2$ configuration. They couple with the fourth d electron, which is in an e_g orbital to $S = 2$. When the latter hops to a neighboring site, the matrix element depends on the angle θ_{ij} between the spins $\mathbf{S}_i$ and $\mathbf{S}_j$ of the two sites involved. More specific

$$t(i,j) = t \cos(\theta_{ij}/2) \quad. \tag{12.20}$$

This form of the hopping matrix elements gives preference to a ferromagnetic coupling of neighboring sites. The latter is forced by the kinetic energy, which is optimal when $\theta_{ij} = 0$. As pointed out before, we speak here of double exchange.

The degeneracy of the e_g doublet, a result of the local site symmetry, is lifted by the Jahn-Teller (J–T) effect. The latter consists of a distortion of the oxygen octahedron in the xy-plane and results in a lowering of the energy. This is seen as follows. From degenerate perturbation theory we know that the energy gain due to a symmetry-breaking perturbation, which here is a strain, is linear in that perturbation. Yet the energy loss due to the accompanying elastic energy is only quadratic in the strains. Therefore an orbital symmetry breaking by a lattice deformation will always lead to a gain in energy. The deeper the energy minimum, the larger the size of the deformation of the octahedra.

For a more detailed discussion of the J–T effect, we introduce the orbitals

$$|\varphi_1\rangle = |d_{3x^2-r^2}\rangle \quad, \quad |\varphi_2\rangle = |d_{3y^2-r^2}\rangle \tag{12.21}$$

as basis for the effective isospin. The two orbitals are obtained from the original two e_g orbitals $d_{3z^2-r^2}$ and $d_{x^2-y^2}$ (see Fig. 12.16) by a rotation in the two-dimensional space, i.e.,

$$|\varphi(\vartheta)\rangle = \cos\vartheta/2\, |d_{3z^2-r^2}\rangle + \sin\vartheta/2\, |d_{x^2-y^2}\rangle \tag{12.22}$$

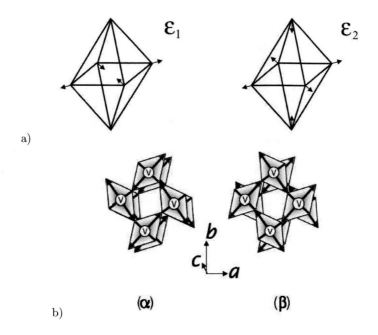

Fig. 12.20. (a) Jahn-Teller (J-T) distortions of Γ_3-symmetry in the form of ϵ_1, ϵ_2 strains on an octahedron; (b) two forms of cooperative J-T distortions corresponding to (α) C-type and (β) G-type ordering, see Fig. 12.19. (From [385])

when for ϑ the values $4\pi/3$ and $-4\pi/3$ are chosen. The deformations of the octahedra i in the ab-plane are of Γ_3-symmetry and described by the Hamiltonian

$$H_{\rm JT} = -g \sum_i \left(\epsilon_1(i)\tau_z(i) + \epsilon_2(i)\tau_x(i) \right) \quad , \tag{12.23}$$

with the strains ϵ_1 and ϵ_2 given by

$$\epsilon_1 = \frac{1}{\sqrt{2}} \left(\epsilon_{xx} - \epsilon_{yy} \right) \quad ; \quad \epsilon_2 = \frac{1}{\sqrt{6}} \left(2\epsilon_{zz} - \epsilon_{xx} - \epsilon_{yy} \right) \quad . \tag{12.24}$$

The Pauli matrices τ_μ act on the pseudospin space spanned by $|\varphi_1\rangle$ and $|\varphi_2\rangle$. The corresponding deformations are depicted in Fig. 12.20a. While ϵ_1 favors either the orbital $|d_{3x^2-y^2}\rangle$ or $|d_{3y^2-z^2}\rangle$ depending on the sign of g, the strain ϵ_2 mixes the two.

Cooperative J–T transitions can be of different type as shown in Fig. 12.20b. Thereby the total volume change must be kept as small as possible in order not to loose additional elastic energy. One way to achieve this is by an antiferro-orbital arrangement of the distorted octahedra. In addition to the J–T effect the octahedra of the perovskite structure are often tilted like in La_2CuO_4. This tilting may affect considerably the hopping matrix elements of the electrons.

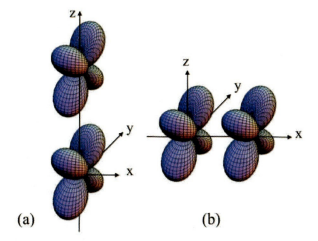

Fig. 12.21. Hybridization between two d_{yz} orbitals; (a) when the bond is in z direction; (b) when it is in x-direction.

Now we proceed to derive an effective Hamiltonian by replacing the virtual hopping processes by effective intersite interactions similar as done before for the t-J model at half filling or as described more generally in Sect. 10.2.2. To be specific, we consider as an example two neighboring sites i and $j(=i+\delta)$ with two electrons in orthogonal orbitals d_{xz} and d_{yz}, a situation which applies to the d electrons of the perovskite Sr_2VO_4. Furthermore, we discuss first the nearest-neighbor interaction along the z axis (see Fig. 12.21a). We assume that the two electrons have opposite spins because of an AF coupling. This allows for singlet as well as triplet states. We introduce again an isospin matrix $\hat{\tau}_z$ for each of the two sites. In the present case we assign the eigenvalue $\tau_z = 1$ to the d_{xz} orbital and $\tau_z = -1$ to the orbital d_{yz}. The state of a d electron at site i is denoted by $|\tau_z;\sigma_z\rangle_i$ and includes the orbital index τ_z as well as the spin index σ_z. We want to reduce the Hamiltonian for sites i and j to an effective one acting on a 8-dimensional Hilbert space. The latter is spanned by states of the form $|\tau_z;\sigma_z\rangle_i \otimes |\tau'_z;-\sigma_z\rangle_j$. The aim is to find the matrix elements $H^{\text{eff}}_{\alpha\beta}$ of the effective Hamiltonian in this reduced Hilbert space.

First we consider the matrix elements in the subspace with $\tau_z = \tau'_z$. By applying (10.15 - 10.17b) to this problem we obtain as intermediate state a doubly occupied orbital with energy U in the part of the Hilbert space perpendicular to the reduced one. Therefore, like in the t-J model the effective Hamiltonian is of the form $J_{\text{ex}}\mathbf{s}_i\mathbf{s}_j$ with $J_{\text{ex}} = 4t^2_{\nu\nu}/U$. The subscript ν refers here to the d_{xz} or d_{yz} orbital. The determination of the remaining matrix elements with $\tau'_z = -\tau_z$ is more involved. In the intermediate state the two electrons are in different orbitals at one of the site. They can be in a singlet state with energy $E_s = U' + J$ or in a triplet state with energy $E_t = U' - J$. The energies U' and J have been defined in (11.1). In accordance with Hund's rules $E_t < E_s$. The electron return process, i.e., H_{PQ} in (10.17a) leads to four final states labeled 1 to 4, i.e.,

$$|\tau_z; \sigma_z\rangle_i \otimes |-\tau_z; -\sigma_z\rangle_j \quad ; \quad |-\tau_z; -\sigma_z\rangle_i \otimes |\tau_z; \sigma_z\rangle_j \quad \text{and}$$

$$|-\tau_z; \sigma_z\rangle_i \otimes |\tau_z; -\sigma_z\rangle_j \quad ; \quad |\tau_z; -\sigma_z\rangle_i \otimes |-\tau_z; \sigma_z\rangle_j \quad . \quad (12.25)$$

The first state is the original one and the second state has the electrons on site i and j interchanged. The remaining two states correspond to an isospin flip and to a spinflip of the original state, respectively.

The Hamiltonian matrix elements are of the following form: $H_{11}^{\text{eff}} = -(t^2/2)(1/E_s + 1/E_t)$ and $H_{12}^{\text{eff}} = -H_{11}^{\text{eff}}$ because of the intersite exchange of the electrons. The matrix element $H_{13}^{\text{eff}} = -t^2(-1/E_s + 1/E_t)$ because of the isospin exchange and $H_{14}^{\text{eff}} = -H_{13}^{\text{eff}}$ because of the spin exchange. The different signs in the matrix elements are immediately obtained by using for the singlet and triplet states the analog of the first and third line of (10.7). The hopping matrix element t stands for $t_{\nu\nu}$ with $\nu = 1(d_{xz})$ and $\nu = 2(d_{yz})$. Since i and j are neighboring sites along the z axis it is $t_{11} = t_{22}$. We notice that spin and isospin remain uncoupled, at least as long as spin-orbit interactions are neglected, as done here. The spin couplings are rotational invariant but the isospin couplings are *not*. From this property of the coupled spins and the known matrix elements, we can construct the effective Hamiltonian in terms of the spin and isospin operators $\mathbf{s}(i)$ and $\boldsymbol{\tau}(i)$, respectively. The matrix elements for $(\tau_z)_i = (\tau_z)_j$ were obtained before. We rewrite it here so that the isospin appears explicitly

$$H_{\text{eff}_1}^{(1)}(i,j) = \frac{8t^2}{U}\left(\mathbf{s}_i \mathbf{s}_j - 1/4\right)\left(\tau_z(i)\tau_z(j) + 1/4\right) \quad . \quad (12.26)$$

The Hamiltonian differs from zero only when the spins couple antiferromagnetically.

The effective Hamiltonian for antiferro-orbital ordering, i.e., when $(\tau_z)_i = (-\tau_z)_j$ follows from the above matrix elements and is found to be

$$H_{\text{eff}}^{(2)}(i,j) = \frac{4t^2 U'}{(U')^2 - J^2}\left(\boldsymbol{\tau}(i)\boldsymbol{\tau}(j) - 1/4\right)\left(\mathbf{s}_i \mathbf{s}_j + 3/4\right)$$

$$+ \frac{4t^2 J}{(U')^2 - J^2}\left(\boldsymbol{\tau}(i)\boldsymbol{\tau}(j) - 1/4\right) \quad . \quad (12.27)$$

One can check easily that this form reproduces the previously derived matrix elements of H_{eff}. The second term is independent of the spin coupling, while the first term favors ferromagnetic spin alignment. This is due to Hund's-rule correlations. The ground-state wavefunction is therefore a product of a spin triplet times an antisymmetric, i.e., singlet isospin part. The ground-state energy is

$$E_0 = -\frac{2t^2}{(U'-J)} \quad . \tag{12.28}$$

Next we consider for the same two orbitals d_{xz} and d_{yz} the effective interaction between neighboring sites in x instead of z direction. The positioning of two d_{yz} orbitals is shown in Fig. 12.21b. It is apparent that the hybridization matrix element $-t_{\nu\nu}$ in (12.18) is largest for two d_{yz}-orbitals, i.e., $t_{22} \gg t_{11}$ and furthermore that $t_{12} = 0$. To lowest approximation one often accounts only for $t_{22} = t$ and neglects all other hopping terms. An analysis analogous to the previous one yields an effective Hamiltonian of the general form

$$H_{\text{eff}}(i,j) = [a + b\left(\tau_z(i) + \tau_z(j)\right) + c\ \tau_z(i)\tau_z(j)]$$
$$+ \mathbf{s}(i)\mathbf{s}(j)\left[a' + b'\left(\tau_z(i) + \tau_z(j)\right) + c'\ \tau_z(i)\tau_z(j)\right] \quad . \tag{12.29}$$

The constants can be expressed again in terms of U, U' and J but we do not reproduce their explicit forms. A point of interest is that only the z component of the isospin appears. This implies that there is no dynamics with respect to the orbital degrees of freedom, i.e., there are no isospin flips. The ground state of the two-site, two-electron system turns out to be a spin triplet multiplied by an orbital singlet. The ground-state energy is $E_0 = -t^2/(U'-J)$ and therefore is higher than for a bond in z direction (see (12.27)). With $H_{\text{eff}}(i,j)$ defined for the z and x and similarly for the y direction one can apply a mean-field approximation and determine a phase diagram involving the different types of ground states. Examples are given below. A similar analysis can be applied to different orbital degeneracies like the two-fold degenerate e_g orbitals.

The situation simplifies very much if we set $t_{\nu\nu'} = t\delta_{\nu\nu'}$, a rather drastic assumption in view of the previous discussion. Then we have as well rotational invariance in isospin space and the effective Hamiltonian reduces to the simple form

$$H_{\text{eff}} = \sum_{\langle ij \rangle} \left(J_1 \mathbf{s}(i)\mathbf{s}(j) + J_2 \boldsymbol{\tau}(i)\boldsymbol{\tau}(j) + J_3 \left(\mathbf{s}(i)\mathbf{s}(j)\right)\left(\boldsymbol{\tau}(i)\boldsymbol{\tau}(j)\right) \right) \quad . \tag{12.30}$$

In the following we want to discuss some material-specific properties.

12.2.1 Manganites: LaMnO$_3$ and related Compounds

The structure of LaMnO$_3$ consists of MnO$_6$ octahedra with Mn^{3+} ($3d^4$) in the center. The octahedra share corners and form infinite, stacked layers. As such LaMnO$_3$ serves as a parent compound of other systems which differ in the number of layers formed periodically. The double layer compound La$_2$Mn$_2$O$_7$ is an example, which is discussed later. The manganites have attracted immense attention. It was triggered by the discovery of a colossal magnetoresistance (CMR) which some of the systems show[5]. The latter is intimately connected with the strong correlations which prevail in these materials.

---
[5] see, e.g., [78, 221]

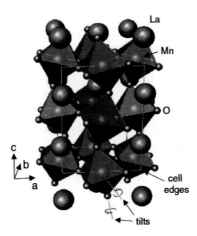

Fig. 12.22. Distortion of the MnO$_6$ lattice of GdFeO$_3$-type due to tilting of the octahedra. The resulting structure is orthorhombic. (From [143])

The ground state of LaMnO$_3$ is insulating. It is the result of a delicate balance between different interactions of orbital-, spin- and lattice-degrees of freedom. As pointed out before, the $3d^4$ electrons of Mn occupy the t_{2g} orbitals in form of a high-spin state with $S = 3/2$, the fourth electron is in a e_g-state and couples ferromagnetically to form a $S = 2$ state. The crystal structure deviates considerably from the ideal perovskite structure, because the MnO$_6$ octahedra are strongly tilted (see Fig. 12.22). This tilting is related to a J–T distortion of the octahedra. The Jahn-Teller effect is cooperative and antiferro-distortive. Spin ordering is of A-type (see Figs. 12.19 and 12.23) while orbital ordering is of C-type. The angle ϑ in (12.22) is not known precisely, because it is difficult to determine it experimentally. One technique which has been applied is anomalous X-ray scattering associated with the K-edge absorption [337]. A distortion of the octahedra respective orbital order shows up here in an otherwise forbidden (3, 0, 0) reflection. A theoretical analysis is found, e.g., in Ref. [444].

The stacking of the ferromagnetic planes is shown in Fig. 12.23. In the z direction electrons on neighboring sites are in the same orbital. Their spin coupling is antiferromagnetic via superexchange and mediated by oxygen sites. This is also what is obtained from LDA+U calculations [282]. It is worth noticing that the transition temperatures for spin ordering T_{cs} and orbital ordering T_{co} are quite different, i.e., $T_{cs} = 145$ K while $T_{co} = 780$ K.

When La^{3+} is partially replaced by Sr^{2+} we deal with a hole doped system. At half filling of the lower e'_g orbital, i.e., for La$_{0.5}$Sr$_{0.5}$MnO$_3$ charge order of the Mn^{3+} and Mn^{4+} takes place. One sublattice is occupied by Mn^{3+} and

Fig. 12.23. Orbital order in LaMnO$_3$ in form of an antiferro-orbital arrangement of $d_{3x^2-r^2}$ and $d_{3y^2-r^2}$ orbitals. The precise value of ϑ in (12.21) is not yet known. (From [451])

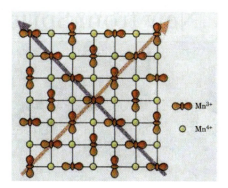

Fig. 12.24. Charge- and orbital ordering in La$_{0.5}$Sr$_{0.5}$MnO$_3$ as confirmed by resonant X-ray scattering. (From [451])

the other by Mn^{4+} ions. In addition there is orbital ordering of the Mn^{3+} sites as confirmed by resonant X-ray scattering [337] (see Fig. 12.24). Explanation of charge ordering requires the inclusion of electron repulsions between neighboring sites. Charge order of the form shown in Fig. 12.24 reduces those repulsions to a minimum.

In the following we want to discuss briefly the double-layer manganite systems La$_{2(1-x)}$Sr$_{1+2x}$Mn$_2$O$_7$ in order to demonstrate possible reentrant charge ordering caused by electron correlations. The structure consists of double layers of MnO$_6$ octahedra with rock-salt like MnO slabs in between. It is shown schematically in Fig. 12.25. When $x = 0.5$ we deal with LaSr$_2$Mn$_2$O$_7$, an antiferromagnetic compound with equal number of Mn^{3+} and Mn^{4+} ions. This

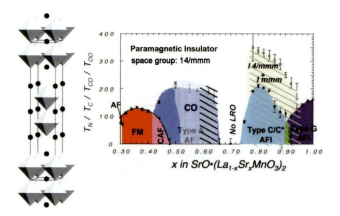

Fig. 12.25. Left panel: Crystal structure of LaSr$_2$Mn$_2$O$_7$ and La$_{2-2x}$Sr$_{1+2x}$Mn$_2$O$_7$ consisting of MnO$_6$ octahedra and (La,Sr)-cations (circles). Lattice constants are ($x = 0.5$) $a = 3.874$Å, $c = 19,972$Å. Right panel: Structural and magnetic phase diagram of bilayer La$_{2-2x}$Sr$_{1+2x}$Mn$_2$O$_7$ in the doping range $0.3 < x < 1.0$. Charge order (CO) occurs in the range $0.5 \leq x \leq 0.65$. Near $x = 0.7$ no long-range order is observed. CAF: canted AF order, AFI: antiferromagnetic insulator. (From [383])

suggests charge, and orbital ordering, which has in fact been observed. The phase diagram of La$_{2(1-x)}$Sr$_{1+2x}$Mn$_2$O$_7$ is very complicated. In Fig. 12.25 we show this to exemplify how electron correlations can lead to a rich variety of phases by slight modifications of materials. It also shows that the materials we understand reasonably well are obviously just the beginning of a vast manifold, with various competing interactions and instabilities.

Replacing in La$_2$SrMn$_2$O$_7$ the La^{3+} ions by Sr^{2+} implies hole doping. Then the following Hamiltonian seems appropriate for a description of charge- and spin-degrees of freedom,

$$H = \sum_{ij\sigma} t_{ij} \left(c_{i\sigma}^+ c_{j\sigma} + h.c. \right) + U \sum_i n_{i\uparrow} n_{i\downarrow} + V \sum_{\langle ij \rangle} n_i n_j$$
$$- J_H \sum_i \mathbf{S}_i \mathbf{s}_i + J \sum_{\langle ij \rangle} \mathbf{S}_i \mathbf{S}_j \quad . \tag{12.31}$$

The compact index $i = (\ell, \lambda)$ includes a layer index $\lambda = 1, 2$ and a site index ℓ within a layer. The first three terms of the Hamiltonian describe the kinetic energy of the e_g electrons, their on-site Coulomb repulsion and the repulsion when they are placed on nearest-neighbor sites. The latter repulsion plays an important role in charge order. The term before the last one describes the Hund's-rule energy of e_g electrons by their coupling to the high-spin $S = 3/2$ configuration of the localized t_{2g} electrons. Finally, there is an intersite spin-spin interaction via superexchange of neighboring $S = 3/2$ spins associated with the t_{2g} orbitals. The orbital e_g degeneracy is still missing in (12.31) but the problem would become too complex to be treated in a reasonably transparent way here. Therefore we concentrate on charge order only and disregard at this stage possible additional orbital order.

For a discussion of charge order we want to limit ourselves to a single layer instead of treating the double layer and assume that charge order is the same in both layers. So one might wonder why we fail to consider $La_{1-x}Sr_xMnO_3$ rather than $La_{2(1-x)}Sr_{1+2x}Mn_2O_7$. The answer is that $LaMnO_3$ has a cubic structure in which charge order in the presence of hole doping has not yet been observed.

Furthermore, for a study of charge degrees of freedom, the last two terms in (12.31) may be neglected. We assume that $U \gg t > V$. In order to describe charge order, we introduce two sublattices A and B and treat V on a mean-field level. The on-site interaction U is treated in CPA as outlined in Chapter 9[6]. For a square lattice with four nearest neighbors we may write the first three terms in (12.31) for a paramagnetic system in the form

$$H = V \sum_{i \in A, \sigma} \sum_{\delta, \sigma'}^{4} a_{i\sigma}^+ a_{i\sigma} b_{i+\delta\sigma'}^+ b_{i+\delta\sigma'} + U \sum_{\ell \in A,B} n_{\ell\uparrow} n_{\ell\downarrow}$$
$$+ t \sum_{\langle ij \rangle \sigma} \left(a_{i\sigma}^+ b_{j\sigma} + b_{j\sigma}^+ a_{i\sigma} \right) \quad . \quad (12.32)$$

Here we have replaced the $c_{\ell\sigma}^+$ operators by the ones $a_{i\sigma}^+, b_{j\sigma}^+$ for the sublattices A and B, respectively. Hopping processes have been limited to nearest neighbors. In CPA the above Hamiltonian is replaced by a single-particle Hamiltonian in the presence of disorder, i.e.,

$$H = \sum_{i \in A_1 \sigma} E_{A\sigma} a_{i\sigma}^+ a_{i\sigma} + \sum_{j \in B_1 \sigma} E_{B\sigma} b_{j\sigma}^+ b_{j\sigma} + t \sum_{\langle ij \rangle \sigma} \left(a_{i\sigma}^+ b_{j\sigma} + b_{j\sigma}^+ a_{i\sigma} \right)$$
$$- 2NV n_A n_B \quad . \quad (12.33)$$

Here $n_{A/B}$ are the average occupation numbers of sites on sublattice A and B, respectively. As before, N is the total number of sites. The orbital energies $E_{A\sigma}$ and $E_{B\sigma}$ are given by

$$E_{A\sigma} = 4V \begin{cases} n_B & \text{with probability} \quad 1 - \frac{1}{2} n_A \\ n_B + U & \text{with probability} \quad \frac{1}{2} n_A \end{cases}, \quad (12.34)$$

by assuming that $\langle a_{i\sigma}^+ a_{i\sigma} \rangle = \langle a_{i-\sigma}^+ a_{i-\sigma} \rangle$ and similar for b. The expression for $E_{B\sigma}$ is analogous. Solving this Hamiltonian by the CPA is almost identical

---

[6] see, e.g., [185]

to the case discussed in Sect. 9.1, except that we deal here with two sublattices. They are connected by the hopping matrix element t. Therefore, for the averaged Green's function $\bar{G}_A(\mathbf{k}, \omega)$ of sublattice A, we may write

$$\bar{G}_A(\mathbf{k}, \omega) = \bar{G}_A^{(0)}(\omega) + \bar{G}_A^{(0)}(\omega) t_\mathbf{k} \bar{G}_B^{(0)}(\omega) t_\mathbf{k} \bar{G}_A(\mathbf{k}, \omega) \tag{12.35}$$

and similar for $\bar{G}_B(\mathbf{k}, \omega)$. Here $t_\mathbf{k} = -2t(\cos k_x + \cos k_y)$ is the energy dispersion caused by nearest neighbor hopping given in units of the lattice constant. The $\bar{G}_{A/B}^{(0)}(\omega)$ are averaged Green's functions in the absence of any hopping and (9.15) applies here. Therefore

$$\bar{G}_A(\mathbf{k}, \omega) = \frac{1}{\omega - \Sigma_A(\omega) - t_\mathbf{k}^2 / (\omega - \Sigma_B(\omega))} \tag{12.36}$$

and similar for $\bar{G}_B(\mathbf{k}\omega)$. The CPA requires that on average the t-matrix vanishes (see (9.12)). This yields two equations of the form of (9.13). More specifically, we obtain

$$\Sigma_A(\omega) = 4V n_B + U n_A/2$$
$$- (4V n_B - \Sigma_A(\omega)) \bar{G}_A^{(0)}(\omega) (4V n_B + U - \Sigma_A(\omega)) \tag{12.37}$$

and similar for $\Sigma_B(\omega)$. These equations have to be solved for $\Sigma_A(\omega)$ and $\Sigma_B(\omega)$ under the constrain that $n_A + n_B = 2n$ with n denoting the average number of electrons per site. We are interested in computing the changes in charge ordering with temperature. Therefore we go over to the temperature Green's functions discussed in Sect. 7.1.2. The occupation numbers n_α ($\alpha = A, B$) are related to the average Green's functions $\bar{G}_\alpha(\mathbf{k}, i\omega_n)$ through

$$n_\alpha = \frac{2T}{N} \sum_{\mathbf{k},n} \bar{G}_\alpha(\mathbf{k}, i\omega_n) \ . \tag{12.38}$$

These two equations have to be solved self-consistently.

Whether solution with $n_A \neq n_B$ can be found depends on the size of V which enters the equations. We show in Fig. 12.26 the phase diagram for different values of $n = 1 - x$ as function of temperature T and V. The on-site repulsion is $U = 2$ in units of t. One notices a reentrant charge order for an interval of V values which depends on n. In that regime the ground state is homogeneous. However, at finite temperatures charge order sets in, and disappears again at even higher temperatures. The absolute temperatures in Fig. 12.26 for reentrant behavior are much too high for possible observation. This is a consequence of having assumed a nearest-neighbor interaction only. Reentrant charge order is not obtained in a self-consistent field approximation and therefore a true correlation effect. However, it can be shown that also polaron formation can lead to reentrant charge order.

In case of a ratio $Mn^{3+}/Mn^{4+} = 1$, i.e., for $x = 0.5$ one would expect charge ordering with a vector $\mathbf{Q} = (\frac{1}{2}, \frac{1}{2}, 0)$. Instead, a superstructure with

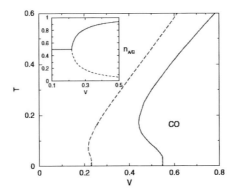

Fig. 12.26. T-V phase diagram for one layer of the double-layer compound $La_{2(1-x)}Sr_{1+2x}Mn_2O_7$ with $n = 1 - x = 0.3$ (solid line) and 0.5 (dashed line). Inset: occupancies n_A (solid line) and n_B (dashed line) as function of V (all energies in units of t) for $n = 0.5$ and $U \to \infty$. (From [185])

$\mathbf{Q} = (\frac{1}{4}, \frac{1}{4}, 0)$ is observed [240]. This is due to a Jahn-Teller distortion as a result of which orbital ordering takes place. As previously discussed the electrons of the Mn^{3+} ions order alternatingly in $d_{3x^2-r^2}$ and $d_{3y^2-r^2}$ orbitals in a staggered fashion. Above 100 K the superstructure reflections in X-ray scattering disappear, indicating that charge and orbital order have been destroyed by melting. Charge order for a ratio of $Mn^{3+}/Mn^{4+} = 1$ is also observed in other systems like $La_{0.5}Ca_{0.5}MnO_3$ or $Pr_{0.5}Sr_{0.5}MnO_3$ [452].

When hole doping is large the correlation energy decreases while the average distance between electrons increases. Then charge order is disfavoured (see Fig. 12.25). The magnetic interaction terms in (12.31) become important then. The double exchange term $-J_H \sum_i \mathbf{S}_i \mathbf{s}_i$ favors ferromagnetic (FM) coupling of Mn sites in the metallic state while the last term in (12.31) due to superexchange favors an antiferromagnetic arrangement of the two layers at $x = 0.5$. In the latter case we deal with an insulating state since the unit cell has doubled. Both types of order are shown in Fig. 12.27.

The spin-wave spectrum of the double layer has been measured[7] and calculated[8]. It consists of an acoustic and an optical branch. The dispersion in z direction is found to be smaller by a factor of 10^{-2} as compared with the one in the $x - y$ plane. The weak exchange coupling between layers results from the large separation of the double layers from each other. In the doping regime $0.4 < x < 0.5$ AF superexchange competes with FM double exchange with the result that a canted AF phase is formed. It can be considered as a superposition of an AF and FM component. For doping $x \leqslant 0.4$ the e_g conduction band turns more and more $d_{3z^2-r^2}$ like. Therefore double exchange

---

[7] see [63, 183, 365]
[8] see [62, 407]

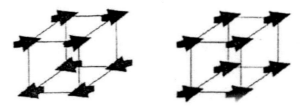

Fig. 12.27. Left side: AF coupling of the two layers in stoichiometric ($x = 0.5$) charge ordered $LaSr_2Mn_2O_7$ due to superexchange in z direction. Right side: FM coupling between layers induced by double exchange ($0.3 < x < 0.4$). Within a plane the FM spin arrangement is also due to double exchange, the mechanism of which is shown in Fig. 12.18.

along the z axis becomes dominant. In that case the superexchange term in (12.31) may be neglected. Before closing we want to reemphasize that the formation of polarons, which is only indirectly related to electron correlations and therefore has been discarded here, has a strong influence on the phase diagram.

12.2.2 Vanadates: $LaVO_3$

In the vanadate $LaVO_3$ the V^{3+} ions are in a $3d^2$ configuration. Therefore only t_{2g} orbitals are partially occupied and the e_g orbitals need not be considered. The structure is tetragonal, i.e., the octahedra are not tilted. Yet, they are J–T distorted and the distortion is of type G (see Fig. 12.28). Since Hund's-rule coupling is strong, the two d electrons are in a high-spin state $S = 1$. The system is a Mott-Hubbard insulator. Electronic structure calculations show that at every V site the d_{xy} orbital is singly occupied [327]. The second d electrons occupy a d_{xz} or d_{yz} orbital in an alternating fashion (see Fig. 12.28). The orbital order is of G type and follows from the J–T transition. The magnetic ordering is of C type for the following reason. The singly occupied d_{xy} orbitals cause an AF coupling within the ab planes. Along the z axis an alternating occupation of a d_{xz} and d_{yz} orbital is favored by double exchange; hence as observed experimentally ferromagnetic order is established. The transition temperatures for orbital ordering T_O and magnetic long-range order T_N are nearly the same, i.e., $T_O = 141$ K (1st order transition) and $T_N = 143$ K (2nd order transition). This is quite distinct from other orbitally ordered systems where T_O is usually much higher than T_N.

12.2.3 Ladder Systems: $\alpha'-NaV_2O_5$

The interest in the layered quasi-one dimensional (1D) perovskite $\alpha'-NaV_2O_5$ originates from the observation of a spin gap in the low-temperature suscepti-

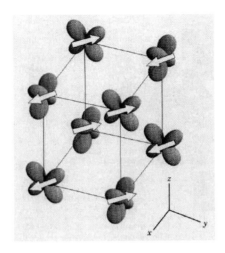

Fig. 12.28. Spin and orbital order in $LaVO_3$. Spin ordering is of type C while orbital ordering is of type G (compare with Fig. 12.19). Not shown is the singly occupied d_{xy}-orbital on each lattice site. (From [451])

bility of that compound. In order to understand the significance of this finding some background information is required.

In Sect. 10.2.2 we have discussed the appearance of a SDW in a metal due to nesting properties of the Fermi surface. The description given there applied to weakly correlated systems. A one-dimensional *metal* is insofar special as the Fermi surface has here the form of a slab. The nesting condition is therefore fulfilled at each point of the Fermi surface. Consequently, a charge density wave (CDW) forms and a gap opens over the whole surface. If the conduction band is half filled, this results in a dimerization, i.e., the CDW is accompanied with a lattice distortion. The unit cell is doubled. This is known as *Peierls transition*. It occurs at a wave vector $q = 2k_F$ and at a transition temperature controlled by the interchain coupling. In Chapter 14 we discuss the case of trans-polyacetylene, which is a prototype for a Peierls transition with an alternating change occurring in the C-C-bond length.

Surprisingly a similar dimerization can occur in a 1D-*insulator*, but here with respect to the spins. This is a spin-Peierls transition. A prerequisite for it is an antiferromagnetic interaction between spins on neighboring sites[9]. There are a number of organic spin-chain systems in which a spin-Peierls transition has been observed, but there exist hardly any inorganic chain systems which show that phenomenon. Noticeable exceptions are $CuGeO_3$ and $\alpha'-NaV_2O_5$ [172, 206]. The reason for the rareness of spin-Peierls transitions in inorganic chain systems is presumably that they are overruled by magnetic interchain coupling.

---

[9] for early work see [74, 381]

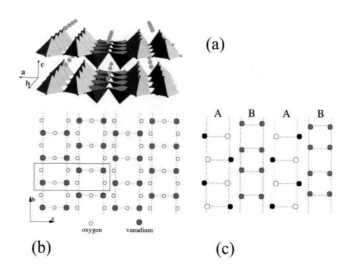

Fig. 12.29. (a) Layered perovskite structure of α'–NaV_2O_5 consisting of chains of oxygen pyramids that contain the V atoms aligned along the crystal b axis. The layers are stacked along c axis. Na atoms (grey spheres) are centered above the ladder plaquettes. (b) ab-plane of the Trellis lattice structure consisting of V-V ladders alternatingly shifted along b by half a lattice constant. This leads to a quasi-'triangular' structure for V-V-rung units. An orthorhombic high-temperature unit cell is indicated. (c) A ladders with zig-zag charge order and B ladders with dimerization. (From [29])

In α'–NaV_2O_5 a spin-Peierls transition has been observed at $T_c \simeq 33$ K. The systems consist of double chains of oxygen pyramids. The V atoms in the center of the base of the pyramid are aligned along the b-axis (see Fig. 12.29). In the homogeneous high-temperature phase the V atoms have valency $V^{4.5+}$ and hence an average of $3d^{0.5}$ per site. Yet the system is not metallic. The low-temperature monoclinic structure can be understood only, if in addition to spin dimerization a charge ordering transition of V^{4+} and V^{5+} ions takes place. For $T < T_c$ the magnetic susceptibility shows a behavior typical for spin chains, i.e., the spin excitations are isotropically gapped.

In order to understand the system, one has to take into account that around 33 K there are actually two phase transition very close to each other, i.e., a first-order transition at $T_{c_1} = 33$ K and a second-order transition at $T_{c_2} = 32.7$ K. Experiments suggest that at T_{c_1} charge ordering takes place on the V sites, i.e., there are lattice sites with ordered V^{4+} ions and other sites with V^{5+} ions. However, that does not imply that all lattice sites participate in charge ordering. Some experiments indicate also that in the Trellis lattice structure shown in Fig. 12.29 only every other ladder charge orders [295]. This finding has been questioned, so that at present uncertainty remains. However, the measured magnetic excitations with **q** vector perpendicular to the ladders are in accord with the assumption of alternating ordered and disordered ladders. At T_{c2} an isotropic spin gap opens which suggests a dimerization of spins, here in the charge ordered A chains (see Fig. 12.29). Associated with the dimerization is a lattice distortion which shortens the distances between sites forming dimers and elongates the remaining intersite distances. Therefore α'-NaV$_2$O$_5$ is another example of an intricate interplay between charge ordering and gapped spin excitations. In order to obtain better insight into the underlying physics, the system was studied by a LDA+U calculation [497]. However, despite their usefulness the calculations do not explain why the system remains insulating even above T_{c_1} as experimentally observed. Therefore, a description in terms of an extended Hubbard model seems more appropriate.

The model Hamiltonian is the one of (11.1) but extended to the following form

$$H = \sum_{\langle ij \rangle_R} t_R \left(a^+_{i\sigma} a_{j\sigma} + h.c. \right) + \sum_{\langle ij \rangle_L} t_L \left(a^+_{i\sigma} a_{j\sigma} + h.c. \right)$$
$$+ \sum_{\langle ij \rangle_I} t_I \left(a^+_{i\sigma} a_{j\sigma} + h.c. \right) + \sum_{\langle ij \rangle_D} t_D \left(a^+_{i\sigma} a_{j\sigma} + h.c. \right)$$
$$+ \sum_{\langle ij \rangle_R} V_R n_i n_j + \sum_{\langle ij \rangle_L} V_L n_i n_j + \sum_{\langle ij \rangle_I} V_I n_i n_j + U \sum_i n_{i\uparrow} n_{i\downarrow} \quad .$$
(12.39)

The indices R, L, D and I refer to V ions on a rung, a leg and a diagonal between two rungs of a ladder, while the index I refers to nearest neighbor V sites belonging to different ladders (see Fig. 12.29). The interactions and hopping matrix elements are illustrated in Fig. 12.30. The problem can be considerably simplified if we limit the configurations to those with one electron per rung. In that case the Hamiltonian can be reduced to one of a pseudospin. It ensures insulating behavior even above the charge ordering temperature. The pseudospin $\tau(i) = \frac{1}{2}$ describes a $3d_{xy}$ electron in rung i with $\tau_z(i) = +\frac{1}{2}$ when it occupies the left and $\tau_z(i) = -\frac{1}{2}$ when it occupies the right V atom of the rung. The Hamiltonian projected onto the Hilbert space of the pseudospins is of the form

$$H = \sum_{\langle ij \rangle_L} K(i,j) \tau_z(i) \tau_z(j) + \sum_{\langle ij \rangle_{LL'}} I(i,j) \tau_z(i) \tau_z(j)$$
$$+ \sum_i 2\tilde{t}_R(i) \tau_x(i) \quad .$$
(12.40)

In the first term the sum refers to neighboring pairs i, j within a ladder. There are three of them per V site. This term favors configurations in which the nearest neighbors of an occupied V site within the same leg are unoccupied. The second term refers to nearest neighbors belonging to different ladders and

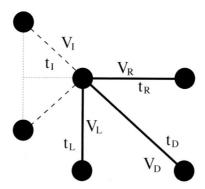

Fig. 12.30. Square segment of a ladder of V ions with different, i.e., frustrated intersite repulsions used in a model Hamiltonian. Dashed lines connect to nearest neighbor sites on a leg belonging to a neighboring ladder. Solid lines connect sites on the same ladder.

minimizes the electron repulsions between them, while the last term describes hopping between two V ions forming a rung. The Hamiltonian (12.40) has the form of an Ising model in a transversal magnetic field.

The spin interactions between all neighboring V ions are of the superexchange type. They take place via oxygen ions. The couplings constant differ when the ions i, j are on the same ladder and when they belong to neighboring ladders of type A and B, respectively. Therefore we want to couple the spin interactions with the charge degree of freedom. We write for the modified parameters in (12.40)

$$K(i,j) = 2V_L + \delta K(\mathbf{S}_i, \mathbf{S}_j)$$

$$I(i,j) = V_I + \delta I(\mathbf{S}_i, \mathbf{S}_j)$$

$$\tilde{t}_R = t_R + \sum_{\langle ij \rangle_R} \delta t_R(\mathbf{S}_i, \mathbf{S}_j) \quad (12.41)$$

without specifying here the adjustable parameters $\delta K, \delta I$ and δt_R. Due to these modifications charge (τ) and spin (**S**) degrees of freedom are coupled. As a consequence the optical conductivity which probes charge excitations of rungs shows signatures of coupled spin excitations even above T_{c_2}.

From LDA+U calculations values of $\tilde{t}_R \simeq -0.2$eV and $K \simeq I = 0.7$eV are obtained [29]. We want to draw attention to the frustration caused by the Coulomb repulsions and also by the antiferromagnetic spin interactions between V ions belonging to different ladders (see Fig. 12.29). Considering a single ladder only and discarding the spin interactions, we obtain a zig-zag charge order when the ratio $K(i,j)/4\tilde{t}_R = \lambda$ exceeds a critical value $\lambda_c = 1$. When the interladder interaction $I(i,j)$ becomes equal or larger than $K(i,j)$, the zig-zag order melts, and the charge order changes to one of V^{4+} ions on one side of the ladder and V^{5+} ions on the other side. Estimates from LDA+U calculations show that $K \approx I$. Therefore, due to frustration the system has many states with nearly the same energy. This quasi-degeneracy can be lifted by distortions of the lattice.

The driving mechanism for spin dimerization is the spin interaction energy obtained from superexchange. Due to the quarter filling of the lattice the intermediate states do not only consist of doubly occupied sites as we found it to be the case in the Hubbard model at half filling (see (10.99)). Instead they include also rungs with two or no electrons instead of one. Furthermore, the effective exchange constant depends on the form of charge order, e.g., it is different for zig-zag order and order with all V^{4+} sites on one side of the ladder. On a given ladder there is a competition between dimerization of rungs and zig-zag charge ordering. This can be optimized by a different order in alternating ladders. While in ladders of type A electrons are mainly ordered in a zig-zag order, in ladders of type B they mainly dimerize in rungs. This takes place at T_{c_1}. Due to the frustration which has to be counterbalanced, the transition temperature is low compared with other charge-ordered transition-metal oxides. Just below T_{c_1} there is a spin gap resulting from dimerization in type-B ladders but no gap in A-type ladders. With increasing charge ordering below T_{c_1} the zig-zag order in the type A ladders increases and the interrung hopping decreases. At this stage the spin excitations in the type A ladders are gapless like in a spin chain. Below T_{c_2} a dimerization of rungs takes also place in type A ladders. Associated with it is a spin-Peierls transition with a spin-gap opening in type A ladders. This is seen in Knight shift experiments [479] and supports the assumption of alternating type A and B ladders. From the above it is seen that an understanding of the properties of α'-NaV_2O_5 requires quite an advanced theoretical modeling. Despite of this we have included that system here as an example of how advanced modern material sciences have become and of the important role electron correlations play hereby.

12.2.4 Other Oxides

Of particular interest are ruthenate perovskites, one reason being that Sr_2RuO_4 seems to be a spin-triplet superconductor. The Ru^{4+} ions are in a $4d^4$ configuration. Hund's-rule correlations are not strong enough to dominate the CEF and therefore all d electrons are in t_{2g} orbitals. We deal here with a $S = 1$ state. In distinction to Sr_2RuO_4 the compound Ca_2RuO_4 is a Mott-Hubbard insulator. Localization of electrons takes place below $T_{MH} = 350$ K, while AF magnetic order sets in at $T_c = 110$ K. The metal-insulator transition is understood best by assuming that it is the d_{xy} orbital of the t_{2g} manifold which is doubly occupied [10,188]. This may be due to distortions of the RuO_6 octahedra in c direction. The spins of electrons in d_{xz} and d_{yz} orbitals at neighboring sites are coupled antiferromagnetically through superexchange.

In the titanate LiTiO$_3$ the Ti^{3+} ions are in a $3d^1$ configuration and therefore only t_{2g} orbitals have to be considered. Experimentally Néel order is observed below $T_N = 130$ K [233]. However, neither J–T distortions nor orbital ordering is observed at any temperature. Apparently the hopping matrix elements are sufficiently large so that J–T distortions do not form. When the lattice constant is increased by replacing La by Y, hopping is reduced and orbital order sets in. At low temperature the system is a ferromagnet with $T_c = 30$ K due to double exchange.

13
Heavy Quasiparticles

One characteristic feature of strong electron correlations is the appearance of low-energy scales. In Sect. 10.2.1 a simple model was used to demonstrate this phenomenon. In addition, that simple model contains key ingredients of the Kondo effect[1].

The Kondo temperature T_K quantifies the low-energy scale which appears when a magnetic impurity with strongly correlated valence electrons is placed into a metal. In a system in which the magnetic ions form a lattice, such as $CeAl_3$, $CeRu_2Si_2$, $CeCu_2Si_2$, Yb_4As_3, $YbAl_3$, UBe_{13}, $NpBe_{13}$, UPt_3 or UPd_2Al_3, the low-energy excitations may form coherent states. As a result the effective mass of the quasiparticles can be several hundred times the free electron mass; thus, it may become larger than, e.g., the mass of μ mesons. When the aforementioned situation occurs, these particles are classified as heavy quasiparticles or heavy fermions.

Although most of the heavy fermion systems contain $4f$ or $5f$ ions, heavy quasiparticles are also found in LiV_2O_4, where d electrons are involved. That system is, however, special insofar as it has a geometrically frustrated lattice structure. Strongly correlated electrons on frustrated lattice structures are an interesting problem in itself and therefore we will examine the subject separately in Sect. 14.3.

Heavy-fermion systems with heavy quasiparticles satisfy the following requirements:

(a) The low-temperature specific heat $C = \gamma T$ has a coefficient γ of order 1 J mol^{-1} K^{-2}, rather than 1 mJ mol^{-1} K^{-2} as, e.g., in the case of sodium or copper metal;
(b) the Pauli paramagnetic susceptibility χ_s is similarly enhanced as γ, and;
(c) the ratio $R = \pi^2 k_B^2 \chi_s / (3\mu_{\text{eff}}^2 \gamma)$ is of order unity. Here μ_{eff} is the effective magnetic moment of the quasiparticles. It is obtained, e.g., from measurements of the spin susceptibility at elevated temperatures.

---

[1] see Sect. 10.3

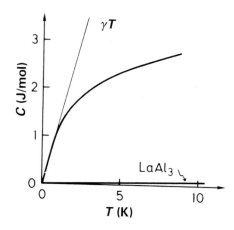

Fig. 13.1. Qualitative plot of the specific heat C(T) of a heavy-fermion system like CeAl$_3$. Also shown is the specific heat of LaAl$_3$, a system without $4f$ electrons.

Sometimes a condition for the resistivity $\rho(T) = AT^2$ with large prefactor A is added. The quantities γ and χ_s are both proportional to the quasiparticle density of states at the Fermi level $N^*(0)$. The latter is proportional to m^*, i.e., the effective mass of the fermionic excitations. Large values of γ and χ_s can therefore be interpreted by ascribing a large m^* to the quasiparticles. When the ratio R (Sommerfeld-Wilson ratio) is calculated, the density of states $N^*(0)$ drops out since χ_s is also proportional to $N^*(0)$. For free electrons $R = 1$. Therefore, when conditions (a)-(c) are met, we may assume a one-to-one correspondence between the quasiparticle excitations of the complex metallic system with strong electron correlations and those of a free-electron gas, provided we use the effective mass m^* instead of the free-electron mass. A ratio $R \neq 1$ indicates that quasiparticle interactions are not negligible. As the temperature increases to values above T^*, the excitations lose their heavy-fermion character; the specific heat levels off as indicated in Fig. 13.1, and the susceptibility changes from Pauli- to Curie-like behavior. With increasing temperature the rare earth or actinide ions behave more and more like ions with well-localized f electrons.

Another interesting phenomenon is the following. With respect to the thermodynamics at low temperatures, the f electrons of constituents like Ce seem to be placed right at the Fermi energy, giving rise to the large density of states (Fig. 13.2). However, it takes approximately 2 eV in a photoemission experiment to excite a $4f$ electron of Ce into an unoccupied conduction electron state above the Fermi energy (see Fig. 10.9). What at first sight seems to be a contradiction proves in fact to be none. The two parts in Fig. 10.9 merely demonstrate that there are two types of electronic excitations: namely, low-energy excitations involving predominantly spin degrees of freedom and

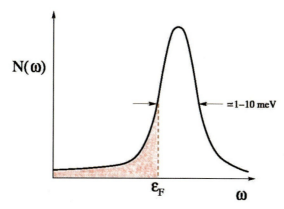

Fig. 13.2. Quasiparticle density of states as obtained from the low-temperature thermodynamics. The high value near ϵ_F is due to the f electrons and is absent when, for example, Ce is replaced by La.

high-energy excitations involving charge degrees of freedom of the $4f$ electrons.

It is important to realize that there are quite different microscopic origins of heavy quasiparticle excitations. Originally it was thought that the Kondo effect is the only source of such behavior. That turned out to be incorrect. At present, a number of different model systems are known. The most widespread is, indeed, the Kondo-lattice model which applies to Ce systems and to a number of Yb compounds as well. But heavy quasiparticles may also form due to partial charge order in a system. The semimetal Yb_4As_3 is such an example. In U intermetallic systems an orbital selective partial localization of $5f$ electrons is responsible for the heavy quasiparticle; UPt_3 as well as UPd_2Al_3 serve as examples. Nearness to a quantum critical point like in YMn_2 or special lattice structures like the pyrochlore lattice with geometric frustration (for example, LiV_2O_4) may also result in the formation of heavy quasiparticles. Finally the Zeeman effect may cause heavy quasiparticle-like behavior as in $Nd_{2-x}Ce_xCuO_4$.

Apparently, the low-lying excitations characterizing systems with heavy quasiparticles (so called heavy-fermion systems) involve predominantly spin degrees of freedom. Direct evidence of this is seen in the entropy associated with the excess specific heat at low temperatures. It is of order $S \simeq k_B ln\nu$ per f or d site. Here ν denotes the degeneracy of the ground-state of the atomic $f(d)$ shell. As in the case of a Kondo impurity, one spin excitation is associated with each (e.g., Ce) site. The excitation energy is of order $k_B T^*$ and defines a characteristic low-energy scale of the system. The lower T^* is, the smaller is the change in the f or d charge associated with the excitation (see the model discussed in Sect. 10.2.1) Eventually we may speak of an approximate separation of excitations with spin and with charge degrees of freedom. This does *not* imply that, for example, for a Kondo lattice system T_K and

T^* are the same temperatures. Instead, one expects in most cases $T^* < T_K$. This difference originates in the magnetic interactions between different Ce or actinide ions. They are coupled by the RKKY interaction via the conduction electrons and that interaction can favor ferromagnetic as well as antiferromagnetic spin alignment. We lose this interaction energy when nonmagnetic singlets are formed: model calculations for two Kondo impurities [218] show that antiferromagnetic correlations between the magnetic sites weaken the energy gain caused by singlet formation.

This argument suggests that singlets should *not* form when the magnetic interaction energy per site exceeds the singlet formation energy (Doniach criterion). In systems like $CeAl_2$, $CePb_3$, and $NpBe_{13}$ this seems to be the case, since at low temperatures they become antiferromagnets. The difficulties in dealing with Kondo lattices result from the two regimes where the formation of local singlets dominates on one side, and where the magnetic interactions dominate on the other. For both regimes separate mean-field theories are available which convert the many-body quantum problem into a classical one. But there is no unifying approach available yet which incorporates the two limiting cases.

We will be mainly dealing with the heavy quasiparticle phase that forms below a temperature $T_{\text{coh}} < T^*$ when the single-site excitations lock together. The formation of heavy quasiparticles does not only show up in the thermodynamic properties of a system, but also in de Haas-van Alphen experiments[2] and photoemission studies[3]. Heavy quasiparticles may lead to strong mass anisotropies at the Fermi surface, a result which raises the question of how one can calculate the Fermi surface and the anisotropic masses for a system of strongly correlated electrons. For Kondo lattice systems, renormalized bandstructure calculations have proven a successful computational scheme for doing so; we can calculate Fermi surfaces as well as the strongly anisotropic effective masses for those systems with only one adjustable parameter.

13.1 Kondo Lattice Systems

Heavy fermion behavior in Ce intermetallic compounds is, to our knowledge, always due to the Kondo effect. In the following we will often refer to $CeRu_2Si_2$ which is a prototype for Kondo lattice systems. As pointed out before, coherent excitations in the form of heavy quasiparticles build up below a temperature T_{coh}. This temperature can be approximately determined by measuring the temperature dependent resistivity $\rho(T)$ (Fig. 13.3). In dilute magnetic alloys it increases rapidly below T^* due to the Kondo effect. Yet below T_{coh} the function $\rho(T)$ decreases strongly because of the formation of coherent Bloch-like states, and one observes a behavior $\rho(T) = AT^2$ typical for Fermi liquids.

---

[2] see, e.g., [12, 290, 423]
[3] see [88]

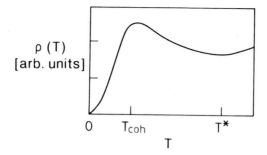

Fig. 13.3. Schematic plot of the temperature dependent resistivity $\rho(T)$ of a Kondo lattice system. Below T_{coh} a typical Fermi liquid behavior is observed, i.e., $\rho(T) \propto T^2$, while near T^* the behavior resembles that of a metal with Kondo impurities.

When T increases above T_{coh}, the mean-free path of the (spin-dominated) excitations of the f electron system becomes so short that coherence can no longer be maintained and the heavy quasiparticles disappear gradually. For $T_{\text{coh}} \leq T \leq T^*$ the specific heat contains large contributions from the incoherent parts of the f electron excitations. This temperature interval can be nicely described by a theory based on a Noncrossing Approximation (NCA). A discussion of that theory would however go beyond the introduction to the correlation problem we intend to give here. For more details of this NCA-based theory, see [131], or for an extensive account of the method see [35] which also lists the original literature.

When $T \gg T^*$ the f electrons can be treated as localized and their local moments are seen to interact weakly with the spins of the conduction electrons (effective $s-f$ exchange interactions). The system can then be treated as a collection of Kondo impurities and coherent heavy quasiparticles no longer exist. The Fermi surface encloses a volume in momentum space given by the conduction electrons only, i.e., excluding the $4f$ electrons. In the case of $CeRu_2Si_2$ it is therefore the same as the one of $LaRu_2Si_2$. This has been confirmed by photoemission spectroscopy (see Fig. 13.4). The situation differs for $T < T_{\text{coh}}$, where the $4f$ electrons of $CeRu_2Si_2$ contribute to the Fermi surface.

13.1.1 Renormalized Band Theory

We want to show that the Fermi surface of a system with heavy quasiparticles can be calculated by Renormalized Band Theory. But in order to understand this approach it is instructive to consider first the case of a single, e.g., Ce impurity embedded in a sea of conduction electrons.

When a magnetic impurity forms a singlet with the conduction electrons, it acts like a nonmagnetic scattering center. A nonmagnetic scattering potential can be characterized by energy-dependent phase shifts $\eta_l(\epsilon)$ [242], where l

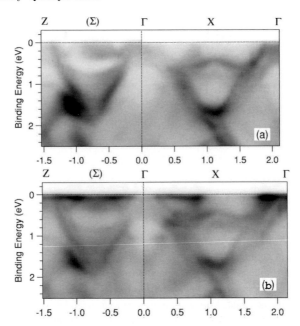

Fig. 13.4. Photoemission results for (a) LaRu$_2$Si$_2$ in comparison to (b) CeRu$_2$Si$_2$ at $T = 25K$, i.e., above the Kondo temperature $T^* = 15K$ of that system. Band structures are very similar for both compounds. In CeRu$_2$Si$_2$ there are still signatures to be seen near zero energy of the heavy quasiparticles [89].

denotes the angular momentum quantum number of the scattered electron. Friedel's sum rule relates the phase shifts at the Fermi energy, i.e., $\eta_l(\epsilon_F)$, to the charge $-Ze$ bound by the scattering potential,

$$Z = \frac{2}{\pi} \sum_l (2l+1)\, \eta_l\,(\epsilon_F) \quad . \tag{13.1}$$

The energy dependence of the phase shifts near ϵ_F is directly related to the density of low-lying excitations. The local enhancement of the density of states (per spin direction) due to the presence of the impurity, $\delta N(\epsilon)$, follows from (13.1) as

$$\delta N(\epsilon) = \frac{1}{\pi} \sum_l (2l+1) \frac{d\eta_l(\epsilon)}{d\epsilon} \quad . \tag{13.2}$$

If we start from the Anderson Hamiltonian (10.23), it becomes evident that the f phase shift ($l = 3$) is of particular importance. In order for the ground state of a Ce impurity to have an f electron number close to $n_f = 1$, the phase shift $\eta_{l=3}(\epsilon_F)$ would have to be in the vicinity of $\pi/[2(2l+1)]$, provided all f orbitals are degenerate. This is, however, never the case. The lowest J multiplet of a Ce ion is $J = 5/2$ and the crystalline electric field splits the $(2J+1)$-fold degenerate multiplet into a sequence of Kramers' doublets. In

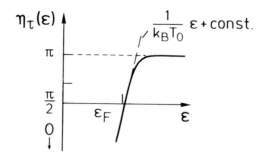

Fig. 13.5. Qualitative plot of the f phase shift $\eta_\tau(\epsilon)$ near ϵ_F. At ϵ_F, $\eta_\tau(\epsilon_F) \simeq \pi/2$. The dashed line has a slope of $1/k_B T_0$. The remaining phase shifts of the s, p, and d electrons are practically constant over energies of the order of $k_B T_0$.

the ground state only the lowest doublet is occupied; we characterize it by a *pseudospin* index $\tau = \pm 1$. Therefore, among the different f electron scattering channels, only those with the symmetry of the crystal-field ground state have a non-vanishing phase shift labeled $\eta_\tau(\epsilon)$. According to the above sum rule, $\eta_\tau(\epsilon_F)$ must be close to, but slightly less than $\pi/2$ in order to bind nearly one f electron. We show a schematic plot of $\eta_\tau(\epsilon)$ in Fig. 13.5. Close to ϵ_F, the following expansion holds for the $\eta_\tau(\epsilon)$:

$$\eta_\tau(\epsilon) = \eta(\epsilon_F) + \frac{1}{k_B T_0}(\epsilon - \epsilon_F) + \sum_{\epsilon'} \phi_{\tau,-\tau}(\epsilon, \epsilon') \delta n_{-\tau}(\epsilon') \qquad (13.3)$$

with $\eta(\epsilon_F) \simeq \pi/2$. The linear term defines a characteristic temperature T_0 and leads to an excess density of states at ϵ_F per pseudospin

$$\delta N_\tau(0) = \frac{1}{\pi k_B T_0}, \qquad (13.4)$$

where T_0 is closely related to the Kondo temperature T_K. From Fig. 13.5 we learn that the excess density of states is limited to an energy range $k_B T_0$ around the Fermi surface. It corresponds to the singlet-triplet excitation energy, which here is smeared out over an interval of order $k_B T_0$.

The last term on the right-hand side of (13.3) describes the effect on the phase shift $\eta_\tau(\epsilon)$ of a distribution $\delta n_{\tau'}(\epsilon')$ of quasiparticles which might be present, i.e., it represents quasiparticle interactions. For completeness we have included it within the spirit of Landau's Fermi liquid theory and it is absent when only one quasiparticle is considered. Since we want to set the stage for computing the Fermi surface we will presently ignore it, but we will show interesting quasiparticle interaction effects in Appendix D.

After having formulated the single impurity Kondo problem in terms of phase shifts, we want to apply a similar procedure to the Kondo lattice case.

A gas of noninteracting quasiparticles near the Fermi energy is parameterized by the direction dependent Fermi wavevector $\mathbf{k}_F$ and Fermi velocity $\mathbf{v}_F$, i.e.,

$$E(\mathbf{k}) = \mathbf{v}_F(\hat{\mathbf{k}})(\mathbf{k} - \mathbf{k}_F) \quad . \tag{13.5}$$

Here $\hat{\mathbf{k}}$ denotes the direction on the Fermi surface. The renormalized band method determines the band structure for a given effective potential. As pointed out before, this potential can be completely described by a set of energy-dependent phase shifts $\{\eta_l^A(\epsilon)\}$. Here A denotes the different atoms in the unit cell, and l is the orbital angular momentum quantum number. The phase shifts contain all necessary information about the periodic potential. Consider, for example, CeRu$_2$Si$_2$. The phase shifts at the Fermi energy ϵ_F, i.e.,

$$\{\eta_l^A(\epsilon_F)\} = \left\{\eta_l^{Ce}(\epsilon_F), \eta_l^{(Ru)\nu}(\epsilon_F), \eta_l^{(Si)\mu}(\epsilon_F)\right\} \quad ; \quad \nu, \mu = 1, 2 \quad , \tag{13.6}$$

determine the Fermi surface of the material. The partial electronic densities n_l^A are given by[4]

$$n_l^A = \frac{2}{\pi}(2l + 1)\eta_l^A(\epsilon_F) \quad , \tag{13.7}$$

and knowing all the η_l^A is equivalent to knowing the Fermi surface.

The Fermi velocities and hence effective masses are obtained from the derivatives

$$\{\dot{\eta}_l^A(\epsilon_F)\} = \left\{\left(\frac{d\eta_l^A(\epsilon)}{d\epsilon}\right)_{\epsilon=\epsilon_F}\right\} \quad . \tag{13.8}$$

There is an important constraint which must be observed by the phase shifts: the volume Ω_F in reciprocal space enclosed by the Fermi surface must equal half the number of valence electrons n_{val} (including the f electrons) per volume of the unit cell (Luttinger's theorem), i.e.,

$$n_{\text{val}} = \frac{2}{(2\pi)^3}\Omega_F\left(\eta_l^A(\epsilon_F)\right) \quad . \tag{13.9}$$

This condition reduces the number of parameters $\eta_l^A(\epsilon_F)$ by one.

From the treatment of a single Kondo impurity we know that the strong correlations are linked to the phase shift for $l = 3$ of the Ce site, i.e., to $\eta_{l=3}^{Ce}(\epsilon)$. All other phase shifts ought to remain essentially unaffected by the correlations of f electrons and thus are taken from an LDA calculation. After all phase shifts except $\eta_{l=3}^{Ce}(\epsilon)$ have been determined from LDA by treating the f electron of Ce as part of the core, the Fermi surface is already fixed to a considerable extent; only the phase shift $\eta_{l=3}^{Ce}(\epsilon)$ remains to be determined. After the discussion related to (13.3) it is apparent that only the phase shifts $\eta_\tau^{Ce}(\epsilon_F)$ among the different $l = 3$ channels differ from zero. We present an

---
[4] see [242]

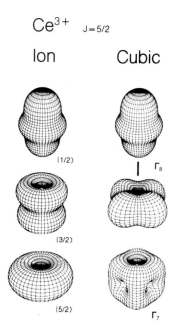

Fig. 13.6. Contours of equal $4f$ charge density for a free Ce^{3+} ion and for Ce^{3+} ion in cubic symmetry (Courtesy of U. Walter).

example of the $4f$ charge density in Fig. 13.6, where the reader can see that it reflects the cubic symmetry. Close to the Fermi energy, the phase shift $\eta_\tau^{Ce}(\epsilon)$ can be parameterized by the resonant form

$$\eta_\tau^{Ce}(\epsilon) = \arctan \frac{\tilde{\Gamma}}{\tilde{\epsilon} - \epsilon} \quad . \tag{13.10}$$

The parameters $\tilde{\epsilon}$ and $\tilde{\Gamma}$ denote the center of the narrow f-like band and its width, respectively (see Fig. 13.7). Thus $\tilde{\epsilon}$ differs slightly from ϵ_F. Instead of using the parameterized form (13.10), we can alternatively expand $\eta_\tau^{Ce}(\epsilon)$ near the Fermi surface in the form of

$$\eta_\tau^{Ce}(\epsilon) = \eta_\tau^{Ce}(\epsilon_F) + \frac{1}{k_B T^*}(\epsilon - \epsilon_F) + \sum_{\tau',i,\epsilon'} \Phi_{\tau\tau'}^i(\epsilon,\epsilon') \delta n_{\tau'}^i(\epsilon') \quad . \tag{13.11}$$

Comparing this expansion with (13.3), we notice a difference in the prefactor of the linear term as well as in the interaction term due to the presence of a lattice. We use the slope of the phase shift at ϵ_F to define a characteristic temperature T^*. Its value fixes the width of the resonance at ϵ_F (Fig. 13.7) and the effective mass of the quasiparticles of f character as well. The last term in (13.11) describes the effect of other quasiparticles, which may be present either on neighboring sites i of the Ce site or on the site itself ($i = 0$).

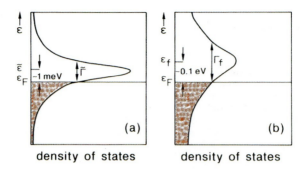

Fig. 13.7. Density of states multiplied by the respective bandwidth versus energy (schematic): (a) when a renormalized band calculation is done, (b) from a LDA calculation for a Ce compound like CeRu$_2$Si$_2$. In a system with Yb^{3+} ($4f^{13}$) ions like YbRh$_2$Si$_2$ the role of electrons and holes is interchanged. Note that the energy scales in (a) and (b) differ [508].

The Pauli principle prevents two f electrons from occupying a Ce site with the same quantum number τ. Since the quasiparticles have predominantly $4f$ character, we have as a consequence $\Phi^0_{\tau\tau} = 0$. Here we are interested only in the energy dispersion $\epsilon_{qp}(\mathbf{k})$ of a single quasiparticle, i.e., when no other quasiparticles are present, and the last term in (13.11) is zero. The theory then contains the parameters $\eta^{Ce}_\tau(\epsilon_F)$ and T^* only, which can be expressed in terms of $\tilde{\epsilon}$ and $\tilde{\Gamma}$ if desired.

The partial density of states derived from (13.10) is proportional to $\tilde{\eta}^{Ce}_\tau(\epsilon)$ and is shown in Fig. 13.7a. A band calculation within LDA would yield a qualitatively similar picture as indicated in Fig. 13.7b with $\tilde{\Gamma}$ and $\tilde{\epsilon}$ replaced by Γ_f and ϵ_f, respectively. The calculated LDA values of ϵ_f and Γ_f are, however, of order 0.1 eV and far too large due to an inadequate treatment of the strong f electron correlations within that approximation (Sect. 4.3). Here we determine $\tilde{\Gamma}$ and $\tilde{\epsilon}$ by using relation (13.10) and by requiring that the γ value in the specific heat be correctly reproduced when the calculated quasiparticle dispersions are employed for its determination. Because of the constraint (13.9), renormalized band calculations constitute a one-parameter theory. A schematic summary of the different computational steps is given in Table 13.1.

Within the frame of renormalized band calculations the calculated f bands are modified as follows:

(a) the crystal-field splitting is adequately taken into account
(b) the f resonance width is reduced
(c) the center of gravity of the f resonance moves closer to the Fermi energy.

After the bands have been determined, we can calculate the Fermi surface cross sections measured in a de Haas-van Alphen experiment as well as the

Table 13.1. List of the different computational steps taken in renormalized band-structure calculations. (From [508])

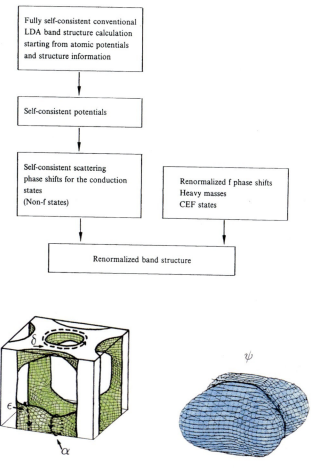

Fig. 13.8. Plot of some of the calculated parts of the Fermi surface of CeRu$_2$Si$_2$ [512]. Of the four closed hole sheets only the heavy hole sheet ψ is shown. (Courtesy of G. J. McMullan)

effective masses. Consider CeRu$_2$Si$_2$ as an example. The Fermi surface consists of five separate sheets: four of them are closed hole surfaces while the remaining one is a multiply-connected electronic surface with extremal orbits of rather different character (Fig. 13.8). Listed in Table 13.2 are some of the measured extremal areas of the Fermi surface with the corresponding effective masses. We compare these experimental findings with the results of conventional bandstructure calculations based on the LDA and with those of renormalized bandstructure theory. As far as the measured Fermi-surface topology is concerned, the deviations between LDA calculations and renormalized band calculations are relatively small. However, in contrast to the former, the renormalized band calculations reproduce well also the large measured mass anisotropies. The contributions from the large hole surface ψ with a measured $m^* \simeq 120\, m_0$ dominate the specific-heat coefficient $\gamma \simeq 350\, mJ\, mol^{-1} K^{-2}$ to which T^* (or $\tilde{\Gamma}$) have been fitted. We may thus conclude that we can obtain the anisotropic mass *ratios* without any additional fit parameter.

Of interest is also the Fermi surface of CeCu$_2$Si$_2$ as obtained by *Zwicknagl* from renormalized band theory. The main sheet of the very heavy quasiparticles with $m^*/m \simeq 500$ is shown in Fig. 13.9. It differs considerably from LDA results and has a characteristic nesting vector $\mathbf{Q} = (0.23, 0.23, 0.52)$ in units of reciprocal lattice vectors. The latter connects flat, i.e., nesting parts of the Fermi surface and gives rise to a spin-density wave (SDW) formed by the heavy quasiparticles. Theory provides here a simple explanation for the so-called A phase in that material, which has been investigated for almost twenty years. The SDW phase with $\mu \simeq 0.1\mu_B$ and T$_{SDW} \simeq 0.7K$ has been identified by neutron scattering experiments. Small changes in the f electron count induce noticeable changes in the heavy quasiparticle sheet. This explains the extreme sensitivity of that material to deviations from the ideal stoichiometry.

It is instructive to leave the quasiparticle picture for a moment and shift our focus to the microscopic picture of f electrons in rare-earth intermetallic Ce systems. Their effective hybridization with the conduction electrons is strongly renormalized. We can explain this easily. A conduction electron can hop onto a $4f$ orbital of a Ce ion only when the latter is empty; otherwise, the large Coulomb repulsion between the $4f$ electrons comes into play. For simplicity we assume only one f orbital per Ce site. But when the f electron number n_f is close to one, i.e., $n_f \lesssim 1$, the f orbital is unoccupied only with probability $(1-n_f)$. The effective hybridization is thus strongly reduced. This results in the large density of low-lying excitations or, alternatively, in a self-

Table 13.2. Comparison of de Haas-van Alphen data for CeRu$_2$Si$_2$ [290] with theoretical results. Shown are some of the extremal areas of the Fermi surface (areas in megagauss) and the effective mass ratios m^*/m_0. Unlike the LDA, the renormalized band theory (RB) reproduces well the large observed mass anisotropies. (From [512])

CeRu$_2$Si$_2$		Experiment		LDA		RB	
Orbit	Field	Area [MG]	Mass ratio	Area [MG]	Mass ratio	Area [MG]	Mass ratio
α	(110)	4.7	12.3	≈ 10	≈ 1.5	≈ 10	≥ 10
ϵ	(110)	25,0	19.7	23	1.2	20	≈ 20
δ	(001)	12.2	4.0	24	1.5	26	2.1
ψ	(100)	53.6	120	70		≈ 62	≈ 100

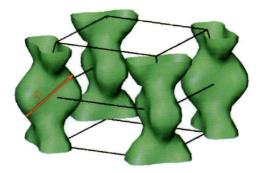

Fig. 13.9. CeCu$_2$Si$_2$: Main Fermi surface sheet of heavy quasiparticles ($m^*/m \simeq 500$) calculated with the renormalized band method. It consists of modulated columns which are oriented parallel to the tetragonal axis. The calculations adopt the CEF scheme of Ref. [150] consisting of a doublet ground state separated from an excited quartet by a CEF splitting $\delta \simeq 330K$. Therefore $\delta \gg T^* \simeq 10K$ (obtained from the γ-value). The nesting vector $\mathbf{Q} = (0.23, 0.23, 0.52)$ connects flat ("nesting") parts of the Fermi surface. (From [430, 511])

energy $\Sigma(\omega) = -A\omega$ with $A \gg 1$. In fact, we know from (4.35) that $A \simeq m^*/m$. The large coefficient A is closely related to the large slope $(k_\mathbf{B} T^*)^{-1}$ of the phase shift $\eta_\tau^{Ce}(\epsilon)$ of the quasiparticles at ϵ_F.

13.1.2 Large Versus Small Fermi Surface

In the following discussion we want to compare the de Haas-van Alphen measurements of CeRu$_2$Si$_2$ with those of CeRu$_2$Ge$_2$. The only difference between the two systems is that the distance between nearest neighbors Ce ions is larger in CeRu$_2$Ge$_2$ than in CeRu$_2$Si$_2$. As a consequence, the hybridization matrix element V between the $4f$ electron of Ce with its surrounding is much smaller in the first case. In fact, there is no Kondo temperature for CeRu$_2$Ge$_2$; instead, the material is ferromagnetic with a Curie temperature of $T_c \simeq 8K$. Therefore, the $4f$ electron of Ce is well localized and must be treated as part of the core. Since it does not take part in the formation of the Fermi surface, the volume in phase space enclosed by the Fermi surface must be less by one electron per unit cell than in CeRu$_2$Si$_2$, where $T^* \simeq 15K$. This picture has been confirmed by a series of very successful de Haas-van Alphen experiments[5].

---

[5] see [241, 290]

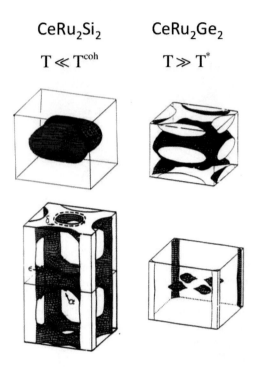

Fig. 13.10. Comparison of the Fermi surface for CeRu$_2$Si$_2$ and CeRu$_2$Ge$_2$ as derived from de Haas-van Alphen measurements [241, 290]. In CeRu$_2$Ge$_2$ the 4f electron of Ce is well localized, while in CeRu$_2$Si$_2$ it participates in the Fermi surface. The volume enclosed by the Fermi surface therefore differs by one electron in the two cases. While the hole sheet is enlarged in CeRu$_2$Ge$_2$, the electronic part is shrunk in comparison with CeRu$_2$Si$_2$.

A comparison of the two Fermi surfaces is shown in Fig. 13.10. One notices there that the hole part of the Fermi surface has increased in CeRu$_2$Ge$_2$ as compared with CeRu$_2$Si$_2$, while the electron part has shrunk. The difference in the enclosed volumes of the two Fermi surfaces is just one electron. The observed effective band masses for CeRu$_2$Ge$_2$ are larger by a factor of 1.3 - 4.5 than the ones calculated by band theory within the LDA, a result which is to be expected. The virtual excitation of higher CEF levels of the $J = 5/2$ multiplet of Ce^{3+} as well as spin-wave excitations are not included in the calculation and will certainly lead to an enhancement of the band masses. The measured γ coefficient of the low-temperature specific heat agrees well with the one determined from the measured masses m^* of the different parts of the Fermi surface. This implies that there are no other sizeable contributions to the linear specific heat term.

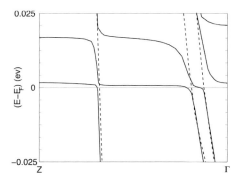

Fig. 13.11. Band dispersion for CeCu$_2$Si$_2$ along $Z - \Gamma$ for low temperatures $T \ll T^*$ (full lines) and high temperatures (dashed lines). The formation of the heavy quasiparticles leads to a characteristic bending in the occupied part of the spectrum (Courtesy of G. Zwicknagl).

One might speculate what happens when sufficiently high pressure is applied to CeRu$_2$Ge$_2$. With increasing pressure the hybridization of the f electrons should increase and they should start delocalizing giving rise to a Kondo temperature T^*. Denote with T the temperature at which a de Haas-van Alphen measurement is done. When T^* increases so that $T^* > T > T_{\text{coh}}$ the Fermi surface should become blurred. Landau's Fermi-liquid theory should no longer apply because the one-to-one correspondence of the excitations with those of nearly free electrons is lost. Finally, when at even higher pressures $T_{\text{coh}} > T$, a new enlarged Fermi surface should emerge which includes the f electrons. The system is again a Fermi liquid and the large specific heat coefficient γ should match the measured quasiparticle masses.

The transition from a large to a small Fermi surface takes also place when T is increased so that $T > T^*$. This has been verified for CeRu$_2$Si$_2$ by photoemission experiments and was discussed before when the photoemission results for LaRu$_2$Si$_2$ and CeRu$_2$Si$_2$ were compared (see Fig. 13.4). At this point a comment is in order on the reflection of heavy-quasiparticle formation in angular resolved photoemission spectroscopy (ARPES). Here one must realize that photoemission experiments only probe the occupied electron states while the most pronounced changes in the spectrum when going from $T \gg T_{\text{coh}}$ to $T \ll T_{\text{coh}}$ occur in the unoccupied part (for an example see Fig. 13.7). Yet the characteristic band bending close to ϵ_F which we show in Fig. 13.11 for CeCu$_2$Si$_2$ has indeed been observed in CeCoIn$_5$[6].

We conclude by discussing briefly the form of the f electron Green's function $G_f(\mathbf{k},\omega)$ close to the Fermi surface. It is of the general form of (7.22). If we assume one band of heavy quasiparticles with dispersion $\epsilon_{qp}(\mathbf{k})$ only – then in accordance with the self-energy discussed above – this form reduces to

---

[6] see [247]

$$G_f(\mathbf{k},\omega) = \frac{1-n_f}{\omega - \epsilon_{qp}(\mathbf{k}) + i\eta \text{ sgn } \omega} + G_{\text{inc}}(\mathbf{k},\omega) \quad . \tag{13.12}$$

The first part contains the quasiparticle pole, whereas the second part $G_{\text{inc}}(\mathbf{k},\omega)$ describes an incoherent, weakly structured background. In agreement with the discussion in Sect. 10.3 the f spectral weight near ϵ_F vanishes like $(1-n_f)$ for small values of T^*. The form of (13.12) should be seen in contrast to the Green's function of the quasiparticles. According to Landau's Fermi-liquid theory, there is a one-to-one correspondence between the low-energy excitations of a heavy-fermion system and those of a noninteracting electron gas, provided that parameters like the band masses are renormalized. Therefore, the quasiparticle Green's function takes the form

$$G_{qp}(\mathbf{k},\omega) = \frac{1}{\omega - \epsilon_{qp}(\mathbf{k}) + i\eta \text{ sgn } \omega} \quad , \tag{13.13}$$

as it does for free electrons. The renormalized parameters are contained in the dispersion $\epsilon_{qp}(\mathbf{k})$.

13.1.3 Mean-Field Treatment

As we have seen before, in a rare-earth Kondo-lattice system the strong correlations of $4f$ electrons lead to reduced hybridization matrix elements with the surroundings. Renormalized band theory takes into account these reductions and this way is able to compute the strong mass anisotropies found in de Haas-van Alphen experiments.

In the following we want to show how reduced hybridization matrix elements can be derived from a mean-field treatment of a Hamiltonian with strong local interactions. As such we choose the Anderson lattice Hamiltonian, which is a generalization of (10.23) to the lattice

$$\begin{aligned} H = & \sum_{\mathbf{k}n\sigma} \epsilon_n(\mathbf{k}) a^+_{\mathbf{k}n\sigma} a_{\mathbf{k}n\sigma} + \sum_{mi} \epsilon_{fm} f^+_m(i) f_m(i) \\ & + \frac{1}{\sqrt{N_0}} \sum_{imkn\sigma} V_{m\sigma}(\mathbf{k},n) \left[a^+_{\mathbf{k}n\sigma} f_m(i) e^{-i\mathbf{k}\cdot\mathbf{R}_i} + h.c. \right] \\ & + \frac{U}{2} \sum_{i,m\neq m'} n^f_m(i) n^f_{m'}(i) \quad . \end{aligned} \tag{13.14}$$

The index i labels the N_0 f-sites at positions $\mathbf{R}_i$. Furthermore, the conduction electron creation operators are denoted here by $a^+_{\mathbf{k}n\sigma}$, where n is a band index; otherwise the notation is the same as in (10.23). In most treatments we consider the limit of large Coulomb repulsion U of the f-electrons. Then the f electron number at a site is either 1 or 0; double occupancies of f sites are strictly excluded. This condition can be accounted for by introducing an auxiliary bosonic field $b^+(i), b(i)$ as done in Sect. 10.6, where the boson operator $b^+(i)$ creates an empty f state at site i.

13.1 Kondo Lattice Systems

Accordingly, the operator $n_b(i) = b^+(i)b(i)$ counts the number of empty f sites i. In a slight generalization of the subsidiary condition (10.139) we have here in the limit of large U to require that

$$Q(i) = \sum_m f_m^+(i) f_m(i) + b^+(i) b(i)$$
$$= 1 \; . \tag{13.15}$$

In terms of these boson operators and by inclusion of the subsidiary condition (13.15) with a Lagrange multiplier Λ_i, the Hamiltonian (13.14) reads

$$H = H_{\text{band}} + \sum_{mi} (\epsilon_{fm} + \Lambda_i) f_m^+(i) f_m(i) + \sum_i \Lambda_i \left[b^+(i) b(i) - 1 \right]$$
$$+ \frac{1}{\sqrt{N_0}} \sum_{im\mathbf{k}n\sigma} \left[V_{m\sigma}(\mathbf{k}, n) b^+(i) a_{\mathbf{k}n\sigma}^+ f_m(i) e^{-i\mathbf{k}\cdot\mathbf{R}_i} + h.c. \right] \; , \tag{13.16}$$

where H_{band} denotes the conduction-electron part of H. Note that U has disappeared from the Hamiltonian, but consequently we must deal with subsidiary conditions (13.15).

In the mean-field approximation, the condition $Q(i) = 1$ is replaced by a weaker one:

$$\langle Q(i) \rangle = 1 \; . \tag{13.17}$$

This is achieved by replacing $b^+(i)$ by the site-independent mean value of the field operator, i.e.,

$$b^+(i) \to \langle b^+(i) \rangle = r \; . \tag{13.18}$$

Thus we end up with the mean-field Hamiltonian

$$H_{\text{MF}} = \sum_{\mathbf{k}n\sigma} \epsilon_n(\mathbf{k}) a_{\mathbf{k}n\sigma}^+ a_{\mathbf{k}n\sigma} + \sum_{mk} \tilde{\epsilon}_{fm} f_{\mathbf{k}m}^+ f_{\mathbf{k}m}$$
$$+ \sum_{nm\mathbf{k}\sigma} r V_{m\sigma}(\mathbf{k}, n) \left(a_{\mathbf{k}n\sigma}^+ f_{\mathbf{k}m} + h.c. \right) + \Lambda N_0 \left(r^2 - 1 \right) \tag{13.19}$$

with $\tilde{\epsilon}_{fm} = \epsilon_{fm} + \Lambda$. The Fourier transform $f_{\mathbf{k}m}^+$ of the operators $f_m^+(i)$ has been introduced and the condition (13.15) has been replaced by $1 - n_f^{\text{op}}(i) = r^2$. The Hamiltonian H_{MF} is a one-particle Hamiltonian and as such can be easily diagonalized as a function of the two unknowns Λ and r. We rewrite it in the diagonalized form as

$$H_{\text{MF}} = \sum_{\mathbf{k}\ell\tau} E_\ell(\mathbf{k}) c_{\ell\tau}^+(\mathbf{k}) c_{\ell\tau}(\mathbf{k}) + \Lambda N_0 \left(r^2 - 1 \right) \; , \tag{13.20}$$

where the $c_{\ell\tau}^+(\mathbf{k})$ denote the creation operators of quasiparticles in band ℓ with pseudospin τ. We may speak of quasiparticles because the complex many-body problem has been mapped onto a one-particle problem as before, when phase shifts were used. The $E_\ell(\mathbf{k})$ are the quasiparticle energies, which depend on Λ and r. In terms of the $c^+_{\ell\tau}(\mathbf{k})$ the ground state $|\Phi_0\rangle$ of H_{MF} is written as

$$|\Phi_0\rangle = \prod_{\substack{\ell,\tau \\ |\mathbf{k}|<k_F}} c^+_{\ell\tau}(\mathbf{k})|0\rangle \quad . \tag{13.21}$$

For the special case of one conduction electron band only and an f pseudo spin-orbital degeneracy of $\nu_f = 2$, we find two quasiparticle bands with a two-fold pseudospin degeneracy and with energies

$$E_\ell(\mathbf{k}) = \frac{1}{2}\{[\epsilon(\mathbf{k}) + \tilde{\epsilon}_f] \mp W(\epsilon(\mathbf{k}))\}$$
$$W(\epsilon(\mathbf{k})) = \sqrt{[\epsilon(\mathbf{k}) - \tilde{\epsilon}_f]^2 + 4\tilde{V}^2} \quad . \tag{13.22}$$

The $-$, $+$ signs refer to the bands $\ell = 1$ and 2, respectively. Furthermore $\tilde{V} = rV$. If we require that $E_\ell(k_F) = \mu$, where μ is the chemical potential, we obtain for the renormalized energy of the f level $\tilde{\epsilon}_f$

$$\tilde{\epsilon}_f = \mu + \frac{\tilde{V}^2}{\epsilon(k_F) - \mu} \quad . \tag{13.23}$$

Because $\tilde{V}$ is very small, the renormalized f electron energy $\tilde{\epsilon}_f$ lies slightly above the Fermi energy. This is shown schematically in Fig. 13.12.

We sketch briefly the derivation of (13.22–13.23). First $a^+_{\mathbf{k}\sigma}$ and $f^+_{\mathbf{k}m}$ are expressed in terms of the $c^+_{\ell\tau}(\mathbf{k})$ as

$$f^+_{\mathbf{k}m} = \sum_{\ell\tau} y_m(\mathbf{k};\ell\tau) c^+_{\ell\tau}(\mathbf{k}) \quad ,$$
$$a^+_{\mathbf{k}\sigma} = \sum_{\ell\tau} x_\sigma(\mathbf{k};\ell\tau) c^+_{\ell\tau}(\mathbf{k}) \quad . \tag{13.24}$$

The coefficients $x_\sigma(\mathbf{k};\ell\tau)$ and $y_m(\mathbf{k};\ell\tau)$ must satisfy the normalization condition

$$\sum_m |y_m(\mathbf{k};\ell\tau)|^2 + \sum_\sigma |x_\sigma(\mathbf{k};\ell\tau)|^2 = 1 \quad . \tag{13.25}$$

We insert (13.24–13.25) into (13.19) and require that the off-diagonal matrix elements vanish since the $c^+_{\ell\tau}(\mathbf{k})$ generate eigenstates of H_{MF}. This gives us two equations for $x_\sigma(\mathbf{k};\ell\tau)$ and $y_m(\mathbf{k};\ell\tau)$.

The $x_\sigma(\mathbf{k};\ell\tau)$ can be eliminated with the help of (13.25) and the remaining equation for $y_m(\mathbf{k};\ell\tau)$ is

$$[\tilde{\epsilon}_{fm} - E_\ell(\mathbf{k})] y_m(\mathbf{k};\ell\tau) + \sum_{\sigma m'} \frac{\tilde{V}_{m\sigma}(\mathbf{k})\tilde{V}^*_{m'\sigma}(\mathbf{k})}{E_\ell(\mathbf{k}) - \epsilon(\mathbf{k})} y_{m'}(\mathbf{k};\ell\tau) = 0 \quad . \tag{13.26}$$

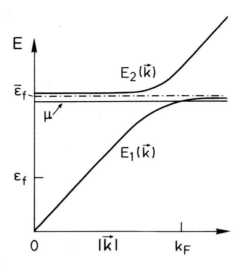

Fig. 13.12. Quasiparticle energies $E_{1(2)}(\mathbf{k})$ given by (13.22). Also shown is the original f orbital energy ϵ_f and the renormalized energy $\tilde{\epsilon}_f \gtrsim \mu$.

The $y_m(\mathbf{k}; \ell \tau)$ are the probability amplitudes of finding a quasiparticle with band index ℓ and pseudospin τ in a f state with quantum number m. From the last equation and $\tilde{V}_{m\sigma}(\mathbf{k}) = \tilde{V}$ (13.22) follows immediately.

For $T = 0$ the unknowns r and Λ can be determined by minimizing the ground-state energy with respect to r, i.e., from

$$\left\langle \Phi_0 \left| \frac{\partial}{\partial r} H_{MF} \right| \Phi_0 \right\rangle = 0 \tag{13.27}$$

and from the condition

$$r^2 = 1 - n_f \quad . \tag{13.28}$$

Without going into details we merely mention that, after Λ and r have been determined, one may define a characteristic temperature T^* through

$$k_B T^* = \mu \exp\left(-\frac{\Lambda}{\nu_f N(0) V^2} \right) \tag{13.29}$$

in terms of which the energy gain per site due to hybridization is

$$\Delta E = -k_B T^* \tag{13.30}$$

while the f electron count becomes

$$n_f = 1 - \frac{k_B T^*}{\nu_f N(0) V^2} \quad . \tag{13.31}$$

The temperature T^* plays the role of a Kondo temperature for a lattice. Its dependence on microscopic parameters resembles closely that of (10.44), since to leading order Λ is given by $\Lambda = |\epsilon_f|$ when we count the energy from the Fermi energy ϵ_F. The above considerations were limited to $T = 0$. However, we can easily extend them to finite temperatures. The averages which appear in the above expressions are then thermodynamic averages with respect to H_{MF}. Only for temperatures T less than an artificial mean-field critical temperature T_c, which is on the order of T^*, does one find a solution of (13.27,13.28) with $r \neq 0$. For $T > T_c$ we find that $r = 0$ and the conduction electrons decouple completely from the f electrons. The fluctuations $\delta b(i) = b(i) - \langle b(i) \rangle$, neglected in a mean-field theory, prevent such a decoupling for $T > T_c$. With increasing orbital degeneracy ν_f the influence of these fluctuations decreases, and in the limit $\nu_f \to \infty$ the mean-field theory becomes exact.

Now we want to establish a connection between the mean-field approach and the renormalized band-structure calculations presented before, since the resemblance between the two approaches is apparent. The reduction of the bare hybridization V to $\tilde{V}$ has its equivalent in the large slope $(k_\mathbf{B} T^*)^{-1}$ of the f phase shift $\eta_\tau^{Ce}(\epsilon)$ at ϵ_F[7]. Similarly, the positioning of $\tilde{\epsilon}_f$ just above ϵ_F corresponds to a value of $\eta_\tau^{Ce}(\epsilon_F)$ slightly less than $\pi/2$. In fact, if we start from the secular equation for the renormalized bands, we can show quite rigorously that by a "downfolding" or reduction procedure we obtain an effective Hamiltonian of the form of (13.19) with the only difference that the f electron energy is also $\mathbf{k}$ dependent, i.e., $\tilde{\epsilon}_{fm} \to \tilde{\epsilon}_{fm}(\mathbf{k})$ [136]. If the CEF ground state is only twofold degenerate and if spin-orbit effects for the conduction electrons can be neglected, the effective Hamiltonian reduces to

$$H_{\text{eff}} = \sum_{\mathbf{k}\tau} \tilde{\epsilon}_f(\mathbf{k}) f_{\mathbf{k}\tau}^+ f_{\mathbf{k}\tau} + \sum_{\mathbf{k}n\tau} \left[\tilde{V}_{n\tau}(\mathbf{k}) f_{\mathbf{k}\tau}^+ a_{\mathbf{k}n\tau} + h.c. \right] + H_{\text{band}}^{\text{eff}} \quad , \qquad (13.32)$$

where $H_{\text{band}}^{\text{eff}}$ is an effective conduction-electron Hamiltonian not coupling to the f electrons. The evaluation of this Hamiltonian corresponds precisely to the renormalized band-structure method.

13.2 Charge Ordering in Yb$_4$As$_3$: an Instructive Example

As will be shown in this section, charge ordering can also be the origin of heavy quasiparticles. Here the system Yb$_4$As$_3$ serves as an excellent example. We will discuss it at some length for the aforementioned reason, and because Yb$_4$As$_3$ is also an example of a generalized Wigner crystal.

Connected with charge ordering are changes in the lattice. Therefore, we are dealing here with a case where heavy quasiparticles, charge ordering and

---

[7] see Eq. (13.11)

13.2 Charge Ordering in Yb_4As_3: an Instructive Example

lattice degrees of freedom are intimately connected. Since the compound has also been extensively studied experimentally, we can gain here good insight into the manifold of microscopic processes which may result in heavy fermions.

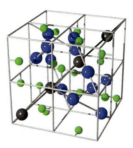 

Fig. 13.13. Left panel: Anti-Th_3P_4 structure of Yb_4As_3. Large and small spheres symbolize Yb and As ions, respectively. The Yb ions are residing on four interpenetrating families of chains oriented along cubic space diagonals. Right panel: Dense rod packing presentation of the Yb-chains. In the charge-ordered structure only *one* family of chains carries Yb^{3+} ions with pseudo-spin $S = 1/2$ whereas the other three families are occupied with Yb^{2+} ions having a filled $4f$ shell.

We begin by showing that at low temperatures Yb_4As_3 may be considered an example of a generalized Wigner crystal. As demonstrated before, a homogeneous electron gas becomes unstable with respect to a lattice formation when the electron density is sufficiently low. In that case the Coulomb repulsion of electrons dominates the kinetic energy gain caused by itinerancy and therefore an electron lattice becomes energetically more favourable than an electron liquid. However, as shown in Sect. 3.3 the critical mean distance between electrons is rather large before lattice formation takes place. This is different when an inhomogeneous system, i.e., a real crystal is considered. Here the energy gain of electrons due to delocalization may be rather small to start with, depending on the type of electrons we are dealing with, i.e., on the size of their specific hybridization matrix elements. Therefore, Coulomb repulsion may prevail already at much higher densities. This holds particularly true for $4f$ electrons, because of their closeness to the nuclei and corresponding small hopping matrix elements to neighboring sites. Note that in this case charge ordering, or the formation of an electronic lattice by the f electrons, is superimposed on the underlying atomic lattice structure.

A good example is the intermetallic compound Yb_4As_3, where below a temperature of $T_c = 292K$ charge order of $4f$ holes takes place [349]. The system has a cubic anti-Th_3P_4 structure. The Yb ions are aligned along four families of interpenetrating chains pointing along the (shifted) diagonals of a cube. The structure is shown in Fig. 13.13. It is often referred to as body-centered cubic rod packing.

An important point to notice is that the distance between two Yb ions within a chain is larger than the one between ions belonging to different chains. Thus nearest neighbor Yb ions belong to different families of chains. Counting valence electrons we notice that As has a valency of -3 which is adding up to -9. Therefore, three of the four Yb ions must have a valency of $+2$ while the remaining one has valency $+3$, i.e., $Yb_4As_3 \to (Yb^{2+})_3(Yb^{3+})(As^{3-})_3$. However, Yb^{2+} has a filled $4f$ shell. Therefore one $4f$ hole remains per formula unit. At sufficiently high temperatures, i.e., for temperatures exceeding 300 K the f holes move freely between sites and the system is metallic. This is confirmed by measurements of the Hall coefficient R_H, which yields a carrier concentration of approximately one hole per formula unit in that temperature regime. The situation is different at low temperatures. Here the measured Hall coefficient has a value of $(ecR_H)^{-1} = 7 \cdot 10^{18}$ cm^{-3}, implying approximately one itinerant hole per 10^4 Yb ions (see Fig. 13.14a).

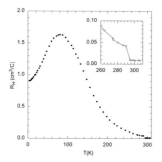 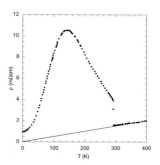

Fig. 13.14. Left panel: Hall coefficient $R_H(T)$ for Yb_4As_3 (pro Coulomb). The inset shows the change at the phase transition temperature T_c. Right panel: resistivity $\rho(T)$. At $T_c = 292K$ a phase transition due to charge ordering is taking place. Solid line: extrapolation of $\rho(T) \sim T$. (From [349])

Thus the system changes from metallic to semi-metallic as temperature decreases. This is reflected in the measured resistivity $\rho(T)$ as seen in Fig. 13.14b. While for $T > 300K$ the resistivity $\rho(T)$ decreases nearly linearly with T, it increases again below the first order phase transition at $T_c \simeq 292K$. At low temperatures $\rho(T) = \rho_0 + AT^2$ showing that Yb_4As_3 has become a semimetallic Fermi liquid. Low temperature thermodynamic properties show typical heavy-quasiparticle behavior. The γ coefficient of the low temperature specific heat is $\gamma \simeq 200 mJ/(mol K^2)$. The spin susceptibility χ_s is equally enhanced and the Sommerfeld-Wilson ratio $R_W = 4\pi^2 k_B^2 (\chi_s/(3(g\mu_{eff})^2\gamma)$ is close to unity. Here μ_{eff} denotes the effective moment of a Yb^{3+} site and g is the gyromagnetic factor. The resistivity is $\rho(T) = \rho_0 + AT^2$ and the Kadowaki-Woods ratio A/γ^2 is similar to that of other systems with heavy quasiparticles. Note

13.2 Charge Ordering in Yb$_4$As$_3$: an Instructive Example

that the large linear specific heat coefficient is found despite the fact that there are hardly any conduction electrons present at low temperatures.

Here we concentrate first on the nature of the phase transition. It is caused by the strong correlations of the $4f$ electrons (holes) and it is crucial for understanding the heavy quasiparticles which appear at low temperatures. An analysis of structural data shows that the phase transition is accompanied by a trigonal distortion and a change of the space group. The structural transition is volume conserving. It is triggered by charge ordering of the $4f$ holes. The angle between rods which are orthogonal in the cubic phase changes to $\alpha = 90.8^o$ in the trigonal phase. Associated with this change is a spontaneous elastic strain which is proportional to the charge order parameter. The structural instability is accompanied by a softening of the c_{44} elastic mode as T approaches T_c from above (see Fig. 13.15). This implies that the trigonal elastic strain $\epsilon_{xy}, \epsilon_{xz}, \epsilon_{yz}$ with Γ_5 symmetry of the lattice plays a significant role. A detailed group-theoretical analysis of different elastic constants is found in [153].

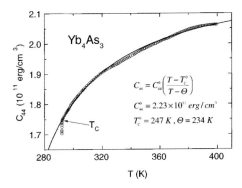

Fig. 13.15. Temperature dependence of the elastic constant $c_{44}(T)$. Above the structural phase transition temperature T_c a strong softening is observed described by (13.36) and caused by coupling of the c_{44} mode to the Γ_5 type charge order parameter. Due to the first-order nature of the transition the theoretical mean-field transition temperature $T_{c0} = 247K$ is smaller than the actual T_c. (From [153])

After the description of the structural phase transition we want to discuss its origin. It turns out to be based on strong correlations between the $4f$ holes. As pointed out before, at high temperatures the f holes are moving freely through the system thereby avoiding configurations with two f holes on a Yb site. At low temperatures however, repulsion between f holes on neighboring sites comes into consideration. As previously pointed out, nearest neighbor Yb sites always belong to *different* families of chains. By avoiding nearest-neighbor Yb^{3+} - Yb^{3+} sites and thus minimizing the short-ranged intersite repulsion of f holes, the latter are accumulating in one family of chains, i.e., they charge order. Therefore at $T=0$ one expects that in the idealized case all Yb^{3+} ions are situated in one family of chains and that the Yb^{2+} ions are in the remaining three ones. Since Yb^{3+} ions are smaller than Yb^{2+} ions, the crystal contracts in the direction of the chains with the Yb^{3+} ions. In order to keep the volume of a unit cell constant (otherwise the elastic energy would increase) the crystal must expand in the direction of the remaining three families of chains filled with Yb^{2+} ions. The above arguments explain the origin of the structural phase transition including the change in the unit cell angle.

If each of the sites in the short chains is occupied by a Yb^{3+} ion, why is the system a semimetal and not a Mott insulator, since one of the chain families is having one hole per site? The reason is found when LDA+U calculations are done. They show that a small number of As $4p$ holes remains at the Γ point. They must be compensated by removing the same number of $4f$ holes, i.e., by filling additionally some of the $4f$ shells. The chains containing the Yb^{3+} ions are therefore not perfect spin $1/2$ chains but have instead in one out of 10^4 sites a spin missing (i.e., a Yb^{2+} ion). This accounts for having one charge carrier per 10^4 Yb ions.

Before discussing a microscopic model for the phase transition driven by electron correlations, we discuss briefly the physics of the softening of the elastic constant c_{44}.

The temperature dependence of the c_{44} mode is obtained from a Ginzburg-Landau expansion of the free energy in terms of the strains ϵ_{ij} and the components Q_{ij} of the charge order parameter. The latter is defined by expanding the changes in the electronic charge distribution $\Delta\rho$ caused by the structural changes in terms of the important charge fluctuation modes $\rho_{ij}(\Gamma_5)$ of Γ_5 symmetry,

$$\Delta\rho = Q_{yz}\rho_{yz}(\Gamma_5) + Q_{zx}\rho_{zx}(\Gamma_5) + Q_{xy}\rho_{xy}(\Gamma_5) \quad . \tag{13.33}$$

For $T < T_c$ the order parameter $Q_{ij} \neq 0$. The free energy contains three contributions. The first (F_Q) results from the electronic charge order parameter, the second (F_{el}) is due to the elastic energy of the lattice and the third (F_{Q-el}) describes the interactions of the order parameter with the lattice strains. Thus we obtain (see [152]):

$$F_Q = F_0 + \frac{\alpha}{2}\left(Q_{xy}^2 + Q_{xz}^2 + Q_{yz}^2\right)$$
$$+ \frac{\beta}{4}\left(Q_{xy}^4 + Q_{xz}^4 + Q_{yz}^4 - \frac{3}{5}\left(Q_{xy}^2 + Q_{xz}^2 + Q_{yz}^2\right)^2\right)$$
$$F_{el} = \frac{c_{44}^0}{2}\left(\epsilon_{xy}^2 + \epsilon_{xz}^2 + \epsilon_{yz}^2\right)$$
$$F_{Q-el} = -g\left(Q_{xy}\epsilon_{xy} + Q_{xz}\epsilon_{xz} + Q_{yz}\epsilon_{yz}\right) \quad . \tag{13.34}$$

The softening of the c_{44} mode is due to the term F_{Q-el}. At a phase transition with an unrenormalized transition temperature Θ the coefficient α vanishes

13.2 Charge Ordering in Yb$_4$As$_3$: an Instructive Example

like $\alpha = \alpha_0(T - \Theta)$. The fourth-order terms in Q_{ij} stabilize the ordered state of the system. For $T > \Theta$ the softening of the elastic constant is obtained by minimizing $F = F_Q + F_{el} + F_{Q-el}$ with respect to the Q_{ij}, which are induced here by an external strain, i.e., sound wave. Hereby the terms $\sim Q_{ij}^4$ in the free energy are neglected. When $\beta > 0$ a trigonal charge order parameter $Q_t = \frac{1}{\sqrt{3}}(Q_{xy}, Q_{xz}, Q_{yz})$ characterises the ordered phase. In addition Q_{ij} and ϵ_{ij} are proportional to each other. The expression for the free energy simplifies therefore to the form

$$F = F_0 + \frac{1}{2}\left(c_{44}^0 - \frac{g^2}{\alpha_0(T-\Theta)}\right)(\epsilon_{xy}^2 + \epsilon_{xz}^2 + \epsilon_{yz}^2) \quad . \tag{13.35}$$

The renormalized elastic constant is

$$c_{44} = c_{44}^0 \left(\frac{T - T_{c0}}{T - \Theta}\right) \quad , \quad \text{where } T_{c0} = \Theta + \frac{g^2}{\alpha_0 c_{44}^0} \tag{13.36}$$

denotes the mean-field transition temperature in the presence of the strain interaction F_{Q-el}. An explanation of the observed first-order phase transition at $T_c < T_{c0}$ requires the inclusion of higher order terms in Q_{ij} in (13.34).

Having described the physics of the phase transition, we want to provide a microscopic model description of it. Thereby we neglect hopping matrix elements between different chains as well as the small number of As 4p holes. We write for the effective Hamiltonian

$$H = -t\sum_\mu \sum_{\langle ij \rangle \sigma} \left(f_{i\mu\sigma}^+ f_{j\mu\sigma} + h.c.\right) + U\sum_\mu \sum_i n_{i\mu\uparrow} n_{i\mu\downarrow}$$
$$+\epsilon_\Gamma \sum_\mu \sum_{i\sigma} \Delta_\mu n_{i\mu\sigma} + \frac{N}{4} c_\Gamma \epsilon_\Gamma^2 \quad . \tag{13.37}$$

The first term describes effective 4f-hole hopping from site i to a nearest neighbor site j *within* a chain of a family $\mu = 1 - 4$. From LDA calculations one can deduce that $4t \simeq 0.2$ eV. The second term is due to the on-site Coulomb repulsion of 4f holes with $n_{i\mu\sigma} = f_{i\mu\sigma}^+ f_{i\mu\sigma}$ and ensures that in the large U limit Yb^{4+} states with $4f^{12}$ configurations are excluded. The third term describes the volume conserving coupling of the f bands to the trigonal strain $\epsilon_\Gamma > 0$ with $\Gamma = \Gamma_5$. It leads to a deformation potential of the form

$$\Delta_\mu = \frac{\Delta}{3}(4\delta_{\mu 1} - 1) \tag{13.38}$$

for 4f holes situated in chains, e.g., in [111] direction denoted by $\mu = 1$.

As previously pointed out, the origin of the deformation potential is the short-range repulsion between holes on neighboring sites. It is treated here as an effective attraction V_{eff} between holes on next-nearest neighbor sites, i.e., nearest-neighbor sites of a chain. The fourth term in (13.37) is the elastic energy in the presence of a trigonal distortion, where N is the number of sites and c_Γ is the background elastic constant. A reasonable value is $c_\Gamma/\Omega = 4 \cdot 10^{11} \text{erg/cm}^3$ where Ω denotes the volume of a unit cell with a lattice constant of $a_0 = 8.789 \text{Å}$. In accordance with the above one may eliminate ϵ_Γ in (13.37) and obtain instead an effective interaction term

$$H_{\text{int}} = -V_{\text{eff}} \sum_{\mu\langle ij\rangle} (n_{i\mu} - \bar{n})(n_{j\mu} - \bar{n}) \quad , \tag{13.39}$$

which in mean-field approximation becomes

$$H_{\text{int}}^{\text{MF}} = -2V_{\text{eff}} \sum_\mu (\bar{n}_\mu - \bar{n}) \left[\sum_i (n_{i\mu} - \bar{n}_\mu) - \frac{N}{8} (\bar{n}_\mu - \bar{n}) \right] \quad . \tag{13.40}$$

Here $\bar{n}$ and $\bar{n}_\mu$ are the f hole occupation number averaged over all sites and all sites of chain family μ, respectively. By relating

$$\frac{4}{3}\epsilon_\Gamma \Delta = -2V_{\text{eff}}(\bar{n}_1 - \bar{n}) \quad , \quad \frac{\Delta^2}{c_\Gamma} = \frac{9}{4} V_{\text{eff}} \tag{13.41}$$

the ϵ_Γ dependent terms in (13.37) become equivalent to the interaction (13.40). When a lattice deforms, the hopping matrix elements usually change, because the overlap of atomic wavefunctions of neighboring atoms changes. But this is not important here and therefore is neglected.

The lattice distortion caused by ordering of the f holes shows similarities to a band Jahn-Teller effect [263], which occurs here as a consequence of electron interactions. The four-fold degeneracy of the one-dimensional f band is lifted by a symmetry-breaking trigonal strain. For a rough estimate consider the case of $U = 0$. This neglects the effects of strong on-site hole repulsions on the band Jahn-Teller effect. This simplification is not too bad in the high temperature phase, where only one Yb site out of four contains a $4f$ hole. However, it is very poor at low temperatures when the $\mu = 1$ family of chains contains nearly one hole per site. The condition for a band Jahn-Teller effect is $\Delta^2/(t\epsilon_F) > 3$ [138]. In that case the symmetry broken solution has a lower energy than the symmetric one with a four-fold degenerate f band.

The strain $\epsilon_\Gamma(T)$ splits the four one-dimensional f bands into one lower and three upper bands. For Yb_4As_3, generalized Wigner crystallization and band Jahn-Teller effect of correlated electrons are therefore alternative points of view. In passing we note that the Yb^{3+} sites are not centers of inversion. This allows for a Dzyaloshinsky-Moriya interaction with interesting consequences regarding the magnetic field dependence of the heavy quasiparticles [410].

The above discussion provides the key for an understanding of the heavy quasiparticles in the system. As pointed out before, in the charge ordered state nearly all Yb sites with a $4f$ hole are in one of the four families of chains while the remaining three families contain almost no holes. The interacting crystal-field ground state doublets of Yb^{3+} behave like an isotropic Heisenberg spin model. This is not immediately obvious but was shown in [410, 458]. Therefore we are dealing here with almost perfect Heisenberg spin chains. It is well known from the work of *Bonner* and *Fisher* [40] that the specific heat $C = \gamma T$ and the spin susceptibility χ_s of a Heisenberg chain is given by

$$\gamma = \frac{2}{3}\frac{k_B R}{J} \quad , \quad \chi_s = \frac{4\mu_{\text{eff}}^2 R}{\pi^2 J} \quad . \tag{13.42}$$

Here $J > 0$ is the coupling constant of nearest-neighbor effective $S = 1/2$ spins:

$$H = J \sum_{\langle ij \rangle} \mathbf{S}_i \mathbf{S}_j \tag{13.43}$$

and R is the gas constant. Note that the Sommerfeld-Wilson ratio is $R_W = 2$ in that case. Thus the large γ coefficient in the specific heat results from spin excitations in the spin chains. It is present even if Yb_4As_3 would become insulating. This picture is confirmed by beautiful inelastic neutron scattering (INS) experiments (see Fig. 13.16). The measured dispersion $\omega(\mathbf{q})$ of the spin excitations agree with old calculations of *Cloizeaux* and *Pearson*[8]. Note that these are no sharp spin-wave excitations. Rather they represent the lower bound of a two-spinon continuum. The latter has a square root singularity in the dynamic structure factor which enters the INS cross-section. Therefore the two-spinon spectrum is strongly peaked at the lower bound with an asymmetric tail leading up to much higher energies. This is precisely what is observed experimentally and proves the one-dimensional character of the spin excitations. From the data one can deduce a coupling constant $J/k_B = 25K$. When set into (13.42) the observed size of the γ coefficient is reproduced. In agreement with this is the observation that $Yb_4(As_{1-x}P_x)_3$ with $x = 0.3 - 0.4$ has a γ coefficient of similar size although that material is an insulator.

The above results show that the microscopic origin of the heavy quasiparticle in Yb_4As_3 is quite different from that in the Ce compounds. The spin excitations in the Heisenberg chains are nearly decoupled from the charge excitations. The effective mass of the As $4p$ holes is not heavy. The large A coefficient in $\rho(T)$ is due to scattering of the light quasiparticles by spin excitations.

We are faced here with a breakdown of Landau's Fermi liquid theory although this is not immediately obvious from the low temperature thermodynamic data. In fact, we are dealing here with *two* Fermi liquids, i.e., one with charge-neutral heavy quasiparticles (spin excitations in chains) and the other with charged light quasiparticles (As $4p$ holes). The light quasiparticles which are observed in cyclotron resonance experiments are scattered off by the neutral heavy quasiparticles. This results in a large A coefficient in $\rho(T) = AT^2$. While only the light quasiparticles contribute to the electrical resistivity, the thermal conductivity is dominated by the neutral, i.e., heavy ones. There is

---

[8] see [68]

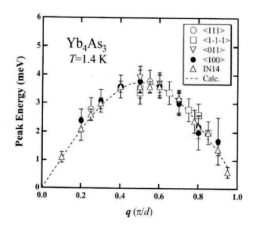

Fig. 13.16. Dispersion of magnetic excitations in Yb_4As_3 in the low-temperature charge ordered phase measured by INS experiments. Here q is the projection of momenta on the $\langle 111 \rangle$ chain direction. All data fall onto the theoretical curve $\omega(q) = \frac{\pi}{2} J \sin dq$ with $J/k_B = 25 K$ and d is denoting the interatomic distance in a chain. This proves the one-dimensional character of the excitations. (From [244, 245])

no longer a one-to-one correspondence between the electronic excitations in Yb_4As_3 and those of a nearly free electron system as required by Landau's Fermi liquid theory.

13.3 Partial Localization: Dual Role of 5f Electrons

In some of the actinide compounds $5f$ electrons show an orbital-selective localization. This concept is known to play an important role for d electrons in transition metal oxides (see Chapter 12). Its extension to $5f$ system is relatively new and has considerable consequences[9]. In both types of materials Hund's rule or intra-atomic correlations play a crucial role. Nevertheless, the physics of partial localization in transition metal oxides and in $5f$ systems is quite different. In $3d$ systems the large crystalline electric field (CEF) set up by the atomic surroundings of a transition metal ion lifts partially orbital degeneracies and causes splitting energies which are often larger than the bandwidths. A much discussed example are the manganites (see Sect. 12.2). In a cubic lattice the five d orbitals are split into a t_{2g} triplet and e_g doublet with well separated subbands. When the Hund's rule energy is larger than the $t_{2g} - e_g$ splitting and when the orbital energy of the t_{2g} multiplet is lower than of the e_g multiplet, the first three d electrons will occupy t_{2g} orbitals and form a high spin $S = 3/2$ state. These electrons remain localized. Additional

---

[9] see [510, 514]

13.3 Partial Localization: Dual Role of 5f Electrons

electrons enter the e_g orbitals with spin parallel to the high-spin state and are delocalized. The situation differs when the CEF splitting is larger than Hund's rule coupling. In that case the t_{2g} subband will accommodate up to six electrons. When the d electron count n_d per transition metal ion is $n_d > 6$, only $(n_d - 6)$ of the d electrons are itinerant and contribute to metallic behavior.

In $5f$ compounds we are confronted with a different situation. The $5f$ atomic wavefunctions are closer to the nuclei than d electron wavefunctions are and therefore CEF splittings are smaller and less important. Yet, Hund's rule energies for $5f$ electrons are larger than for d electrons. Thus when dealing with a situation where the $5f$ count per actinide ion n_f exceeds two, i.e., $n_f > 2$ only those $5f$ electrons will delocalize which enable the remaining ones to form a Hund's rule state. Otherwise the increase in Coulomb energy would overcompensate the energy gain due to delocalization. This results in orbital-selective localization. In UPt$_3$ with $n_f \simeq 2.5$ the Hund's rule ground state of a $5f^2$ configuration has total angular momentum $J = 4$. The differences in Coulomb energies for different J multiplets (only even values of J are allowed by Pauli's principle) can be calculated and are found to be

$$\Delta U_4 = U_{J=4} - U_{J=0} = -3.79 \ eV$$
$$\Delta U_2 = U_{J=2} - U_{J=0} = -2.72 \ eV \quad . \tag{13.44}$$

Although these are bare Coulomb integrals, i.e., unscreened ones, their differences are less affected by screening and to first approximation can be regarded as remaining unchanged. The differences are larger than a typical hopping matrix element of a $5f$ electron which is of order $t \simeq 0.5$ eV. By partially suppressing hopping processes, Hund's rule correlations may strongly enhance anisotropies in the kinetic energy and eventually lead to the coexistence of band-like itinerant $5f$ states with localized atomic-like ones. Direct evidence for the coexistence of $5f$-derived quasiparticles and local magnetic excitations has been obtained from neutron scattering experiments[10]. As will be shown, the dual model provides for a natural explanation of the heavy quasiparticle excitations found, for example, in UPt$_3$ and UPd$_2$Al$_3$. Furthermore, there is clear evidence that in superconducting UPd$_2$Al$_3$ Cooper-pair formation is due to virtual intra-atomic excitations of localized $5f$ electrons rather than phonons[11]. More details are found in Sect. 15.3. Here we provide first a microscopic justification for the dual model and afterward demonstrate the emergence of heavy quasiparticles.

In order to study the role of intra-atomic correlations we consider a model Hamiltonian for the $5f$ subsystem, where the hybridization with the conduction electrons is accounted for by effective $5f$ hopping matrix elements. Note that the model should be considered as an effective Hamiltonian for low-energy excitations, typical less than 10 meV. The high-energy processes have been integrated out. The conjecture is that in $5f$ systems the hybridization

---
[10] see [182]
[11] see [317]

between conduction electrons and $5f$ states is effectively renormalized to zero for some orbitals, while it remains finite for others. The simplest example for demonstrating orbital selection consists of two actinide ions at sites a and b. In order to model uranium compounds, where LDA calculations often give a $5f$ count of $n_f \simeq 2.5$, we shall assume a total of five $5f$ electrons for the two sites. We assume that the on-site Coulomb repulsion U is large as compared with the hybridization matrix elements. In this case the ground state is a superposition of different $|a; f^3\rangle|b; f^2\rangle$ and $|a; f^2\rangle|b; f^3\rangle$ configurations. They are coupled by hopping matrix elements. The states at a site are characterized by the total angular momenta $J(a)$ and $J(b)$, respectively. The differences between different J states at fixed n_f are of the order of the $5f$ exchange constant, i.e., 1 eV. Since the spin-orbit interaction is large, we use $j-j$ coupling and limit ourselves to $j = 5/2$ single particle states. For the low-energy excitations the relevant states are $|a; f^3, J(a) = 9/2\rangle|b; f^2, J(b) = 4\rangle$ and $|a; f^2, J(a) = 4\rangle|b; f^3, J(b) = 9/2\rangle$. When an electron is transferred from site a to b and vice versa, the final state is a mixture of different excited multiplets, i.e., $|a; f^3, J(a) = 9/2\rangle|b; f^2, J(b) = 4\rangle \to |a; f^2, J'(a)\rangle|b; f^3, J'(b)\rangle$. The energy loss due to the excitation of different J multiplets must be balanced by the gain in kinetic energy. The crucial point is that the energy loss depends on the atomic orbital from which the electron is leaving site a. It also depends on the relative orientation of $\mathbf{J}(a)$ and $\mathbf{J}(b)$. By requiring that the overall energy gain due to hopping is as large as possible, anisotropies in the hopping matrix elements are enlarged and orbital selection is obtained. With these considerations as background we write down the following model Hamiltonian for N sites

$$H = H_{\text{kin}} + H_{\text{coul}} \quad , \tag{13.45}$$

where the local Coulomb term is

$$H_{\text{coul}} = \frac{1}{2} \sum_n \sum_{j_{z_1} \ldots j_{z_4}} U_{j_{z_1} j_{z_2} j_{z_3} j_{z_4}} c^+_{j_{z_1}}(n) c^+_{j_{z_2}}(n) c_{j_{z_3}}(n) c_{j_{z_4}}(n) \quad . \tag{13.46}$$

Here $c^+_{j_z}(n)$ creates an electron at site n in state $j = 5/2$ with j_z. The Coulomb matrix elements are given in terms of Clebsch-Gordan coefficients and Coulomb integrals U_J:

$$U_{j_{z_1} j_{z_2} j_{z_3} j_{z_4}} = \sum_J U_J C^{J J_z}_{5/2, j_{z_1}; 5/2, j_{z_2}} C^{J J_z}_{5/2, j_{z_3}; 5/2, j_{z_4}} \quad . \tag{13.47}$$

Pauli's principle limits J to $J = 0, 2, 4$. Furthermore $J_z = \sum_{\nu=1}^{4} j_{z_\nu}$. The first term in (13.45) is

$$H_{\text{kin}} = -\sum_{j_z} t_{j_z} \left(c^+_{j_z}(a) c_{j_z}(b) + h.c. \right) + \epsilon_f \sum_n c^+_{j_z}(n) c_{j_z}(n) \quad , \tag{13.48}$$

where ϵ_f is the orbital energy. We have assumed that the hopping matrix elements are diagonal in j_z.

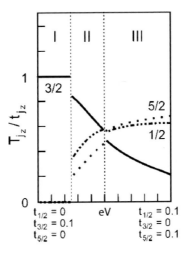

Fig. 13.17. Values T_{j_z}/t_{j_z} for a two-site cluster with 5 electrons along a line connecting linearly the t_{j_z} values written below the figure. The numbers 1/2, 3/2, 5/2 refer to different j_z values. The regions I, II, III have a ground state with $J_z = 15/2$, 5/2 and 1/2, respectively. The U_J are in accordance with (13.44). (From [107])

The Hamiltonian can be diagonalized numerically and the ground state can be determined. The degree of localization is given by the reduction of the hopping matrix element for a given j_z orbital by intra-atomic correlations, i.e., by the ratio of the j_z projected kinetic energy T_{j_z} and the bare hopping t_{j_z}

$$\frac{T_{j_z}}{t_{jz}} = \langle \psi_{GS} | \left(c_{j_z}^+(a) c_{j_z}(b) + h.c. \right) | \psi_{GS} \rangle \quad . \tag{13.49}$$

Here $|\psi_{GS}\rangle$ is the ground state for a given f electron number. In Fig. 13.17 results are shown for a cluster consisting of two sites and five electrons. The parameters resemble the ones for UPt$_3$. One notices strong renormalization effects and orbital selective localization in some parameter regimes. For example, in region I only electrons with $j_z = 3/2$ delocalize. The results agree qualitatively with those of larger clusters.

After this introduction we point out the origin of heavy quasiparticle excitations in systems like UPt$_3$ and UPd$_2$Al$_3$ and demonstrate the predictive power of the theory. We start by keeping $5f$ electrons in orbitals with $j_z = \pm 5/2$, $\pm 1/2$ as localized. Only those with $j_z = \pm 3/2$ are kept itinerant. Since experiments don't indicate any Kramers' degeneracy of the localized $5f$ electrons, we keep *two* of them as localized. Note that electrons with $|j_z| = 5/2$ should remain localized since otherwise no $J = 4$ state could form when going over from $5f^3 \to 5f^2$. It turns out that best results are obtained by constructing the localized $5f^2$ states from $|j_z| = 5/2$ and $1/2$ instead of $|j_z| = 5/2$ and $3/2$ states.

The first step is the determination of the self-consistent LDA potentials, e.g., for UPd$_2$Al$_3$ by keeping two localized $5f$ electrons as part of the core. Since UPd$_2$Al$_3$ is an antiferromagnet with $T_N = 14.5K$ and $\mu \simeq 0.83$ μ_B per U ion, LDA calculations use the observed AF structure. Next the localized $5f^2$ subsystem is diagonalized, thereby excluding $j_z = \pm 3/2$ states. Six states of the $5f^2$ system can be formed with $j_z = \pm 5/2, \pm 1/2$. The Coulomb matrix elements are evaluated by assuming jj coupling and $5f$ radial functions of the ab initio band structure potential. The result is a two-fold degenerate ground state with $J_z = \pm 3$ implying $J = 4$. The overlap with the Hund's rule ground state $^3H_4$ derived from the LS-coupling scheme is 0.865. The two states are split by the hexagonal crystalline electric field (CEF) of the surroundings of a U ion. The CEF eigenstates $|\Gamma_3\rangle$ and $|\Gamma_4\rangle$ are

$$|\Gamma_{3(4)}\rangle = \frac{1}{\sqrt{2}} \left(|J = 4; J_z = +3\rangle \, (\pm) \, |J = 4; J_z = -3\rangle \right) \quad . \tag{13.50}$$

The coupling matrix element of the itinerant and localized $5f$ electrons is directly obtained from the expectation values of the Coulomb interactions in the $5f^3$ states. For the latter the product states $|f^2; J = 4, J_z = \pm 3\rangle \otimes |f^1, j = 5/2, j_z = \pm 3/2\rangle$ are being used. From the energy difference $\langle f^1; 5/2, 3/2| \otimes \langle f^2; 4, 3|U_{\text{coul}}|f^2; 4, 3\rangle \otimes |f^1; 5/2, 3/2\rangle - \langle f^1; 5/2, 3/2| \otimes \langle f^2; 4, -3|U_{\text{coul}}|f^2; 4, -3\rangle \otimes |f^1; 5/2, 3/2\rangle$ of approximately -0.4 eV we are able to determine the transition matrix element between $|\Gamma_3\rangle$ and $|\Gamma_4\rangle$, i.e.,

$$M = \left\langle f^1; \frac{5}{2}, \frac{3}{2} \right| \otimes \langle \Gamma_4 |U_{\text{coul}}| \Gamma_3\rangle \otimes \left| f^1; \frac{5}{2}, \frac{3}{2} \right\rangle$$
$$\simeq -0.2 eV \quad . \tag{13.51}$$

The itinerant f electrons couple with an RKKY type of interaction the CEF excitations $|\Gamma_3\rangle \to |\Gamma_4\rangle$ with excitation energy δ at different sites i and j. The Hamiltonian for these coupled excitations is

$$H_{\text{CEF}} = \delta \sum_i |\Gamma_4(i)\rangle\langle\Gamma_4(i)| + \sum_{\langle ij \rangle} J_{\text{ex}}(i-j) J_z(i) J_z(j) \quad . \tag{13.52}$$

The coupling $J_{\text{ex}}(i-j)$ is the one caused by the itinerant electrons. It is indicated in Fig. 13.18. In UPd$_2$Al$_3$ this coupling is strong enough that it induces an AF ground state with $\mathbf{Q} = (0, 0, \pi/c)$, where c is the lattice constant perpendicular to the u planes. This is an example of an induced moment system, a phenomenon well known from $4f$ rare-earth systems. We refer the reader who wants to learn more about these systems to various contributions to the Handbook on the Physics and Chemistry of the Rare Earth [159][12]. The AF CEF excitations, otherwise called magnetic excitons, have been observed by inelastic neutron scattering[13]. We show the results in Fig. 13.19 since we will

---
[12] see, e.g., [130]
[13] see [30, 315]

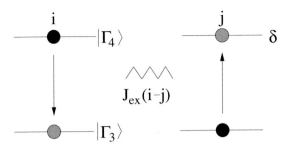

Fig. 13.18. Coupling of CEF excitations at sites i and j via itinerant electrons, i.e., RKKY interactions. Due to the interaction, the discrete excitation energy δ goes over into a band and eventually into induced magnetic order if the coupling is strong enough. Magnetic excitons are associated with the resulting band of excitations.

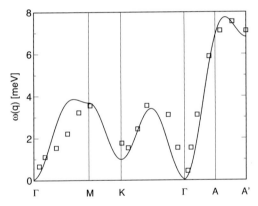

Fig. 13.19. Magnetic excitons in UPd$_2$Al$_3$ observed by inelastic neutron scattering [30, 315] together with a fit given by (13.53). (From [448])

refer to them repeatedly in Chapter 15. They demonstrate nicely the dual character of $5f$ electrons. The system UPd$_2$Al$_3$ behaves like an intermetallic compound with Pr^{3+} ions, i.e., a $4f^2$ system. The magnetic-exciton dispersion parallel to the hexagonal c*-axis can be modeled by

$$\omega_{\mathrm{ex}}(q_z) = \omega_{\mathrm{ex}}\left[1 + \beta \cos(cq_z)\right] \quad, \tag{13.53}$$

with $\omega_{\mathrm{ex}} = 5$ meV, and $\beta = 0.8$[14]. This shows us that the original CEF splitting is indeed of order 5 meV. The heavy quasiparticle masses result from a dressing of the conduction electrons with intra-atomic CEF excitations, or stated more accurately, with induced AF excitons. The situation resembles that in Pr metal, where a mass enhancement of conduction electrons by a factor of 5 is obtained from virtual CEF excitations of localized $4f^2$ electrons[15].

---

[14] see [448]
[15] see [134]

Fig. 13.20. Self-energy of an electron (solid line) due to the emission and reabsorption of a magnetic exciton (wavy line).

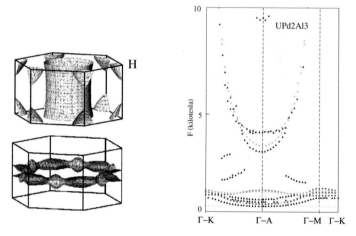

Fig. 13.21. Left panel: Fermi surface of UPd$_2$Al$_3$ calculated within the dual model. The cylindrical main part and the H centered ellipsoid denoted by γ and β have effective masses with $m^*/m = 33, 19$. The highest masses are found on the torus (only the most important sheets of the FS are shown). Right panel: Comparison of experimental dHvA frequencies F (black symbols) (From [204]) and calculated frequencies (open symbols). The large parabola corresponds to the main FS cylinder. (From [510])

For quantitative results the self-energy $\Sigma(\mathbf{k},\omega)$ has to be calculated. The microscopic process is shown in Fig. 13.20. The wavy line represents the susceptibility $\chi(\mathbf{q},\omega)$ of the magnetic excitons. It turns out to be sufficient to neglect the dispersion of the $5f^2$ low-energy CEF excitations, in which case the self-energy becomes $\Sigma(\omega)$. This provides for the following, rather complete picture: The Fermi surface of UPd$_2$Al$_3$ is obtained from the band structure calculation with $5f^2$ kept as part of the core. The band masses m_b of the f-like parts of the conduction electrons are multiplied isotropically by a factor $m^*/m_b = 1 - (\partial \Sigma/\partial \omega)_{\omega=0}$. The enhancement of the bandmass m_b by the emission and reabsorption of magnetic excitons reduces in this case to

$$\frac{m^*}{m} = 1 + 8a^2 N(0)\frac{|M|^2}{\bar{\bar{\delta}}} \quad . \tag{13.54}$$

The prefactor a^2 is the weight of the $5f$ contribution to the conduction electrons close to the Fermi energy. From band structure calculations it is known that this weight is nearly 40 %. For an estimate of (13.54) we need to know $N(0)$ and $\bar{\delta}$ which is the averaged energy of a magnetic exciton. We extract a value of $N(0) = 2.76$ states/(eV cell spin) from LDA calculations with two $5f$ electrons in the core and choose a value of $\bar{\delta} = 7$ meV from the INS data[16]. Results for the Fermi surface and for de Haas-van Alphen frequencies are shown in Fig. 13.21. Also shown are experimental results. The strongly anisotropic quasiparticle masses are listed in Table 13.3. The agreement with experiments is surprisingly good. It should be noticed that once $\bar{\delta}$ is obtained from experiment, the theory does *not* contain any adjustable parameters. Similar findings were obtained for UPt$_3$, where the strong mass anisotropies are again well explained. In this case the theory contains one adjustable parameter, for which an adjustment of the γ coefficient of the low temperature specific heat is used.

13.4 Heavy d Electrons: LiV$_2$O$_4$

The spinel LiV$_2$O$_4$ is considered the first d-electron system with heavy quasiparticle excitations [248]. Note that (YSc)Mn$_2$, which is close to a magnetic instability shows similar features[17]. The γ coefficient of the low temperature specific heat of LiV$_2$O$_4$ is strongly enhanced, i.e., $\gamma \simeq 0.4 J$ mol$^{-1}K^{-2}$ and so is the spin susceptibility. The latter shows for $T > \theta$ a Curie-Weiss like behavior

$$\chi(T) = \chi_0 + \frac{c}{T+\theta} \quad ; \quad \theta = 63K \tag{13.55}$$

and for $T \ll \theta$ a Fermi-liquid behavior with a similar enhancement as γ. No magnetic ordering has been observed down to 0.02 K. The Sommerfeld-Wilson ratio is of order unity. Furthermore, the resistivity is $\rho(T) = \rho_0 + AT^2$

Table 13.3. Quasiparticle mass m^* in unit of free electron mass for UPd$_2$Al$_3$. Experimental values from de Haas-van Alphen measurements with $H||c$. (From [204] and [510])

FS sheet	m^* (exp.)	m^* (theory)
ζ	65	59.6
γ	33	31.9
β	19	25.1
ϵ_2	18	17.4
ϵ_3	12	13.4
β	5.7	9.6

---

[16] see [397]
[17] see [120, 475]

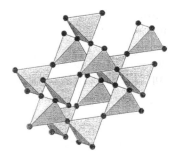

Fig. 13.22. Pyrochlore lattice. The V atoms of LiV_2O_4 occupy corner sharing sites (solid dots) of tetrahedra.

with a large coefficient $A = 2 \ \mu\Omega cm K^{-2}$. When the entropy is calculated from the specific heat data, it is found that $S(T = 60K) - S(T = 2K) = 10 \ J mol^{-1} K^{-1}$. This is close to $2R \ln 2$ where R is the gas constant and implies that at 60 K there is nearly one excitation per V ion present. This proves that correlations are strong in the system, since in a conventional band description of $3d$ electrons only a fraction of $k_B T/\epsilon_F$ of the conduction electrons takes part in the excitations.

Spinels have the composition AB_2O_4. The B ions are surrounded by an octahedron of O^{2-} ion each, i.e., they form a BO_6 complex. They are positioned at the sites of a pyrochlore lattice (see Figs. 13.22), i.e., each B site belongs to two tetrahedra. The pyrochlore structure belongs to a class of lattices which are called geometrically frustrated and are discussed in Sect. 14.3. By this we mean that if spins with a nearest neighbor AF interaction are attached to the different sites, then an AF spin alignment is frustrated by the lattice structure.

Of particular interest are spinels in which the B sites have half-integer valency. Here LiV_2O_4 is an example. With O^{2-} and Li^+ the two V atoms have a valency of $V^{+3.5}$, i.e., they fluctuate between $3d^1$ and $3d^2$ configurations. Another, much studied example is Fe_3O_4, i.e., magnetite. One Fe^{3+} ion occupies an A site while the remaining Fe^{2+} and Fe^{3+} occupy the B sites. Magnetite shows a metal-insulator transition (Verwey transition) of Mott-Hubbard type at $T_v = 120K$ [468]. It demonstrates that correlations are strong in this system. The metal-insulator transition is accompanied by charge ordering, the precise form of which has been repeatedly a subject of debate.

In addition to LiV_2O_4 and Fe_3O_4, there are a number of other spinels with half-integer valency of the B atoms. We list some of them in Table 13.4.

Different models have been proposed to explain the existence of heavy quasiparticles in LiV_2O_4. The geometrical frustration of the pyrochlore lattice is an important ingredient in those theories. We want to discuss two of them, with quite different starting points. One starts from a conventional LDA type of band calculation and treats the strong correlations by a RPA. With the

13.4 Heavy d Electrons: LiV$_2$O$_4$

assumption that we are close to a magnetic instability, we can explain the neutron inelastic experiments as well as γ and $\chi(T)$. The second approach starts from the strong correlation limit and can explain parameter free the correct order of magnitude of γ and $\chi(T)$. The latter approach, however, has not been sufficiently developed in order to explain the neutron data. The hope is that the two models with their opposite starting points can eventually be brought to convergence in a regime which is in between the two cases.

It is worth pointing out that the band structure of a pyrochlore lattice has special features. They are seen best when only one orbital per site and nearest-neighbor hopping processes with matrix element t are taken into account. The Hamiltonian is then of the diagonal form

$$H_0 = \sum_{\mathbf{k}\alpha\sigma} (\epsilon_\alpha(\mathbf{k}) - \mu) c^+_{\mathbf{k}\alpha\sigma} c_{\mathbf{k}\alpha\sigma} \quad , \tag{13.56}$$

where $\alpha = 1, \ldots, 4$ is a subband index due to the four atoms per unit cell. This assumes that there is one orbital per site. The $\epsilon_\alpha(\mathbf{k})$ are given by

$$\epsilon_\alpha(\mathbf{k}) = \begin{cases} 2t & \alpha = 3, 4 \\ -2t \left[1 \pm (1 + \eta_\mathbf{k})^{\frac{1}{2}}\right] & \alpha = 1, 2 \end{cases}$$

$$\eta_\mathbf{k} = \cos(2k_x)\cos(2k_y) + \cos(2k_y)\cos(2k_z) + \cos(2k_z)\cos(2k_x) \quad , \tag{13.57}$$

where the k_ν are expressed in reciprocal lattice units $2\pi/a$. The band structure is shown in Fig. 13.23. Going beyond nearest neighbor hopping does not result in any dispersion of the flat band. A finite dispersion is obtained only when we deal with several orbitals per site and different hopping matrix elements. This feature is still visible in the calculated density of states for LiV$_2$O$_4$ shown in Fig. 13.24. The peak at $\simeq 1$ eV corresponds to the flat band in Fig. 13.23. We notice that the $3d$ electrons are in the t_{2g} subbands which are well separated from the empty e_g bands. They are also well separated from the oxygen $2p$ bands.

The first model calculation which we discuss starts from the t_{2g} bands. More precisely, a weak trigonal distortion splits the t_{2g} orbitals into $a_{1g} + e'_g$ orbitals. Starting from weakly or nearly uncorrelated electrons, the strong correlations are treated within a RPA[18]. One aim is to explain available quasielas-

Table 13.4. Spinels with a half-integer valency of d ions.

	M =	Ti	V	V(Cr)	Mn
Li(Al)M$_2$O$_4$		LiTi$_2$O$_4$	LiV$_2$O$_4$	AlV$_2$O$_4$ (LiCr$_2$O$_4$)	LiMn$_2$O$_4$
average d-electron count per M-atom		$d^{0.5}$	$d^{1.5}$	$d^{2.5}$	$d^{3.5}$

---

[18] see [494, 505]

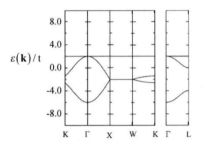

Fig. 13.23. Energy bands for a pyrochlore lattice with nearest neighbor hopping t and one orbital per site. The dispersionless band is twofold degenerate. (From [207])

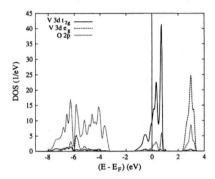

Fig. 13.24. Partial densities of states (DOS) for LiV_2O_4 obtained from LDA calculations. The sharp peak around 1 eV corresponds to the flat band in Fig. 13.23. (From [112])

tic neutron scattering experiments[19]. For that purpose the wavenumber and frequency dependent spin susceptibility $\chi(\mathbf{q}, \omega)$ for LDA multiband electrons has to be calculated. This is a generalization of the calculation outlined in Sect. 11.3.2, where $\chi(\mathbf{q}, \omega)$ was determined for a one-band system (see (11.106) and (11.109)). As a first step the unenhanced susceptibility $\chi^{(0)}(\mathbf{q}, \omega)$ must be determined, which is the analogue of $u(\mathbf{q}, \omega)$ in (11.106). The frustrated lattice structure only enters here in form of the energy dispersion $\epsilon_{\mathbf{k}\alpha}$. It has the effect that $Re\chi^{(0)}(\mathbf{q})$ has broad peaks in different symmetry directions. They are in the same region $0.4 \text{Å}^{-1} \lesssim q \lesssim 0.8 \text{Å}^{-1}$ of q-space in which the main quasielastic neutron scattering is observed. The structures in $\chi^{(0)}(\mathbf{q})$ are amplified when $\chi(\mathbf{q}, \omega)$ is computed (compare with (11.109)). We then solve the integral equation

---

[19] see [251, 275]

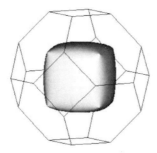

Fig. 13.25. Surface in momentum space representing the lower bound of the critical regime of strongly enhanced slow spin fluctuations. The upper bound is given by a larger equivalent surface which differs by a momentum $|\delta q| = 0.3 a^{-1}$ from the one shown where a is the lattice parameter. (From [505])

$$\chi(\mathbf{r},\mathbf{r}';\omega) = \chi^{(0)}(\mathbf{r},\mathbf{r}';\omega) + \int d\mathbf{r}'' \chi^{(0)}(\mathbf{r},\mathbf{r}'';\omega) K(\mathbf{r}'') \chi(\mathbf{r}'',\mathbf{r}';\omega) \ , \quad (13.58)$$

where $K(\mathbf{r}'')$ is the spin-dependent part of the exchange-correlation potential $v_\sigma^{xc}(\mathbf{r})$ (see (4.21)). Taking the Fourier transform results in $\chi_{\mathbf{GG}'}(\mathbf{q},\omega)$ where $\mathbf{G}$ denotes reciprocal lattice vectors. The peak structure in $\chi^{(0)}(\mathbf{q})$ is found to be enhanced in $\chi(\mathbf{q})$, especially when $\mathbf{G} = \mathbf{G}' = 0$. What enters the calculations is an average of $K(\mathbf{r})$ over the local d electron density. At a critical coupling K_c the energy denominator in $\chi(\mathbf{q},\omega)$ vanishes and the system becomes magnetic. If we would use the unrenormalized $K(\mathbf{r})$ from LDA calculations, LiV$_2$O$_4$ would be indeed magnetic with a range of possible $\mathbf{q}$ values, in contradiction with experiments. Thus the LDA potential is apparently strongly renormalized. In fact, the geometrically frustrated pyrochlore structure prevents the system from becoming an antiferromagnet.

Assuming that $K \lesssim K_c$ one finds a broad range of $\mathbf{q}_c$ values at which $\chi(\mathbf{q}) = $ max. The system does not find a single $\mathbf{q}$ vector at which the free energy of antiferromagnetic spin fluctuations is minimized. The competition between different spin structures is a consequence of geometric frustration. In Fig. 13.25 we show a surface in $\mathbf{k}$ space which defines the lower bound for the critical regime of strongly enhanced slow spin fluctuations. The upper bound is given by a larger surface with a difference $a|\delta q| \approx 0.3$ from the one in that figure. The mean radius of the region of enhanced fluctuations is $q_c \approx 0.6 \text{Å}^{-1}$. This is in good agreement with inelastic neutron scattering experiments on polycrystalline LiV$_2$O$_4$ samples at $T \leq 2$ K. Here short-range AF correlations with a relaxation rate of $\Gamma(\mathbf{q}) \simeq 1$ meV were observed in a broad region of $\mathbf{q}$ vectors around $|\mathbf{q}| \simeq 0.6 \text{Å}^{-1}$. A small relaxation rate is precisely what is expected near a critical point, i.e., when K is close to K$_c$.

We use (11.123) in order to write for low frequencies

$$\text{Im}\chi(\mathbf{q},\omega) = z_{\mathbf{q}} \chi(\mathbf{q}) \omega \frac{\Gamma(\mathbf{q})}{\omega^2 + (\Gamma(\mathbf{q}))^2} \ , \quad (13.59)$$

where $z_\mathbf{q}$ is a weighting factor of order unity. By adjusting the value K $\lesssim$ K$_c$, we obtain a $\Gamma(q_c)$ of the observed size, i.e., of approximately 1 meV. It reproduces also the large observed linear specific heat coefficient γ. The relation is

$$\gamma = \frac{k_B^2 \pi}{N} \sum_\mathbf{q} \frac{z_\mathbf{q}}{\Gamma(\mathbf{q})} \quad (13.60)$$

and the summation is over the cubic Brillouin zone.

In summary, the strong point of the RPA based theory is the finding of a broad region in $\mathbf{k}$ space around $q_c \simeq 0.6 \text{Å}^{-1}$ in which, low-energy spin fluctuations are strong. This broad region is intimately related to the frustrated pyrochlore lattice structure. The nearness to a AF transition is again related to the frustrated structure which prevents long-range AF order[20].

The second approach to heavy quasiparticle formation in LiV$_2$O$_4$ starts from the strong correlation limit. This seems justified in particular in view of an observed charge ordering which takes place under pressure [127]. The analogy to the Verwey transition in Fe$_3$O$_4$ is apparent and is another earmark of strong correlations. The model Hamiltonian for the d electrons of vanadium is

$$H = -\sum_{\langle ij \rangle \nu} t_\nu \left(c_{i\nu\sigma}^+ c_{j\nu\sigma} + h.c. \right) + U \sum_{i\nu} n_{i\nu\uparrow} n_{i\nu\downarrow} + U \sum_{i;\nu>\mu} n_{i\nu} n_{i\mu}$$

$$+ \tilde{J} \sum_{i\nu\mu} \mathbf{s}_{i\nu} \mathbf{s}_{i\mu} + V \sum_{\langle ij \rangle} n_i n_j + \sum_{\langle ij \rangle} J_{ij}(S_i, S_j) \mathbf{S}_i \mathbf{S}_j \quad . \quad (13.61)$$

Here i is the site index and $\nu = 1, 2, 3$ labels the different t_{2g} bands. The first term is the kinetic energy term while the next three terms describe the intra-atomic Coulomb repulsion (compare with (11.1)). For simplicity, the differences in Coulomb repulsions between different orbitals have been neglected. The last two terms are due to the Coulomb interactions and spin-spin interactions between neighboring sites, respectively. Here $\mathbf{S}_i = \sum_\nu \mathbf{s}_{i\nu}$.

Let us assume that $U \to \infty$. Because the average d electron number is $3d^{1.5}$ the V ions are either in a $3d^1$ or $3d^2$ configuration. Other configurations like $3d^0$ or $3d^3$ are excluded by the large U value. Due to Hund's rule correlations the $3d^2$ ions are in a high spin $S = 1$ state. Furthermore, let us first set the hopping matrix elements $t_\nu = 0$. This corresponds to the classical limit. In order to minimize the Coulomb repulsions V between $3d^1$ and $3d^2$ ions, we must have on each tetrahedron two V ions in a $3d^1$ and two ions in a $3d^2$ configuration (tetrahedron rule). It is easily seen that any other distribution of the $3d^1$ and $3d^2$ sites increases the total intersite electron repulsion. The ground-state is in this case macroscopically degenerate because there are different configurations of order $(3/2)^{N/2}$, which satisfy the tetrahedron rule where N is the number of sites. We show one of these configurations in Fig.

---

[20] For further details, as well as extensions of the theory to finite temperatures we refer to the original literature [504, 505]

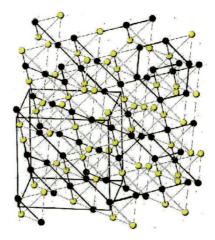

Fig. 13.26. Pyrochlore lattice: Example of a configuration satisfying the tetrahedron rule. Occupied sites with $S = 1/2$ (black dots) are connected by thick solid lines which form chains or rings. The same may be done for sites with $S = 1$ (yellow dots).

13.26. When we connect all sites with $S = 1/2$ they form loops or rings, provided periodic boundary conditions are applied. The same holds true for the $S = 1$ sites. From LDA calculations we can determine the nearest neighbor coupling constants $J_{ij}(S_i S_j)$. One finds $J(1/2, 1/2) = 3$ meV, $J(1,1) = 24$ meV. Because $J(1,1) \gg J(1/2, 1/2)$ the spins in the $S = 1$ chains are much more strongly coupled than the spins in the $S = 1/2$ chains.

Note that a Heisenberg $S = 1$ chain has a gap in the excitation spectrum (Haldane gap) [168]. Therefore, the low energy excitations are within the $S = 1/2$ subsystem. Due to the geometrical frustrations the spin 1/2 loops or chains are essentially decoupled from the spin 1 chains. They are also decoupled from each other. A coupling can take place only via the $S = 1$ chains respective rings, which require excitation energies of order $J(1,1)$. The specific heat of a Heisenberg chain is linear in temperature T. The γ coefficient as well as the spin susceptibility of a spin chain are given according to (13.42) by

$$\gamma = \frac{2}{3} \frac{k_B R}{J(1/2, 1/2)} \quad , \quad \chi_s = \frac{4\mu_{\text{eff}}^2 R}{\pi^2 J(1/2, 1/2)} \quad , \tag{13.62}$$

where R is the gas constant. The experimental γ coefficient would require a value of $J(1/2, 1/2) = 1.2$ meV instead of the calculated 3 meV. However, this strong coupling model gives nearly the right density of low-energy excitations required by experiments.

Up to now we have set all kinetic energy terms equal to zero. In Chapter 14 we shall discuss in detail how the macroscopic degeneracy of the ground state configurations is lifted by t_ν. It turns out that for small hopping matrix elements this occurs only to order t_ν^3/V^2. Therefore, when $J(1/2, 1/2) > t_\nu^3/V^2$ then the above classical limit is a reasonable starting point for including the dynamics of the system. A disadvantage of the above considerations is that it is difficult to explain the neutron scattering experiments, which show enhanced spin fluctuations in the regime $0.4\text{Å}^{-1} \lesssim q \lesssim 0.8\text{Å}^{-1}$. This was found possible in the RPA based model, although the γ coefficient requires corrections by a factor of 30 when the starting point is an uncorrelated system.

14
Excitations with Fractional Charges

Condensed matter physicist became acquainted for the first time with the concept of fractional charges when *Su* and *Schrieffer* [439] discussed excitations in heavily doped trans-polyacetylene $(CH)_n$. They found that the magnitude of the fractional charge is always a rational of e and depends on the degree of doping in this case. Before that the two authors together with *Heeger* had shown that in undoped trans-polyacetylene kink excitations may exist which have only a spin and no charge or only a charge $\pm e$ but no spin [440]. This phenomenon is known as spin-charge separation. When the electron system is fully spin polarized it behaves like a system of spinless fermions, because the spin degrees of freedom are frozen out. If we consider a one dimensional π electron system similar to $(CH)_n$ but with *one* π electron per *two* carbon sites only (half filled band of spinless fermions), then the kink excitations have fractional charge $\pm e/2$. They are related to an excitation with fermion number 1/2 in relativistic field theory discovered by *Jackiw* and *Rebbi* [212].

A characteristic feature of spin-charge separation or of fractional charges in trans-polyacetylene is that both appear within the independent electron approximation, i.e., they have little to do with electron correlations. Instead, they require the inclusion of atomic displacements, i.e., lattice degrees of freedom. That is completely different in the fractional quantum Hall (FQH) effect. Two-dimensional (2D) semiconducting inversion layers in a perpendicular magnetic field can support excitations with fractional charges, most notable with $\pm e/3$, which are solely due to electron correlations [455]. A fascinating aspect of them is that they neither obey Fermi- nor Bose statistics but instead are anyons[1]. The name indicates that an exchange of two of these excitations multiplies the wavefunction by a phase $e^{i\varphi}$ where φ is neither π like for fermions nor 2π like for bosons but instead a rational of it. A stringent connection between fractional charges and fractional, i.e., anyonic statistics would exclude fractionally charged excitations in three dimensions. It is well known, that in three dimensions only fermions or bosons may exist. Therefore it is of interest to

---
[1] see [272, 279, 487]

know whether or not a model Hamiltonian can be found which supports deconfined excitations with fractional charges in three dimensions [416]. This is indeed the case. What is required is a system of strongly correlated electrons on a three-dimensional geometrically frustrated lattice. An example is the pyrochlore lattice (see Fig. 13.23). When this lattice is half-filled with strongly correlated electrons, excitations with fractional charges $\pm e/2$ may result. We will discuss a number of different phenomena which occur when electrons with strong nearest-neighbor repulsions occupy frustrated lattices in two and three dimensions. This includes the checkerboard as well as the kagome lattice. Interesting features are relations to *Pauling*'s ice model [355] as well as to gauge theories. A fascinating aspect is the appearance of magnetic monopoles in spin ice.

We start with trans-polyacetylene despite the fact that correlations play no role as far as the basic effect is concerned. However, good insight is gained into the phenomenon of fractional charges.

14.1 Trans-Polyacetylene

Polyacetylene forms chains $(CH)_x$ with x typically of the order of 10 - 40. Two configurations of $(CH)_x$ are known: trans- and cis-polyacetylene. We are interested here in the trans configuration which is depicted in Fig. 14.1.

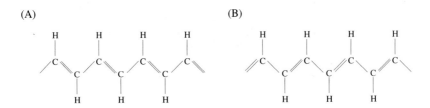

Fig. 14.1. Chemical structure of the two ground states of *trans*-polyacetylene.

Note that an infinite chain has two ground states which differ in the arrangement of single and double C-C bonds. The structure is produced by a sp^2 hybridization of the valence electrons of carbon. Three of its four valence electrons participate in σ bonding, while one electron does so in π bonding. The π bond is formed from electrons with $2p_z$ symmetry (Fig. 14.2). There are two C atoms per unit cell. If the C atoms were equally spaced in *trans*-$(CH)_x$, the π band would be degenerate at the edge of the Brillouin zone due to a C_2 symmetry. If this were the case, the π band would be half-filled in an extended zone scheme, and the system would exhibit metallic behavior when the chains become infinitely long. We do not, however, observe this; instead, infrared absorption experiments reveal a gap in the electronic excitation spectrum of approximately 2 eV.

Fig. 14.2. Schematic picture of the $2p_z$ orbitals which form π bonds.

This gap results from bond alternation. In fact, consecutive C-C bond lengths d_n and d_{n+1} are found experimentally to differ by a dimerization length $\xi = (1/2)|d_n - d_{n+1}| \simeq 2.7$ pm. This difference results in a loss of the C_2 symmetry and hence in a gap in the π band at the edge of the reduced Brillouin zone (Fig. 14.3).

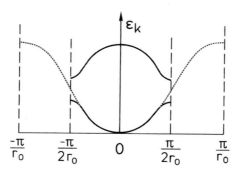

Fig. 14.3. Energy dispersion for a π bond of an equidistant (dotted lines) and a dimerized (solid lines) chain. The latter requires a doubling of the unit cell and therefore a reduction of the Brillouin zone. The quantity r_0 is closely related to the average C-C bond length.

The phenomenon of dimerization in polymers has a long-standing history in quantum chemistry[2]. It represents a special case of a general instability first suggested by *Peierls* for one-dimensional systems with a half-filled conduction band. We notice from Fig. 14.3 that doubling the unit cell leads to a gain in kinetic energy, because the states near the edge of the new Brillouin zone have a lower energy than in the undimerized chain. *Peierls* showed that this gain in kinetic energy is always larger than the loss in elastic energy associated with

---

[2] see, e.g., [257, 289]

a distortion. *Su, Schieffer* and *Heeger* demonstrated the latter by means of a model Hamiltonian for the π electron system:

$$H = -\sum_{(i,j)\sigma} (t_0 \pm 2\alpha\xi)\, a^+_{i\sigma} a_{j\sigma} + 2N_0 K\xi^2 \quad . \quad . \quad (14.1)$$

The operators $a^+_{i\sigma}$, $a_{i\sigma}$ create and destroy a π electron at the carbon site i. The first term takes into account that the hopping matrix elements between π electrons at neighboring sites (i,j) depend in the dimerization length ξ, i.e., it is larger than t_0 for the shorter bond and smaller than t_0 for the longer bond. The second term describes the elastic energy which must be associated with the dimerization of a chain of N_0 carbon atoms. We can relate the parameters α and K to the electron-phonon coupling strength and to an elastic constant, respectively. The dimerization length acts like an order parameter. It vanishes for an undimerized chain. A dimerized $(CH)_x$ chain has a twofold degenerate

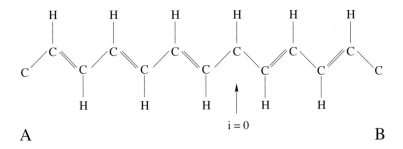

Fig. 14.4. Soliton or domain wall between an (A) domain to the left and a (B) domain to the right. (From [440])

ground state denoted by A and B. In state B the double bonds are shifted by one C site as compared with the ones in state A.

Consider now a soliton or domain wall with the A phase to the left and the B phase to the right of the wall as illustrated in Fig. 14.4. Denote with u_i the displacements of the C atoms projected onto the chain axis relative to an undistorted chain. If the C atom at site $i = 0$ marks the center of the domain wall than $u_0 = 0$. Because of symmetry $u_j = -u_{-j}$. A trial function for the u_j is

$$u_j = (-1)^{j+1} \frac{\xi}{2} \tanh \frac{j}{2l} \quad , \quad j = \pm 1, \pm 2, \ldots \quad (14.2)$$

and similarly for the bond length d_n. In passing we mention that an energy minimization of the soliton or kink excitation gives $l \simeq 7$ for reasonable parameters t_0, α and K [440].

The kink gives raise to a midgap state as indicated in Fig. 14.5. It can accommodate at most two electrons due to Pauli's principle. The appearance

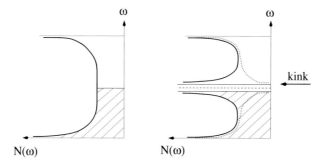

Fig. 14.5. Density of states for a half-filled undimerized chain and a dimerized one with a soliton or kink state inside the gap. Dotted lines indicate the states contributing to the midgap state.

of a midgap state can be understood simply by the fact that the center of the kink is undistorted and therefore its energy must be in the center of the band[3]. Because of symmetry half of the oscillator strength of the midgap state must come from the upper (unoccupied) band and half of it from the lower (occupied) band. Thus when the kink state is empty one electron is missing from the lower band. Therefore the system is no longer charge neutral but has a charge $-e$. However, its total spin is zero, because each state of the lower band has as an equal amount of electrons in both spin directions. Thus we may speak of spin-charge separation because the system has a charge but no spin. If the kink state is singly occupied the system is charge neutral but now has a spin. Similarly, when the midgap state is doubly occupied there is a net charge e but the total spin is zero. Note that when we close a trans-polyacetylene chain to a ring then generating a kink implies automatically generation of an anti-kink. The latter has two double bonds next to each other. When the kink state is empty the anti-kink state must be doubly occupied in order to ensure charge neutrality.

Next let us assume that electrons are fully spin polarized so that spin degrees of freedom do not play any role or, what is equivalent, that we deal with spinless fermions. We want to start again from a half filled conduction band, which implies here one spinless fermion per two C atoms. In that case the kink state can accommodate at most one spin-polarized (or spinless) electron. Again half of the oscillator strength comes from the lower band and half from the upper one. Therefore when the kink state is empty one half of an electron is missing and the charge is $-e/2$. On the other hand, when the kink state is occupied there is a surplus charge $e/2$. Thus we are dealing here with excitations of fractional charge. Thereby the degeneracy of the ground state is a necessary prerequisite. The topologically protected excitation with fractionalized charge, i.e., the kink or domain wall connects degenerate ground

---

[3] a more sophisticated phase-shift argument comes to the same conclusion [440]

states. This is the main message of this Section. Note that kink and antikink are identical in the present case. A topological protection of fractionalized charges applies also to higher dimensions. This will be seen in the following.

14.2 Fractional Quantum Hall Effect

The fractional quantum Hall effect (FQHE) was discovered by *Tsui, Störmer* and *Gossard* shortly after the discovery of the integer quantum Hall effect (IQHE) by *von Klitzing*. While the IQHE can be understood within an independent electron theory of a disordered system the FQHE is based on strong electron correlations. Both effects are found in $GaAs/Ge_xGa_{1-x}As$ heterostructures or silicon field-effect transistors when a magnetic field B is applied perpendicularly to the two-dimensional electron liquid. The experimental arrangement is schematically shown in Fig. 14.6. At certain magnetic field strengths the Hall resistance R_H shows pronounced plateaus at which the resistance R drops to zero.

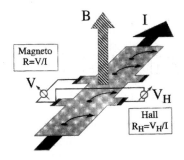

Fig. 14.6. Experimental arrangement for measuring the quantum Hall effect.

The physics of the quantum Hall effect is very rich and has developed into a large field of its own. It has been dealt with in a number of good text books. Examples are Refs. [60, 113, 378, 502] of which [502] is especially recommended to students. Here we shall restrict ourselves to discuss how the strong correlations result in a FQHE. The first question to answer is why are we dealing here with strong correlations? Let us begin with the Hamiltonian of the system of interacting electrons with band mass m in a magnetic field:

$$H = \sum_j \left[\frac{1}{2m} \left[\frac{\hbar}{i}\nabla_j + \frac{e}{c}\mathbf{A}_j \right]^2 + V(\mathbf{r}_j) \right] + \frac{1}{2}\sum_{j\neq k} \frac{e^2}{|\mathbf{r}_j - \mathbf{r}_k|} \quad . \quad (14.3)$$

14.2 Fractional Quantum Hall Effect

The charge e is chosen to be positive. The potential $V(\mathbf{r})$ is due to the background and ensures charge neutrality. First consider the Hamiltonian for a single electron in an external field

$$H_e = \frac{1}{2m}\left[\frac{\hbar}{i}\nabla + \frac{e}{c}\mathbf{A}\right]^2 \quad . \tag{14.4}$$

We choose for $\mathbf{A}(\mathbf{r})$ the symmetric gauge

$$\mathbf{A} = \frac{B}{2}(y\hat{x} - x\hat{y}) \tag{14.5}$$

where $\hat{x}$ and $\hat{y}$ are unit vectors in the plane of the layer. We are interested in the low-energy physics only. The spin degree of freedom is discarded here, because the magnetic field is freezing it out. The eigenfunctions of the Hamiltonian are of the form

$$\psi_{m,n}(\mathbf{r}) = e^{\frac{1}{4}(x^2+y^2)}\left(\frac{\partial}{\partial x} + i\frac{\partial}{\partial y}\right)^m \left(\frac{\partial}{\partial x} - i\frac{\partial}{\partial y}\right)^n e^{\frac{1}{2}(x^2+y^2)} \quad . \tag{14.6}$$

They are related to Laguerre polynomials. All lengths have been expressed in units of the magnetic length

$$l_m = \sqrt{\frac{\hbar c}{eB}} \tag{14.7}$$

which is the basic scale in quantum Hall physics. For a magnetic field of 1T it is approximately 250 Å. The eigenvalues are

$$E_{mn} = (n + 1/2)\hbar\,\omega_c \quad , \quad n = 0, 1, ... \tag{14.8}$$

where ω_c is the cyclotron frequency given by

$$\omega_c = \frac{eB}{mc} \quad . \tag{14.9}$$

It is noticed that the eigenvalues depend on n only, which labels Landau levels with equally spaced energies. Therefore, when we deal with a system of noninteracting electrons the ground state will be highly degenerate because of the index m in (14.8). The degree of degeneracy is obtained as follows. The density of states per spin direction and unit area of a two-dimensional, homogeneous electron gas is constant and given by $N_E = m/(2\pi\hbar^2)$. If the system is in a magnetic field with energy levels (14.8) the averaged density of state is $\overline{N}_E = 1/(\hbar\omega_c)$. Therefore the degeneracy per level is $n_B = N_E/\overline{N}_E = \frac{eB}{hc}$. Since

$$\phi_0 = \frac{hc}{e} \tag{14.10}$$

is the flux quantum in quantum mechanics, we find that n_B is also the number of flux quanta per unit area. By using (14.7) the number of states per unit area within a Landau level can also be written as

$$n_B = \frac{1}{2\pi \ell_m^2} \ . \tag{14.11}$$

The FQHE is most pronounced when the magnetic field is so strong that the lowest Landau level is partially filled only. In that case, the ground state of noninteracting electrons is highly degenerate, a prerequisite for the appearance of fractional charges. But instead of a noninteracting system, we are dealing here with a strongly correlated system. The kinetic energy of the electrons is practical reduced to zero as noticed from (14.8) where for $n = 0$ only the zero-point fluctuations $\hbar\omega_c/2$ of the cyclotron frequency remain. Therefore the Coulomb repulsion of the electrons dominates and the ground state must minimize it. The electronic wavefunctions in the lowest Landau level are

$$\psi_{\ell 0}(\mathbf{r}) = \left(2^{\ell+1}\pi\ell!\right)^{-1/2} z^\ell e^{-|z|^2/4} \tag{14.12}$$

where we have introduced the coordinate $z = x + iy$ in units of the magnetic length. For a given value of ℓ the wavefunction has a density $|\psi_{\ell 0}(\mathbf{r})|^2$ as shown in Fig. 14.7. The quantum number ℓ denotes the angular momentum, yet the radius of the orbit in that figure should not be mistaken as the cyclotron orbit. This is seen, when the classical analogue is considered (see Fig. 14.7(b)). Since all $\psi_{\ell 0}(\mathbf{r})$ have the same energy, any function

$$\psi(\mathbf{r}) = f(z) e^{-|z|^2/4} \tag{14.13}$$

with an arbitrary polynomial $f(z)$ in z is also an eigenfunction of (14.4). Note that when the electronic system is confined to a finite area, ℓ may not exceed a maximum value.

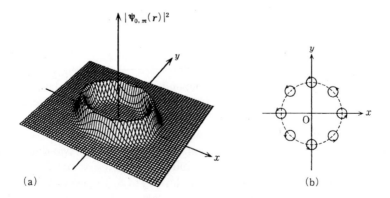

Fig. 14.7. (a) Density distribution of an electron in the lowest Landau level with angular momentum ℓ, (b) classical analogue. (From [502])

In order to minimize the Coulomb repulsion of the electrons we have to find a proper superposition of Slater determinants. Because of (14.13) this superposition is of the general form

$$\psi(\mathbf{r}_1, ..., \mathbf{r}_N) = f(z_1, ..., z_N) e^{-\sum_i |z_i|^2/4} \quad (14.14)$$

where $f(z_1, ..., z_N)$ is a polynomial in z_i. The different terms of it can be written as products $\prod_i z_i^{\ell_i}$. Since ℓ_i is the angular momentum of the i-th electron (see (14.12)) the total angular momentum of the product is $L = \sum_i \ell_i$. But L is a constant of motion and therefore only terms in the polynomial with the same L may be used in constructing $\psi(\mathbf{r}_1, ..., \mathbf{r}_N)$. The main purpose of $f(z_1, ..., z_N)$ is to keep electrons well apart in order to minimize their mutual repulsion. This can be achieved by a Jastrow-type ansatz discussed in Sect. 5.4.4. Yet here the Jastrow prefactor is not applied to a Hartree-Fock ground state as in (5.105) but to a Hartree ground state given by the exponential function in (14.14). The latter is not antisymmetric with respect to particle exchanges and therefore the Jastrow prefactor must here take care of Pauli's principle. This is done by replacing the pair wavefunction in (5.105) by a direct product

$$f(z_1, ..., z_N) = \prod_{i>j} f(z_i - z_j) \quad . \quad (14.15)$$

The expression must change sign when electrons i and j are interchanged and

$$f(z_i - z_j) = (z_i - z_j)^m \quad , \quad m > 0, \text{ odd integer} \quad (14.16)$$

fulfills all requirements. The wavefunction is strongly suppressed when z_i approaches z_j and it is antisymmetric. Thus we end up with the total wavefunction suggested by *Laughlin*

$$\psi_L(\mathbf{r}_1, ..., \mathbf{r}_N) = \prod_{i<j} (z_i - z_j)^m e^{-\sum_i |z_i|^2/4} \quad , \quad m = 1, 3, 5, ... \quad (14.17)$$

It constitutes a major accomplishment of condensed matter theory.

Next we have to find the proper value of m for a given filling factor of the lowest Landau level, or vice versa, the right filling factor ν for a given value of m. For that purpose we use a heuristic argument. In a system of N electrons with wavefunction (14.17) the largest exponent an electron with coordinate z_i can have is $(N-1)m$. Therefore its maximal angular momentum is L. It gives raise to a circle of radius $R = \sqrt{2(L+1)}\ell_m$ (see Fig. 14.7(a)) and area $F = 2(L+1)\pi\ell_m^2$. Given a homogeneous electron distribution, we may write $F = (L+1)/n_B$ because of (14.11). But $L \simeq Nm$ for large electron number N. Therefore $N = n_B F/m$ and $\nu = 1/m$ is found to be the corresponding filling factor of the lowest Landau level. At filling factors $1/3$, $1/5$ etc. the system is particularly able to keep electrons apart and minimize their repulsions. This is seen from (14.16 - 14.17). A Laughlin wavefunction with $m = 1$ corresponds to a completely filled lowest Landau level. It can be shown that in this case the wavefunction corresponds to a single Slater determinant formed with functions

(14.12) and $1 \le \ell \le N$. In passing we mention that the Slater determinant can be written in the form of a Vandermonde determinant

$$\psi(\mathbf{r}_1, ..., \mathbf{r}_N) = \frac{1}{\sqrt{N!}} \begin{vmatrix} 1 & z_1 & z_1^2 & ... & z_1^N \\ 1 & z_2 & z_2^2 & ... & z_2^N \\ 1 & z_3 & z_3^2 & ... & z_3^N \\ \vdots & \vdots & \vdots & & \vdots \\ & & ... & & \end{vmatrix} e^{-\sum_i |z_i|^2/4} \quad . \qquad (14.18)$$

Describing the system by one Slater determinant only, neglects correlation effects by which electron pairs are virtually excited into the next Landau level. This corresponds to the configuration interactions discussed in Sect. 5.1. However, because of the large magnetic length ℓ_m or low carrier concentration those effects are very small here.

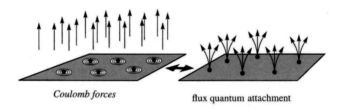

Fig. 14.8. Composite bosons consisting of electrons with three flux quanta attached each. (From [431])

Returning to the case of filling factor $\nu = 1/3$ we notice that there are three flux quanta per electron. So let us attach three flux quanta to each electron. The new objects, i.e., an electron plus three flux quanta attached to it is a boson, more precisely a composite boson. This is shown schematically in Fig. 14.8. The boson character follows from the behavior under particle interchange. Exchanging two electrons with each of them having a flux ϕ attached gives raise to a phase factor $\exp(i\pi \left(1 + \frac{\phi}{\phi_0}\right))$. Thus, when an odd number of flux quanta are attached to an electron it behaves like a boson, while attachment of an even number of flux quanta results in fermions. Since all flux quanta are used up when composite bosons are formed these bosons are in a field-free plane. Note that we could have as well associated three flux quanta of an fictitious opposite magnetic field to an electron with the same result, i.e., obtaining composite bosons in a field free system. In mean-field approximation the composite bosons interact solely through a short-range repulsive interaction. Therefore, like any interacting boson system in two dimensions, they Bose condense at low enough temperatures. Condensation of charged bosons is known from superconductivity where electron pairs, i.e., Cooper pairs act nearly like boson particles. Characteristic for a superconductor are a dissipationless current and the Meissner effect. A dissipationless current prevents the build up of a voltage in the direction of the current. At the same time, the motion of the flux quanta, which are part of the bosons, sets up via induction an electric field perpendicular to the current. Indeed, a current per unit area $\mathbf{j} = n_B e \mathbf{v}$ in a field $\mathbf{B} = n_B \cdot 3\phi_0 \hat{\mathbf{e}}_z$ where $\hat{\mathbf{e}}_z$ is a unit vector perpendicular to the plane, gives rise to an induced electric field $\mathbf{E} = [\mathbf{v} \times \mathbf{B}]$. This results in

$$e\mathbf{E} = 3\phi_0 \, [\mathbf{j} \times \hat{\mathbf{e}}_z] \tag{14.19}$$

and a Hall conductivity $\sigma_{xy} = -\frac{e}{3\phi_0}$ as experimentally observed. Needless to say that similar arguments apply to filling factors $\nu = 1/5, 1/7$ etc.

Consider a system with filling factor $\nu = 1/3$ and assume that the field is increased by such a tiny amount that there is just one additional flux unit present. There is no electron available for including this flux quant as part of a composite boson. Let us move this flux quantum to the origin. Thus we may think of a tiny solenoid at the origin containing one unit of flux. This implies keeping an area $2\pi\ell_m^2$ at the origin free of electrons. This is achieved by shifting all electron states $\psi_{\ell 0}(\mathbf{r}_i)$ to $\psi_{\ell+1,0}(\mathbf{r}_i)$ (see (14.12)). We do this by multiplying each function $\psi_{\ell 0}(\mathbf{r}_i)$ by z_i. This factor causes an additional phase shift of 2π when electron i is moved in a circle around the origin and encircles the solenoid. The total wavefunction is therefore of the form

$$\psi_{\text{ex}}(\mathbf{r}_1, \cdots, \mathbf{r}_N) = \prod_{i}^{N} z_i \psi_L(\mathbf{r}_1, \cdots, \mathbf{r}_N) \tag{14.20}$$

or, when we move the extra flux to position z_0

$$\psi_{\text{ex}}(\mathbf{r}_1, \cdots, \mathbf{r}_N) = \prod_{i} (z_i - z_0) \prod_{j<n} (z_j - z_n)^m e^{-\sum_\ell |z_\ell|^2/4} \; . \tag{14.21}$$

It follows from the above that the zero in the wavefunction at z_0 reduces the probability of finding an electron close to that point. Therefore, there is a net positive background charge which we associate with a quasihole. Its charge is $-e/3$. This is seen as follows. Assume that we have three flux quanta at z_0 to which we add an extra electron so that the total electron number is $N+1$. Then the system is again charge neutral. This implies a charge of $-e/3$ for one quasihole. Similar arguments can be applied for the construction of a quasiparticle excitation, which is obtained when the magnetic field is slightly decreased. When the number of flux quanta is reduced by one, there is an electron left with two flux quanta only, so that it can not form a composite boson. By removing three flux quanta and one electron the system is again charge neutral, implying that the removal of one flux quantum creates a quasiparticle or quasi-electron with charge $e/3$.

The argument can be made more rigorous by moving z_0 and with it the associated positive, yet undertermined charge e^* of a quasihole adiabatically around a circle which encloses a flux ϕ. By calculating the associated geometrical phase (generalized Aharonov-Bohm or Berry phase) one finds

$$\frac{e^*}{\hbar c} \oint d\mathbf{r} A(\mathbf{r}) = 2\pi \frac{e^*}{e} \frac{\phi}{\phi_0} \tag{14.22}$$

where e^* is still undetermined. On the other hand, the geometrical phase change $d(\gamma(z_0))$ under an adiabatic motion of the quasihole is [31]

$$d(\gamma(z_0)) = i \langle \psi_{\text{ex}}(z_0) | d\, \psi_{\text{ex}}(z_0) \rangle \quad . \tag{14.23}$$

From (14.21) it follows that

$$\frac{d\psi_{\text{ex}}}{dz_0} = \sum_i \left(\frac{d}{dz_0} \ln(z_i - z_0) \right) \psi_{\text{ex}} \tag{14.24}$$

and furthermore that

$$\frac{d\gamma}{dz_0} = \left\langle \psi_{\text{ex}} \left| \frac{d}{dz_0} \sum_i \ln(z_i - z_0) \right| \psi_{\text{ex}} \right\rangle \quad . \tag{14.25}$$

We use that

$$\left\langle \psi_{\text{ex}} \left| \frac{d}{dz_0} \sum_i \ln(z_i - z_0) \right| \psi_{\text{ex}} \right\rangle = \int d^2z \frac{d}{dz_0} \ln(z - z_0)$$

$$\cdot \left\langle \psi_{\text{ex}} \left| \sum_i \delta(z - z_i) \right| \psi_{\text{ex}} \right\rangle \tag{14.26}$$

and identify the expectation value with the density, i.e.,

$$\rho(z) = \left\langle \psi_{\text{ex}} \left| \sum_i \delta(z - z_i) \right| \psi_{\text{ex}} \right\rangle \quad . \tag{14.27}$$

Therfore, when z_0 is moved adiabatically around a circle of radius R we obtain

$$\gamma = i \int \rho(z) d^2z \oint dz_0 \frac{d}{dz_0} \ln(z - z_0) \quad . \tag{14.28}$$

All z values contribute to the z-integration which are inside the circle. For them the dz_0 integral gives a value of $2\pi i$. Note that the electron density is $\rho = n_B/3$ since the lowest level is filled by one third. With (14.11) this gives us $\gamma = -\pi R^2/(3\ell_m^2)$. When this is compared with (14.22) we obtain $e^*/e = 1/3$ or more generally for a filling factor ν of the lowest Landau level $e^*/e = -\nu$. In close analogy one finds for quasi-electrons a charge νe.

Let us repeat the adiabatic motion of a quasihole at z_0 along a circle of radius R, but this time with an additional quasihole positioned at the center of the circle. Then all arguments up to (14.28) still hold except that $\pi R^2 \rho$ must be replaced by $(\pi R^2 \rho - \nu)$, since a fraction ν of an electron is missing. This is so because of the positive background associated with the additional hole. This change in γ would seem to imply a change in the effective charge e^* of the quasihole. However, this is unphysical, because e^* should not depend on the presence or absence of other quasiholes. Thus the extra phase shift of $\Delta\gamma = 2\pi\nu$ must come from exchanging the two quasiholes twice when the quasihole is moved along the circle. Remember that exchanging two particles (or holes) is equivalent to moving one particle on a *semicircle* around the second one. When the encirclement of a quasihole by a second one gives a phase change of $\Delta\gamma = 2\pi\nu$, the exchange of two quasiholes gives raise to a phase factor of $e^{i\varphi} = e^{i\nu\pi} (= e^{i\Delta\gamma/2})$, i.e.,

$$\psi(\mathbf{r}_1, \mathbf{r}_2) = e^{i\nu\pi} \psi(\mathbf{r}_2, \mathbf{r}_1) \quad . \tag{14.29}$$

This is distinct from the case of bosons where $\varphi = 0$ as well as fermions for which $\varphi = \pi$. The quasiholes therefore obey fractional statistics and are called anyones [279, 487]. The reason for the fractional statistics is, of course, found in the association of a flux line with a quasihole. So when two quasiholes are interchanged one has to account for the Aharonov-Bohm phase, which is building up when a quasihole is moved on a semicircle around another one.

We hope that from the above considerations the origin of fractional charges in the FQHE has become clear. There is a wealth of additional new phenomena related to the FQHE. They go beyond the scope of this book, though. For further studies, we have to refer therefore to the original literature collected in the following volumes [97, 177, 436].

14.3 Correlated Electrons on Frustrated Lattices

Strongly correlated electrons on geometrically frustrated lattices lead to a number of new physical phenomena. For most of them experimental realizations are still missing. Nevertheless, it is very instructive to look at these systems in more details. Geometrically frustrated lattices have the property that when a spin is attached to each lattice site with antiferromagentic interactions between neighboring sites, then spins are frustrated. An antiferromagnetic spin arrangement is only partially possible on these lattice structures. Simple examples are the triangular, pyrochlore, checkerboard and kagome lattice shown in Fig. 14.9. But there are many other frustrated lattice forms. An extensive list of them is found, e.g., in Ref. [171].

Later, when we establish a connection between strongly correlated electrons on a frustrated lattice structure and a U(1) gauge theory, it will turn out advantageous to work not with the original lattices but with the medial ones. They are obtained by connecting the centers of the tetrahedra (pyrochlore lattice) or criss-crossed squares (checkerboard lattice) or triangles (kagome

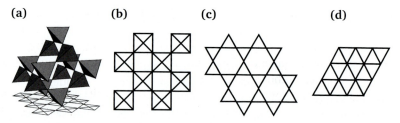

Fig. 14.9. (a) Pyrochlore, (b) checkerboard, (c) kagome and (d) triangular lattice. The checkerboard lattice can be viewed as a projection of a pyrochlore lattice onto a plane. Therefore, in both lattices each site has six nearest neighbors connected by solid lines.

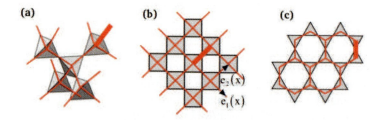

Fig. 14.10. Medial or dual lattice (red) of the pyrochlore, checkerboard, and kagome lattice. Particles are occupying here links instead of sites. This is indicated for one particle by a thick solid line on each medial lattice. The vector **x** labels a square.

lattice) with each other. It is easy to see that the resulting medial lattices are the diamond, square and honeycomb lattice, respectively. They are shown in Fig. 14.10. In distinction to the original lattices, where particles are occupying sites, the latter are occupying links of the medial lattices. Of main concern will be the pyrochlore lattice, because of its three-dimensional nature and because this lattice structure is quite common in nature. As pointed out in Sect. 13.4, in spinels of chemical composition AB_2O_4 the B sites form a pyrochlore lattice. Examples belonging to this class of materials are LiV_2O_4 or magnetite, i.e., Fe_3O_4, of which the former has been discussed at length in Sect. 13.4. But a number of physical features are much easier to visualize when we consider a checkerboard rather than a pyrochlore lattice. The checkerboard lattice may be considered a projection of a pyrochlore lattice onto a plane. Therefore we will often call the criss-crossed squares simply tetrahedra. The kagome lattice will also be used in order to demonstrate some interesting general properties of electrons on frustrated lattices.

There is considerable literature available on spins on frustrated lattices. Perhaps the most interesting result is the prediction and subsequent finding of

14.3 Correlated Electrons on Frustrated Lattices

magnetic monopoles in spin ice[4]. Here we will be mainly concerned with charge degrees of freedom of strongly correlated electrons on those lattices[5]. This suggest to eliminate the spin degrees of freedom either by assuming spinless fermions, or what is equivalent, fully spin polarized electrons. Therefore, from now on we will forget about spin degrees of freedom.

Using spinless fermions implies that a lattice site is either empty or singly occupied. Double occupancies are ruled out due to Pauli's principle. Thus there are two states possible for a lattice site in distinction to the Hubbard model in the large U limit, where we deal with three states per site. Considering spinless fermions is not as unrealistic as it might seem. Consider, e.g., magnetite $Fe^{3+}(Fe^{2+}Fe^{3+})O^4$. Here the $Fe^{2+}Fe^{3+}$ ions occupy a pyrochlore lattice. The Fe^{3+} ions are in a $3d^5$ and $S = 5/2$ configuration while the Fe^{2+} ions with $3d^6$ and $S = 2$ have one electron with opposite spin direction. Thus in first approximation only one spin direction may be considered.

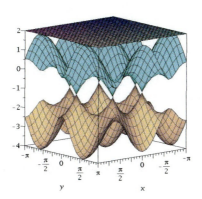

Fig. 14.11. The three bands of the kagome lattice. The first Brillouin zone is of hexagonal shape. The highest band is flat. At 1/3 filling the Fermi energy coincides with the Dirac points between the first and second bands. At those points the energy depends linearly on momentum. Note similar features, especially a flat band, in the energy dispersions of a pyrochlore lattice (see Fig. 13.23). (From [348])

The following Hamiltonian will be studied

$$H = -t \sum_{\langle ij \rangle} \left(c_i^+ c_j + h.c. \right) + V \sum_{\langle ij \rangle} n_i n_j \quad . \tag{14.30}$$

The operator c_i^+ (c_i) creates (annihilates) a spinless fermions on site i. The kinetic energy term is limited to nearest-neighbor hopping processes and we

---
[4] see [44, 56]
[5] see [137]

assume that $t > 0$. It is worth noticing that frustrated lattices with nearest-neighbor hopping have often flat electronic bands, implying local excitations. This was already seen in Fig. 13.23 where the energy bands of a pyrochlore lattice are shown. The same feature is found for the kagome lattice. The energy bands are drawn in Fig. 14.11 where it is noticed that the highest band is flat. The interaction term V with $n_i = c_i^+ c_i$ is limited to nearest neighbor repulsions. We are interested in the case of strong correlations and therefore assume that $V \gg t$. At special fillings of the lattice the strong correlations lead to strong subsidiary conditions. In the following we shall concentrate on their effects.

14.3.1 Loop Models

Let us consider a pyrochlore lattice with twice as many sites than particles (half-filled case). We start out by first assuming that $t = 0$. This is the classical limit with no fluctuations. Then the repulsion V between the particles is minimized when on each tetrahedron two of the four sites are occupied and two sites remain empty. Any deviation from this prescription, which we call the *tetrahedron rule*, will result in an increase of the repulsive energy. This is easily

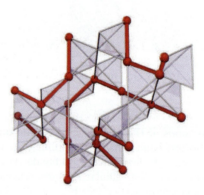

Fig. 14.12. Part of the pyrochlore lattice with two particles on each tetrahedron. When occupied sites are connected, open and closed chains cover the finite lattice cluster.

seen since each tetrahedron with three occupied sites requires a corresponding tetrahedron with one occupied site only. The difference is $\Delta E = V$ per such a pair. Satisfying the tetrahedron rule implies a macroscopic degeneracy of the ground state. The degeneracy has an entropy $S_0 \simeq k_B \ln(3/2)$ per tetrahedron

14.3 Correlated Electrons on Frustrated Lattices

associated with it. This estimate is due to *Pauling* who applied it to water ice [356]. The ice rule states that each O^{2-} ion at the center of a tetrahedron has two H^+ ions with a short bond and two with a long bond associated with it. The equivalence of the ice rule and tetrahedron rule is obvious. One specific example of the many ground-state configurations obeying the tetrahedron rule is shown in Fig. 14.12 for a cluster. When we connect the occupied sites of neighboring tetrahedra by solid lines, interpenetrating open and closed chains result. In fact, when we require periodic boundary conditions we obtain for each configuration a complete loop covering of the lattice.

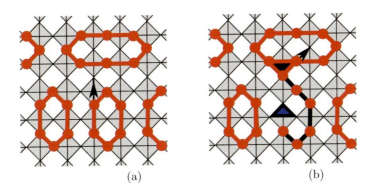

Fig. 14.13. (a) One of the many ground-state configurations which cover the plane with loops. By moving a particle as indicated by an arrow we break a loop and end up with the configuration in (b). The two tetrahedra with three occupied and three empty sites are connected by a string of occupied sites. Further displacement of a particle as indicated by an arrow does not increase the repulsion energy. (After [137])

All this can be better visualized when we go over from the pyrochlore lattice to the checkerboard lattice which we consider from now on. Here the macroscopic degeneracy of the ground state is $N_{\text{deg}} = (4/3)^{\frac{3}{4}N}$ where N is the number of sites. The degenerate ground-state configurations consist of all the different loop coverings of the plane. One example is shown in Fig. 14.13a. Assume that we add an energy V to that configuration. This suffices to break up a loop by moving a particle as indicated by an arrow in that figure. At the ends of the string formed by the opened loop there is one tetrahedron with three particles and another with one particle only (see Fig. 14.13b). Both are connected by a string consisting of an even number of occupied sites. Note that the repulsion energy does not change if we continue to shift a particle as indicated by the arrow in that figure.

We want to show that the two tetrahedra at the ends of the string carry a charge of $e/2$ and $-e/2$, respectively. This is seen best by adding a particle of charge e to the ground state as shown in Fig. 14.14a. There are now two neighboring tetrahedra with three particles each. Again, the repulsion energy

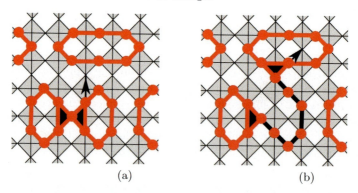

Fig. 14.14. (a) Adding a particle costs an energy $4V$ and results in two neighboring tetrahedra with three occupied sites each. Shifting a particle as indicated by an arrow leads to configuration (b) with the same repulsion energy. Continued displacement of a particle as indicated by an arrow leaves the repulsion energy unchanged.

remains unchanged when a particle is shifted as indicated by an arrow with the resulting configuration shown in Fig. 14.14b. If we continue to shift a particle as indicated by the arrow in that figure the repulsion energy remains unchanged. The two tetrahedra are connected here by a string consisting of an odd number of occupied sites. Since nothing is distinguishing the two special tetrahedra from each other, each of them must carry a charge $e/2$ in view of the fact that a charge e has been added to the system. But then the assignment of charges $\pm e/2$ to the two ends of the broken loop in Fig. 14.13b is natural. Analyzing the origin of the fractional charges one realizes that it is due to backflow of charge $e/2$ when a particle is moved. This backflow results from the requirement that the tetrahedra which are left behind by a moving particle must again fulfill the tetrahedron rule. It is therefore a consequence of the strong electron correlations. The same break-up of a charge e into charges $e/2$ is found for a half filled pyrochlore lattice. This is shown in Fig. 14.15 where it cannot be as easily visualized as for the checkerboard lattice.

Up to this point we have not included any dynamics in our model, since we had set $t = 0$. This is what we want to change now by allowing for $t \neq 0$. Thereby we assume that $t \ll V$. Of particular interest is to study, how the macroscopic degeneracy of the ground state is reduced by dynamical processes caused by t. To order t^2/V the degeneracy remains unaffected since the only possible process is that a particle is breaking a loop as in Fig. 14.13 and returns again to its original position. This results in a constant energy contribution which is equal for all ground-state configurations. We want to point out that ring hopping processes of order t^2/V do not contribute for spinless fermions considered here. Clockwise and counter clockwise hopping processes cancel in that order. But in order t^3/V^2 the ground-state degeneracy is nearly completely lifted. Ring-hopping processes involving three particles do connect

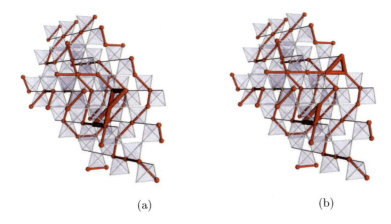

Fig. 14.15. The analogue of Fig. 14.14 for a pyrochlore lattice. (a) a particle has been added to the system; (b) the charge e has split into two charges of $e/2$.

different ground-state configurations. In that order the effective Hamiltonian is

$$H_{\text{eff}} = \frac{12t^3}{V^2} \sum_{\{\bigcirc\}} c_{j_6}^+ c_{j_4}^+ c_{j_2}^+ c_{j_5} c_{j_3} c_{j_1} \tag{14.31}$$

and the sum is over all hexagons formed from sites of the lattice. For the checkerboard lattice we rewrite this expression in a pictorial form as

$$H_{\text{eff}} = g \sum_{\{\bigcirc\}} \left(|\bowtie\rangle\langle\bowtie| - |\bowtie\rangle\langle\bowtie| + \text{h.c.} \right)$$

$$= g \sum_{\{\bigcirc,\bigcirc\}} \left(|A\rangle\langle\bar{A}| + |\bar{A}\rangle\langle A| - |B\rangle\langle\bar{B}| - |\bar{B}\rangle\langle B| \right) \tag{14.32}$$

with $g = 12t^3/V^2 > 0$. The sum is taken over all hexagons of the checkerboard lattice to which the following considerations apply.

The difference between the B and A processes is that in the former case the site in the center of the hexagon is unoccupied while in the latter case it is occupied. The different sign results from a different number of commutations of fermion operators that have to be performed when $\langle j|H_{\text{eff}}|i\rangle$ is evaluated. B and A hopping processes act quite differently on a configuration $|i\rangle$. While B processes merely shift loops around, so that $\langle j|$ has the same topological structure as $|i\rangle$ has, this is different for A processes. Here different loops are reconnected by $H_{\text{eff}}|i\rangle$ so that $\langle j|$ has a different topological structure. This is seen from Fig. 14.16 where both types of processes are shown. In passing we want to mention that the sign in (14.32) can be removed by a simple gauge transformation so that the Hamiltonian is amenable to a Monte Carlo treatment [375]. This does not hold true for the pyrochlore lattice though, were we are facing the typical sign problem of fermionic systems.

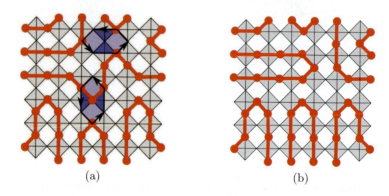

Fig. 14.16. Ring hopping of type B (lower hexagon) and of type A (upper hexagon): (a) before and (b) after ring hopping. In the first case (type B) the loops merely shift while in the second case (type A) they are differently connected by ring hopping.

The ground state of H_{eff} for the half-filled checkerboard lattice consists of a superposition of configurations $|\phi_n\rangle$ of the degenerate classical ground state, i.e.,

$$|\psi_0\rangle = \sum_n \alpha_n |\phi_n\rangle \quad . \tag{14.33}$$

Numerical diagonalization of H_{eff} for clusters up to 72 sites shows that $|\psi_0\rangle$ is charge ordered. The largest contributions to $|\psi_0\rangle$ are made by configurations with the largest number N_{fl} of flipable hexagons. For the afore mentioned cluster this is a configuration, called squiggle, with $N_{fl} = 24$. Because of the large unit cell the ground state is 10-fold degenerate. In Fig. 14.17 we show the values of $|\alpha_i|^2$ for configurations with a decreasing number N_{fl}. We are dealing here with the phenomenon of order by disorder. It is the dynamics which gives the ordered state a lower energy than the disordered ones.

In a charge-ordered ground state a pair of charges $\pm e/2, \pm e/2$ generates a string of disorder when the fractional charges separate. This causes an energy increase linear in the length of the string which is connecting the two charges. Thus the fractional charges are confined by a linear potential similar to quarks. Before continuing we want to point out that the situation seems to be different for the 3D pyrochlore lattice. Here the ground state of a related model remains a strongly correlated liquid and the fractional charges are deconfined (see Sect. 14.3.4) [408].

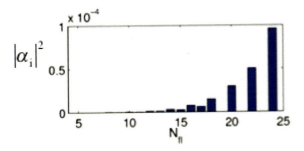

Fig. 14.17. Weighting factor of ground-state configurations with different numbers of flipable hexagons N_{fl} (Courtesy of F. Pollmann).

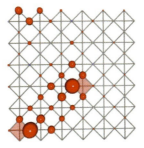

Fig. 14.18. Loss of kinetic energy in proportion to the size of the circles in the presence of two fractional charges $e/2$ (plaquettes marked in red). (From [376])

Returning to the checkerboard lattice we want to determine the tension of the confining string. We can calculate it by determining the kinetic energy ϵ_i at a site i in the presence of two tetrahedra (i.e., criss-crossed squares) with three particles each. They are positioned at plaquettes $\mathbf{0}$ and $\mathbf{r}$ and the wavefunction is $|\bar{\psi}_0(\mathbf{0},\mathbf{r})\rangle$. It is

$$\varepsilon_i = -\frac{1}{6}\sum_{\hexagon \ni \circ} \langle \bar{\psi}_0(\mathbf{0},\mathbf{r})|H_{\text{eff}}|\bar{\psi}_0(\mathbf{0},\mathbf{r})\rangle \tag{14.34}$$

and the sum is over all hexagons containing site i. The loss of kinetic energy at different sites is obtained from cluster calculations and is shown in Fig. 14.18. The total kinetic energy change is found to increase linearly with r. Thus the restoring force is independent of r, unlike in the case of a spring. The string tension τ_{st} is defined as the total increase in kinetic energy divided by r (in units of the lattice constant). It can be determined from the numerical data and it is found that $\tau_{st} = 0.2g$. The same result holds true when a loop is breaking up with charges $e/2$, $-e/2$ at its ends.

384   14 Excitations with Fractional Charges

When the two fractional charges are pulled apart so that the confining energy $\Delta E = 0.2gr$ is larger than $\Delta E \geq V$ then it becomes energetically favourable to create a new pair $e/2, -e/2$ which is cutting the string into two pieces. The situation reminds us in many aspects of quark confinement in QCD. Also quarks experience a constant confining force. When they are separated too far, quark-antiquark pairs, i.e., mesons are generated. The prevailing picture in that field is depicted in Fig. 14.19. We want to reemphasize the finding that the confinement of fractionally charged particles in our model is a direct consequence of the symmetry-broken, i.e., charge ordered vacuum state. This vacuum is here the ground state $|\psi_0\rangle$ given by (14.33).

For completeness we want to mention that between two fractional charges the vacuum is polarized, i.e., there is an alternating increase and decrease of the local net charge. It is obtained by evaluating

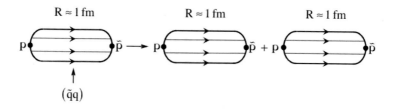

($\tilde{q}q$)

Fig. 14.19. When quark q and antiquark $\tilde{q}$ are separated too far (i.e., by more than 1 fm) a new $q\tilde{q}$ pair (μ-meson) is created. In our model a pair of fractional charges $e/2, -e/2$ corresponds to $q\tilde{q}$. (From [102])

$$\delta n_i = \langle \bar{\psi}_0(\mathbf{0}, \mathbf{r})|n_i|\bar{\psi}_0(\mathbf{0}, \mathbf{r})\rangle - \langle \psi_0|n_i|\psi_0\rangle \quad . \tag{14.35}$$

Therefore we may argue as well that the confining force is a consequence of the vacuum polarization caused when fractional charges are separated. The loss of kinetic energy as well as the vacuum polarization follow directly from the property that the string consists of occupied sites connecting the fractional charges. In the vicinity of the string, ring hopping is reduced, since it requires alternating empty and occupied sites. Similarly, the string modifies the charge distribution in its surrounding as compared with its average value.

Assume that we start from the ground- or vacuum state $|\psi_0\rangle$ and that we pump an increasing amount of energy into the system. Then more and more loops are broken and we obtain an increasing number of pairs of fractional charges $e/2, -e/2$ linked by strings. When these pairs come close to each other they can recombine differently and form also $e/2, e/2$ and $-e/2, -e/2$ pairs. Eventually we end up with a plasma consisting of particles with fractional charges $e/2, -e/2$ and of strings confining them in different combinations (see

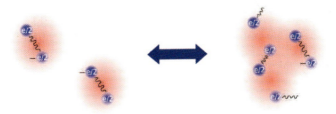

Fig. 14.20. As the number of broken loops increases, pairs with different combinations of $\pm e/2$ may form. Finally a plasma is obtained of fractionally charged particles with attached strings acting like a glue.

Fig. 14.20). The system, which was previously an insulator has then turned into a metal.

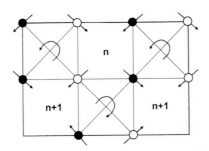

Fig. 14.21. Height representation. Filled circles denote occupied sites with vectors pointing in the orientational direction. This direction alternates for the two sublattices. Empty circles denote empty sites with vectors pointing against the orientational direction. On the criss-crossed squares curl $f = 0$. The height of neighboring squares changes by one in the direction of the arrows.

The insulator-metal transition can also be studied by using a *hight representation*. As we will see the break up of a loop corresponds here to the generation of a vortex-antivortex pair. The transition to a metal corresponds to a proliferation of these pairs (Berezinskii–Kosterlitz–Thouless transition). In the hight representation each ground-state configuration of the checkerboard lattice is uniquely represented by a vector field $\mathbf{f}(i)$ where i is the site index. The discretized lattice version of curl $\mathbf{f}$ vanishes. The vector field is obtained by using the property that the checkerboard lattice is bipartite. This allows for assigning alternating orientational directions (i.e., clockwise and counter clockwise) to the different criss-crossed squares (see Fig. 14.21). Each site has associated with it a vector which points into the orientational direction when the site is occupied and into opposite direction when it is empty. The tetrahedron rule requires that on each criss-crossed square curl $\mathbf{f} = 0$ since two sites are occupied and two sites are empty. Since curl $\mathbf{f} = 0$, the vector field vanishes, it can be represented in terms of a scalar or hight field as

$$\mathbf{f} = \mathrm{grad}\, h \ . \tag{14.36}$$

For a finite lattice with $N_x \cdot N_y$ criss-crossed squares and periodic boundary conditions, the height at opposite boundaries $h(-N_x/2)$ and $h(+N_x/2)$ can differ only by an integer $\kappa_{x(y)}$. It varies from $-N_x \leq \kappa_x \leq N_x$ and similar for κ_y. The quantum numbers $\kappa_{x(y)}$ are topologically protected, i.e., they are not affected by ring-hopping processes [393]. Application of H_{eff} merely changes the height of two neighboring plaquettes by ± 2. States with $(\kappa_x, \kappa_y) \neq (0,0)$ are charge ordered. When, for example, $\kappa_x > 0$ this implies a charge modulation along a diagonal stripe.

In the height representation it is easily seen that breaking up a loop corresponds to the creation of a vortex-antivortex pair. We merely have to exchange an occupied site with a neighboring empty site (see the filled and empty circle in Fig. 14.21). Then two neighboring plaquettes obtain a vorticity of opposite sign. The vortex and antivortex can separate in analogy to Fig. 14.13. This motion can be described by a Hamiltonian

$$H_{t-g} = H_{\mathrm{eff}} - t \sum_{\langle ij \rangle} P\left(c_i^+ c_j + \mathrm{h.c.}\right) P \ . \tag{14.37}$$

P is here a projector which projects onto the subspace of configurations with two fractional charges $e/2$ and $-e/2$, or in height presentation with one vortex and one antivortex. We consider the two parameters in (14.37), i.e., g which appears in H_{eff} and t as being independent. The above Hamiltonian does not take into account that with increasing ratio t/V ring hopping on rings larger than hexagons becomes important. A more advanced theory should also take 8-site, 10-site etc. hopping processes into account and not limit itself to 6-site rings. Thus a transition from an insulator to a metal can have two origins. One is within the model described by H_{eff} by means of proliferation of vortex-antivortex pairs caused by energy input. The other is by extending H_{eff} to larger ratios of t/V and including hopping on larger rings.

14.3.2 Dimer Models

At this stage we want to come back to the medial lattice representation (see Fig. 14.10) in order to introduce dimer models. Until now we have concentrated on half-filled pyrochlore and checkerboard lattices. We have seen that they lead to loop coverings of the lattices. If we consider instead those two lattices at quarter filling, we end up with a dimer model on the respective medial lattice. The tetrahedron rule is replaced here by requiring that each

14.3 Correlated Electrons on Frustrated Lattices

lattice site of the medial lattice must have attached to it one occupied link, i.e., one dimer. The same holds true for a kagome lattice at 1/3 filling. In the following we want to concentrate on that particular system.

Looking for ground-state configurations of the Hamiltonian (14.30) with $t = 0$ it is obvious that the mutual nearest-neighbor repulsion vanishes when each triangle contains one occupied site (triangle rule). The effective Hamiltonian, which lifts most of the degeneracy of the ground-state configurations is written here in the form

$$H_{\text{eff}} = g \sum_{\bigcirc} (|\bigcirc\rangle\langle\bigcirc| + \text{h.c.}) , \qquad (14.38)$$

and refers to the medial lattice representation (see Fig. 14.10). Ring hopping takes place on hexagons of the kagome lattice. On the medial lattice, which here consists of hexagons, it corresponds to flips of dimers. Note that $g = 12t^3/V^2$ like for the pyrochlore lattice. In addition, there is a constant energy shift ΔE, which for a system of N sites has the form

$$\Delta E = -\frac{N}{3}\left(\frac{4t^2}{V} + \frac{2t^3}{V^2}\right) + O\left(t^4/V^3\right) . \qquad (14.39)$$

Although the kagome lattice is not bipartite it turns out that the sign of g is irrelevant. It can be changed by a simple gauge transformation. This is achieved by multiplying each configuration $|C\rangle$ by a phase factor

$$|C\rangle \to i^{\nu(C)}|C\rangle \qquad (14.40)$$

where $\nu(C)$ is the number of particles in $|C\rangle$ on any of the three sublattices of the hexagon lattice. More details are found in [348].

Often a term first introduced by *Rokhsar* and *Kivelson* is added to H_{eff}. In the medial lattice representation it is of the form

$$\delta H = \mu \sum_{\bigcirc} (|\bigcirc\rangle\langle\bigcirc| + |\bigcirc\rangle\langle\bigcirc|) \qquad (14.41)$$

and counts all flipable hexagons. It is used to counteract the dimer flipping in H_{eff} and has the advantage that for the resulting quantum dimer model

$$H_{\text{QDM}} = H_{\text{eff}} + \delta H \qquad (14.42)$$

an exact ground state can be found when $\mu = g > 0$. At this point, often referred to as Rokhsar-Kivelson point, all ground-state configurations satisfying the triangle rule have the same weight. This is seen by rewriting H_{QDM} for $\mu = g$ in the form

$$H_{\text{QDM}} = g \sum_{\bigcirc} (|\bigcirc\rangle - |\bigcirc\rangle) \times (\langle\bigcirc| - \langle\bigcirc|) \qquad (14.43)$$

14 Excitations with Fractional Charges

which is a sum of projectors. When $H_{\rm QDM}$ is applied to a superposition of all ground-state configurations with equal prefactor we obtain zero. Therefore the above statement follows. Knowing the ground state for $\mu = g$, we can search for the ground state in the vicinity of that point. This way one can prove that for $\mu = 0$ the ground state consists of resonating plaquettes or hexagons, and has the form of a plaquette phase [328]. The latter is schematically shown in Fig. 14.22.

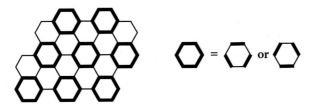

Fig. 14.22. Plaquette phase of a kagome lattice at 1/3 filling. Ring hopping takes place on the black plaquettes of the medial lattice. The ground state is threefold degenerate.

Moving a particle from one triangle of the kagome lattice to another (quantum fluctuation) or adding or removing a particle from the lattice produces fractionally charged excitations with $e/2, -e/2$ in close analogy to the checkerboard lattice case. They are confined because of the symmetry broken vacuum, i.e., charge ordered ground state. The "string" tension is found to be $\tau_{\rm st} = 0.2g$. However, due to the low 1/3 lattice filling it is not possible to identify a single string of occupied sites connecting the two fractional charges as found to be the case at half-filling. Instead, in a given configuration there exist several paths of alternating occupied and empty sites connecting the two.

Of particular interest is a study of the spectral density. It is expected to show fingerprints of the fractional charges. Since fractional charges are confined by a linear potential, we expect to see excited bound states similar to the ones known in elementary particle physics between a charmed quark and antiquark, i.e., charmonium[6]. The spectral function is written as $A(\mathbf{k}, \omega) = A^+(\mathbf{k}, \omega) + A^-(\mathbf{k}, \omega)$ with

$$A^+(\mathbf{k}, \omega) = -\frac{1}{\pi} \operatorname{Im} \left\langle \psi_0 \left| c_{\mathbf{k}} \frac{1}{\omega + \mathcal{E}_0 - H + i\eta} c_{\mathbf{k}}^+ \right| \psi_0 \right\rangle$$

$$A^-(\mathbf{k}, \omega) = -\frac{1}{\pi} \operatorname{Im} \left\langle \psi_0 \left| c_{\mathbf{k}}^+ \frac{1}{\omega - \mathcal{E}_0 + H + i\eta} c_{\mathbf{k}} \right| \psi_0 \right\rangle \qquad (14.44)$$

---

[6] see, e.g., [13, 392]

14.3 Correlated Electrons on Frustrated Lattices

where $\mathcal{E}_0$ is the ground-state energy of the N particle system when H is given by (14.30) (compare with (7.103)). The resulting **k** integrated spectral density is

$$D^{\pm}(\omega) = -\frac{1}{N} \sum_{\mathbf{k}} A^{\pm}(\mathbf{k}, \omega) \quad (14.45)$$

where N is the number of lattice sites.

When $D^+(\omega)$ and $D^-(\omega)$ are calculated for a cluster of 27 sites dramatic changes appear when V/t varies from $0 \le V/t \le 30$. Some of them we want to discuss.

When V/t is large, or $g/t \ll 1$ the confined charges $e/2$, $\pm e/2$ are on average far apart because of the small string tension. In that limit bound states cannot be obtained from a cluster calculation. But we may confine in that limit calculations to the reduced Hilbert space onto which H_{t-g} (see (14.37)) acts. That enables us to treat large clusters. Results for $g = 0.01t$ using a 108 sites cluster are shown in Fig. 14.23.

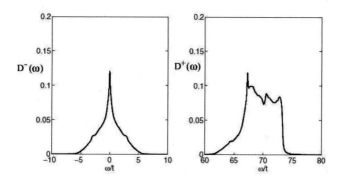

Fig. 14.23. Hole- and particle spectral density for a cluster of 108 sites of a kagome lattice at 1/3 filling. It is $g = 0.01t$. (From [348])

Most noticeable is the bandwidth of the excitations which is approximately twice as large as for $V = 0$. In the hole density of states $D^-(\omega)$ a δ-function like peak is seen at $\omega = 0$. Its origin is that the operator $c_{\mathbf{k}=0} = (3N)^{-1/2} \sum_j c_j$, when applied on $|\psi_0\rangle$, gives an approximate eigenstate of the Hamiltonian (14.30). A proof is found in Appendix J. The spectrum $D^+(\omega)$ is shifted by $2V$, i.e., by the energy it takes to add a particle.

We can study bound states by formally increasing g/t so that eventually the first excited state of a pair of fractional charges is inside, e.g., a 75 site cluster. This is seen in Fig. 14.24 and an estimate when this will happen is obtained by solving the Schrödinger equation for a particle in a linear potential. As expected the criterion is that $g \simeq t$. In future experiments the above features may serve as fingerprints of excitations with fractional charges.

14 Excitations with Fractional Charges

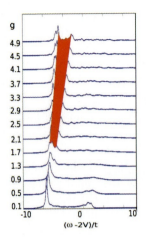

Fig. 14.24. Spectral function $A^+(\mathbf{k} \simeq 0, \omega)$ for different values of g/t ranging from 0.1 - 4.9. One notices a split-off peak for $g \gtrsim 1.3$ which is due to an excited state of a confined pair $e/2, -e/2$. The cluster includes 75 sites. (From [348])

Before we close this Section we want to discuss briefly which changes we expect when the spin is introduced. We still want to exclude double occupancies of sites. Therefore, we add a large on-site, Hubbard type of repulsion $U \sum_i n_{i\sigma} n_{i-\sigma}$ to the Hamiltonian. Let us assume that we add to a half-filled checkerboard or pyrochlore lattice an electron with charge e and spin $1/2$. As we have seen before, the charge is breaking up into two pieces $e/2$ and $e/2$. Yet we would like to know what happens to the spin. There will be always a spin-spin interaction between particles on neighboring sites and only an antiferromagnetic, i.e., Heisenberg type of coupling is of interest here. Like in the case of spinless fermions the string connecting the two charges $e/2$ consists of an odd number of sites (see Fig. 14.14b). A Heisenberg chain with an odd number of sites has a two-fold degenerate, i.e., doublet ground state. This doublet is representing the spin degree of freedom. Thus the latter is distributed over the chain as a whole, i.e., the spin degree of freedom is delocalized. In agreement with this is the observation that a pair $e/2, -e/2$ of fractional charges (see Fig. 14.13b) is connected by a string consisting of an even number of sites. The ground state of such a Heisenberg chain is a singlet and there is no spin degree of freedom connected with such a particle-hole like pair created out of the vacuum.

14.3.3 Mapping to a U(1) Gauge Theory

As it turns out, strongly correlated electrons on a half-filled pyrochlore or checkerboard lattice, or on a 1/3-filled kagome lattice can be mapped onto a local U(1) gauge theory. The basis of this mapping is the observation that H_{eff}

given by (14.32) and (14.38) conserves the number of occupied links attached to a site of the medial lattice. This number is two for the half-filled pyrochlore and checkerboard lattice and one for the 1/3-filled kagome lattice. Therefore different phases may be attached to the different sites. The invariance of H_{eff} against (local) gauge changes manifests the strong relations, which exist between the requirement of local gauge invariance and the form of the interactions of the particles. Here they enforce the tetrahedron or triangular rule, respectively. Relations of this kind are a corner stone of elementary particle physics.

The mapping which we want to demonstrate for the checkerboard lattice at half filling is similar to the one derived by *Fradkin* [122] for the dimer model on a square lattice [348]. We start from the medial square lattice (see Fig. 14.10) with unit vectors $\mathbf{e}_j$ $(j = 1, 2)$. For each link ab between neighboring lattice sites a and b of the square lattice we define a particle number operator $\hat{n}_{ab}$ with integer eigenvalues. In order to limit the number of particles on a link to zero and one, the effective Hamiltonian must include a term of the form

$$H_{\text{lim}} = \lim_{U \to \infty} U \sum_i \left((\hat{n}_i - 1/2)^2 - \frac{1}{4} \right) \ . \tag{14.46}$$

All configurations with more than one particle on a link i are eliminated by attaching to them an infinite energy. Before we can express H_{eff} in terms of the $\hat{n}_i$ operators we have to introduce their canonical conjugates, which are the phase operators $\hat{\phi}_i$ on links. They satisfy the commutation relations

$$\left[\hat{n}_i, \hat{\phi}_j \right] = i \delta_{ij} \ . \tag{14.47}$$

Since the eigenvalues of the $\hat{n}_i$ are integers, the spectrum of the $\hat{\phi}_i$ is in the range $[0, 2\pi]$ modulo 2π. The phase operator acts like a shift operator on the number operator, i.e., $\exp(im\hat{\phi}_j)|n_j\rangle = |n_j + m\rangle$ for integer values of m. This relation follows from the identity

$$e^{-im\hat{\phi}} \hat{n} \, e^{im\hat{\phi}} = im \sum_{p=0}^{\infty} \frac{1}{p!} \left[\hat{n}, \hat{\phi} \right]_p = \hat{n} + m \tag{14.48}$$

where $[\hat{n}, \hat{\phi}]_p = [\hat{n}[\hat{n} \ldots [\hat{n}, \hat{\phi}]] \ldots]$, i.e., the iteration is performed p times. The identity of both sides is obtained when $\hat{n}$ and $\exp(im\hat{\phi})$ are commuted. From (14.48) we obtain

$$\hat{n} e^{im\hat{\phi}} |0\rangle = m e^{im\hat{\phi}} |0\rangle$$
$$= m|m\rangle \ . \tag{14.49}$$

We are now able to express the effective Hamiltonian in terms of the $\hat{n}_i$ and $\hat{\phi}_i$. For the medial, i.e., square lattice it is of the form

$$H_{\text{eff}} = g \sum_{\{\square \cdot \square\}} \left(|\square\!\!\!\rightarrow\rangle\langle\square\!\!\!\rightarrow| + |\square\!\!\!\leftarrow\rangle\langle\square\!\!\!\leftarrow| + h.c. \right)$$

$$= \lim_{U \to \infty} U \sum_i \left[\left(\hat{n}_i - \frac{1}{2} \right)^2 - \frac{1}{4} \right] + 2g \sum_{\{\square \cdot \square\}} \cos\left(\hat{\phi}_1 - \hat{\phi}_2 + \hat{\phi}_3 - \hat{\phi}_4 + \hat{\phi}_5 - \hat{\phi}_6 \right) \quad .$$

(14.50)

The dots in the first equation reemphasize occupied links. The last term results from the shift operator and is a sum over all double squares, each of which contain six links (1, ..., 6) involved in a ring-hopping process. The $\hat{\phi}_\nu$ alternate in sign, since in a ring-hopping process an empty site goes over into an occupied one, implying $+\hat{\phi}_\nu$ while the neighboring site changes from occupied to empty $(-\hat{\phi}_{\nu\pm1})$.

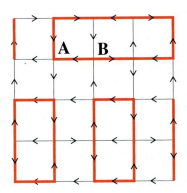

Fig. 14.25. Medial bipartite lattice with sublattices A and B. Occupied links always point towards A lattice sites while unoccupied links point in opposite direction. The vectors are used to define an electric field. The configuration corresponds to the one in Fig. 14.13a. It is noticed that according to (14.53) $\text{div}\mathbf{E}(\mathbf{x}) = 0$ at each site.

When dealing with a bipartite lattice as is the case here, a vector field can be associated with each configuration. This was previously done in connection with the height representation (see Fig. 14.21). Thus the links obtain a direction with occupied links always pointing from a site of sublattice B to one of sublattice A. The opposite holds true for empty links (see Fig. 14.25). This suggest the introduction of an electric field $\hat{E}_j(\mathbf{x})$ where links j are specified by two indices, i.e., $\mathbf{x}$ and $j = 1, 2$. Here $\mathbf{x}$ is labeling a square of the lattice and j refers to the two vectors $\mathbf{e}_j$ shown in Fig. 14.10. With this notation we introduce two vector fields

$$\hat{E}_j(\mathbf{x}) = (-1)^{x_1+x_2} \left(\hat{n}_j(\mathbf{x}) - \frac{1}{2} \right)$$

$$\hat{A}_j(\mathbf{x}) = (-1)^{x_1+x_2} \hat{\phi}_j(\mathbf{x}) \quad .$$

(14.51)

The electric field has been defined so that its average vanishes. The $\hat{A}_j(\mathbf{x})$ have here the same sign in a ring with alternating occupied and empty links. We must also incorporate the constraint, that each lattice site is touched by exactly two occupied links. This follows from the (tetrahedron) rule that each criss-crossed square contains two particles. Here the constraint for any physical state reads

$$\left(\mathrm{div}\hat{\mathbf{E}}(\mathbf{x}) - \rho(\mathbf{x})\right)|\mathrm{Phys}\rangle = 0 \quad, \tag{14.52}$$

where the lattice divergence of the electric field and $\rho(\mathbf{x})$ are defined by

$$\mathrm{div}\mathbf{E}(\mathbf{x}) = \hat{E}_1(\mathbf{x}) - \hat{E}_1(\mathbf{x} - \mathbf{e}_1) + \hat{E}_2(\mathbf{x}) - \hat{E}_2(\mathbf{x} - \mathbf{e}_2)$$

$$\rho(\mathbf{x}) = n_1(\mathbf{x}) + n_1(\mathbf{x} - \mathbf{e}_1) + n_2(\mathbf{x}) + n_2(\mathbf{x} - \mathbf{e}_2) - 2 \quad. \tag{14.53}$$

For all ground-state configurations $\rho(\mathbf{x}) = 0$. Then $\mathrm{div}\hat{\mathbf{E}}(\mathbf{x}) = 0$ as seen in Fig. 14.25. It is noticed that the constraint has the form of Gauss' law. A system obeying a constraint of this form is said to be in a Coulomb phase [178]. In terms of $\hat{E}_j(\mathbf{x})$ and $\hat{A}_j(\mathbf{x})$ the effective Hamiltonian H_{eff} becomes

$$H_{\mathrm{eff}} = \lim_{U \to \infty} U \sum_{\mathbf{x}j} \left(\hat{E}_j^2(\mathbf{x}) - \frac{1}{4}\right) + 2g \sum_{\mathbf{x}j} \cos\left(\sum_{\Box} \hat{A}_j(\mathbf{x}) + \sum_{\Box} \hat{A}_j(\mathbf{x} - \hat{\mathbf{e}}_j)\right) \quad. \tag{14.54}$$

The oriented sum of the vector potential around one plaquette is

$$\sum_{\Box} \hat{A}_\ell(\mathbf{x}) = \hat{A}_1(\mathbf{x}) + \hat{A}_1(\mathbf{x} + \mathbf{e}_2) + \hat{A}_2(\mathbf{x}) + \hat{A}_2(\mathbf{x} + \mathbf{e}_1) \quad. \tag{14.55}$$

It is noticed that ring hopping is equivalent to the presence of a **B** field via $\sum_\Box \hat{A}_j(\mathbf{x}) + \sum_\Box \hat{A}_j(\mathbf{x} - \mathbf{e}_j)$ and therefore a flux. Thus the Hamiltonian (14.54) resembles that of compact electrodynamics in 2+1 dimensions [377]. That model is known to yield confinement of charges under the assumption that the vacuum state is field free. This is not the case here, since the vacuum state contains a staggered **E** field (see (14.51)). Yet, as we have seen before, also in our model fractional charges are confined, because of a charge-ordered vacuum or ground state. This state is produced by the dynamics in the system, i.e., by vacuum fluctuations in form of ring-hopping processes. The confinement of fractional charges is caused by an increase of disorder in the vacuum state when the charges are separated.

14.3.4 Magnetic Monopoles

Despite of extensive search, magnetic monopoles originally proposed by *Dirac* in 1931 have not been found until now. Dirac had suggested them, because they give the basic Maxwell's equations of electromagnetism a more symmetric form than they have presently. It is of considerable interest, that for certain magnetic solids, i.e., spin ice, corresponding excitations have been predicted by *Castelnovo, Moessner* and *Sondhi*, which set up a magnetic field like Dirac's magnetic monopoles do [56]. The difference is that Dirac's monopoles refer to the vacuum of Maxwell's theory while in the solids discussed here, monopoles refer to a vacuum which is the highly degenerate ground state of spin ice. The advantage is, that monopoles in spin ice have actually been observed and widened our physical understanding not only of spin ice but also of monopoles in general [44]. $Dy_2Ti_2O_7$, the material in which magnetic monopoles have been detected, is a magnetic insulator. The incomplete $4f$ shell of Dy^{3+} contains strongly correlated electrons, yet the large field of magnetism of localized spins is outside the scope of this book. Nevertheless, it seems justified to describe briefly this interesting field of research, in particular, since it is related to the one of fractional charges, i.e., the topic we are considering in this Chapter.

The Dy ions of $Dy_2Ti_2O_7$ occupy the sites of a pyrochlore lattice. They have a large magnetic moment of nearly $10\mu_B$. It results from the incomplete $4f$ shell which contains nine $4f$ electrons. The lowest J multiplet is according to Hund's rules $J = 15/2$. It is split by a strong crystalline electric field into a ground-state which is nearly a pure $|J_z = \pm 15/2\rangle$ Kramers' doublet and excited states with an energy higher than 30 meV. They are neclected here. The CEF forces the magnetic moment $\boldsymbol{\mu}_i$ to point along the lines, which connect the centers of neighboring tetrahedra. In the medial diamond lattice the moments point therefore along the links of the lattice. The interaction Hamiltonian is

$$H_{\text{int}} = \frac{J}{3} \sum_{\langle ij \rangle} S_i S_j + Da^3 \sum_{i,j} \left[\frac{\hat{\mathbf{e}}_i \hat{\mathbf{e}}_j}{|\mathbf{r}_{ij}|^3} - \frac{3 (\hat{\mathbf{e}}_i \mathbf{r}_{ij}) (\hat{\mathbf{e}}_j \mathbf{r}_{ij})}{|\mathbf{r}_{ij}|^5} \right] S_i S_j \quad , \quad (14.56)$$

with $S_i = \pm 1$ describing the directions of the moments (Ising case). Here $a = 3.5 Å$ is the nearest-neighbor distance, and the $\hat{\mathbf{e}}_i$ are local Ising unit vectors with relative orientation $\hat{\mathbf{e}}_i \hat{\mathbf{e}}_j = -1/3$. Furthermore, $\boldsymbol{\mu}_i = \mu S_i \hat{\mathbf{e}}_i$ with $S_i = \pm 1$ and $D = 1.4$ K. Since we have discarded higher CEF eigenstates and do not deal with multipolar interaction either, the two states of the doublet remain disconnected by H_{int} and quantum fluctuations are absent. The system has converted into a classical one. Note that because of the large moment μ the dipolar and exchange interactions are nearly of equal size. An interesting feature of the dipolar interaction term is that the state which minimizes it is the same, which minimizes a nearest neighbor repulsion. This is seen as follows:

Let us represent the magnetic dipole moment $\boldsymbol{\mu}_i$ by a pair of magnetic charges $\pm Q$ separated by a vector $\mathbf{d}_i$ so that $\boldsymbol{\mu}_i = Q\mathbf{d}_i$. Normally one represents a dipole by fixing $\boldsymbol{\mu}$ while taking $Q \to \infty$ and $d = |\mathbf{d}_i| \to 0$. Yet, here it is advantageous and in fact an excellent approximation to fix $d = \sqrt{3/2}a$ which is the distance between the centers of two neighboring tetrahedra or alternatively, the bond length in the medial diamond lattice. Depending on the directions of the four magnetic moments of a tetrahedron we have at the center of tetrahedron n a total magnetic charge q_n. The dipole interaction term in (14.56) is the same as the interaction energy of magnetic charges q_n of the different tetrahedra. This follows from Coulomb's law. It is minimized when all $q_n = 0$, i.e., when on each tetrahedron two of the $\boldsymbol{\mu}_i$ point towards the center and two of them point away from it. This is a configuration in which the tetrahedron rule is fulfilled everywhere, like in the case of a nearest neighbor repulsive interactions. Because of the similarity to Pauling's model for water ice we speak of the spin-ice rule rather than of a tetrahedron rule. Therefore, as far as the ground state of H_{int} is concerned, the problem reduces to the one of (14.30) with $t = 0$. Indeed, measurement of the entropy of $Dy_2Tb_2O_7$ have revealed that up to the lowest temperatures a huge ground-state degeneracy remains with an entropy of $S = 1/2 \log 3/2$ per site as envisaged by *Pauling*. As pointed out above, this degeneracy is not lifted here, because of the absence of quantum fluctuations.

Assume that we flip a dipole at site i, which costs an energy of order J or respective D. In that case the neighboring sites on the diamond lattice acquire a magnetic charge $Qd = q_m$, since in one tetrahedron three moments point away and one points towards the center, while on the neighboring tetrahedron three moments point in and one points out. The two magnetic charges on the medial diamond lattice can be separated at the cost of an energy

$$V(\mathbf{r}_{ij}) = \frac{\mu_0}{4\pi} \frac{q_m^2}{r_{ij}} \quad , \quad r_{ij} > d \tag{14.57}$$

according to the magnetic Coulomb law. This is depicted in Fig. 14.26. Moving one of the two charges to infinite results in a magnetic monopole with a Dirac string attached to it. Note that when the magnetic charges obtain a dynamics, i.e., kinetic energy due to either the inclusion of higher CEF levels or multipolar interactions, then bound states between a monopole and an antimonopole become an issue. They can be computed by solving Schrödinger's equation with a lattice version of the potential (14.57), supplemented by the nearest-neighbor potential when $r_{ij} = d$.

Magnetic monopoles are sparse at low temperatures. It takes an energy of order J to flip a spin and therefore the density of monopoles drops exponentially fast with decreasing temperature. One way to look for monopoles is by the string of flipped spins that connects them (see Fig. 14.26). A *Dirac string* emanates from each of the monopoles and effects many more spins than there are monopoles. These strings can be oriented by a magnetic field [334]. In this setting, deconfinement shows up by the strings fluctuating freely in the directions perpendicular to the applied field. These fluctuations take a form which is mathematically identical to a random walk (note that tilting the field can bias this random walk). The resulting neutron scattering pattern is diffuse – the absence of long-range order in the strings means that they do not cause Bragg peaks – and as such a characteristic feature which has been observed experimentally.

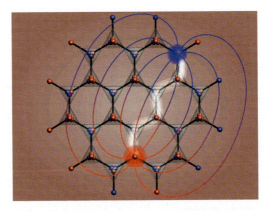

Fig. 14.26. Two magnetic charges $\pm q_m$ on the medial diamond lattice at a distance r. Pulling them apart requires an energy $(\mu_0/4\pi)q_m^2/r$. Therefore the magnetic charges are deconfined. Note the similarity of the evolving Dirac string with Fig. 14.13. (From [56])

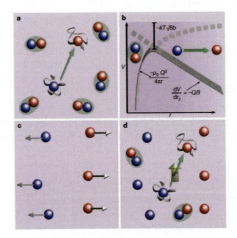

Fig. 14.27. Temperature induced spin flips with an associated generation of magnetic charges $\pm q_m$ at neighboring sites of the dual diamond lattice (a). The magnetic charges can be separated by an energy proportional to q_m^2/r (magnetic Coulomb law). They can also recombine again, so that an equilibrium is attained. An external magnetic field B along $\hat{z}$ provides a field energy $-q_m B r_z$ which counter balances the magnetic Coulomb interaction and causes dissociation of the magnetic charges (b) and (c). New spin flips are required to reestablish equilibrium. The magnetic fluctuations with associated local fields due to the magnetic charges and the associated local fields can be detected by implanted μ^+ muons (d). (From [44])

Monopoles also give rise to a new form of collective behavior. They form a magnetic Coulomb liquid [56]. The resulting magnetolyte physics – in analogy to electrolytes formed by dilute electric charges – is a rich field indeed. In application of physical chemistry to spin ice, *Bramwell* et al. [44] have pointed out an interesting analogy between the creation of separated magnetic charges and the autoionization of water, i.e., of the decomposition

$$2H_2O = [H_3O^+OH^-] = H_3O^+ + OH^- \quad . \tag{14.58}$$

It corresponds to the transition of two charge neutral tetrahedra into ones with magnetic charges $\pm q_m$. The dissociation constant K which describes this process increases nonlinearly when an electric field is applied. This leads in turn to an increase in the electrical conductivity of water, the so-called Wien effect. *Onsager* has developed a theory for this increase which can be directly transferred to the present case, i.e., the break-up of a magnetic dipole into a monopole and an antimonopole. The electric field is here replaced by an external magnetic field B. The result of the transfer of Onsager's theory is that in the weak field limit the magnetic field dependence of the dissociation constant is given by

$$K(B) = K(0)\left(1 + b + b^2/3\right) \quad , \quad b = \frac{\mu_0 Q^3 B}{8\pi k_B^2 T^2} \quad . \tag{14.59}$$

Following a disturbance, the density of monopoles will depend on time in a characteristic manner which is expected to show up in dynamical measurements such as magnetization curves or relaxation processes. An unequivocal observation of such an effect poses a challenge to experimentalists.

15
Superconductivity

Superconductivity, one of the most fascinating phenomena in solid-state physics, was discovered in 1911 by *Kamerlingh Onnes*, but it took 47 years for a satisfactory microscopic theory of the effect to become available via *Bardeen, Cooper* and *Schrieffer* (BCS) [21]. The major obstacle which theorists were facing before the appearance of the BCS theory can be summarized as follows. The superconducting transition temperature T_c is usually of the order 10 K, which corresponds to an energy of order 1 meV (we are not considering, for the moment, the more recent high-temperature superconductors). Provided that superconductivity is based on electron correlations, and taking into account that the correlation energy of electrons in a metal is of the order 1 eV per electron, is it then necessary to compute that energy to an accuracy of order 1 ‰ in order to find a superconducting ground state? If so, this would indeed be an impossible task and would eliminate any hope for a microscopic theory of superconductivity in the foreseeable future. Fortunately, a very special *pair correlation* leads to the phenomenon of superconductivity and a detailed treatment of the remaining correlation contributions is not required in order to understand the phenomenon as such. All the correlations that are difficult to treat are left out and enter the theory only in the form of renormalized parameters. This explains why reliable calculations of the superconducting transition temperature have so far remained an unsolved problem. They require the microscopic calculation of those parameters and therefore a detailed treatment of the correlations, which goes beyond the special pair correlations.

The presence of electron attractions may lead to the formation of electron pairs (Cooper pairs) [71]. These pairs may act like bosons, i.e., they can condense, although their commutation relations deviate from those of bosons. The transition from Cooper-pair condensation to true Bose-Einstein condensation can be studied in systems with ultracold fermionic atoms.

It was a breakthrough when BCS realized that the superconducting ground state can be written in the form of a coherent state. The pair state with zero total momentum is macroscopically occupied. This implies that it is not an eigenstate to the electron number operator. But in the limit of large electron number N the ground state is nearly an eigenstate of both the particle-number operator $\hat{N}$ as well as its conjugate, the phase operator $\hat{\alpha}$. The relative deviation is $\Delta N/N \simeq \Delta\alpha/\alpha \simeq N^{-1/2}$ and therefore becomes negligible. It is interesting to note that the results of the BCS theory can also be obtained by working with a ground state of fixed particle number (see, e.g., [277]).

A very fruitful period with many new experimental and theoretical findings followed the appearance of the BCS theory. One milestone was the realization that superconductivity does not require a gap in the excitation spectrum and that gapless superconductors do exist, a feature which is now common knowledge. Crucial is the presence of an order parameter and, associated with it, of a broken symmetry. In the BCS theory it is global gauge invariance which is broken by the superconducting ground state due to the nonconserved particle number. The Josephson effect is a prominent and important example of the many consequences of that symmetry breaking. A summary of that period up until the late sixties is found in the two volumes of "Superconductivity" [354] which came out in 1969 and has since been used as a reference in many laboratories worldwide. After a number of years of relatively slow but steady progress, which included the discovery of superconductivity in organic conductors [219] and in compounds with heavy quasiparticles (for reviews see, e.g., [156, 427]), the field gained immense impetus from the discovery in 1986 of the high-temperature superconducting perovskites by *Bednorz* and *Müller* [27]. The subsequent development has raised the transition temperature T_c to values as high $T_c = 125$ K. Examples of the new high-T_c materials are $La_{2-x}Sr_xCuO_4$ ($T_c \simeq 40$ K), $YBa_2Cu_3O_7$ ($T_c \simeq 92$ K), $Bi_2Sr_2Ca_2Cu_3O_{10}$ ($T_c \simeq 110$ K), and $Tl_2Ca_2Ba_2Cu_3O_{10}$ ($T_c \simeq 125$ K)[1].

Soon it was realized and particularly emphasized by *Anderson* [6] that systems like $La_{2-x}Sr_xCuO_4$ are hole-doped Mott- or, more precisely, charge-transfer insulators. Note that the antiferromagnet La_2CuO_4 remains an insulator above the Néel temperature despite having one unpaired electron per unit cell. The important structural element of the high-T_c cuprates are Cu-O planes with strong correlations which have been discussed at length in Sect. 12.1. Therefore, research on strongly correlated electrons became very significant and gained tremendously from high-temperature superconductivity.

Two features of superconductivity obtained special attention; the form of the pair state, and the origin of electron-electron attractions leading to Cooper pair formation. Before the discovery of the high-T_c cuprates s-wave, spin singlet pairing was assumed to be the general rule, although there were indications that in some of the superconductors with heavy quasiparticles the pair state was of a more complex form. However, in the high-T_c materials electrons turned out to pair in a d-wave state. The symmetry of the pair state is here lower than the point symmetry of the lattice. Therefore, we speak of unconventional pairing. Subsequently, pairing states in other superconductors were reexamined. It was found that scattering of electrons by impurities av-

---

[1] see, e.g., [67]

erages anisotropies of the order parameter over the Fermi surface. Therefore, one would expect that unconventional pairing, e.g., in the form of a d-wave is destroyed when the mean-free path becomes shorter than the superconducting coherence length. However, this argument is obviously not valid in general. Exceptions may be caused by strongly anisotropic potential scattering. The second topic, namely that of the origin of electron attractions and hence Cooper-pair formation also initiated many studies.

In principle, pairs may also form when the electron interactions are purely repulsive, but then they must meet certain stringent requirements. For example, the order parameter must change sign in different parts of the Fermi surface and the interaction must be less repulsive for electrons near the Fermi surface than away from it. While in the BCS theory phonons are assumed to be responsible for electron attractions and hence superconductivity, this has been seriously questioned in the case of the high-T_c superconductors. The idea that electron-electron interactions might provide the glue for Cooper-pairs formation in those materials finds support from the observation that superconductivity often does occur near a magnetic phase transition when parameters of the system are changed. However, quantitative, verifiable facts are still scarce.

The situation is different in some superconductors with low transition temperatures T_c involving f electrons. Here we have overwhelming experimental evidence that pair formation can primarily be due to intra-atomic low-energy excitations. This is the case in $PrOs_4Sb_{12}$ and UPd_2Al_3.

More recently, new surprises in the field of superconductivity have arisen. The long known material MgB_2 turned out to be a superconductor with a $T_c = 39$ K [340] and most recently iron pnictides [65, 340] like RE FeAsO with RE = La, Sm were, against all expectations, also found to be superconductors with rather high transition temperatures, i.e., up to $T_c = 55$ K. The ruthenate Sr_2RuO_4 was found to be superconducting at temperatures below $T_c = 1.5$ K. It is a layered perovskite like the cuprates but with Ru taking the positions of the Cu ions [300]. Probably pairing takes place here in a spin triplet state. In fullerenes, superconductivity was observed. Furthermore, ferromagnetic superconductors have been found, such as UGa_2 where superconductivity is observed in the vicinity of a quantum critical point, i.e., near an external pressure where the ferromagnetic transition temperature goes to zero [398].

Not only can Cooper pairs form but they can also be broken by external perturbations. Here it plays a role whether or not the perturbation acting on the electronic system conserves time reversal symmetry. Usually electrons are paired in time-reversed states like $(\mathbf{k}, \sigma)$ and $(-\mathbf{k}, -\sigma)$. When a perturbation changes sign when a time-reversal transformation is applied to the conduction electron system, then this implies that the perturbation acts differently on the two partners of a pair. It, therefore, breaks them if it is sufficiently strong. Examples are magnetic impurities which interact through an exchange poten-

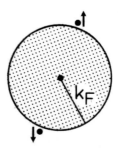

Fig. 15.1. Two interacting electrons outside a filled Fermi sphere. When the interaction is attractive they form a bound state (Cooper pair).

tial with the conduction electrons or an external magnetic field. Pair breaking plays a significant role in the theory of superconductivity.

Another important aspect is Cooper pairing of electrons with unbalanced spin populations. The latter are generated, e.g., by a Zeeman term in the Hamiltonian. It is of interest that the same topic applies to pairing of other particles, like quarks, or of different species of ultracold atoms. In a certain parameter range this imbalance may lead to inhomogeneous superconducting states. Its practical application is found in π junctions in superconducting circuitry.

Superconductivity has become such a large field of research that it cannot be our aim to cover it in any detail, in particular since there are a number of good textbooks already available[2]. Instead we will concentrate on those aspects of superconductivity, which widen our view on correlations in solids.

15.1 The Superconducting State

The essence of the phenomenon of superconductivity is an instability of the normal state with respect to the formation of electron pairs, known as Cooper pairs [71]. They have a boson-like character and condense thereby forming a superfluid. Consider a filled Fermi sphere in momentum space with radius k_F and two extra electrons outside of it (Fig. 15.1). These two electrons are assumed to attract each other through a potential $V(\mathbf{r}_1 - \mathbf{r}_2)$. The origin of this attraction is usually the electron-phonon interaction. But this need not always be the case and is not important at this stage. The center of mass is assumed to be at rest. None of the other electrons participate in the interactions. Their only role is to block the interior of the Fermi sphere for the orbital part $\phi(\mathbf{r}_1 - \mathbf{r}_2)$ of the wavefunction of the two extra electrons.

When we take the Fourier transform

---

[2] for examples, see Ref. [82, 403, 449]

15.1 The Superconducting State

$$\phi(\mathbf{r}_1 - \mathbf{r}_2) = \sum_{\mathbf{k}} g(\mathbf{k}) e^{i\mathbf{k}\cdot(\mathbf{r}_1 - \mathbf{r}_2)} \quad , \tag{15.1}$$

this implies that

$$g(\mathbf{k}) = 0 \quad \text{for} \quad |\mathbf{k}| < k_F \quad . \tag{15.2}$$

The function $g(\mathbf{k})$ is the probability amplitude that one electron is in momentum state $\mathbf{k}$ and the other is in state $-\mathbf{k}$. When the two electrons are in a spin-singlet state, antisymmetry of the wavefunction requires that $g(\mathbf{k}) = g(-\mathbf{k})$. The wavefunction $\phi(\mathbf{r}_1 - \mathbf{r}_2)$ satisfies the Schrödinger equation

$$\left(-\frac{1}{2m}(\nabla_1^2 + \nabla_2^2) + V(\mathbf{r}_1 - \mathbf{r}_2)\right)\phi(\mathbf{r}_1 - \mathbf{r}_2) = \left(E + \frac{k_F^2}{m}\right)\phi(\mathbf{r}_1 - \mathbf{r}_2) \quad . \tag{15.3}$$

In Fourier space this equation takes the form

$$\frac{k^2}{m} g(\mathbf{k}) + \sum_{\mathbf{k}'} g(\mathbf{k}') V_{\mathbf{k}\mathbf{k}'} = (E + 2\epsilon_F) g(\mathbf{k}) \quad , \tag{15.4}$$

where

$$V_{\mathbf{k}\mathbf{k}'} = \frac{1}{\Omega} \int d^3 r V(\mathbf{r}) e^{i(\mathbf{k}-\mathbf{k}')\cdot\mathbf{r}} \tag{15.5}$$

is the Fourier transform of the attractive potential. We have used $\epsilon_F = k_F^2/2m$.

In order to study (15.4), it is advantageous to use a form for $V_{\mathbf{k}\mathbf{k}'}$ which is as simple as possible. The following one is easy to handle:

$$V_{\mathbf{k}\mathbf{k}'} = \begin{cases} -\dfrac{V}{\Omega} & \text{for } \epsilon_F < \dfrac{k^2}{2m}, \dfrac{k'^2}{2m} < \epsilon_F + \omega_D, \\ 0 & \text{otherwise} \end{cases} \tag{15.6}$$

We notice that the attraction is limited to an energy shell of size ω_D above ϵ_F and that the initial ($\mathbf{k}$) and final ($\mathbf{k}'$) states must both be within that interval in order for the attraction to become effective. With this choice of $V_{\mathbf{k}\mathbf{k}'}$ we have

$$\left(-\frac{k^2}{m} + E + 2\epsilon_F\right) g(\mathbf{k}) = -\frac{V}{\Omega} {\sum_{\mathbf{k}'}}' g(\mathbf{k}')$$
$$= C \quad . \tag{15.7}$$

The prime on the summation symbol implies that $\mathbf{k}'$ must satisfy the inequality

$$\epsilon_F < \frac{k'^2}{2m} < \epsilon_F + \omega_D \quad . \tag{15.8}$$

From (15.7) we obtain

$$g(\mathbf{k}) = \frac{C}{-k^2/m + E + 2\epsilon_F} \tag{15.9}$$

and the self-consistency condition

$$C = \frac{V}{\Omega} C \sum_{\mathbf{k}'}{}' \frac{1}{k'^2/m - E - 2\epsilon_F} \quad . \tag{15.10}$$

With the abbreviation

$$\epsilon' = \frac{k'^2}{2m} - \epsilon_F \tag{15.11}$$

the density of states (per spin and unit volume) is

$$N(\epsilon') = \frac{1}{(2\pi)^3} 4\pi k'^2 \frac{dk'}{d\epsilon} \quad . \tag{15.12}$$

The self-consistency equation (15.10) can then be written in the form

$$1 = V \int_0^{\omega_D} d\epsilon \, N(\epsilon) \frac{1}{2\epsilon - E} \quad . \tag{15.13}$$

Provided that $\omega_D \ll \epsilon_F$, the density of states can be replaced by its value for $\epsilon = 0$, i.e., $N(0) = mk_F/2\pi^2$. After integration we find that

$$1 = \frac{N(0)}{2} V \ln\left(\frac{E - 2\omega_D}{E}\right) \quad . \tag{15.14}$$

For weak attraction $[N(0)V \ll 1]$ this expression simplifies to

$$E = -2\omega_D \, e^{-2/N(0)V} \quad . \tag{15.15}$$

A solution of the Schrödinger equation with an eigenvalue $E < 0$ implies a bound state of the two extra electrons in the presence of the filled Fermi sphere. The surprising fact is that here a bound state always exists, *regardless how weak the attractive potential V is*. This differs from the case of an electron in a three-dimensional potential well. There a bound state exists only if the depth V_0 of the well exceeds a threshold value V_c which depends on the diameter a of the well, i.e., $V_c = (2ma^2)^{-1}$. The difference compared to the one-electron problem lies in the blocking of the states within the Fermi sphere which are unavailable for the two interacting electrons and lead to the condition (15.2). The formation of a bound state is therefore a true many-body phenomenon. The electrons within the Fermi sphere influence the bound-state formation through the Pauli principle.

Another important point has to do with the form of the binding energy E, which cannot be obtained by a perturbation expansion with respect to V owing to the exponential dependence of E on V. The above calculations show that in the presence of weak electron net attractions the normal state of a metal becomes unstable with respect to the formation of pairs.

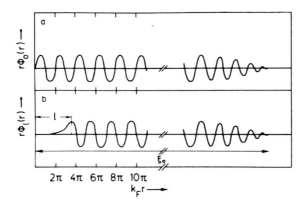

Fig. 15.2. Qualitative behavior of the Cooper-pair wavefunction for (a) isotropic and (b) anisotropic pair states. $\phi_l(r)$ has its first maximum at $r \simeq l/k_F$.

15.1.1 Pair States

Before discussing the ground-state wavefunction of the total electron system (BCS wavefunction) we ought to consider the form of the function $g(\mathbf{k})$ for a nonspherical Fermi surface. Instead of $g(\mathbf{k})$ we may write $g(\hat{\mathbf{k}}, \xi)$, where the unit vector $\hat{\mathbf{k}}$ specifies a point on the Fermi surface and ξ denotes the energy measured from ϵ_F. Here, we are assuming that $g(\hat{\mathbf{k}}, \xi = 0)$ has the same symmetry properties as the Fermi surface itself (conventional pairing). Later we will discuss the case when the function $g(\hat{\mathbf{k}}, \xi = 0)$ has a lower symmetry than the Fermi surface (unconventional pairing).

We expand $g(\hat{\mathbf{k}})$ in terms of a set $\theta_l(\hat{\mathbf{k}})$ of orthonormal functions classified according to the angular momentum l, which have the full symmetry of the lattice, i.e.,

$$g(\hat{\mathbf{k}}) = g_0 + \sum_{l>0} g_l \, \theta_l(\hat{\mathbf{k}}) \quad . \tag{15.16}$$

More precisely, $\theta_l(\hat{\mathbf{k}})$ is usually expressed in terms of a linear combination of spherical harmonics $Y_{lm}(\hat{\mathbf{k}})$ and is fully symmetric under the operations of the symmetry group of the lattice.

Returning to $\mathbf{r}$ space, we expand the pair function $\phi(\mathbf{r})$ where $\mathbf{r}$ is the relative coordinate,

$$\phi(\mathbf{r}) = \phi_0(r) + \sum_{l>0} \phi_l(r) \theta_l(\hat{\mathbf{r}}) \tag{15.17}$$

in close analogy to (15.16). The $\theta_l(\hat{\mathbf{r}})$ are the Fourier transforms of the $\theta_l(\hat{\mathbf{k}})$, and $\hat{\mathbf{r}}$ is a unit vector. Of interest is the radial dependence of $\phi(\mathbf{r})$, i.e., of $\phi_l(r)$. It is shown in Fig. 15.2 for the isotropic case, in which only $\phi_0(r)$ contributes, and for an anisotropic one with a given value of l. In systems with strongly correlated electrons, like heavy-fermion systems or the high-T_c superconducting materials, the two electrons (or, more generally, quasiparticles) which form a pair have only a small probability of occupying the same site, as this would imply a large Coulomb repulsion between the two electrons. Therefore, correlations suppress these configurations. The isotropic component $\phi_0(r)$ of $\phi(\mathbf{r})$ must then be small or even vanish. In the heavy-fermion system UBe$_{13}$, the two quasiparticles forming a Cooper pair cannot be closer than a U-U distance, which is $d = 510$ pm. The first maximum of $\phi(\mathbf{r})$ should be at this distance. With $k_F^{-1} \simeq 100$ pm this requires $l = 4$ or 6 when ℓ is even. We therefore expect the pair wavefunction to be strongly anisotropic [136].

The expulsion of a magnetic field from a sample when it is cooled below the superconducting transition temperature (Meissner effect) is extremely important, because it proves that a superconductor is not simply a metal with infinite conductivity but rather a new thermodynamic state. As such, it is characterized by an order parameter, different to zero in the superconducting state and vanishing in the normal state. The particular order in a superconductor must obviously be related to the formation of Cooper pairs. An elegant formulation of the order parameter is obtained from the two-particle density matrix

$$\left\langle \mathbf{r}_1\sigma_1; \mathbf{r}_2\sigma_2 \left| \rho^{(2)} \right| \mathbf{r}_3\sigma_3; \mathbf{r}_4\sigma_4 \right\rangle = \langle \psi_{\sigma_1}^+(\mathbf{r}_1)\psi_{\sigma_2}^+(\mathbf{r}_2)\psi_{\sigma_3}(\mathbf{r}_3)\psi_{\sigma_4}(\mathbf{r}_4) \rangle \quad . \quad (15.18)$$

The single-electron field operators $\psi_\sigma(\mathbf{r})$ are the same as in (2.2, 2.3). In the superconducting state, the two-particle density matrix remains finite in the limit of large distances between pairs of points $\mathbf{r}_1, \mathbf{r}_2$ and $\mathbf{r}_3, \mathbf{r}_4$,

$$\left\langle \mathbf{r}_1\sigma_1; \mathbf{r}_2\sigma_2 \left| \rho^{(2)} \right| \mathbf{r}_3\sigma_3; \mathbf{r}_4\sigma_4 \right\rangle \to \langle \psi_{\sigma_1}^+(\mathbf{r}_1)\psi_{\sigma_2}^+(\mathbf{r}_2) \rangle \times \langle \psi_{\sigma_3}(\mathbf{r}_3)\psi_{\sigma_4}(\mathbf{r}_4) \rangle$$
$$= \Phi_{\sigma_1\sigma_2}(\mathbf{r}_1, \mathbf{r}_2)\, \Phi^*_{\sigma_3\sigma_4}(\mathbf{r}_3, \mathbf{r}_4) \quad . \quad (15.19)$$

This property of $\rho^{(2)}$ characterizes the superconducting states and is called off-diagonal long-range order (ODLRO) [496]. The function $\Phi_{\sigma_1\sigma_2}(\mathbf{r}_1, \mathbf{r}_2)$ behaves like a two-fermion wavefunction and can be identified with the one in (15.1), when only the orbital part is considered; however, it now deals with quasiparticles instead of bare electrons and we will refer to it as the order parameter. Generally this order parameter is antisymmetric with respect to particle interchange, i.e.,

$$\Phi_{\sigma_1\sigma_2}(\mathbf{r}_1, \mathbf{r}_2) = -\Phi_{\sigma_2\sigma_1}(\mathbf{r}_2, \mathbf{r}_1) \quad . \quad (15.20)$$

In homogeneous systems it depends only on the relative coordinate $(\mathbf{r}_1 - \mathbf{r}_2)$.

Being a two-particle wavefunction, the order parameter has the form of a 2×2 matrix which can be decomposed into an antisymmetric part proportional to the Pauli matrix τ_2 and a symmetric part. It is therefore of the general form

$$\Phi = \phi(\mathbf{r}_1 - \mathbf{r}_2)i\tau_2 + \sum_{\mu=1}^{3} d_\mu(\mathbf{r}_1 - \mathbf{r}_2)\tau_\mu i\tau_2 \quad , \quad (15.21)$$

15.1 The Superconducting State

where $\phi(\mathbf{r})$ and $d_\mu(\mathbf{r})$ are four complex functions [276]. From (15.20) it follows that

$$\phi(\mathbf{r}_1 - \mathbf{r}_2) = \phi(\mathbf{r}_2 - \mathbf{r}_1)$$
$$d_\mu(\mathbf{r}_1 - \mathbf{r}_2) = -d_\mu(\mathbf{r}_2 - \mathbf{r}_1) \quad . \tag{15.22}$$

The corresponding relations for the Fourier transforms are $\phi(\hat{\mathbf{k}}) = \phi(-\hat{\mathbf{k}})$ and $d_\mu(\hat{\mathbf{k}}) = -d_\mu(-\hat{\mathbf{k}})$ respectively. The unit vector $\hat{\mathbf{k}}$ defines a point on the Fermi surface.

When the crystal lattice of the superconductor has an inversion center, the order parameter (i.e., the pair wavefunction) can be classified according to its parity. From (15.20) we see that $\phi(\mathbf{r})$ has even parity, while the $d_\mu(\mathbf{r})$ are odd-parity states.

When the spin-orbit interaction is sufficiently small and may be neglected, then the spin S of an electron pair is a good quantum number. It is either $S = 0$ implying a singlet state or $S = 1$ (triplet state). The order parameter may be also written in the form of a state vector as

$$\left| \Phi(\hat{\mathbf{k}}) \right\rangle = \phi(\hat{\mathbf{k}})|0\rangle + \sum_{\mu=1}^{3} d_\mu(\hat{\mathbf{k}})|x_\mu\rangle \quad , \tag{15.23}$$

where

$$|0\rangle = \frac{1}{\sqrt{2}} (|\uparrow\downarrow\rangle - |\downarrow\uparrow\rangle) \quad , \tag{15.24a}$$

denotes the singlet state and

$$|x_1\rangle = \frac{-1}{\sqrt{2}} (|\uparrow\uparrow\rangle - |\downarrow\downarrow\rangle)$$
$$|x_2\rangle = \frac{-1}{i\sqrt{2}} (|\uparrow\uparrow\rangle + |\downarrow\downarrow\rangle)$$
$$|x_3\rangle = \frac{1}{\sqrt{2}} (|\uparrow\downarrow\rangle + |\downarrow\uparrow\rangle) \quad , \tag{15.24b}$$

is the triplet state [276]. The $|s_z, s'_z\rangle$ denote states in which the quasiparticles have spins with z components s_z and s'_z. While the singlet state is invariant under spin rotations, the triplet states $|x_\nu\rangle$ transform like the three components of a vector. This choice proves more advantageous than using the three eigenfunctions $|S = 1, S_z\rangle$ with $\mathbf{S} = \mathbf{s} + \mathbf{s}'$.

Consider a superconductor with an order parameter

$$\Phi(\hat{\mathbf{k}}) = \sum_{\mu=1}^{3} d_\mu(\hat{\mathbf{k}}) \tau_\mu i \tau_2$$
$$= \left(\mathbf{d}(\hat{\mathbf{k}}) \cdot \boldsymbol{\tau} \right) i\tau_2 \quad . \tag{15.25}$$

Noticing that
$$\Phi\Phi^+ = |\mathbf{d}|^2 + i\,(\mathbf{d}\times\mathbf{d}^*)\cdot\boldsymbol{\tau} \quad. \tag{15.26}$$
we compute its norm
$$\frac{1}{2}\text{Tr}\left|\Phi(\hat{\mathbf{k}})\right|^2 = \left|\mathbf{d}(\hat{\mathbf{k}})\right|^2 \quad. \tag{15.27}$$

We see that $|\mathbf{d}(\hat{\mathbf{k}})|^2$ measures the magnitude of the pair condensate at point $\hat{\mathbf{k}}$ on the Fermi surface. The direction of the expectation value of the spin $\mathbf{S}$ of an electron pair is given by that of $(\mathbf{d}^*\times\mathbf{d})$, which can be verified by explicitly evaluating $\frac{1}{2}\text{Tr}(\Phi^+(\hat{\mathbf{k}})\mathbf{S}\Phi(\hat{\mathbf{k}}))$.

For a given crystal structure, e.g., cubic or hexagonal, the spin-singlet and spin-triplet pair function can be expanded in terms of the basis functions $\Phi_{\Gamma_i}(\hat{\mathbf{k}})$ of the different irreducible representations Γ_i of the symmetry group, where i is a degeneracy index. We write

$$\phi(\hat{\mathbf{k}}) = \sum_{(\Gamma_i)} A_{\Gamma_i}\Phi_{\Gamma_i}(\hat{\mathbf{k}}) \tag{15.28}$$

for the spin-singlet and

$$\mathbf{d}(\hat{\mathbf{k}}) = \sum_{(\Gamma_i)} B_{\Gamma_i}\mathbf{d}_{\Gamma_i}(\hat{\mathbf{k}}) \tag{15.29}$$

for the spin-triplet order parameter. A list of the different basis functions for various symmetries is found in [473].

An example of a spin-singlet order parameter in a hexagonal crystal structure is a d-wave pair state corresponding to the two-fold degenerate irreducible representation $\Gamma = E_{1g}$. In the case, the basis function are

$$\begin{aligned} i &= 1 : \Phi_{\Gamma_1}(\hat{\mathbf{k}}) = \hat{k}_x\hat{k}_z \\ i &= 2 : \Phi_{\Gamma_2}(\hat{\mathbf{k}}) = \hat{k}_y\hat{k}_z \quad. \end{aligned} \tag{15.30}$$

Specific examples of spin-triplet order parameters are found in [473]. A prominent one is a p-wave pair state in a system of cubic symmetry with basis functions
$$d_{\Gamma_1}(\hat{\mathbf{k}}) = \hat{k}_x, \quad d_{\Gamma_2}(\hat{\mathbf{k}}) = \hat{k}_y, \quad d_{\Gamma_3}(\hat{\mathbf{k}}) = \hat{k}_z \quad. \tag{15.31}$$

The superconducting order parameter has a lower symmetry than the Hamiltonian. The symmetry group G_t of the Hamiltonian consists of the space group G of the crystal, the time-reversal symmetry group with the operation T_R, and the gauge group $U(1)$. Adding a phase α to each electron by multiplying $\psi_\alpha^+(\mathbf{r})$ with $e^{i\alpha}$ (i.e., changing the gauge) leaves the Hamiltonian invariant. Thus it is
$$G_t = G \otimes T_R \otimes U(1) \quad. \tag{15.32}$$

When the spin-orbit interaction may be neglected, the Hamiltonian is additionally invariant under the spin-rotation group $SU(2)$.

15.1 The Superconducting State

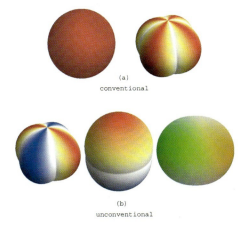

Fig. 15.3. Amplitude of the order parameter for conventional and unconventional pairing. The crystal is assumed to have tetragonal symmetry. In both cases the order parameter can vanish at the Fermi surface. Red sections: positive sign of the order parameter, blue sections: negative sign. Green section: order parameter of predominantly p_x symmetry and orange section: of p_y symmetry. In (a) the symmetry of the order parameter agrees with the point symmetry of the lattice while in (b) it is lower than the point symmetry (Courtesy of T. Takimoto).

Systems with heavy quasiparticles involve, in most cases, $4f$ or $5f$ electrons and therefore spin-orbit interactions are strong. In this case, point group transformations also rotate the spin. In the limit of strong spin-orbit coupling we consider the spin variables in (15.24a,b) as those of pseudospins, reflecting Kramers' degeneracy.

When superconductivity sets in, some of these symmetries are broken. One symmetry always broken is gauge symmetry. This is intimately connected with the formation of Cooper pairs and is in fact obvious from the definition of the order parameter through $\langle \psi^+_{\sigma_1}(\mathbf{r}_1)\psi^+_{\sigma_2}(\mathbf{r}_2) \rangle$. When the $U(1)$ symmetry is the only one broken in the superconducting state, it would be called *conventional* pairing. Should the order parameter also break symmetries of the crystal, one speaks of *unconventional* pairing. An example is an odd-parity pair state in a crystal with inversion symmetry. The order parameter has lower symmetry than the lattice because it does not remain invariant under inversion. In Fig. 15.3 we show examples of the order parameter for conventional and unconventional pairing; in *both* cases, the amplitude may vanish at points, lines, or on parts of the Fermi surface.

Unconventional pairing is realized in the high-T_c cuprates, in the ruthenate SrRu$_2$O$_4$, where p-wave pairing takes place, in some of the heavy-fermion superconductors (e.g., CeCoIn$_5$ and UPt$_3$) and presumably in some of the organic superconductors. The superconducting fullerenes seem to have a conventional pair state. A comprehensive review of the subject is found in [415].

15.1.2 BCS Ground State

In the following discussion we neglect spin-orbit interactions and consider conventional spin-singlet pairing. The important step of generalizing the theory of a single pair, discussed previously, to the ground-state wavefunction of a superconductor was done by *Bardeen*, *Cooper* and *Schrieffer*. In contrast to the previous calculations, all electrons are treated on an equal level. An appropriate ansatz for the ground state of N electrons would seem to be

$$\tilde{\psi}(\mathbf{r}_1,\ldots,\mathbf{r}_N) = A(N)\left[\phi(\mathbf{r}_1-\mathbf{r}_2)S(1,2)\phi(\mathbf{r}_3-\mathbf{r}_4)S(3,4)\ldots\right.$$
$$\left.\phi(\mathbf{r}_{N-1}-\mathbf{r}_N)S(N-1,N)\right] \quad . \tag{15.33}$$

This is a wavefunction of independent pairs. The antisymmetrizing operator $A(N)$ ensures that Pauli's principle is satisfied. The function $S(i,j)$ denotes a spin singlet (15.24a) formed by electrons i and j. The wavefunction is a natural generalization to pairs of the SCF wavefunction, an antisymmetrized product of one-particle states (independent electrons). Equation (15.33) has been written down for an even electron number. For large electron numbers like in a solid (i.e., $N \simeq 10^{23}$) it should not make any difference whether an even or odd number of electrons is considered.

Although the ansatz (15.33) can, in principle, be used to calculate expectation values of operators [277, 338], it is generally not convenient to do so. The theory has be cast by BCS into a very elegant form by working with an alternative form of the ground-state wavefunction. For that purpose we write for the creation operator of a single pair

$$\phi_0^+ = \sum_{\mathbf{k}} g(\mathbf{k}) c_{\mathbf{k}\uparrow}^+ c_{-\mathbf{k}\downarrow}^+ \quad . \tag{15.34}$$

The subscript 0 indicates that the total momentum of the pair is zero. The operator ϕ_0^+ is boson-like because it involves two fermions. However, some features distinguish it from a true boson. In particular the commutation relations, which are easily derived, differ from those of a true boson. With ϕ_0^+ one can construct a coherent state of the form

$$\left|\tilde{\psi}_0\right\rangle = e^{\phi_0^+}|0\rangle$$
$$= \exp\left(\sum_{\mathbf{k}} g(\mathbf{k}) c_{\mathbf{k}\uparrow}^+ c_{-\mathbf{k}\downarrow}^+\right)|0\rangle \quad , \tag{15.35}$$

where $|0\rangle$ is the vacuum state. Coherent states have the property that they are eigenstates of the corresponding bosonic creation operator, i.e., in the present case $\phi_0^+|\tilde{\psi}_0\rangle = \nu|\tilde{\psi}_0\rangle$ with $\nu \simeq N^{1/2}$. They are "almost" eigenstates of the total particle number operator $\hat{N}$ *and* its conjugate, i.e., the phase operator $\hat{\alpha}$. The uncertainties in the two eigenvalues are of order

$$\frac{\Delta N}{N} \simeq \frac{1}{\sqrt{N}} \quad , \quad \Delta\alpha \simeq \frac{1}{\sqrt{N}} \tag{15.36}$$

and vanish in the limit of large N. Note that the uncertainty principle requires $\Delta N \Delta \alpha \geqslant 1$.

By expansion of (15.35) one obtains

$$|\psi_0\rangle = \prod_{\mathbf{k}} \left[1 + g(\mathbf{k})c^+_{\mathbf{k}\uparrow}c^+_{-\mathbf{k}\downarrow}\right]|0\rangle \qquad (15.37)$$

(remember that $c^+_{\mathbf{k}\sigma}c^+_{\mathbf{k}\sigma} = 0$). It is customary to set

$$g(\mathbf{k}) = v_{\mathbf{k}}/u_{\mathbf{k}} \;, \quad \text{with} \quad u^2_{\mathbf{k}} + v^2_{\mathbf{k}} = 1 \;. \qquad (15.38)$$

In this notation the normalized ground state is written as

$$|\psi_0\rangle = \prod_{\mathbf{k}} \left(u_{\mathbf{k}} + v_{\mathbf{k}} c^+_{\mathbf{k}\uparrow}c^+_{-\mathbf{k}\downarrow}\right)|0\rangle \;. \qquad (15.39)$$

We can easily check that the condition (15.36) for a coherent state is indeed satisfied. We have

$$\langle N_{\mathrm{op}}\rangle = \left\langle \psi_0 \left| \sum_{\mathbf{k}\sigma} c^+_{\mathbf{k}\sigma} c_{\mathbf{k}\sigma} \right| \psi_0 \right\rangle$$
$$= 2\sum_{\mathbf{k}} v^2_{\mathbf{k}} = \frac{2\Omega}{(2\pi)^3}\int d^3k\, v^2_{\mathbf{k}} = N \qquad (15.40)$$

and similarly

$$(\Delta N)^2 = \left\langle \hat{N}^2 \right\rangle - \left\langle \hat{N} \right\rangle^2$$
$$= \frac{4\Omega}{(2\pi)^3}\int d^3k\, v^2_{\mathbf{k}} u^2_{\mathbf{k}} \;. \qquad (15.41)$$

We notice that the last expression is also proportional to the volume Ω, and therefore to N. Thus $\Delta N \simeq \sqrt{N}$ in agreement with (15.36). The BCS wavefunction $|\psi_0\rangle$ can be decomposed into a sum of states (15.33), i.e., normalized eigenstates $|\tilde{\psi}_N\rangle$ of the electron number operator $\hat{N}$,

$$|\psi_0\rangle = \sum_N e^{i\alpha N} w^{1/2}_N |\tilde{\psi}_N\rangle \;. \qquad (15.42)$$

Here α is an arbitrary phase which we may attach to each electron by a gauge transformation. This form is typical for a coherent or Glauber state[3]. The coefficients w_N depend on N as indicated in Fig. 15.4. Writing the superconducting ground state in the form of a coherent state has the advantage that we can perform calculations with it. It does not suffer from the shortcomings of the wavefunction (15.33), i.e., that in practice expectation values can hardly be calculated with it. As mentioned before, in the limit of large N a coherent

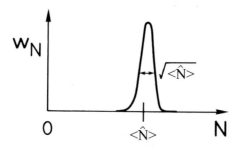

Fig. 15.4. Dependence of the coefficients w_N on the electron number N. The function is strongly peaked at the average value $\langle \hat{N} \rangle$.

state $|\psi_0\rangle$ is an eigenstate of the phase operator, i.e., one may associate a fixed phase with it.

The function u_k and v_k in (15.38) have remained undetermined. They are obtained by minimization of the energy expectation value

$$W = \langle \psi_0 | H | \psi_0 \rangle \quad . \tag{15.43}$$

For H we use a reduced Hamiltonian of the form

$$H_{\text{red}} = \sum_{k\sigma} \epsilon(k) c^+_{k\sigma} c_{k\sigma} + \sum_{k,q} V(q) c^+_{k+q\downarrow} c^+_{-k-q\uparrow} c_{-k\uparrow} c_{k\downarrow} \quad . \tag{15.44}$$

Working with a state $|\psi_0\rangle$ which is not an eigenstate of the electron number operator requires a Hamiltonian with a coupling to a particle reservoir. We achieve this by taking the zero of the energy $\epsilon(\mathbf{k})$ at the chemical potential μ. We have kept only those interaction matrix elements which scatter electrons of opposite spins and which have zero total momentum. One can easily check that

$$\langle \psi_0 | H_{\text{red}} | \psi_0 \rangle = 2 \sum_{k} \epsilon(\mathbf{k}) v_k^2 + \sum_{k,q} V(\mathbf{q}) u_k v_k u_{k+q} v_{k+q} \quad . \tag{15.45}$$

At this stage, it is useful to introduce $h_\mathbf{k} = v_\mathbf{k}^2$ and to set $\mathbf{q} = \mathbf{k}' - \mathbf{k}$ and $V(\mathbf{q}) = V_{\mathbf{kk}'}$. The last equation is then written in the form

$$W = 2 \sum_{\mathbf{k}} \epsilon(\mathbf{k}) h_\mathbf{k} + \sum_{\mathbf{k},\mathbf{k}'} V_{\mathbf{kk}'} \sqrt{h_\mathbf{k}(1-h_\mathbf{k}) h_{\mathbf{k}'}(1-h_{\mathbf{k}'})} \quad . \tag{15.46}$$

Minimizing W with respect to $h_\mathbf{k}$ leads to

$$\frac{\partial W}{\partial h_\mathbf{k}} = 0 = 2\epsilon(\mathbf{k}) + \sum_{\mathbf{k}'} V_{\mathbf{kk}'} \sqrt{h_{\mathbf{k}'}(1-h_{\mathbf{k}'})} \frac{1 - 2h_\mathbf{k}}{\sqrt{h_\mathbf{k}(1-h_\mathbf{k})}} \quad , \tag{15.47}$$

---

[3] see, e.g., [320]

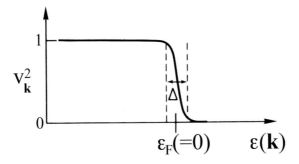

Fig. 15.5. Plot of v_k^2 as a function of ϵ_k. The function decreases to zero over an energy interval of order Δ.

where we have used that $V_{\mathbf{kk'}} = V_{\mathbf{k'k}}$. By using the notation

$$\Delta_{\mathbf{k}} = -\sum_{\mathbf{k'}} V_{\mathbf{kk'}} \sqrt{h_{\mathbf{k'}}(1 - h_{\mathbf{k'}})} \qquad (15.48)$$

we can write (15.47) in the form

$$2\epsilon(\mathbf{k}) = \Delta_{\mathbf{k}} \frac{1 - 2h_{\mathbf{k}}}{\sqrt{h_{\mathbf{k}}(1 - h_{\mathbf{k}})}} \qquad . \qquad (15.49)$$

From this equation, $h_{\mathbf{k}}$ is determined as

$$h_{\mathbf{k}} = \frac{1}{2}\left(1 - \frac{\epsilon(\mathbf{k})}{E(\mathbf{k})}\right) \qquad (15.50)$$

with $E(\mathbf{k})$ given by

$$E(\mathbf{k}) = \sqrt{\epsilon^2(\mathbf{k}) + \Delta_{\mathbf{k}}^2} \qquad . \qquad (15.51)$$

Accordingly, we obtain $v_{\mathbf{k}}$ and $u_{\mathbf{k}}$ as

$$v_{\mathbf{k}} = \sqrt{\frac{1}{2}\left(1 - \frac{\epsilon(\mathbf{k})}{E(\mathbf{k})}\right)} \quad , \quad u_{\mathbf{k}} = \sqrt{\frac{1}{2}\left(1 + \frac{\epsilon(\mathbf{k})}{E(\mathbf{k})}\right)} \qquad . \qquad (15.52)$$

We plot the function v_k^2 in Fig. 15.5, which describes the occupation probability of states $\mathbf{k}$ in momentum space; compare it to (15.40). It is noticed that the Fermi surface is no longer sharp. Instead it is smeared out over an energy interval of order Δ.

Equation (15.48) has the form of a self-consistency condition, because $h_{\mathbf{k}}$ depends on $\Delta_{\mathbf{k}}$. If we make use of (15.50,15.51), this condition can be written as

$$\Delta_{\mathbf{k}} = -\frac{1}{2}\sum_{\mathbf{k'}} V_{\mathbf{kk'}} \frac{\Delta_{\mathbf{k'}}}{E(\mathbf{k'})} \qquad . \qquad (15.53)$$

In order to extract from it an expression for Δ in terms of the attractive potential $V_{\mathbf{kk'}}$ one again chooses the form (15.6). The region in momentum space within which $V_{\mathbf{kk'}} \neq 0$ is given by $|\epsilon(\mathbf{k})|, |\epsilon(\mathbf{k'})| < \omega_D$, which implies

$$\Delta_{\mathbf{k}} = \begin{cases} \Delta & : \quad |\epsilon(\mathbf{k})| < \omega_D \ , \\ 0 & : \quad \text{otherwise} \ . \end{cases} \qquad (15.54)$$

From

$$\Delta = \Delta \frac{V}{\Omega} \sum_{\mathbf{k'}} \frac{1}{2E(\mathbf{k'})} \ , \qquad (15.55)$$

when we compare it to (15.10), it follows that

$$1 = N(0)V \int_0^{\omega_D} \frac{d\epsilon}{\sqrt{\epsilon^2 + \Delta^2}} \ . \qquad (15.56)$$

Provided that $N(0)V \ll 1$ one finds that

$$\Delta = 2\omega_D e^{-1/N(0)V} \ . \qquad (15.57)$$

If we compare this result with (15.15), we notice that the only difference between the two is a factor of 2 in the exponent. The inclusion of all electrons in the Cooper-pair formation increases the binding energy when compared with that of a single pair.

The superconducting condensation energy is obtained by subtracting from (15.45) the energy of the normal state, i.e., that of a filled Fermi sphere. Specifically

$$\Delta E_S = \sum_{\mathbf{k}} |\epsilon(\mathbf{k})| \left(1 - \frac{|\epsilon(\mathbf{k})|}{\sqrt{\epsilon^2(\mathbf{k}) + \Delta^2}}\right) - \frac{1}{2} \sum_{\mathbf{k}} \frac{\Delta^2}{\sqrt{\epsilon^2(\mathbf{k}) + \Delta^2}} \ . \qquad (15.58)$$

After converting the sum into an integral we obtain

$$\Delta E_S = 2N(0) \int_0^{\omega_D} d\epsilon \left(\epsilon - \frac{1}{2}\frac{2\epsilon^2 + \Delta^2}{\sqrt{\epsilon^2 + \Delta^2}}\right) \ . \qquad (15.59)$$

After integration, thereby taking into account that $\omega_D \gg \Delta$, we obtain

$$\Delta E_S = -\frac{N(0)}{2}\Delta^2 \ . \qquad (15.60)$$

The excited states of the system are described by the two types of wavefunctions

$$|\psi_{\text{ex}}(\mathbf{k})\rangle = c_{\mathbf{k}\sigma}^+ \prod_{\mathbf{k'} \neq \mathbf{k}} \left(u_{\mathbf{k'}} + v_{\mathbf{k'}} c_{\mathbf{k'}\uparrow}^+ c_{-\mathbf{k'}\downarrow}^+\right) |0\rangle \qquad (15.61)$$

and

$$|\psi_{\text{ex}}^{\text{pair}}(\mathbf{k})\rangle = \left(v_{\mathbf{k}} - u_{\mathbf{k}} c_{\mathbf{k}\uparrow}^{+} c_{-\mathbf{k}\downarrow}^{+}\right) \prod_{\mathbf{k}' \neq \mathbf{k}} \left(u_{\mathbf{k}'} + v_{\mathbf{k}'} c_{\mathbf{k}'\uparrow}^{+} c_{-\mathbf{k}'\downarrow}^{+}\right) |0\rangle \quad . \tag{15.62}$$

We notice here that

$$\langle \psi_{\text{ex}}(\mathbf{k})|\psi_0\rangle = 0 \ ; \quad \langle \psi_{\text{ex}}^{\text{pair}}(\mathbf{k})|\psi_0\rangle = 0 \quad , \tag{15.63}$$

i.e., the states (15.61,15.62) are orthogonal to the ground state. The $|\psi_{\text{ex}}(\mathbf{k})\rangle$ describes single-particle excitations, while the states $|\psi_{\text{ex}}^{\text{pair}}(\mathbf{k})\rangle$ describe pair excitations.

The excitation energies are calculated from

$$E_{\text{ex}}(\mathbf{k}) = \langle \psi_{\text{ex}}(\mathbf{k})|H_{\text{red}}|\psi_{\text{ex}}(\mathbf{k})\rangle - \langle \psi_0|H_{\text{red}}|\psi_0\rangle \tag{15.64}$$

and similarly for the pair excitations. The evaluation of the expectation values is straightforward as demonstrated before for ΔE_S. One finds that the energy of a single-particle excitation is simply

$$E_{\text{ex}}(\mathbf{k}) = E(\mathbf{k}) = \sqrt{\epsilon^2(\mathbf{k}) + \Delta^2} \quad , \tag{15.65}$$

whereas that of a pair excitation is

$$E_{\text{ex}}^{\text{pair}}(\mathbf{k}) = 2E(\mathbf{k}) \quad . \tag{15.66}$$

The excitations have a gap in their spectrum, the size of which is given by Δ. For a more detailed derivation of the excitation energies see, e.g., [82, 403, 449]. A distinction between single-particle and pair excitations gives valuable insight but has become obsolete by the introduction of a linear transformation due to *Bogoliubov* and *Valatin*.

$$\begin{aligned} \gamma_{\mathbf{k}\uparrow}^{+} &= u_{\mathbf{k}} c_{\mathbf{k}\uparrow}^{+} - v_{\mathbf{k}} c_{-\mathbf{k}\downarrow} \\ \gamma_{\mathbf{k}\downarrow}^{+} &= u_{\mathbf{k}} c_{\mathbf{k}\downarrow}^{+} + v_{\mathbf{k}} c_{-\mathbf{k}\uparrow} \quad . \end{aligned} \tag{15.67}$$

A special feature of the transformation is the superposition of creation and annihilation operators. This is closely related to the nonconservation of the particle number in the BCS theory. By applying the above γ-operators to the BCS ground state $|\psi_0\rangle$ it is very simple to verify that

$$\begin{aligned} |\psi_{\text{ex}}(\mathbf{k})\rangle &= \gamma_{\mathbf{k}\sigma}^{+} |\psi_0\rangle \\ |\psi_{\text{ex}}^{\text{pair}}(\mathbf{k})\rangle &= \gamma_{\mathbf{k}\sigma}^{+} \gamma_{-\mathbf{k}-\sigma}^{+} |\psi_0\rangle \quad . \end{aligned} \tag{15.68}$$

In terms of the γ operators it is therefore not necessary to distinguish between single-particle and pair excitations. Since in addition $\gamma_{\mathbf{k}\sigma}|\psi_0\rangle = 0$ we may consider the $\gamma_{\mathbf{k}\sigma}^{+}$ operators as quasiparticle operators.

Associated with the excitation energy is a quasiparticle density of states. It is defined by the number N of excited states per spin and per energy interval dE_σ

$$N_\sigma(E) = \frac{dN}{d\epsilon} \frac{d\epsilon}{dE_\sigma} \quad . \tag{15.69}$$

Here $(dN/d\epsilon) = N(0)$ is the density of states per spin direction in the normal state.

The absence of a Fermi surface and the presence of the gap in the excitation spectrum are clear evidence that superconductors are not normal Fermi liquids. Nevertheless, the excitations can be treated as well-behaved fermionic quasiparticles with a dispersion given by (15.65). Thus, their distribution is expected to be given by the Fermi function

$$f(\mathbf{k}) = \frac{1}{1 + e^{\beta E(\mathbf{k})}} \quad , \tag{15.70}$$

a result which can be rigorously derived when the free energy of a superconducting system is constructed. Note that $\mathbf{k}$ states filled with a quasiparticle are blocked for virtual pair scattering via $V_{\mathbf{k}\mathbf{k}'}$ (see (15.48) and (15.53)). Therefore, the self-consistency condition (15.53) changes at temperatures $T \neq 0$ to

$$\Delta_{\mathbf{k}} = -\frac{1}{2} \sum_{(\mathbf{k}')} V_{\mathbf{k}\mathbf{k}'} \frac{\Delta_{\mathbf{k}'}}{E(\mathbf{k}')} (1 - 2f(E(\mathbf{k}'))) \quad . \tag{15.71}$$

The factor 2 in front of $f(E(\mathbf{k}))$ accounts for blocking $\mathbf{k}$ and $-\mathbf{k}$ for pair scattering. For the BCS interaction and using (15.54), the last equation reduces to

$$1 = N(0)V \int_0^{\omega_D} \frac{d\epsilon}{\sqrt{\epsilon^2 + \Delta^2}} \left(1 - 2f\left(\sqrt{\epsilon^2 + \Delta^2}\right)\right) \tag{15.72}$$

and yields a relation between Δ and T. For T = 0 one recovers (15.56). The superconducting transition temperature T_c is obtained by letting Δ go to zero. This results in

$$\frac{1}{N(0)V} = \int_0^{\omega_D} \frac{d\epsilon}{\epsilon} \tanh \frac{\epsilon}{2k_B T_c}$$

$$= \ln \frac{2\omega_D e^\gamma}{\pi k_B T_c} \tag{15.73}$$

where $\gamma = 0.5772$ is Euler's constant [2]. The integral is evaluated, e.g., in [354]. Therefore,

$$k_B T_c = 1.14 \, \omega_D e^{-1/N(0)V} \quad . \tag{15.74}$$

The equation may be used to derive the size of $N(0)V$ for given values of the Debye frequency and transition temperature. Typical values are 0.18 for Al or 0.4 for Pb. In the latter case we speak of strong coupling superconductivity.

15.2 Cooper Pair Breaking

The concept of pairing as represented in the ansatz (15.39) for the ground state can be generalized to systems which contain nonmagnetic scattering centers (impurities). Via spin-orbit scattering the electrons may relax not only their momenta but also their spins. Yet, the Hamiltonian of the electronic system still has time-reversal symmetry. It is useful to recall the concept of time inversion.

The time-dependent Schrödinger equation for a single electron

$$i\frac{\partial \psi(t)}{\partial t} = H\psi(t) \qquad (15.75)$$

remains unchanged under the replacement $t \to -t$, provided we take the complex conjugate of the wavefunction and replace i by $-i$. If we include the spin of the electrons, we have to require that the spin operators $\mathbf{s}$, like any angular momentum operator, anticommutes with the time-reversal operator $T_\mathbf{R}$, i.e.,

$$\mathbf{s}T_\mathbf{R} = -T_\mathbf{R}\mathbf{s} \quad . \qquad (15.76)$$

We set $\mathbf{s} = \frac{1}{2}\boldsymbol{\sigma}$, which implies that s_x and s_z are real operators, while s_y is purely imaginary. This follows from the form of the Pauli matrices. Equation (15.76) is not fulfilled for s_x and s_z if $T_\mathbf{R}$ is identified with the operator K, which changes a function into its complex conjugate. However, the equation is satisfied by the form

$$T_\mathbf{R} = -i\sigma_y K \quad , \qquad (15.77)$$

as can easily be checked.

Consider a one-electron Hamiltonian H_0 which has time reversal symmetry, i.e., for which $[H_0, T_\mathbf{R}] = 0$. When ψ_n is an eigenstate of H_0,

$$H_0 \psi_n = \epsilon_n \psi_n \quad , \qquad (15.78)$$

so is $\psi_{\bar{n}} = T_\mathbf{R} \psi_n$. The subscript n stands for the four quantum numbers of an electron whereas $\bar{n}$ denotes the time-reversed quantum numbers. The eigenvalue of $\psi_{\bar{n}}$ is the same as that of ψ_n, i.e.,

$$H_0 \psi_{\bar{n}} = \epsilon_n \psi_{\bar{n}} \quad . \qquad (15.79)$$

The generalized ansatz for the superconducting ground state is then

$$|\psi_0\rangle = \prod_n \left(u_n + v_n c_n^+ c_{\bar{n}}^+\right) |0\rangle \quad , \qquad (15.80)$$

i.e., the electrons are paired in time-reversed states (Anderson's theorem). It is intuitively clear that this concept breaks down in the presence of interactions, which break the time-reversal symmetry of the conduction electron system, but *not* of the total Hamiltonian. Examples are an external magnetic field

H or magnetic impurities with spin **S**. In the former case, the interaction Hamiltonian is

$$H_{\text{int}} = \frac{e}{2mc}\sum_i (\mathbf{p}_i\mathbf{A} + \mathbf{A}\mathbf{p}_i) - \mu_B \sum_i \boldsymbol{\sigma}_i\mathbf{H} \quad , \tag{15.81}$$

where **A** is the vector potential, while the $\mathbf{p}_i$ and $\boldsymbol{\sigma}_i$ are the electronic momenta and spins respectively. We notice that changing $\mathbf{p}_i \to -\mathbf{p}_i$ and $\boldsymbol{\sigma}_i \to -\boldsymbol{\sigma}_i$ does not leave any of the terms in H_{int} unchanged. The same holds true for the interaction

$$H_{\text{int}} = -J_{\text{ex}} \sum_i \boldsymbol{\sigma}_i \mathbf{S} \delta(\mathbf{R} - \mathbf{r}_i) \quad , \tag{15.82}$$

between a magnetic impurity with spin **S** at position **R** and conduction electrons at positions $\mathbf{r}_i$. Note that $T_\mathbf{R}$ is neither applied to **A** nor **H** nor to **S**.

Interactions which break time-reversal symmetry destroy superconductivity, provided they are sufficiently strong. If they become operative, either the net pairing interaction is reduced (pair weakening) or the Cooper pairs are partially broken (pair breaking). Determining which of the two cases prevails will depend on the long-time behavior of the correlation function $\langle T_\mathbf{R}^+(0)T_\mathbf{R}(t)\rangle$.

15.2.1 Ergodic vs. Nonergodic Perturbations

De Gennes has shown that the equation for the superconducting transition temperature can be related to the Fourier transform of the time-reversal correlation function

$$g(\epsilon) = \int \frac{dt}{2\pi} \langle T_R^+(0)T_R(t) \rangle e^{-i\epsilon t} \quad . \tag{15.83}$$

This relation is found in [82] and we state it here without proof

$$1 = N(0)V \int d\epsilon d\epsilon' \frac{1 - f(\epsilon) - f(\epsilon')}{\epsilon + \epsilon'} g(\epsilon - \epsilon') \quad . \tag{15.84}$$

When H_{int} is time-reversal invariant, i.e., when $[H_{\text{int}}, T_\mathbf{R}]_- = 0$, then $T_\mathbf{R}(t) = T_\mathbf{R}(0)$ and $g(\epsilon - \epsilon') = \delta(\epsilon - \epsilon')$. In that case (15.84) reduces to (15.72) with $\Delta = 0$.

When time-reversal symmetry is broken and $[H, T_\mathbf{R}] \neq 0$, the time evolution of the time-reversal operator $T_\mathbf{R}$ is given by

$$\frac{dT_\mathbf{R}}{dt} = i[H, T_\mathbf{R}]_- \quad . \tag{15.85}$$

Two distinct cases can arise with respect to the above correlation function, namely

(a) $\lim_{t\to\infty} \langle T_{\mathbf{R}}^+(0) T_{\mathbf{R}}(t)\rangle = \eta\ ;\quad 1 > \eta > 0\ ,$

(b) $\lim_{t\to\infty} \langle T_{\mathbf{R}}^+(0) T_{\mathbf{R}}(t)\rangle = e^{-2t/\tau_R}$ (15.86)

In case (a) we speak of nonergodic processes which lead to *pair weakening* while in case (b) we deal with ergodic behavior which results in *pair breaking*.

When pair weakening takes place the transition temperature is reduced to

$$k_B T_c = 1.14\ \omega_D e^{-1/(\eta N(0)V)}\ .$$ (15.87)

This is seen when (15.86) is set into (15.83) and (15.84). The effect of a non-time-reversal invariant interaction is just a reduction of the effective electron-electron interaction. If, however, ergodic behavior prevails we obtain from (15.86) and (15.83)

$$g(z) = \frac{1}{2\pi}\ \frac{\tau_R}{1 + z^2 \tau_R^2/4}\ .$$ (15.88)

With this expression the integrations in (15.84) have to be done. Here it is very useful to use the identity

$$\tanh\frac{\epsilon}{2k_B T} - 1 = 2k_B T \sum_n \frac{e^{i\omega_n \eta}}{\epsilon - i\omega_n}\ ,\ \eta \to +0$$ (15.89)

where $\omega_n = 2\pi k_B T(n + 1/2)$ and n runs over all (positive and negative) integers. It can be proven by comparing the poles and residues of both sides in the complex ϵ plane. With that identity we can express the Fermi function in terms of that sum and perform the integration. The result is

$$1 = N(0)V \sum_{n=0}^{\infty} \frac{1}{n + 1/2 + 1/(2\pi T_c \tau_R)}\ .$$ (15.90)

Note that here a cut off at ω_D is still missing so that the sum is divergent. After the cut off is introduced (15.90) becomes

$$\ln(T_c/T_{c0}) + \psi\left(\frac{1}{2} + \frac{1}{2\pi T_c \tau_R}\right) - \psi(1/2) = 0$$ (15.91)

where $\psi(x)$ is the digamma function [2]. The transition temperature T_{c0} is obtained when $H_{\text{int}} = 0$. The last relation was first derived by *Abrikosov* and *Gorkov* [3] when they treated the pair-breaking effect of paramagnetic impurities described by the interaction Hamiltonian (15.82) on superconductivity. One finds from (15.91) that T_c/T_{c0} drops continuously with increasing pair-breaking parameter $1/\tau_R$ and vanishes at a critical value of

$$\left(\frac{1}{\tau_R}\right)_{\text{crit}} = \frac{\pi T_{c0}}{2e^\gamma}\ .$$ (15.92)

A plot of T_c/T_{c0} as function of $(\tau_R T_{c0})^{-1}$ is shown in Fig. 15.6.

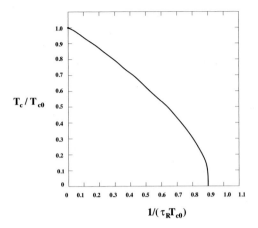

Fig. 15.6. Transition temperature T_c/T_{c0} as function of the pair-breaking parameter $(\tau_R T_{c0})^{-1}$ according to (15.91).

An important aspect of ergodic, time-reversal symmetry breaking interactions is that they may lead to gapless superconductivity. We have seen before that unconventional pairing may result in nodes (e.g., d-wave pairing) or node lines (e.g., p-wave pairing) of the order parameter. At these **k** points or lines the gap is zero. In distinction to unconventional pairing, gapless superconductivity implies a constant density of states at E_F and therefore a specific heat which is linear in T in the low temperature limit.

Gaplessness can be rederived from the behavior of the time-reversal operator T_R. Let us see under which circumstances the excitation energy E_n can be expanded for small order parameter Δ in the form

$$E_n = |\epsilon_n| + \left\langle |\Delta(\mathbf{r})|^2 \right\rangle \sum_m \frac{|\langle n|T_R|m\rangle|^2}{\epsilon_n - \epsilon_m} \tag{15.93}$$

where $|m\rangle, |n\rangle$ are single-electron states. When $[H_{\text{int}}, T_R]_- = 0$ the matrix elements $\langle n|T_R|m\rangle \neq 0$ only when $|m\rangle$ and $|n\rangle$ are time reversed. Then $\epsilon_n = \epsilon_m$ and the expression diverges. An expansion of that form is therefore not allowed. If, however, $[H_{\text{int}}, T_R]_- \neq 0$ then $\langle n|T_R|m\rangle \neq 0$ for a range of $|\epsilon_m - \epsilon_n| \simeq \tau_R^{-1}$. Replacing the quantum numbers n by $\mathbf{p}, \sigma$ we can write (15.93) in analogy to (15.84) as

$$E(\mathbf{p}) = |\epsilon(\mathbf{p})| + \left\langle |\Delta|^2 \right\rangle P \int \frac{d\epsilon'}{\epsilon(\mathbf{p}) + \epsilon'} \, g\left(\epsilon' - \epsilon(\mathbf{p})\right)$$

$$= |\epsilon(\mathbf{p})| + \frac{1}{2} \frac{\left\langle |\Delta(\mathbf{r})|^2 \right\rangle |\epsilon(\mathbf{p})|}{\epsilon(\mathbf{p})^2 + (\tau_R)^{-2}} \, . \tag{15.94}$$

An expansion of the form (15.93) is therefore possible. Note that there is no gap in $E(\mathbf{p})$ in the limit of small Δ. From the last relation we obtain for the density of states

$$N_S(E) = N(0)\left[1 + \frac{\Delta^2}{2}\frac{E^2 - (1/\tau_R)^2}{\left(E^2 + (1/\tau_R)^2\right)^2}\right] \quad (15.95)$$

This expression agrees with the one obtained by the more sophisticated Green's function method of *Abrikosov* and *Gorkov*.

We want to apply the above theory to practical cases and will presently discuss some examples of nonergodic versus ergodic behavior. To illustrate nonergodic behavior of time-reversal symmetry breaking we give two examples.

Consider a thin superconducting film of thickness d with a rough surface but no scattering centers inside the film when a magnetic field is applied parallel to the film [83]. In that case, we obtain from (15.81) and (15.85) for an electron

$$\frac{dT_R}{dt} = \frac{ie}{m}(\mathbf{pA} + \mathbf{Ap})T_R$$

$$= -i\frac{d\phi}{dt}T_R \quad (15.96)$$

where $\phi(t)$ is the phase of the propagating particle. When the electron moves ballistically from one surface of the film to the other, its phase does not change (the vector potential has only a component perpendicular to the film which is symmetric with respect to the film center). At the surface the electron is scattered into all directions because of the surface roughness. Therefore, the only contribution to dT_R/dt comes from the parts of the path before it hits the surface for the first time and after the last hit. Thus $\langle T_R^+(0)T_R(t)\rangle$ remains finite even in the limit $t \to \infty$. This results in pair weakening caused by the applied field and therefore we obtain (15.87).

A second example concerns a staggered field $h_\mathbf{Q}$ imposed onto the conduction electrons, i.e.,

$$H_{int} = -I\sum_{\mathbf{k},\mathbf{Q}} h_\mathbf{Q}\left(c^+_{\mathbf{k}\uparrow}c_{\mathbf{k}+\mathbf{Q}\uparrow} - c^+_{\mathbf{k}\downarrow}c_{\mathbf{k}+\mathbf{Q}\downarrow}\right) \quad . \quad (15.97)$$

The $\mathbf{Q}$'s are reciprocal lattice vectors of the magnetic lattice and are restricted to the first Brillouin zone. This situation occurs in antiferromagnetic superconductors. Superconductivity and antiferromagnetism (AF) need not exclude each other. Early examples were the ternary compounds $(RE)Mo_6S_8$ with RE = Nd, Gd, Tb, Dy, Er; $(RE)Mo_6Se_8$ with RE = Gd, Er and $(RE)Rh_4B_4$ with RE = Nd, Sm, Er, Tm. A representative for that class of materials is $TbMo_6S_8$ with a superconducting transition temperature of $T_c \approx 1.5$ K and a Néel temperature $T_N \approx 0.9$ K.[4] Later, the quaternary compounds of the family $(RE)Ni_2B_2C$ with RE = Ln, Er, Tm, Ho, Dy were added. Here $DyNi_2B_2C$

---
[4] for a review see [311]

with $T_c = 6$ K and $T_N = 10.6$ K is an example of $T_N > T_c$ [57, 341]. In the above cases antiferromagnetism is set up by *localized* $4f$ electrons, i.e., by electrons which differ from the paired conduction electrons. In distinction to the above, in systems in which antiferromagnetism and superconductivity are generated by the same itinerant electrons both phenomena work apparently against each other. Examples are the doped cuprates discussed in Sect. 15.4 and CeCu$_2$Si$_2$, where itinerant SDW antiferromagnetism is expelled by the onset of superconductivity [46, 342].

In the following we consider conduction electrons in an external staggered magnetic field. This corresponds to the first group of AF materials. The time-reversal operator $T_{\mathbf{R}}$ does not commute with (15.97). Instead it is

$$\frac{dT_R}{dt} = 2iH_{\text{int}}T_R \quad . \tag{15.98}$$

While $[H_{\text{int}}, T_R] \neq 0$ we find that the operator $Y = T_R R$ commutes with H_{int}, where R shifts the electron system by a vector connecting the two sublattices. Thus $[H_{\text{int}}, Y]_- = 0$. This is self-evident; after application of T_R the electron spins change sign which increases their energy in the staggered field. By shifting the electrons from one sublattice to the other the spin direction is again in line with the staggered field. This implies that when $\psi_{\mathbf{k}\sigma}(\mathbf{r})$ is an eigenfunction in the staggered field, then so is $Y\psi_{\mathbf{k}\sigma}(\mathbf{r})$ with the same eigenvalue. Therefore, we may pair $\psi_{\mathbf{k}\sigma}(\mathbf{r})$ with $e^{i\varphi}Y\psi_{\mathbf{k}\sigma}(\mathbf{r})$ where the phase φ is chosen by convenience [20]. Electrons forming spin-singlet Cooper pairs are preferably located on different sublattices (see Fig. 15.7). This way they can take advantage of the external staggered magnetic field. Clearly, this requirement affects their mutual attraction, e.g., through phonons. In most cases it leads to a reduction of this attraction. This results in a decreased upper critical magnetic fields H_{c2} below the onset of antiferromagnetic order [333, 509].

The above considerations can be generalized to more complex magnetic structures than common antiferromagnetism. For example, in ErNi$_2$B$_2$C or TmNi$_2$B$_2$C the magnetic structures which coexist or compete with superconductivity are transversely polarized incommensurate spin density waves [297]. It can be shown that the effect of a helical magnetic background on superconductivity like in HoNi$_2$B$_2$C is nearly identical to the effect of antiferromagnetism. The interaction between electrons via phonons is similarly reduced as in the case of pairing in antiferromagnetic Bloch states [5]. Since we are able to pair electrons properly, despite that $[H_{\text{int}}, T_R] \neq 0$, we may conclude that in the presence of the interaction (15.97) Cooper pairs are possibly weakened but not broken.

Now we turn our attention to interactions with ergodic behavior of the time-reversal symmetry correlation function. An example are paramagnetic impurities in a BCS superconductor. The interaction is given by (15.82) and the pair-breaking parameter is here

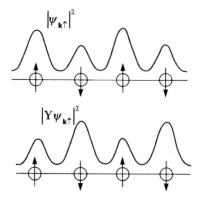

Fig. 15.7. Energetically degenerate electronic wave function in an antiferromagnetic lattice.

$$\frac{1}{\tau_R} = 2\pi n_I N(0) J_{\text{ex}}^2 S(S+1) \tag{15.99}$$

where n_I is the impurity concentration.

A second example is a magnetic field acting on only the electron orbits, i.e., neglecting the effect on the spins. When the mean free path ℓ is much less than the coherence length $\xi_0 = \hbar v_F/\Delta_0$ [82], i.e., when we are in the so-called dirty limit, the pair-breaking theory and therefore (15.91) and (15.95) do apply. The pair-breaking parameter is found to be of the form

$$\frac{1}{\tau_R} = \frac{\tau_{\text{tr}} v_F^2 e H}{3} \tag{15.100}$$

where τ_{tr} is the transport mean free time [81, 83, 305].

Further details concerning pair breaking are found in a comprehensive review by *Maki* [306].

15.2.2 Pairing Electrons with Population Imbalance

Of special interest is the effect of the Zeeman term in (15.81) on superconductivity. It describes the interaction of an internal or external magnetic field with the spins of the conduction electrons. We are faced here with the problem of pairing a different number of spin up and down electrons, i.e., particles with different chemical potential. Pairing of species with imbalanced populations takes also place in other fields of physics like in dense quark matter, nuclear physics or ultracold atoms [55].

Usually the effect of the field on the electron orbits, i.e., the paramagnetic or orbital term proportional to $(e/2mc)(\mathbf{pA}+\mathbf{Ap})$, is dominating the Zeeman effect. When the penetration depth of the magnetic field exceeds the superconducting coherence length ξ_0 (which is a measure of the spatial extent of the Cooper pairs), the effect of the orbital term results in an inhomogeneous superconducting state. The magnetic field penetrates the superconductor in the form of Abrikosov vortices with the latter forming a flux lattice (type II superconductors).

However, the orbital effect is not always dominant. When the mean free path is much less than the coherence length (dirty limit) the pair-breaking effect of the magnetic field due to the orbital term is strongly reduced and the Zeeman term may become more important. Another case is that of a superconducting thin film (i.e., of order 10 nm) in a parallel external magnetic field. Here the effect of the field on the orbits is again strongly reduced because the cyclotron motion of the electrons is limited by the thickness of the film; when the film is thinner, then the limitation on the cyclotron motion of the electrons increases. Finally, in superconductors with heavy quasiparticles the orbital effect is small because of the large quasiparticle mass. The inverse of it enters the interaction term. In all these cases the Zeeman term may become dominant.

We assume in the following that the Zeeman term dominates the paramagnetic one. We will see that also the Zeeman term can lead to an inhomogeneous superconducting state which is quite different from the Abrikosov flux lattice. In the following we limit ourselves to a reduced interaction Hamiltonian of the form

$$H_{\text{int}} = -\mu_{\mathbf{B}} \sum_i \sigma_i \mathbf{H} \quad . \tag{15.101}$$

Then the quasiparticle excitation spectrum is simply

$$E_\sigma(\mathbf{k}) = \sqrt{\epsilon^2(\mathbf{k}) + \Delta_0^2} - \mu_{\mathbf{B}} \sigma_z H \quad . \tag{15.102}$$

The Zeeman interaction (15.101) splits the quasiparticle density of states per spin direction. From (15.102) we find that

$$N_\sigma(E) = N(0)\text{Re}\left\{\frac{|E - \mu_{\mathbf{B}}\sigma H|}{\sqrt{(E - \mu_{\mathbf{B}}\sigma H)^2 - \Delta_0^2}}\right\} \quad ; \quad \sigma = \pm 1 \quad . \tag{15.103}$$

The total density of states is therefore a superposition of the BCS density of states

$$N_{\text{BCS}}(E) = N(0)\text{Re}\left\{\frac{|E|}{\sqrt{(E^2 - \Delta_0^2)}}\right\} \tag{15.104}$$

shifted by $\pm\mu_{\mathbf{B}}H$. This effect has been observed experimentally by tunnelling measurements on thin Al films in a parallel magnetic field [319]. In most cases, the conduction-electron spin direction is not a good quantum number due to the presence of spin-orbit interaction. We characterize the magnitude of the latter by a spin-orbital scattering rate τ_{so}^{-1}. When this rate is small, i.e., when

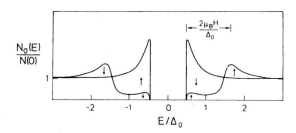

Fig. 15.8. Mixing effect of the spin-orbit scattering rate $\tau_{so}^{-1} \ll \Delta_0$ on the spin-split quasiparticle density of states.

$\tau_{so}\Delta_0 \gg 1$, the two spin-split densities of states are only slightly mixed (Fig. 15.8). In the case of large spin-orbit scattering, i.e., $\tau_{so}\Delta_0 \ll 1$, we expect a form which approaches that of (15.104). The Zeeman term has little influence in that case.

The Zeeman term may also cause reentrant superconductivity. Consider a superconductor containing magnetic impurities with total angular momentum J of concentration n_I. Furthermore, assume that the impurities interact ferromagnetically with each other and that n_I is sufficiently large, so that superconductivity is suppressed by them. Then the Zeeman energy in the presence of an applied magnetic field leads to an interaction term

$$H_{\text{int}} = -\sum_i \sigma_i \left(\mu_B H - n_I(g_J - 1)J_{\text{ex}}\langle J_z\rangle\right) \quad . \tag{15.105}$$

Here g_J is the Landé factor. The external field and the internal field due to the impurity polarization may therefore partially cancel each other with the result that Cooper pairing no longer needs to be suppressed (Jaccarino-Peter effect). The application of a magnetic field may thus cause a transition from a normal into a superconducting state, which is indeed an unexpected feature. However, if the magnetic field becomes too large, superconductivity is again destroyed by the field. The effect has indeed been observed (see Fig. 15.9). In passing we want to mention that reentrant superconductivity has also been observed for different reasons in superconductors with Kondo impurities [310, 336].

Similarly as the orbital term in (15.81) can give rise to an inhomogeneous superconducting state in the form of a flux lattice, so can the Zeeman term also give rise to inhomogeneous superconducting states, often referred to as FFLO states [132, 270]. They require though, that the parameters of the system are appropriate. This is seen as follows.

Consider the BCS ground state in the presence of the Zeeman term (15.101). From (15.102) it is seen that for $\mu_B H < \Delta_0$ the excitation energy is always positive. The ground state cannot take advantage of the Zeeman term because electrons are paired with opposite spins. As long as $\mu_B H < \Delta_0$ they remain paired, i.e., a self-consistent gap function is found. Yet, the normal

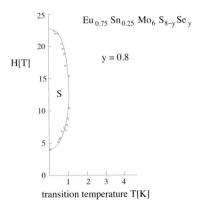

Fig. 15.9. Field induced superconductivity based on the Jaccarino-Peter effect (From [324]). The effect was also observed in $\lambda-(BETS)_2\ FeCl_4$ [459].

state lowers its energy according to

$$\Delta E_H = -\frac{\chi_0}{2}H^2$$
$$= -N(0)\mu_B^2 H^2 \quad (15.106)$$

where χ_0 is Pauli's spin susceptibility. When ΔE_H exceeds the superconducting condensation energy, the energy of the normal state becomes lower than that of the superconducting state; this is given by (15.60). Therefore, at $\frac{\mu H_{Cl}}{\Delta} = \frac{1}{\sqrt{2}}$ the superconducting state is expected to go over by a first-order phase transition into the normal state. The critical field H_{Cl} is called the Clogston limit. Note that this limit holds true in the absence of spin-orbit scattering only and is modified when $\tau_{s0}^{-1} \neq 0$. In the following we want to show that by modifying the Cooper pairing, a superconductor can also respond to the Zeeman term and lower its energy even when $\tau_{s0}^{-1} = 0$. This is possible when the pairing momentum is included in the considerations.

Assume that in the absence of an external field we pair electrons with a pairing momentum $2\mathbf{q}$. This corresponds to pairing electrons on a shifted Fermi sphere (see Fig. 15.10a) and implies a supercurrent flowing. The excitation energies are in this case

$$E(\mathbf{k}) = \sqrt{\epsilon^2(\mathbf{k}) + \Delta^2} - v_F q \cos\vartheta \quad (15.107)$$

where ϑ is the angle between $\mathbf{k}$ and $\mathbf{q}$. When $qv_F/\Delta_0 > 1$ where Δ_0 is the BCS gap, electrons on one side of the shifted Fermi sphere start to depair as indicated in Fig. 15.10b. The $\mathbf{k}$ states with unpaired electrons are blocked for virtual pair scattering in (15.44). Therefore, Δ is reduced and becomes a function of q, i.e., $\Delta(q)$. The depaired region in $\mathbf{k}$ space is defined by requiring that at its surface $E(\mathbf{k}) = 0$.

It turns out that superconductivity is destroyed for $q > q_s$ where $q_s = 1.36\Delta_0/v_F$. We want to stress that even in the presence of unpaired electrons

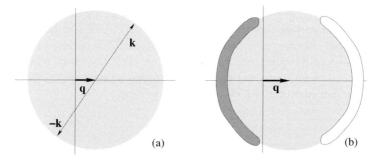

Fig. 15.10. (a) Pairing in the presence of a supercurrent, given by the shifted Fermi sphere. (b) Depairing takes place when $1 < qv_F/\Delta_0 < 1.36$. Black area: unpaired electrons. Light area: unoccupied states. At the rim of those regions $E(\mathbf{k}) = 0$.

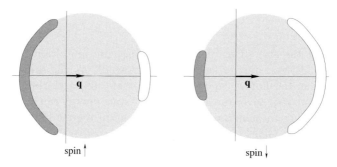

Fig. 15.11. Depairing in the presence of a Zeeman energy. The depaired regions differ for spin up and down. Dark areas: unpaired electrons. Gray areas: occupied states according to (15.52). The total current is zero. The spin current is nonzero.

the corresponding current is still nondissipative since elastic scattering would increase the energy of the partially depaired state.

Next we consider the effect of the Zeeman term in the presence of a finite pairing momentum. The excitation energies are now given by

$$E_\sigma(\mathbf{k}) = \sqrt{\epsilon^2(\mathbf{k}) + \Delta^2} - qv_F \cos\vartheta - \mu_B \sigma H \quad . \tag{15.108}$$

We can again search for a region of depaired electrons in $\mathbf{k}$ space by requiring that at its surface $E_\sigma(\mathbf{k}) = 0$. This region is now depending on q and H and differs for electrons with spin σ equal ± 1 (see Fig. 15.11). The two parameters q and H are reduced to one by requiring that in the ground state the total current must vanish. Such a region does indeed exist and must be excluded from pair scattering (blocking effect). Due to the unpaired electrons, the spin susceptibility of the superconducting state differs from zero and the state can lower its energy in the presence of the Zeeman term. A self-consistent determination of $\Delta(H)$ and computation of the ground-state energy shows that a superconducting state with finite pairing momentum q can have a lower energy than the BCS superconductor as well as the normal state. For a spherical Fermi surface that regime is given by $0.71 < \frac{\mu_B H}{\Delta_0} < 0.76$ and hence is rather small. In one- and two dimensions the stable regime of the FFLO state is considerably extended[5]. For example, in 1D it covers the regime $0.71 < \mu_B H/\Delta_0 < \infty$. A finite pairing momentum implies a position-dependent phase of the order parameter Δ which here is of the form

$$\Delta(\mathbf{r}) = \bar{\Delta} e^{2i\mathbf{q}\mathbf{r}} \quad . \tag{15.109}$$

Thus we are dealing here with a spontaneous translational symmetry breaking and with an inhomogeneous superconducting state. Clearly, there is a priori no preferred direction of $\mathbf{q}$. Therefore, the order parameter is degenerate with respect to the direction of $\mathbf{q}$ and is generally of the form

$$\Delta(\mathbf{r}) = \sum_{\nu} \bar{\Delta}_{\nu} e^{2i\mathbf{q}_{\nu} \mathbf{r}} \quad , \tag{15.110}$$

with $\mathbf{q}_{\nu}$ being the star of $\mathbf{q}$. Strictly speaking, the limitation to the lowest harmonics holds true only when the $\bar{\Delta}_{\nu}$ are small. As the order parameter increases, higher harmonics can become important [69]. Determining which of the combination of $\mathbf{q}_{\nu}$ gives the lowest ground-state energy can be investigated either by studying the nonlinear Ginzburg-Landau regime [55] or by solving the Bogoliubov–de Gennes equations, which are a generalization of the BCS equations (15.45–15.57) to a spatial-dependent order parameter $\Delta(\mathbf{r})$ [82, 298]. A suggestive choice is

$$\Delta(\mathbf{r}) = \bar{\Delta} \cos 2\mathbf{q}\mathbf{r} \quad , \tag{15.111}$$

but triangular, octahedral or cubic states are possible as well to mention some of them [43, 412]. The best combination, i.e., the one leading to the lowest energy will depend on the form of the BCS pairs (s-wave, d-wave or else), on the shape of the Fermi surface and on the specific material.

It is instructive to approach the inhomogeneous superconducting state from the Ginzburg-Landau regime, i.e., when the order parameter is small and given by (15.109). The free energy near a second-order phase transition can be expanded in the form

$$F(\mathbf{r}) = \alpha(T) \left| \Delta(\mathbf{r}) \right|^2 + a(T, H) \left| \boldsymbol{\nabla} \Delta(\mathbf{r}) \right|^2 + \frac{b}{2}(T, H) \left| \boldsymbol{\nabla}^2 \Delta(\mathbf{r}) \right|^2$$
$$+ \text{ higher order terms} \quad . \tag{15.112}$$

As long as only the first three terms are considered, the right-hand side is independent of $\mathbf{r}$. We have included here a gradient term $\boldsymbol{\nabla}\Delta(\mathbf{r})$ without a vector potential in order to allow for a finite pairing momentum. A vector potential is not required because the effect of the magnetic field on the electron orbits is neglected. For a ground state with finite pairing momentum

---

[5] see, e.g., [47, 100, 189, 412]

$a(T, H) < 0$, i.e., the free energy is lowered by a finite value of q. This value is stabilized by the inclusion of a higher-order gradient proportional to $b(T, H)$. Furthermore, we neglect here a weak dependence of the coefficient $b(T, H)$ on the gradient of $\Delta(\mathbf{r})$, which would lift the degeneracy contained in (15.109). The q value which minimizes the free energy is obtained from

$$(2q)^2 = -a(T, H)/b(T, H) \ . \tag{15.113}$$

When $a(T, H) > 0$ the ground state is homogeneous. In that case the coefficient $\alpha(T) = c(T - T_{c0})$ with $c > 0$ where T_{c0} is the transition temperature for zero pairing momentum. The linear dependence of $\alpha(T)$ on $(T - T_{c0})$ follows by minimizing the free energy given by (15.112) when $q = 0$.

In the inhomogeneous phase the transition temperature is modified depending on the value of H. When $\Delta(\mathbf{r})$ is of the form of (15.109) then close to T_c the free energy is minimized by

$$c(T - T_{c0}) + a(2q)^2 + \frac{b}{2}(2q)^4 = c(T - T_c) \tag{15.114}$$

or using (15.113) by

$$c(T - T_{c0}) - \frac{a^2(T_c, H)}{b(T_c, H)} = c(T - T_c) \ . \tag{15.115}$$

Therefore, the increase in transition temperature is

$$T_c - T_{c0} = \frac{a^2}{2b^2 c} \ . \tag{15.116}$$

Explicit expressions for the coefficients $a(T, H)$ and $b(T, H)$ are found, e.g., in [270]. By including terms up to order $|\Delta(\mathbf{r})|^6$ studies of 23 different crystalline structures show that an octahedron with $|\mathbf{q}_m| = q$, $m = 1, \ldots, 8$ gives the lowest energy.

For $T_c/T_{c0} < 0.56$ the free energy is minimized and the highest values of H are obtained for $q \neq 0$. We want to draw attention to the observation that an inhomogeneous superconducting state requires a mean free path which is long when compared with the superconducting coherence length ξ_0. Otherwise the inhomogeneous state with a characteristic wavelength of the order of the coherence length cannot form. Superconductors of low dimension or with heavy quasiparticles are good candidates for its occurrence. As mentioned before, in both cases the effect of a magnetic field on the electron orbit is strongly reduced: in the first case when the field is parallel, e.g., to a layered system, and in the second because the heavy quasiparticles mass appears in the denominator of the vector potential term $(e/mc)\mathbf{p} \cdot \mathbf{A}$ of the Hamiltonian. Superconductors with heavy quasiparticle have in addition a small coherence length ξ_0 since v_F is small. Indeed, FFLO states seem to have been observed in λ-(BETS)$_2$GaCl$_4$ [445], λ-(BETS)$_2$FeCl$_4$ [460], (TMTSF)$_2$ClO$_4$ [501] and

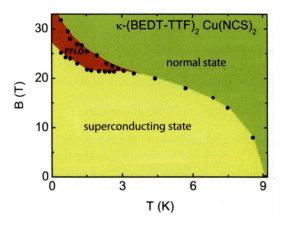

Fig. 15.12. Phase diagram of the organic superconductor κ-(BEDT-TTF)$_2$ Cu(NCS)$_2$ as obtained from specific heat measurements in a magnetic field. (After [291, 390])

most clearly in κ-(BEDT-TTF)$_2$Cu(NCS)$_2$ [291, 417]. The measured phase diagram of the latter material is shown in Fig. 15.12.

An interesting case is CeCoIn$_5$, a d-wave superconductor. This material has a layered structure. In a magnetic field parallel to the layers one finds in the high-field regime signatures of a FFLO phase [34, 384]. Nuclear magnetic resonance (NMR) and neutron-scattering experiments have shown that there is also a magnetic-field induced incommensurate SDW appearing in that regime [235, 258]. The induced staggered magnetization is thereby pointing perpendicular to the layers. It disappears when the field is increased beyond $H_{c2}(T)$, at which the sample goes over into the normal state. This suggests that the superconducting order parameter is the principle one and that the SDW is driven by it. Indeed, it has been shown that via mode coupling a superconductor with a d-wave order parameter in the FFLO state $\Delta_{\mathbf{q}}$ induces an equal spin, odd parity order parameter $\Delta_{-\mathbf{Q}_0}$ with pairing momentum $-\mathbf{Q}_0$ together with an SDW magnetization in c direction $M_{\mathbf{Q}_0+\mathbf{q}}$. Here $\mathbf{Q}_0 = (0.5, 0.5, 0.5)$ in units of the reciprocal lattice vectors [326, 495]. We want to discuss this point in some more detail. The Ginzburg-Landau free energy which includes mode-mode couplings is of the form

$$F = F_0 + \frac{a_M}{2} M^2_{\mathbf{Q}_0+\mathbf{q}} + \frac{b_M}{4} M^4_{\mathbf{Q}_0+\mathbf{q}} + \frac{a_0}{2} \Delta^2_{-\mathbf{Q}_0} + \frac{b_0}{4} \Delta^4_{-\mathbf{Q}_0}$$
$$+ C \Delta_{\mathbf{q}} M_{\mathbf{Q}_0+\mathbf{q}} \Delta_{-\mathbf{Q}_0} \quad . \tag{15.117}$$

The term F_0 is the free energy of the FFLO state. Since $\Delta_{-\mathbf{Q}_0}$ and $M_{\mathbf{Q}_0+\mathbf{q}}$ are induced by the order parameter of a FFLO state, they must vanish when $\Delta_{\mathbf{q}} = 0$. That implies that the prefactors a_M, b_M, a_0, b_0 are all positive. The mode coupling coefficient C can be obtained from the corresponding Feynman diagram [326]. Here we simply take it as given. Minimizing F we obtain

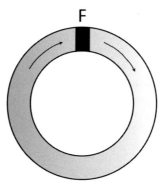

Fig. 15.13. A π junction consisting of a superconducting ring of length L (e.g., of Nb) and a sandwiched thin ferromagnetic field F. The order parameter has a different sign at the two sides of F.

$$0 = \frac{\partial F}{\partial M_{\mathbf{Q}_0+\mathbf{q}}} = a_M M_{\mathbf{Q}_0+\mathbf{q}} + b_M M_{\mathbf{Q}_0+\mathbf{q}}^3 + C\Delta_{\mathbf{q}}\Delta_{-\mathbf{Q}_0}$$

$$0 = \frac{\partial F}{\partial \Delta_{-\mathbf{Q}_0}} = a_0 \Delta_{-\mathbf{Q}_0} + b_0 \Delta_{-\mathbf{Q}_0}^3 + C\Delta_{\mathbf{q}} M_{\mathbf{Q}_0+\mathbf{q}} \quad . \quad (15.118)$$

For sufficiently small a_0 and a_M this leads to

$$M_{\mathbf{Q}_0+\mathbf{q}}^2 = \frac{1}{b_M}\left[\frac{(C\Delta_{\mathbf{q}})^2}{a_0} - a_M\right] \quad (15.119)$$

and explains in a simple way why in CeCoIn$_5$ a SDW with wave vector $\mathbf{Q} = \mathbf{Q}_0 + \mathbf{q}$ is induced by a FFLO state.

It seems that the most important realization of FFLO-like states is in superconducting π junctions [48, 50]. They have considerable potential for applications in superconducting circuitry. The π junctions are produced by sandwiching a thin ferromagnetic layer between two superconductors with s-wave pairing. A π junction has the property that the superconducting order parameter has different signs at the two sides of the junction. Therefore, the phase difference of the order parameter at the two sides is $\pm\pi$ since $-1 = e^{\pm i\pi}$. When such a junction is part of a superconducting ring (see Fig. 15.13) the wavefunction or order parameter within the ring must compensate for this phase difference when a path is taken within the ring. Thus, the order parameter Δ has a phase $\exp(2iqx)$ with $2q = \pi/L$, where L is the circumference of the ring. However, $\Delta(x) \sim \exp(2iqx)$ implies a ground state with a supercurrent, provided that the inductance of the ring is sufficiently small (otherwise setting up a phase-compensating current costs too much energy).

As mentioned earlier a π-shift of the phase of the superconducting order parameter can be achieved when a sufficiently thin ferromagnetic layer is sandwiched between two conventional s-wave superconductors. A combination of

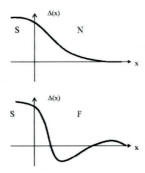

Fig. 15.14. Schematic plot of $\Delta(x)$ near a superconductor-normal interface due to the proximity effect. (a) when the normal metal is nonmagnetic (N) and (b) when it is a ferromagnet (F). The oscillatory behavior results from a FFLO-like state. (From [49])

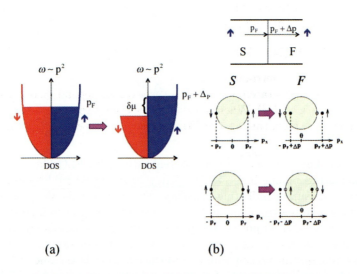

Fig. 15.15. (a) Unbalanced spin population due to a difference $\delta\mu$ in the chemical potential of a ferromagnetic thin film when compared with a paramagnetic state. DOS denotes the density of states. (b) Increase (decrease) in momentum by $\Delta_\mathbf{p}$ as an electron with spin ↑ (↓) moves from left to right into the ferromagnet. (From [87])

a Cu-Ni alloy for the ferromagnet and Nb for the superconductor works particularly well here. In a sandwich of this kind Cooper pairs leak from Nb into the ferromagnetic film. Their amplitude, however, decreases exponentially in the sandwiched film. Therefore, the latter has to meet thickness requirements depending on mean free path, spin-orbit interactions etc. in order to ensure that the superconducting order parameter remains finite in the sandwiched film. The order parameter may change its sign in the ferromagnetic film according to (15.109 - 15.110) because of the exchange field which is present (see Fig. 15.14). This can be seen in more detail with a nice and simple argument. Consider an interface between a superconductor and a ferromagnet, e.g., Nb and a Cu-Ni alloy. Imagine an electron with spin ↑ and momentum $\mathbf{p}$ near p_F passing from Nb through the interface into the ferromagnetic layer. In the ferromagnet its momentum will increase by $\Delta\mathbf{p}$ because of the increase in Fermi momentum due to spin population imbalance (see Fig. 15.15). This electron is paired with a spin ↓ electron of momentum $-\mathbf{p} + \Delta\mathbf{p}$, because when that electron moves from the ferromagnetic layer into Nb it ends up there with momentum $-\mathbf{p}$. On the other hand, when a spin ↓ electron with momentum $\mathbf{p}$ moves from Nb into the ferromagnetic layer it looses momentum $-\Delta\mathbf{p}$ and therefore is paired with a spin ↑ electron of momentum $-\mathbf{p} - \Delta\mathbf{p}$ (see again Fig. 15.15). Consequently, we find

$$\uparrow\downarrow \;\to\; \uparrow\downarrow e^{2i\Delta p x} \quad ; \quad \downarrow\uparrow \;\to\; \downarrow\uparrow e^{-2i\Delta p x} \tag{15.120}$$

where x is normal to the interface. Thus, the singlet $(\uparrow\downarrow - \downarrow\uparrow)$ in Nb goes over in the ferromagnetic film into

$$(\uparrow\downarrow - \downarrow\uparrow) \;\to\; (\uparrow\downarrow - \downarrow\uparrow)\cos 2\Delta p \cdot x + i(\uparrow\downarrow + \downarrow\uparrow)\sin 2\Delta p \cdot x \tag{15.121}$$

indicating that in the ferromagnet an odd-parity spin triplet superconducting component is also generated. Its strength turns out to depend on details of the boundary (see, e.g., [111]). When ϕ_s denotes the even-parity pair wavefunction in the superconductor, then in the ferromagnet the corresponding part is

$$\phi_F(\mathbf{x}) = \alpha \cos 2qx \cdot e^{-x/\xi} \phi_s \tag{15.122}$$

where α depends on the boundary conditions and ξ is a characteristic decay length for Cooper pairs. The momentum $q = \Delta p$ depends on the exchange field in the ferromagnet like $q = \mu_B H_{\text{ex}}/v_F$. A detailed discussion of the modifications caused by potential- and spin-orbit scattering as well as of the proper boundary conditions is found, e.g., in Refs. [87, 211].

The above arguments can be quantified by looking at the linearized equation for $\Delta(x)$. This is possible since the induced superconductivity in the ferromagnetic film is weak. First, we assume a contact at $x = 0$ between a superconductor and a nonmagnetic normal metal. From (15.112) we obtain for the minimum of the free energy and $x > 0$

$$\alpha\Delta(x) - a\frac{d^2\Delta(x)}{dx^2} + \frac{b}{2}\frac{d^4\Delta(x)}{dx^4} = 0 \;. \tag{15.123}$$

Since the film is nonmagnetic, it is $a > 0$ and we find the solution

$$\Delta(x) = \Delta_I e^{-kx} \tag{15.124}$$

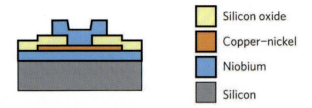

Fig. 15.16. Schematic cross-section through a π-junction consisting of a Nb-Cu/Ni-Nb sandwich. (From [116])

where $k = \sqrt{\alpha/a}$ and $\Delta_I = \Delta(x = 0)$ is the order parameter at the interface. The b term is unimportant here. However, for a ferromagnetic film in a FFLO-like state it is $a < 0$ and we find a complex wave vector $k = k_1 + ik_2$ with

$$k_1^2 = \frac{|a|}{2b}\left[\left(1 + \frac{T - T_c}{T_c - T_{c0}}\right)^{1/2} - 1\right]$$

$$k_2^2 = \frac{|a|}{2b}\left[1 + \left(1 + \frac{T - T_c}{T_c - T_{c0}}\right)^{1/2}\right] \quad . \tag{15.125}$$

If the order parameter in the superconductor is chosen to be real, so is the one in the ferromagnetic film and

$$\Delta(x) = \Delta_I e^{-k_1 x} \cos k_2 x \quad . \tag{15.126}$$

It is seen from (15.126) that when T is close to T_c, then $k_1 \to 0$ while k_2 reduces to $k_2^2 = |a|/b$. A comparison with (15.113) shows explicitly the relation to the FFLO state. The above arguments are qualitative, since we have not included any boundary conditions at the interface [238]. A detailed discussion of this situation is found in the review [49]. In Fig. 15.16 we show schematically the cross-section of a π-junction. Those junctions can be used in superconducting circuitry, e.g., as single-flux quantum cells, frequency binary dividers or self-biased phase qubits.

Before closing we want to draw attention to important progress which has been made by trapping ultracold fermionic atoms like ^{6}Li in optical lattices [236]. With the help of the Feshbach resonance and by populating the two lowest hyperfine levels differently, one can achieve proper conditions for the appearance of an FFLO state [515]. Indeed, there is evidence that in a one-dimensional optical lattice filled with ^{6}Li atoms this state has been observed [281].

Pairing states with finite pairing momentum have also been discussed in QCD where in dense matter (neutron stars) u and d quarks with different chemical potentials are expected to pair. A similar situation can arise in nu-

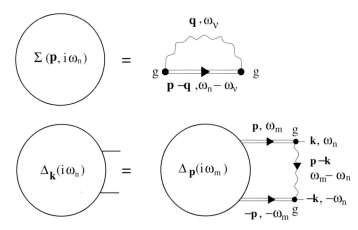

Fig. 15.17. Eliashberg's equations for the determination of T_c. They are set up for the electron self-energy $\Sigma(\mathbf{p}, i\omega_n)$ and for the order parameter $\Delta_\mathbf{k}(i\omega_n)$ in form of diagrams. A double line symbolizes the full Green's function $G(\mathbf{p}, i\omega_n)$ in the normal state. The wavy line is a boson propagator which causes the interactions between electrons.

clear physics where pairing of particles with different isospins can take place. For more details we refer to a number of reviews[6].

15.3 Cooper Pairing without Phonons

Until now, we have not discussed the origin of the attractive electron-electron interaction which leads to Cooper pairing. As briefly mentioned earlier, in most conventional superconductors the electron-phonon interaction causes that attraction. Its significance for superconductivity was realized after the discovery that in many superconductors the transition temperature depends on the isotope mass M of the involved atoms, i.e., $T_c \sim M^{-1/2}$. Originally this was puzzling, since the electron-phonon interaction is known to be small when compared with electron-electron repulsions; how could it lead to an electron attraction? However, careful investigations showed that when screening of the electron and lattice system is included, there remains a net attraction in some domains of space and time [22]. Clearly, the interaction between electrons via phonons is a retarded one. When an electron moves through a lattice it attracts the positively charged ions. Since they are heavier than the electrons, the positive charge accumulation caused by the electron follows the electron motion with retardation. It attracts other electrons and is the source of Cooper-pair formation.

The afore mentioned retardation has been set aside in the BCS theory but is included in Eliashberg's equations. These equations take into account

---

[6] see [49, 55, 513]

that electrons in time-reversed states can exchange phonons (more generally bosons) and also that an electron can emit and reabsorb a phonon, acquiring in this way an enhancement of its effective mass m^*. We can easily write down these equations in a pictorial way, i.e., in terms of diagrams. This is shown in Fig. 15.17. The temperature dependent Green's function formalism is applied here (see Sect. 7.1). We are only interested in the determination of the transition temperature T_c. In only that case does the Green's function $G(\mathbf{k}, \omega_n)$ in the normal state appear, i.e.,

$$G(\mathbf{k}, i\omega_n) = \frac{1}{i\omega_n - \epsilon(\mathbf{k}) - \Sigma(\mathbf{k}, i\omega_n)} \quad . \tag{15.127}$$

The wavy lines describe the boson propagator which in the case of phonons is

$$D(\mathbf{q}, \omega_\nu) = -\frac{\omega_\mathbf{q}^2}{\omega_\nu^2 + \omega_\mathbf{q}^2} \quad ; \quad \omega_\nu = 2\pi T \nu \tag{15.128}$$

where $\omega_\mathbf{q}$ stands for the phonon dispersion. Furthermore, g is the coupling constant, in this case between electrons and phonons and $\Delta(\mathbf{p}, \omega_n)$ is the superconducting order parameter. Due to retardation of the effective electron-electron interactions the order parameter is now frequency dependent. The diagrams in Fig. 15.17 are written in form of equations for T_c as follows,

$$\Sigma(\mathbf{p}, i\omega_n) = g^2 T \sum_\nu \int d^3q\, D(\mathbf{q}, i\omega_\nu) G(\mathbf{p} - \mathbf{q}, i(\omega_n - \omega_\nu))$$

$$\Delta_\mathbf{k}(i\omega_n) = -g^2 T \sum_m \int d^3p\, D(\mathbf{p} - \mathbf{k}, i(\omega_m - \omega_n))$$
$$\times G(\mathbf{p}, i\omega_m) G(-\mathbf{p}, -i\omega_m) \Delta_\mathbf{p}(i\omega_m) \quad . \tag{15.129}$$

The diagram for the electron self-energy $\Sigma(\mathbf{p}, i\omega_n)$ is self-explanatory. The one for the frequency dependent order parameter $\Delta_\mathbf{k}(i\omega_n)$ accounts for the generation of an electron pair in time reversed states out of the vacuum (remember that $\Delta_\mathbf{k} \sim \langle c_{\mathbf{k}\sigma}^+ c_{-\mathbf{k}-\sigma}^+ \rangle$). The pairing interaction is a retarded one and represented by a space and time dependent bosonic propagator. The same one is used for computing the self-energy.

There are computer programs available which solve Eliashberg's equations numerically for given $\epsilon(\mathbf{p})$, $\omega_\mathbf{q}$ and g and find the corresponding value of $T = T_c$. It is noticed that the equation for $\Delta(\mathbf{p}, i\omega_n)$ generalizes (15.71) in the limit $\Delta_\mathbf{k} \to 0$. In the following we show that bosons other than phonons can provide for electron-electron attractions and hence Cooper-pair formation.

Although some of the long-known superconductors show an isotope effect only in a much reduced form, the issue of other than electron-phonon interactions received major attention only after the discovery of high-T_c superconductors [7]. In the CuO_2 planes of the cuprates, electron correlations are strong. Therefore, considerable efforts went into studying the Hubbard model on a square lattice as the simplest realization of a strongly correlated electron system. It was shown by numerical studies on finite systems that a Hamiltonian of the form (9.22) may indeed have a superconducting ground state in a certain parameter range of U, t and particle numbers per site [400]. We will deal with pairing in the high-T_c cuprates in more detail in Sect. 15.5. Those systems are examples where superconductivity occurs in the vicinity of another second-order phase transition. At such an instability either the density susceptibility $\chi_n(\mathbf{q},\omega)$ or spin susceptibility $\chi_s(\mathbf{q},\omega)$ diverges depending on whether the phase transition is structural or magnetic. It is suggestive that under these circumstances density or spin fluctuations are contributing to the Cooper-pair formation.

Above, it has been assumed that the susceptibilities refer to the same electron system which forms Cooper pairs. In the following, we consider examples where the bosonic excitations are provided by localized electrons, in particular f electrons in incomplete $4f$ or $5f$ shells. Here experiments and theory occupy firmer ground so that much more definite statements can be made than for the high-T_c materials.

15.3.1 Filled Skutterudite PrOs$_4$Sb$_{12}$

The filled skutterudites La$_{1-x}$Pr$_x$Os$_4$Sb$_{12}$ are superconductors with transition temperatures $T_c = 0.74$ K for LaOs$_4$Sb$_{12}$ and $T_c = 1.85$ K for PrOs$_4$Sb$_{12}$. The order parameter very likely has isotropic s-wave symmetry with an element of uncertainty remaining. The crystal has tetrahedral site symmetry T_h. The only difference between LaOs$_4$Sb$_{12}$ and PrOs$_4$Sb$_{12}$ are $4f^2$ electrons of Pr^{3+} since La^{3+} has an empty $4f$ shell. Furthermore, PrOs$_4$Sb$_{12}$ has an enhanced mass $m^*/m_b \simeq 2.5$ when compared with the computed band mass m_b and a large jump in the specific heat at T_c, i.e., $\Delta C/T_c \simeq 500$ mJ/(mol K^2), a hallmark of heavy quasiparticles. The increase in T_c by more than a factor of two must be caused by the $4f$ electrons of Pr^{3+} since the lattice vibrations are barely affected by the small mass difference between La and Pr. In order to understand the effect of localized $4f$ electrons on superconductivity consider first a Pr^{3+} impurity in a superconductor. According to Hund's rules the ground-state multiplet of a $4f^2$ system has the total angular momentum J = 4. The two most important interactions with conduction electrons are the isotropic exchange interaction

$$H_{\rm ex} = -2\left(g_J - 1\right) J_{\rm ex} \sum_{\mathbf{k}q\sigma\sigma'} (\mathbf{s}_{\sigma'\sigma}\mathbf{J})\, c^+_{\mathbf{k}-\mathbf{q}\sigma'} c_{\mathbf{k}\sigma} \qquad (15.130)$$

and the aspherical charge scattering

$$H_{\rm AC} = \left(\frac{5}{4\pi}\right)^{1/2} \sum_{kk'\sigma} \sum_{m=-2}^{+2} I_2(k's;kd) Q_2 \left[Y_2^m(\mathbf{J}) c^+_{k's\sigma} c_{kdm\sigma} + h.c. \right] \; . \qquad (15.131)$$

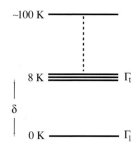

Fig. 15.18. CEF energy levels as obtained from inelastic neutron scattering (see [151, 262]).

In the first equation g_J is the Landé factor and $J_{\rm ex}$ is the exchange coupling constant. In the second equation Q_2 is the quadrupole moment of the Pr^{3+} ions and the definition of the Coulomb integrals $I_2(k's; kd)$ is found, e.g., in [133]. The $c_{kdm\sigma}$ destroy a conduction electron with momentum $k = |\mathbf{k}|$, in a $\ell = 2$ state with azimuthal quantum number m and spin σ while $c^+_{k's\sigma}$ creates an electron with momentum k' in a $\ell = 0$ state. The operators $Y_2^m(J)$ are given by

$$\begin{aligned} Y_2^0 &= (2/3)^{1/2}\left[3J_z^2 - J(J+1)\right]/N_J \\ Y_2^{\pm 1} &= \pm\left(J_z J^\pm + J^\pm J_z\right)/N_J \\ Y_2^{\pm 2} &= \left(J^\pm\right)^2/N_J \end{aligned} \quad (15.132)$$

with $N_J = (2/3)^{1/2}(2J^2 - J)$. The Hamiltonian $H_{\rm AC}$ is of a quadrupolar type. It causes a transfer of angular momentum $\ell = 2$ between the conduction electrons and the $4f^2$ shell.

The two interactions differ in one significant aspect. When the conduction electron system, but not the magnetic ions undergoes a time-reversal transformation (see (15.77)), $H_{\rm ex}$ changes sign while $H_{\rm AC}$ remains invariant. Therefore, their effect on Cooper pairs is quite different. While $H_{\rm ex}$ acts as a pair breaker, $H_{\rm AC}$ supports the formation of Cooper pairs. In a crystalline environment the $(2J+1)$-fold degeneracy of the Hund's rule ground-state multiplet $J = 4$ is split by the crystalline electric field (CEF). The splitting energies are usually of the order of a few meV. The eigenstates are found according to the irreducible representations of the point symmetry group.

Inelastic neutron scattering experiments on $PrOs_4Sb_{12}$ have demonstrated that Pr^{3+} is in a CEF singlet ground state Γ_1 with a low-lying excited triplet $\Gamma_t^{(2)}$ at an energy $\delta = 8$ K [151, 262]. The other CEF levels are much higher in energy and may be neglected (see Fig. 15.18).

The triplet Γ_t of T_h symmetry is a superposition of two triplets Γ_4 and Γ_5 of O_4 symmetry. More specifically it is [411]

$$|\Gamma_t, m\rangle = \sqrt{1-d^2}\,|\Gamma_5, m\rangle + d\,|\Gamma_4, m\rangle \ , \quad m = 1, 2, 3 \ . \quad (15.133)$$

The two triplets are of the form

$$|\Gamma_5, \pm\rangle = \pm\sqrt{\frac{7}{8}}|\pm 3\rangle \mp \sqrt{\frac{1}{8}}|\mp 1\rangle \quad ;$$

$$|\Gamma_5, 0\rangle = \sqrt{\frac{1}{2}}(|+2\rangle - |-2\rangle) \quad ;$$

$$|\Gamma_4, \pm\rangle = \mp\sqrt{\frac{1}{8}}|\mp 3\rangle \mp \sqrt{\frac{7}{8}}|\pm 1\rangle \quad ;$$

$$|\Gamma_4, 0\rangle = \sqrt{\frac{1}{2}}(|+4\rangle - |-4\rangle) \quad , \tag{15.134}$$

where $|n\rangle$ means $|J_z = n\rangle$. Furthermore, from experiments one can deduce that $|d| = 0.26$ implying that $|\Gamma_t, m\rangle$ is mainly of $|\Gamma_5, m\rangle$ character. However, $|\Gamma_1\rangle$ to $|\Gamma_5, m\rangle$ transitions can be induced by H_{AC} but not by H_{ex}, which causes transitions to $|\Gamma_4, m\rangle$ instead. Therefore, H_{AC} is the dominant interaction here and it is pair forming. We specialize H_{AC} to cubic symmetry and write it in a basis of Bloch states in the form

$$H_{AC}(i) = g \sum_{\mathbf{k}\mathbf{q}\sigma} \sum_{\alpha\beta\text{cycl}} O^i_{\alpha\beta} \hat{q}_\alpha \hat{q}_\beta c^+_{\mathbf{k}-\mathbf{q}\sigma} c_{\mathbf{k}\sigma} e^{i\mathbf{k}\mathbf{R}_i} \tag{15.135}$$

where $q_\alpha = q_\alpha/|\mathbf{q}|$ and i is the site index. Furthermore, $O_{\alpha\beta} = \sqrt{3}/2(J_\alpha J_\beta + J_\beta J_\alpha)$ where $\alpha\beta = yz, zx, xy$ denotes the three quadrupole operators of Γ_5 symmetry. The coupling constant g refers here to the quadrupolar coupling of conduction electrons to CEF levels of the Pr ions. It may be determined by experiments.

We are now in the position to solve Eliashberg equations for T_c. The boson propagator due to intra-atomic excitations is given by

$$g^2 D(\mathbf{q}, i\omega_\nu) = \sum_{\alpha\beta r} |\Lambda^r_{\alpha\beta}(\hat{\mathbf{q}})|^2 \frac{2\delta}{\omega_\nu^2 + \delta^2} \tag{15.136}$$

with

$$\Lambda^r_{\alpha\beta}(\hat{\mathbf{q}}) = g\hat{q}_\alpha \hat{q}_\beta \langle \Gamma_1 | O_{\alpha\beta} | \Gamma^r_t \rangle \quad . \tag{15.137}$$

It has to be supplemented by the phonon propagator $K_{ph}(\mathbf{q}, \nu_n)$ to which the electrons couple. The way to proceed is as follows. For LaOs$_4$Sb$_{12}$ where the mechanism described here is not operative, a phonon with average excitation energy $\bar{\omega}_\mathbf{q} = 26$ meV is chosen and the coupling constant is adjusted so that a transition temperature of $T_c \simeq 0.74$ K is obtained when (15.129) is solved. For PrOs$_4$Sb$_{12}$ we use the same phonons and add the propagator (15.136). Since g has not yet been determined by experiments we adjust it so that the right transition temperature is obtained. The required value of $g \simeq 0.04$ eV is very reasonable, since it is of similar size as known from Pr^{3+} ions dissolved in metals. It implies at the same time a mass enhancement through (4.35,13.54) which is of the observed magnitude. Furthermore, we can compute $T_c(x)$ for the alloy $\text{La}_{1-x}\text{Pr}_x\text{Os}_4\text{Sb}_{12}$ without any other adjustable parameter and find good agreement with the observed nonlinear behavior. It is also gratifying that for $\text{Pr}(\text{Os}_{1-x}\text{Ru}_x)_4\text{Sb}_{12}$ a decrease of $T_c(x)$ is predicted in agreement with observations. Neutron data show that here the excitation energy δ of the Γ_t triplet increases with the replacement of Os by Ru. Due to the larger denominator in (15.136) the transition temperature T_c decreases with increasing Ru content.

15.3.2 UPd$_2$Al$_3$: Pairing and Time-Reversal Symmetry Breaking

In Sect. 13.3 we have shown that because of strong intra-atomic or Hund's rule correlations the $5f$ electrons of U ions remain localized in some of the f orbitals while they delocalize in others. The model is supported by numerous experiments. It was pointed out that inelastic neutron scattering experiments show dispersive CEF excitations (magnetic excitons) below the Neél temperature of $T_N = 14.3$ K. Coupling of conduction electrons to these magnetic excitons explains the strongly anisotropic heavy quasiparticle mass in that system. Here we want to show that the same excitations act as bosons leading to Cooper pairing. However, as we shall see, a special form of the order parameter $\Delta(\mathbf{p})$ is required in order that the magnetic excitons can act as a binding agent [317].

We start from the two lowest energy levels $|J = 4, J_z = \pm 3\rangle$ of the two localized $5f$ electrons with a splitting energy due to the CEF of order $\delta = 7$ meV. As explained in Sect. 13.3 interionic interactions lead to induced antiferromagnetism. Low-energy excitations in the form of magnetic excitons have been observed (see Fig. 13.19) providing direct evidence for the dual model of $5f$ electrons. Their dispersion $\omega_{\text{ex}}(\mathbf{q})$ is described by (13.53). We may solve Eliashberg's equations (15.129) by using for the boson propagator the form

$$g^2 D(q_z, i\omega_\nu) = \frac{I^2}{2} \frac{\omega_{\text{ex}}}{\omega_\nu^2 + \omega_{\text{ex}}^2(\mathbf{q})} \quad . \tag{15.138}$$

The coupling constant $I = 0.16$ eV is chosen so that with $N(E_F) = 1$ state / (eV uc) the correct mass enhancement m^* is obtained. The latter was derived in Sect. 13.3 without an adjustable parameter. Therefore, we may use that result here in order to fix the coupling constant for the simplified density of states $N(\epsilon_F)$. We find superconducting order, provided the order parameter has one of the two forms

$$\Delta(\mathbf{p}) = \Delta_0 \cos(cp_z) \quad \text{or} \quad \Delta(\mathbf{p}) = \Delta_0 \sin(cp_z) \quad . \tag{15.139}$$

There is no solution for an s-wave order parameter. The one proportional to $\sin(cp_z)$ requires spin triplet pairing and can be eliminated because it contradicts experiments. For the $\cos(cp_z)$ solution we find $T_c \simeq 3$ K. There is no need to invoke phonons. Instead, the same physics which explains the anisotropic heavy quasiparticle masses also leads to superconductivity, with a specific form of the order parameter. It is reassuring that experiments on UPd_2Al_3 have indeed verified the above spin-singlet form of the order parameter [478].

So why is s-wave pairing excluded while an order parameter of the form of (15.139) gives a solution with a sizable T_c? The answer is simple: the transitions from $|\Gamma_3\rangle$ to $|\Gamma_4\rangle$ are caused by exchange interactions with the conduction electrons. Thus they are of magnetic origin and violate time-reversal symmetry of the conduction electron system as explained in connection with (15.76) and (15.130) (remember that the time-reversal operation is applied only to the conduction electrons and not to the localized $5f$ electrons). Those processes break Cooper pairs when pairing takes place of time-reversed states. However, when the order parameter is of the form $\cos(cq_z)$, electrons are not paired in time-reversed states but rather in time reversed states followed by a lattice translation (see the discussion following (15.98)). The partners of a pair occupy preferably different sublattices. In that case pairing can be achieved with a propagator of the form of (15.138) despite violation of time-reversal symmetry.

15.4 Magnetic Resonances

As we have repeatedly discussed, the glue for the formation of Cooper pairs is provided by the exchange of bosons between the conduction electrons. These bosons are either phonons, like in the original BCS theory; intra-atomic excitations, like in the filled skutterudite $PrOs_4Sb_{12}$; or collective excitations within the same electron systems in which Cooper pairing takes place. Here we want to discuss how the onset of superconductivity acts back on those bosons. Thereby we concentrate on bosons involving magnetic degrees of freedom. The effect of superconductivity on phonons can also be quite dramatic; it may stop the softening of phonons and in this way prevent lattice instability, a situation encountered in V_3Si [90]. However, the effect on magnetic collective excitations seems more general and important. It may lead to new forms of magnetic resonances which are purely due to superconductivity.

The first observation of this kind of excitation was made by *Rossat-Mignod* et al. [391] who found by inelastic neutron scattering a by-now-famous resonance peak below T_c in $YBa_2Cu_3O_{6+\delta}$, a high-temperature superconductor. The peak grows in intensity and shifts in energy with decreasing temperature. Not long thereafter it was suggested that it originates from 2D spin fluctuations in combination with d-wave pairing [273, 424]. Later it was found that the appearance of a magnetic resonance below T_c is a quite general phenomenon. Resonance peaks were observed in UPd_2Al_3 [397], $CeCu_2Si_2$ [429], $CeCoIn_5$ [428] and also in ferropnictides [66]. The common characteristic found in all these systems is that in the superconducting state pairing is unconventional. In fact, it turns out that this is a prerequisite for the appearance of a magnetic resonance induced by superconductivity. The observation of a res-

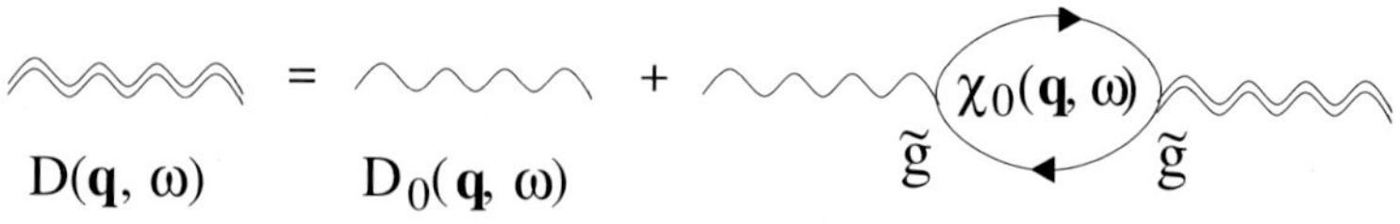

Fig. 15.19. Effect of electron-hole excitations (solid lines) on boson symbolized by wavy lines. The coupling constant of the electrons to bosons is $\tilde{g}$. Double line: renormalized boson, single wavy line: bare boson.

onance structure even enables us to distinguish between different types of unconventional pairings and is therefore helpful in identifying the right pair state.

Generally we have to distinguish between resonances, which are associated with localized and with delocalized electrons. An example of the former kind is UPd_2Al_3. Due to strong intra-atomic correlations the $5f$ electrons are divided into localized and delocalized ones. The dual model was discussed in Sect. 13.3 and 15.3.2. Here the magnetic resonance is associated with *localized* $5f$ electrons, i.e., with electrons in f orbitals with vanishing renormalized hybridization matrix elements. Their interaction with *itinerant* f electrons, i.e., with those in hybridizing orbitals results in a new resonance structure when superconductivity sets in. Examples of the second kind are the high-T_c cuprates or $CeCu_2Si_2$ and $CeCoIn_5$. Here the magnetic resonance is associated with the same electrons which also form Cooper pairs.

In the following we want to discuss one example of each kind. We begin with UPd_2Al_3. As pointed out in Sect. 10.3, in UPd_2Al_3 two $5f$ electrons of a U ion in $j_z = 5/2$ and $1/2$ orbitals remain localized and form a $J = 4$, $J_z = \pm 3$ ground-state doublet. This doublet is split by the CEF. Due to a coupling of the CEF excitations at different sites via the conduction electrons (see Fig. 13.20) an AF ground state is induced. Its excitations, called magnetic excitons have the dispersion shown in Fig. 13.19. As demonstrated in Sect. 15.3.2 they provide for an electron-electron interaction, which results in superconductivity with an unconventional order parameter (see (15.139)). Here we want to discuss the feedback effect of superconductivity on the magnetic excitons.

Any boson described by a propagator $D_0(\mathbf{q},\omega)$ which is interacting with conduction electrons is affected by that interaction. This is shown in Fig. 15.19 in the form of a diagrammatic equation. The electron-hole bubble denotes a susceptibility. Due to the interaction in (13.51) it is, here, the spin susceptibility $\chi_0(\mathbf{q},\omega)$ of the conduction electrons. Let us write the boson propagator in the standard form, i.e., for real frequencies (compare with (15.138))

$$D_0(\mathbf{q},\omega) = -\frac{2\omega_\mathbf{q}}{\omega^2 - \omega_\mathbf{q}^2} , \qquad (15.140)$$

where $\omega_\mathbf{q}$ is here identified with the magnetic exciton dispersion $\omega_{ex}(q_z)$ given by (13.53). Then from

$$D(\mathbf{q},\omega) = D_0(\mathbf{q},\omega) + D_0(\mathbf{q},\omega)\tilde{g}^2\chi_0(\mathbf{q},\omega)D(\mathbf{q},\omega) \qquad (15.141)$$

it follows that

$$D(\mathbf{q},\omega) = -\frac{2\omega_\mathbf{q}}{\omega^2 - \omega_\mathbf{q}^2 + 2\tilde{g}^2 \omega_\mathbf{q} \chi_0(\mathbf{q},\omega)} \quad . \tag{15.142}$$

The zeros of the denominator define the excitations of the bosonic system. In the normal state at low energies, i.e., when ω is of order 10 meV it is Re $\chi_0(\mathbf{q},\omega) = const$, Im $\chi_0(\mathbf{q},\omega) = i\gamma\omega$. Thus, the coupling to the conduction electrons leads merely to a shift of frequencies and a Landau damping in the form of a line width of the excitonic excitations. The measured magnetic excitons modelled by (13.53) already contain this frequency shift and therefore we may discard it here. However, when a superconducting state is formed, $\chi_0(\mathbf{q},\omega)$ changes dramatically due to the appearance of a gap. As we shall see, the form of the order parameter also strongly affects the elelctron susceptibility.

We are interested in $\chi_0(\mathbf{q},\omega)$ at $T = 0$. By absorption of a magnetic exciton of momentum $\mathbf{q}$ and energy ω an electron is moving from a state $\mathbf{k}\sigma$ to a state $\mathbf{k} + \mathbf{q}\sigma$. The initial state is here the BCS ground state $|\psi_0\rangle$ (see (15.39)) while the final state is

$$|\psi_f\rangle = c^+_{\mathbf{k}+\mathbf{q}\sigma} c^+_{-\mathbf{k}-\sigma} \prod_{\mathbf{p}\neq\mathbf{k},\mathbf{k}+\mathbf{q}} \left(u_\mathbf{p} + v_\mathbf{p} c^+_{\mathbf{p}\sigma} c^+_{-\mathbf{p}-\sigma} \right) \quad , \tag{15.143}$$

i.e., we are dealing with two unpaired electrons. The interaction Hamiltonian is

$$H_1 = -\frac{\tilde{g}}{N} \sum_{\mathbf{k},\mathbf{q}} c^+_{\mathbf{k}+\mathbf{q},\alpha} \sigma^z_{\alpha\beta} c_{\mathbf{k}\beta} \left(b_\mathbf{q} + b^+_{-\mathbf{q}} \right) \tag{15.144}$$

where $b_\mathbf{q}$ annihilates a magnetic exciton with momentum $\mathbf{q}$. We apply H_1 for fixed value of $\mathbf{q}$ and evaluate $\left|\left\langle \psi_f \left| c^+_{\mathbf{k}+\mathbf{q}\sigma} c_{\mathbf{k}\sigma} \right| \psi_0 \right\rangle\right|^2$. This gives a coherence factor $(u_{\mathbf{k}+\mathbf{q}} v_\mathbf{k} - u_\mathbf{k} v_{\mathbf{k}+\mathbf{q}})^2$. With the expression (15.52) for $u_\mathbf{k}$ and $v_\mathbf{k}$ we obtain from perturbation theory for the response at finite temperature T

$$\chi_0(\mathbf{q},\omega) = \sum_\mathbf{k} \frac{1}{4} \left[1 - \frac{\epsilon_\mathbf{k} \epsilon_{\mathbf{k}+\mathbf{q}} + \Delta_\mathbf{k} \Delta_{\mathbf{k}+\mathbf{q}}}{E_\mathbf{k} E_{\mathbf{k}+\mathbf{q}}} \right]$$

$$\times \left[\frac{f(E_{\mathbf{k}+\mathbf{q}}) + f(E_\mathbf{k}) - 1}{\omega - E_{\mathbf{k}+\mathbf{q}} - E_\mathbf{k} + i\delta} + \frac{1 - f(E_{\mathbf{k}+\mathbf{q}}) - f(E_\mathbf{k})}{\omega + E_{\mathbf{k}+\mathbf{q}} + E_\mathbf{k} + i\delta} \right]$$

$$+ \sum_\mathbf{k} \frac{1}{2} \left[1 + \frac{\epsilon_\mathbf{k} \epsilon_{\mathbf{k}+\mathbf{q}} + \Delta_\mathbf{k} \Delta_{\mathbf{k}+\mathbf{q}}}{E_\mathbf{k} E_{\mathbf{k}+\mathbf{q}}} \right]$$

$$\times \frac{f(E_{\mathbf{k}+\mathbf{q}}) - f(E_\mathbf{k})}{\omega - (E_{\mathbf{k}+\mathbf{q}} - E_\mathbf{k}) + i\delta} \quad . \tag{15.145}$$

As before, $f(x)$ is Fermi's function. At $T = 0$ and $\Delta_\mathbf{k} \neq 0$ the first term in the second bracket is the only one which is left. It describes the generation of a spin triplet out of a spin-singlet Cooper pair. The second term describes the reverse process while the last term is caused by the scattering of excited quasiparticles due to the external perturbation like in the normal state. It follows that for $\omega > 0$ the function $Im\chi_0(\mathbf{q},\omega)$ is given by

$$Im\chi_0(\mathbf{q},\omega) = \frac{\pi}{4}\sum_{\mathbf{k}}\left[1 - \frac{\epsilon_{\mathbf{k}}\epsilon_{\mathbf{k+q}} + \Delta_{\mathbf{k}}\Delta_{\mathbf{k+q}}}{E_{\mathbf{k}}E_{\mathbf{k+q}}}\right]$$
$$\cdot \delta\left(\omega - E_{\mathbf{k}} - E_{\mathbf{k+q}}\right). \qquad (15.146)$$

It is noticed that when $\mathbf{k}$ and $\mathbf{k+q}$ are on the Fermi surface, i.e., when $\epsilon_{\mathbf{k}} = \epsilon_{\mathbf{k+q}} = 0$, then $Im\chi_0(\mathbf{q},\omega)$ can discontinuously increase at the onset frequency of the quasiparticle-quasihole continuum $\omega_c = \text{Min}\left(|\Delta_{\mathbf{k}}| + |\Delta_{\mathbf{k+q}}|\right)$. However, that requires that $\Delta_{\mathbf{k}} = -\Delta_{\mathbf{k+q}}$, in which case the coherence factor becomes two. This requirement excludes s-wave superconductors.

A discontinuity in $Im\chi_0(\mathbf{q},\omega)$ leads to a logarithmic singularity in $Re\chi_0(\mathbf{q},\omega)$, because the two functions are connected with each other via Kramers-Kronig relations. This in turn results in a resonance as shown below. Since UPd$_2$Al$_3$ is an AF below $T_N = 14.2$ K with $\mathbf{Q} = (0,0,\pi/c)$ we have to modify $\chi_0(\mathbf{q},\omega)$ correspondingly. This implies that we have to replace $E_{\mathbf{k}}$ by $E_{\mathbf{k}}^{\pm} = \sqrt{\left(\epsilon_{\mathbf{k}}^{\pm}\right)^2 + \Delta_{\mathbf{k}}^2}$ where $\epsilon_{\mathbf{k}}^{\pm} = \xi_{\mathbf{k}}^a \pm \sqrt{\left(\xi^b\right)^2 + m^2}$ with $\xi_{\mathbf{k}}^a = \frac{1}{2}\left(\epsilon_{\mathbf{k}} + \epsilon_{\mathbf{k+Q}}\right)$ and $\xi_{\mathbf{k}}^b = \frac{1}{2}\left(\epsilon_{\mathbf{k}} - \epsilon_{\mathbf{k+Q}}\right)$. The two bands $\xi_{\mathbf{k}}^{\nu}$ ($\nu = a,b$) are a consequence of the doubling of the unit cell in the presence of AF long range order. Furthermore, m denotes the effective staggered field in the AF. We do not want to write down the corresponding expression for $\chi_0(\mathbf{q},\omega)$ in an AF superconductor. Instead, we only want to point out some salient features. The Fermi surface of UPd$_2$Al$_3$ has little dispersion along the z-axis, i.e., it looks like a cylinder. On the other hand, the magnetic excitons have little dispersion in the $a-b$ plane and are well described by $\omega_{\text{ex}}(q_z)$ given by (13.53). For both forms of the order parameter shown in (15.139) it is $\Delta_{\mathbf{k+Q}} = -\Delta_{\mathbf{k}}$ for $\mathbf{k}$ values at the Fermi surface. Due to its cylindrical shape we find that for $T = 0$ the coherence factor in $\chi_0(\mathbf{Q},\omega)$ equals two for all k_z momenta.

In order to determine how Cooper-pair formation affects magnetic excitons we have to solve the equation

$$\omega^2 = \omega_{\mathbf{q}}^2 - 2\tilde{g}^2\omega_{\mathbf{q}}\chi_0(\mathbf{q},\omega) \qquad (15.147)$$

which yields the singularities of (15.142). When the form of $\chi_0(\mathbf{q},\omega)$ in the presence of AF order is evaluated numerically and set into that equation, it is found that near $\mathbf{q} \simeq \mathbf{Q}$ two solutions exist with small damping (see Fig. 15.20). One refers to the slightly renormalized magnetic exciton and is positioned at ω_m. The second is within the gap region at ω_r and is due to the strong frequency dependence of $Re\chi_0(\mathbf{Q},\omega)$. In fact, for $\omega = 2\Delta_0$ the real part of $\chi_0(\mathbf{Q},\omega)$ is strongly peaked for reasons discussed earlier (Kramers-Kronig). A resonance peak at ω_r has been clearly observed in UPd$_2$Al$_3$ below T_c near

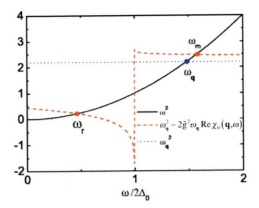

Fig. 15.20. Graphical solution of (15.147) for $\mathbf{q} \simeq \mathbf{Q}$. In addition to the magnetic exciton pole ω_m a second pole of $D(\mathbf{q} \simeq \mathbf{Q}, \omega)$ is found at ω_r with small damping only. It arises from the strong frequency dependence of $\chi_0(\mathbf{q}, \omega)$. The third crossing point is strongly damped and therefore uninteresting. (From [61])

$q_z = \pi/c$. This is seen from Fig. 15.21a [397]. The computed two peaks are shown in Fig. 15.21b in the form of a contour plot of $ImD(q_z, \omega)$.

As previously discussed, the resonance is possible only when $\Delta_{\mathbf{k}+\mathbf{Q}} = -\Delta_{\mathbf{k}}$, i.e., when the order parameter changes sign under translation by $\mathbf{Q}$. This condition is fulfilled by both forms (15.139) of the order parameter; the observation of resonance cannot distinguish between the two forms. However, an s-wave order parameter can be excluded with certainty by this experiment. The resonance in conjunction with the magnetic exciton reconfirms the dual character of $5f$ electrons in UPd_2Al_3.

The magnetic resonance in UPd_2Al_3 serves as a nice example how superconductivity affects bosons with magnetic degrees of freedom. Here the boson, which is a magnetic exciton is caused by localized $5f$ electrons. They differ from the electrons which form Cooper pairs. The latter are conduction electrons with a strong component of itinerant $5f$ electrons.

Next we study the effect of superconductivity on bosons, which are formed by the same electrons which form Cooper pairs. Here we choose $CeCu_2Si_2$ and $CeCoIn_5$ as examples. High-temperature superconductors like $YBa_2Cu_3O_{7-\delta}$ also fit into that category. However, it seems to us that the resonances in the Ce-based heavy quasiparticle systems are simpler to explain.

$CeCu_2Si_2$ has a Fermi surface with nesting properties. It consists of stacked columns along the c direction. This is shown in Fig. 13.9 for the first few Brillouin zones. There are flat parts seen which are connected by a nesting vector $\mathbf{Q}_{SDW} = (0.22, 0.22, 0.52)$ in reciprocal lattice units; indeed, a spin density wave with that $\mathbf{Q}$ vector has been observed by neutron scattering [430]. The Fermi surface with the nesting vector $\mathbf{Q}_{SDW}$ was determined by renormalized bandstructure calculations described in Chapter 13. They contain a single ad-

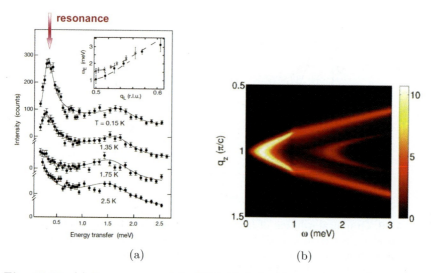

Fig. 15.21. (a) Resonance peak in UPd$_2$Al$_3$ below T_c near $q_z = \pi/c$ as observed by inelastic neutron scattering. The structure near $\omega = 1.5$ meV is the magnetic exciton mode. (From [397])
(b) Computed contour plot of $ImD(q_z,\omega)$ near $Q = \pi/c$ at $T = 0$. The low energy peak is the resonance peak while the one starting near $\omega = 1.5$ meV is the magnetic exciton. Both peaks disperse upwards in energy like the magnetic exciton does in the normal state. Bright colour implies high intensity. (From [61])

justable parameter, i.e., the slope of the $4f$ phase shift at the Fermi energy ϵ_F. With the heavy quasiparticle bands one can calculate the Lindhard spin susceptibility $\chi_0(\mathbf{q},\omega)$ in the normal and superconducting state. Its static part $\chi_0(\mathbf{q})$ is peaked at $\mathbf{Q}_{\mathrm{SDW}}$ due to nesting. The resonance peak in the superconducting state of CeCu$_2$Si$_2$ as well as of CeCoIn$_5$ can be determined within RPA by computing

$$\chi_{\mathrm{RPA}}(\mathbf{q},\omega) = \frac{\chi_0(\mathbf{q},\omega)}{1 - U_{\mathbf{q}}\chi_0(\mathbf{q},\omega)} \qquad (15.148)$$

where $U_{\mathbf{q}}$ is due to the interactions of the heavy quasiparticles. As discussed before, when $\Delta_{\mathbf{k}} = -\Delta_{\mathbf{k+q}}$ for $\mathbf{k},\mathbf{q}$ on the Fermi surface, then $Re\chi_0$ has a logarithmic singularity due to a discontinuity in $Im\chi_0$. In that case the resonance conditions, i.e., $U_{\mathbf{q}}Re\chi_0(\mathbf{q},\omega) = 1$ and $Im\chi_0(q,\omega) = 0$ can both be fulfilled at $\omega_{\mathrm{res}} < \omega_c$, as long as $U_{\mathbf{q}} > 0$. This causes an additional spin excitation below T_c. When a finite lifetime is given to the quasiparticles, then $U_{\mathbf{q}}$ must exceed a critical value U_c for the new mode to appear. Instead of $Im\chi_0(\mathbf{q},\omega) = 0$ we must require that $Im\chi_0(\mathbf{q},\omega)/Re\chi_0(\mathbf{q},\omega) \ll 1$. In CeCu$_2Si_2$ a sharp spin resonance was found below T_c at $\mathbf{q} = \mathbf{Q}_{\mathrm{SDW}}$ by inelastic neutron scattering [429]. One might then ask which forms of $\Delta_{\mathbf{k}}$ are consistent with the observation of the resonance. A detailed analysis shows that a discontinuous jump in $Im\chi_0$

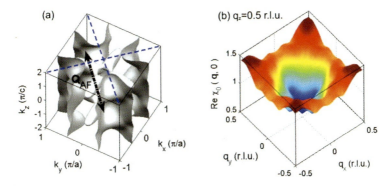

Fig. 15.22. (a) Fermi surface of CeCoIn$_5$ as obtained from renormalized bandstructure theory. The arrows connect points which differ by $\mathbf{Q}_{AF}$. The dashed lines show the node position at which the $d_{x^2-y^2}$ order parameter vanishes. (b) Static susceptibility for $q_z = 0.5$ in reciprocal lattice units. (From [108])

with an associated singularity in $Re\chi_0$ is present for the following order parameters:

$$\Delta_{\mathbf{k}} = \Delta_0 \left(\cos k_x a - \cos k_y a \right) \quad , \quad B_{1g} \text{ irred. representation}$$

$$\Delta_{\mathbf{k}} = \begin{bmatrix} \Delta_0 \sin k_x a \sin k_z a \\ \Delta_0 \sin \tfrac{1}{2}(k_x + k_y) \sin \tfrac{1}{2} k_z a \end{bmatrix} \quad , \quad E_g \text{ irred. representation}$$
(15.149)

with a dispersion of the resonance which is by far the strongest in the B_{1g} channel. This points strongly towards a $d_{x^2-y^2}$ symmetry of the associated order parameter.

The situation is similar in CeCoIn$_5$. Here the Fermi surface obtained from renormalized bandstructure calculations is too complicated to describe by a single band. We can, however, model it by two bands, i.e., a heavy electron f-like band hybridizing with a conduction electron band (see, e.g., Fig. 13.12). The resulting Fermi surface is shown in Fig. 15.22a. It has again nesting properties with a $\mathbf{q}$ vector which coincides with the AF wave vector $\mathbf{Q}_{AF} = (\pi/a, \pi/a, \pi/c)$. This is in agreement with neutron scattering data [428]. As expected, the static Lindhard spin susceptibility of the normal state is found to be peaked at $\mathbf{Q}_{AF}$ (see Fig. 15.22b).

When the real and imaginary part of $\chi_{RPA}(\mathbf{Q}_{AF}, \omega)$ are calculated as previously explained, one finds a resonance peak only when the order parameter is $\Delta_{\mathbf{k}} = (\Delta_0/2)(\cos k_x a - \cos k_y a)$, i.e., of B_{1g} symmetry. Here the resonance forms slightly below ω_c. This is shown in Fig. 15.23. The observation of the resonance excludes other proposals which have been made for the form of the order parameter, such as $d_{xy}(B_{2g})$ symmetry. The downward dispersion of the resonance, when $\mathbf{q}$ deviates from $\mathbf{Q}_{AF}$ resembles the one found in hole-doped

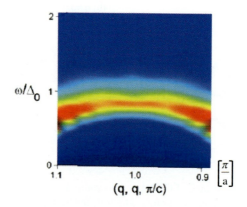

Fig. 15.23. Dispersion of the magnetic resonance in CeCoIn$_5$ along the $(q, q, \pi/c)$ direction calculated for an order parameter of $d_{x^2-y^2}$ symmetry. For the interaction the form $U_\mathbf{q} = U_{\mathbf{Q}_{AF}}[1 - 0.8(\mathbf{q} - \mathbf{Q}_{AF})^2/\mathbf{Q}_{AF}^2]$ was used. (From [108])

superconducting cuprates. The similarity of the resonance in the heavy quasi-particle systems CeCu$_2$Si$_2$ and CeInCo$_5$ and of hole-doped cuprates, and the same $d_{x^2-y^2}$ form of the order parameter found in these three cases gives us hints on the microscopic origin of superconductivity in these strongly correlated systems.

15.5 High-T_c Superconductors

It is customary to associate high-T_c superconductivity with the superconducting cuprates although other materials like the Fe pnictides have also high transition temperatures[7]. Over the last 20 years an enormous amount of work has gone into studying the cuprates. What makes their study difficult is a competition of various instabilities in these systems, which is a characteristic feature of them. Not only may a superconducting- or pairing instability develop, but a magnetic or structural one may develop as well. Thus different order parameters interact and compete with each other. This often causes a sample dependence of experimental results; this can only be avoided by very carefully prepared and characterized samples.

The most important structural element of the cuprates are copper-oxide planes with a unit cell CuO$_2$. These planes are formed from octahedra, pyramids or squares and have been extensively discussed in Sect. 12.1. Here we concentrate on their superconducting properties.

A generic feature of the high-T_c cuprates is that they are antiferromagnetic charge-transfer insulators when they are undoped. With increasing doping, which is most often hole doping, antiferromagnetism is suppressed and

---

[7] see, e.g., [205, 209, 220, 480]

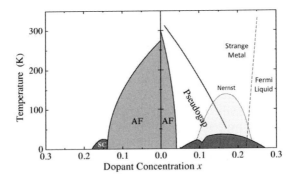

Fig. 15.24. Schematic phase diagram for doped cuprates. Right side: hole-doped $La_{2-x}Sr_xCuO_4$; left side: electron-doped $Nd_{2-x}Cl_xCuO_4$. In parts from [274].

superconductivity starts to appear. The phase diagram looks schematically as indicated in Fig. 15.24. One notices that hole- as well as electron doping destroys AF long range order, and that holes do it more efficiently. As the highest values of T_c have been obtained by hole doping, the by-far-largest amount of research has gone into hole-doped rather than electron-doped systems. In the following we want to restrict ourselves to the former case and we will only discuss the right-hand part of the phase diagram.

It can be seen in Fig. 15.24 that the superconducting region has the shape of a dome, with a small dip at x = 0.12. The highest superconducting transition temperature is achieved at an optimal hole-doping concentration x_{op}. When x < x_{op} we speak of underdoped systems and when x > x_{op} of overdoped systems. If doping becomes too large, superconductivity is suppressed. In the normal state there is a region where in distinction to an ordinary Fermi liquid the system exhibits a pseudogap with respect to spin- as well as density excitations. It shows up experimentally in the form of a reduced spin- and density response to external perturbations. When we speak of a reduction in response, we refer to what is expected if all valence electrons participate in the formation of the Fermi surface. In addition, there is an area in the diagram to the right of the line defining the pseudogap region in which the pseudogap has vanished but where deviations from ordinary Fermi liquid behavior are found. This region is sometimes called that of a *strange metal*. There is also a regime labeled *Nernst*, where it is found that the Nernst effect is of the same size as in the superconducting state. This has been taken as an indication of the presence of preformed electron pairs above the superconducting transition temperature. For large doping, the system shows normal Fermi liquid behavior above T_c. In the following we want to discuss the different parts of that phase diagram. Thereby we have primarily the system $La_{2-x}Sr_xCuO_4$ in mind.

15.5.1 Suppression of Antiferromagnetic Order by Holes

When La^{3+} is replaced by Sr^{2+} and therefore holes are doped into the system, the long-range antiferromagnetic order is destroyed first. In Sect. 10.5 we studied the motion of a hole in an antiferromagnet by means of the $t - J$ model. This model is considered a minimal model to account for the strong correlations, which are prevailing in La_2CuO_4. While the effects of AF order on the hole motion were studied in detail, the effect of holes on antiferromagnetic spin fluctuations was neglected, except for a trivial correction. The latter took into account that holes of concentration x dilute a spin system and modify the dispersion of spin waves by a factor $(1 - x)^2$. This would imply that AF order is destroyed only when $x = 1$, i.e., when no spins are left. Here we want to go beyond that simple estimate and discuss the effect of a finite hole concentration on the dispersion of antiferromagnetic spin waves. Finding the critical doping concentration at which long-range order is destroyed is related to determining the hole concentration at which the spin-wave velocity vanishes, i.e., $v_s(x) = 0$.

When small amounts of holes are present, they form pockets at $(\pm\pi/2, \pm\pi/2)$ in the Brillouin zone. That is where the dispersion $E(\mathbf{k})$ of coherent hole motion in form of a Zhang-Rice singlet has its minimum (see Figs. 12.10 or 10.22). In order to calculate their effect on v_s, one must determine self-consistently the Green's function of spin waves, as well as the Green's function of holes, dressed by spin-wave emission and absorption.

Formally this is done by using the slave-fermion Schwinger boson representation (10.132) and by introducing in accordance with (7.2) the Green's function for holes

$$G_{\mu\nu}(\mathbf{k}, t) = -i\left\langle T\left(f_{\mathbf{k}}^{\mu}(t) f_{\mathbf{k}}^{\nu+}(0)\right)\right\rangle , \quad \mu, \nu = a, b \qquad (15.150)$$

The indices μ and ν refer here to the sublattices a and b of the Néel state, respectively. Similarly, a Green's function for the Schwinger bosons, i.e., spin waves can be defined after a Bogoliubov transformation similar to (10.124) has been performed. The hole propagator is dressed by a self-energy in Born approximation depicted in Fig. 10.20. Spin waves can be absorbed by creating electron-hole pairs and they can be scattered by holes. This is schematically shown in Fig. 15.25. Calculations show that electron-hole excitations produced by spin waves as well as the inverse processes are much more important than spin-wave scattering by holes. The spin-bag degrees of freedom or incoherent part $G_{\text{inc}}(\mathbf{k}, \omega)$ of the hole propagator $G_{\mu\nu}(\mathbf{k}, \omega)$ make the largest contributions here [202]. It leads to the finding that

$$v_s(\hat{\mathbf{q}}) = (1 - \alpha(\hat{\mathbf{q}})x)v_s^{(0)} \qquad (15.151)$$

where $\alpha(\hat{\mathbf{q}})$ depends slightly on the direction of $\mathbf{q}$. It depends also on the ratio J/t, i.e., for $\hat{q} = (1,1)$ one finds $\alpha = 30$ for $J/t = 0.1$ and 6.4 for $J/t = 0.5$. The

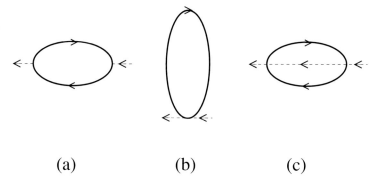

Fig. 15.25. Different contributions to the self-energy of the spin-wave propagator (dashed lines) in schematic form. The most important contribution results from electron-hole generation (a) while the scattering of spin-waves by holes (b) and (c) plays a secondary role only.

values in (1,0) direction differ by less than 5 % from those in (1,1) direction. Antiferromagnetic long-range order is destroyed when the spin-wave velocity goes to zero, i.e., when $\alpha x_{\text{crit}} = 1$. For $J/t = 0.3$, a value usually assumed for the high-T_c cuprates, the theory predicts a critical doping concentration of $x_{\text{crit}} \simeq 0.1$ for the destruction of AF order[8]. As discussed in Sect. 12.1, even when long-range order is destroyed, short-range antiferromagnetic order still prevails and the spin-bag concept remains valid.

15.5.2 Pseudogap Regime

We continue with a discussion of the normal state with a pseudo gap. A pseudogap appears as a strong drop in the normal state spin susceptibility with lowering temperatures. This drop starts below approximately 300 - 400 K, i.e., far above T_c and has been observed, e.g., in $La_{2-x}Sr_xCuO_4$, $YBa_2Cu_3O_{6+x}$ and $YBa_2Cu_4O_8$. Apparently, an increasing fraction of the accessible low-energy excitations freezes out when the temperature decreases, i.e., the system becomes more and more gapped. Pauli's temperature-independent susceptibility $\chi_s = 2\mu_B^2 N(0)$ is based on a Fermi surface with a fixed area $4\pi p_F^2$ and therefore a fixed density of states $N(0)$ at it. A reduction of $\chi_s(T)$ therefore looks like the area of the available Fermi surface is shrinking with decreasing temperature, an interesting feature.

Before we discuss this point in more detail, we first want to consider another possible origin of the pseudogap; preformed Cooper pairs. One might argue that pairs are formed at a much higher temperature than T_c, i.e., at a temperature $T_{\text{MF}} \gg T_c$, but that for $T_{\text{MF}} > T > T_c$ they are not phase locked. Since $\Delta(\mathbf{r}) = |\Delta|e^{i\Theta(\mathbf{r})}$ this implies that $\lim_{r\to\infty} \langle \Theta(\mathbf{r})\Theta(\mathbf{0})\rangle = 0$ instead

---

[8] see, e.g., [202, 237, 368]

of a constant. Thus, the system has an order parameter $|\Delta| \neq 0$ characterizing the pair density, yet it is not a superconductor since the existence of a supercurrent requires phase locking.

We want to understand why the temperatures can differ at the point where $|\Delta| \neq 0$ sets in and at which phase locking occurs. For that purpose assume that the density of electrons participating in pairing is low and that binding is strong. In that case pairs would form which are well separated from each other. Since they are boson-like they would Bose condensate at a temperature T_c, which is quite different from the much higher temperatures at which the pairs form. We recall that this is quite different from the BCS theory where the spatial extent of a pair, given by the coherence length ξ_0 is much larger than the interpair spacing, i.e., the pairs overlap strongly and cannot be separated. In doped Mott-Hubbard insulators we expect that we are between the two limits, i.e., the BCS theory and Bose condensation, as long as the density of the doped holes is small. A proper description of the transition between the two limits is a subject of its own. The key to the problem is the behavior of the chemical potential μ. In the BCS theory $\mu = \epsilon_F$, i.e., it is equal to the Fermi energy. When the attractive potential increases or the electron density decreases μ/ϵ_F decreases. Eventually a self-consistent determination of μ results in $\mu \to -\infty$ which is the limit of separated pairs[9].

A different approach to phase unlocking of boson-like electron pairs is via the kinetic energy increase of the superfluid caused by phase changes of the order parameter. The Hamiltonian describing it can be written in the form

$$H = 1/2 K_p (\nabla \Theta)^2 \tag{15.152}$$

where K_p is the phase stiffness constant. Since the velocity of the superfluid is $v_s = \frac{1}{2m^*} \nabla \Theta$ where m^* is the effective electron mass we find that $K_p = n_s(0)/4m^*$. Here $n_s(0)$ is the density of the superfluid at $T = 0$. Near a Mott-Hubbard transition the stiffness constant is small and phase fluctuations become large. The unlocking of the phase takes place here in the form of a Berezinskii–Kosterlitz–Thouless (BKZ) transition in which vortex-antivortex pairs are created in the 2D system and unbind. The BKZ transition is well understood. It is driven by a competition between the energy of forming vortex-antivortex pairs and the entropy. Eventually both approaches, i.e., the interpolation between the BCS and Bose condensate description and the one based on vortex-antivortex formations should merge into one.

Electron pairs are observed by the Nernst effect in the regime schematically shown in Fig. 15.24. A film with a thermal gradient and a magnetic field applied perpendicular to its surface shows a voltage which is transverse to that gradient. That is the Nernst effect. In a superconductor film, where an applied perpendicular external magnetic field penetrates the film in the form of vortices, the vortices move along the thermal gradient and generate via the Josephson effect an extra large transverse voltage. An interesting observation

---

[9] For more detailed information the reviews [64, 278, 288] should be consulted.

has been made that the large Nernst effect does not only exist below T_c but also in a wide regime in the normal state [477]. This implies that we must have vortices in the normal state as well; however in distinction to the superconducting state vortices in the normal state build up and decay with time. Therefore we expect a reduced density of states for energies less than the binding energy of pairs, i.e., a pseudogap. From Fig. 15.24 it is apparent that preformed or unsynchronized pairs cannot alone explain the phenomenon of a pseudogap, since it appears in a much wider regime of the phase diagram than the enhanced Nernst effect does.

This brings us back to a partial gapping of what would be the large Fermi surface if all valence electrons contributed to it. It is the Fermi surface we would have if correlations were weak enough, such that we would not have to consider a Mott-Hubbard transition. As we have discussed earlier, small hole doping generates lenses of holes in the Brillouin zone at $(\pm\pi/2; \pm\pi/2)$ (see Figs. (12.10)). At their edges the excitation energy goes to zero. The density of states associated with the hole pockets is denoted by $N^*(0)$. The resulting Pauli spin susceptibility is $\chi_s = \mu_B^2 N^*(0)/(1+F_0^a)$ and small where the Landau parameter F_0^a is due to effective quasihole-quasihole interactions. As T increases, larger parts of the gapped portion of the Fermi surface become accessible. Therefore, χ_s increases with T and differs from Pauli's temperature-independent susceptibility. A detailed theoretical analysis has to account for temperature changes in the gapped part of Fermi surfaces caused by spin disorder and associated changes in the quasiparticle dispersion discussed in Sect. 12.1. Remember that with increasing temperature short-ranged AF correlations are reduced, which increases the values of t_{eff} (see (12.5) and the discussion of it). This affects the dispersion of the Zhang-Rice singlets and hence the Fermi surface. In addition, the singlets start to break up and the number of triplets increases as the temperature rises. The system becomes more an more a normal Fermi liquid.

There have been attempts to describe the above scenario by a mean-field approach based on composite operators introduced in Sect. 10.6. The electron operators $\hat{a}_{i\sigma}^+$, $\hat{a}_{i\sigma}$ (see 10.100) are expressed as products of fermionic spinon operators $f_{i\sigma}^+(f_{i\sigma})$ and bosonic holon operators $b_i^+(b_i)$ (see (10.139)). The $t-J$ Hamiltonian, which is considered to be the minimal model for describing the strongly correlated hole-doped CuO_2 planes, is then expressed in terms of the spinons and holons before a mean-field approximation is made. In Sect. 10.6 we discussed in detail the case of half filling, which is that of a Heisenberg antiferromagnet. Here we are interested in the case of hole doping. Then H_0 defined by (10.142) becomes relevant, which in terms of the order parameter χ_{ij} given by (10.146) is written as

$$H_0 = -t\sum_{\langle ij \rangle} \chi_{ij} b_i b_j^+ \quad . \tag{15.153}$$

The spin-spin interaction part (10.147) remains unchanged. Thus, in distinction to half filling, we have to deal additionally with holons, i.e., bosons.

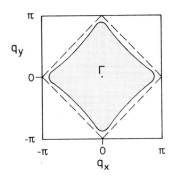

Fig. 15.26. Schematic plot of the Fermi surface of $\text{La}_{2-x}\text{Sr}_x\text{CuO}_4$ for moderate hole doping as found by ARPES.

Their dispersion is governed by a reduced hopping matrix element $-t\chi_{ij}$. It is assumed that at T_c Bose condensation is taking place, which is identified with the formation of a superconducting state. The pseudogap in the normal state is explained by the gapped spinon excitations, which were found in (10.152) when two order parameters $\chi_{ij} \neq 0$ and $\Delta_{ij} \neq 0$ were present.

Although that picture is appealing, a number of open questions remain. In the present mean-field approach, spinons and holons are separate particles. We know, however, that in two dimensions this cannot be the case. The spin bag remains attached to the hole. This was discussed in detail in Sect. 10.5. It is also born out by the wavefunction-based quantum chemical calculations of Sect. 12.1. Therefore, fluctuations neglected in the mean-field approach must bind again the spinons to the holons. Basically the explanation of the pseudogap in mean-field approximation is that it reflects the antiferromagnetic spin excitation spectrum (10.152) of a two-dimensional SDW obtained within that approximation. In view of the discussion given in Sect. 10.6 concerning this spectrum, questions may be raised.

15.5.3 Strange Metal

Next we turn towards the region near optimal doping, where for $T > T_c$ we deal with a strange metal. Deviations from an ordinary Fermi liquid can be seen by the fact that angular resolved photo electron spectroscopy (ARPES) shows, for that doping regime, a large Fermi surface (see Fig. 15.26) in accordance with Luttinger's Theorem, while transport properties like the Drude peak in the optical conductivity remain proportional to the hole-doping concentration δ. Photoelectron spectroscopy experiments show very broad quasiparticle peaks with a large incoherent background. The renormalization constant Z (see (7.20)) seems to be nearly zero. Also the behavior of the resistivity in this doping regime is anomalous, i.e., $\rho(T) \sim T$ and not $\sim T^2$ as expected for a Fermi liquid. As pointed out in Sect. 10.9 the unusual behavior can be explained if we assume that $Im\Sigma(\omega,T)$ has the property

$$Im\Sigma(\omega,T) = \begin{cases} \sim T & \omega \ll T \\ \sim \omega & T \ll \omega \end{cases} \quad (15.154)$$

instead of a usual quadratic dependence on ω and T (see Sect. 7.2). In case of (15.154) we speak of a marginal Fermi liquid which we have discussed in some detail in Sect. 10.9. There we have shown that a one-band Hubbard model on a square lattice can indeed produce marginal Fermi liquid behavior, but only for half filling and possibly near a hole-doping concentration of $x = 0.12$. Therefore, it remains an open question how the resistivity can be explained within that model, since the linear in T behavior extends to higher doping concentrations. It has been speculated that a state with broken time-reversal symmetry but unbroken translational symmetry might be responsible for the strange metal behavior [463].

Within the mean-field scenario the strange metal regime is identified with the RVB regime, in which $\chi_{ij} \neq 0$ is the only nonvanishing order parameter [274]. Certainly, that phase is not a Fermi liquid. The large Fermi surface seen in ARPES experiments is interpreted here as the one of spinons, while the Drude peak in the conductivity is associated with the small hole density δ. Nevertheless, the relation of the spinon Fermi surface to ARPES experiments remains unresolved.

15.5.4 Optical Properties: Drude Peak

It is gratifying that finite cluster calculations for the one-band Hubbard model or $t - J$ model can qualitatively explain the Drude peak as well as a midgap state which is found for moderate doping[10]. It is instructive to have a closer look at these features. Since they are antiferromagnetic charge-transfer type insulators, La_2CuO_4 and Nd_2CuO_4 have a gap of order 1 eV in the frequency-dependent conductivity $\sigma(\omega)$. Nonetheless, when these materials are doped with holes or electrons, they become metallic and therefore a Drude conductivity of the form

$$\sigma(\omega) = \frac{\sigma(0)}{1 - i\omega\bar{\tau}} \quad (15.155)$$

should exist at low enough frequencies. Here $\sigma(0) = ne^2\bar{\tau}/m$ is the static conductivity and it seems appropriate to identify n with the hole or electron doping concentrations. The scattering rate of the charge carriers is $1/\bar{\tau}$. Experiments show[11] two noticeable features: the scattering rate behaves in an unusual manner, i.e., at low temperatures $1/\bar{\tau} \propto \omega$ over a wide range of frequencies, and there is considerable absorption in the gap, called mid-infrared

---

[10] see, e.g., [77]
[11] for reviews see, e.g., [386, 446]

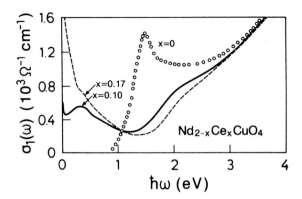

Fig. 15.27. Doping dependence of the optical conductivity $\sigma_1(\omega)$ of $Nd_{2-x}Ce_xCuO_4$. (Redrawn from [457])

absorption. The experimental results for $\sigma_1(\omega) = Re\{\sigma(\omega)\}$ of the electron-doped system $Nd_{2-x}Ce_xCuO_4$ are shown in Fig. 15.27 for different doping concentrations. The results for the hole-doped system $La_{2-x}Sr_xCuO_4$ look qualitatively similar.

Before describing the numerical results, we want to outline some of the properties of the conductivity tensor $\sigma_{\alpha\beta}(\omega)$. We assume a space and time-dependent vector potential acting on the electron system of the form

$$\mathbf{A}(j,t) = \frac{\hat{x}}{2\pi N} \sum_{\mathbf{q}} \int d\omega A_x(\mathbf{q},\omega) e^{i(\mathbf{q}\cdot\mathbf{R}_j - \omega t)} \quad , \quad (15.156)$$

where $\hat{x}$ is a unit vector in x direction and N is the number of sites. Furthermore, we require that $qa_0 \ll 1$ where a_0 is the lattice constant. In the presence of a vector potential, we have to replace the electron momentum $\mathbf{p}$ by $(\mathbf{p} - e\mathbf{A}(j,t))$. As usual, the velocity of light c is set $c = 1$. Due to $\mathbf{A}(j,t)$ an electron acquires an additional phase change when it hops in the x direction by one lattice site. This phase change is incorporated in the nearest-neighbor hopping matrix element t resulting in

$$ta^+_{j\sigma}a_{j+\delta\sigma} \to te^{-iea_0 A_x(j,t)} a^+_{j\sigma} a_{j+\delta\sigma}$$
$$ta^+_{j+\delta\sigma}a_{j\sigma} \to te^{iea_0 A_x(j,t)} a^+_{j+\delta\sigma} a_{j\sigma} \quad , \quad (15.157)$$

where $j + \delta$ denotes the nearest neighbor of site j in the x direction. In the following discussion, we set $a_0 = 1$.

We expand the Hamiltonian to second order in A_x and obtain

$$H_A = H_0 - \sum_i j^p_x(i) A_x(i,t) - \frac{1}{2} \sum_i T_x(i) A^2_x(i,t) \quad , \quad (15.158)$$

with H_0 given by (8.22). The two new terms describe the paramagnetic and diamagnetic current contributions to the energy. The paramagnetic current operator is

$$j_x^p(i) = -ite \sum_\sigma \left(a_{i\sigma}^+ a_{i+\delta\sigma} - a_{i+\delta\sigma}^+ a_{i\sigma}\right) \quad , \tag{15.159a}$$

while the diamagnetic contribution relates to the local kinetic energy $T_x(i)/e^2$

$$T_x(i) = -te^2 \sum_\sigma \left(a_{i\sigma}^+ a_{i+\delta\sigma} + a_{i+\delta\sigma}^+ a_{i\sigma}\right) \quad . \tag{15.159b}$$

To lowest order in $A_x(\mathbf{q},\omega)$ the total induced current is given by the expectation value with respect to the perturbed system $\langle\ldots\rangle_{H_A}$ of the operator $j_x(i,t) = -\partial H_A/\partial A_x(i,t)$. Therefore,

$$\langle j_x(\mathbf{q},\omega)\rangle_{H_A} = -\left[\langle -T_x\rangle + R_{xx}(\mathbf{q},\omega)\right] A_x(\mathbf{q},\omega) \quad , \tag{15.160}$$

where the expectation value $\langle\ldots\rangle$ is with respect to the unperturbed system. We have taken advantage of the fact that $T_x(i)$ is independent of the site index i. The ground state of the unperturbed system does not carry a current, and the paramagnetic current contribution therefore follows from the well-known Kubo relation of linear response theory.

$$R_{xx}(\mathbf{q},\omega) = -\frac{i}{N}\int_0^\infty dt\, e^{i\omega t} \langle [j_x^p(\mathbf{q},t), j_x^p(-\mathbf{q},0)]_-\rangle_{H_A} \tag{15.161}$$

with

$$j_x^p(\mathbf{q},t) = \sum_i e^{-i\mathbf{q}\cdot \mathbf{R}_i} j_x^p(i,t) \quad , \tag{15.162}$$

where the time dependence of $j_x^p(i,t)$ is given according to (7.36).

The optical conductivity $\sigma_{xx}(\omega)$ relates the induced current to the electric field $E_x(\mathbf{q}=0,\omega) = i\omega A_x(\mathbf{q}=0,\omega)$ and is therefore

$$\sigma_{xx}(\omega) = -\frac{1}{i\omega - \eta}\left[\langle -T_x\rangle + R_{xx}(\mathbf{q}=0,\omega)\right] \quad . \tag{15.163}$$

Here η is an infinitesimal damping factor.

One notices that the real (or absorptive) part of $\sigma_{xx}(\omega)$ contains a δ-function contribution and is given by

$$\mathrm{Re}\left\{\sigma_{xx}(\omega)\right\} = 2\pi D \delta(\omega) + \mathrm{Re}\left\{\sigma_{xx}^{\mathrm{reg}}(\omega)\right\} \quad , \tag{15.164}$$

where $\sigma_{xx}^{\mathrm{reg}}(\omega)$ is the finite frequency part of $\sigma_{xx}(\omega)$ and

$$D = \frac{1}{2}\left[\langle -T_x\rangle + R_{xx}(\mathbf{q}=0, \omega\to 0)\right] \quad . \tag{15.165}$$

This expression for D proves rather impractical because the two terms inside the bracket cancel partially. For a static vector potential the induced current $\langle j_x(\mathbf{q}\to 0, \omega=0)\rangle_{H_A} = 0$ and we obtain from (15.163) the relation

$$\langle -T_x\rangle + R_{xx}(\mathbf{q}\to 0,\omega=0)=0 \quad . \tag{15.166}$$

Therefore, D can also be written in the form

$$D = \frac{1}{2}[-R_{xx}(\mathbf{q}\to 0,\omega=0)+R_{xx}(\mathbf{q}=0,\omega\to 0)] \quad . \tag{15.167}$$

This expression turns out to be very useful when D is calculated by means of diagrams, a suitable method when the Hubbard or $t-J$ Hamiltonian H is expressed in terms of auxiliary fields, i.e., slave particle fields.

The δ-function or Drude term in (15.164) is due to free acceleration of the charge carriers. For noninteracting particles $Re\{\sigma_{xx}^{\text{reg}}(\omega)\}=0$, i.e., the real part of the conductivity, has a zero-frequency part only. This is a consequence of $[j_x,H]_-=0$, i.e., the current operator and H can be diagonalized simultaneously, and the expectation value in (15.161) vanishes. Evaluating $\langle -T_x\rangle$ we would find from (15.165) that $D = ne^2/(2m)$ in this case. Thereby it is assumed that each carrier has a kinetic energy of $-2t$, which is reasonable for low carrier concentrations and $t=(2m)^{-1}$. When there are scattering centers in the sample the term $2\pi D\delta(\omega)$ is replaced by

$$Re\{\sigma(\omega)\} = \frac{\sigma(0)}{1+\omega^2\bar{\tau}^2} \quad , \tag{15.168}$$

a consequence of (15.155). According to (15.163), the regular part of the optical conductivity is

$$Re\{\sigma_{xx}^{\text{reg}}(\omega)\} = -\frac{1}{\omega}Im\{R_{xx}(\mathbf{q}=0,\omega)\} \quad . \tag{15.169a}$$

For zero temperature and $\omega>0$, this expression can be rewritten with the help of (15.161,15.162) in the form

$$Re\{\sigma_{xx}^{\text{reg}}(\omega)\} = \frac{1}{N\omega}Im\left\{\left\langle \Phi_0\left|j_x^p\frac{1}{\omega-L_H+i\eta}j_x^p\right|\Phi_0\right\rangle\right\} \quad , \tag{15.169b}$$

where L_H refers to the Hubbard Hamiltonian (8.22), $|\Phi_0\rangle$ denotes the ground state of that Hamiltonian, and $j_x^p = \sum_i j_x^p(i)$. If we wish, we can replace L_H by $(H-E_0)$, where E_0 is the ground-state energy. The expression (15.169b) is suitable for numerical evaluation with the help of the Lanczos algorithm; see Appendix F, especially (F.9.). Note that we may also write

$$Re\{\sigma_{xx}^{\text{reg}}(\omega>0)\} = -\frac{\pi}{N\omega}\sum_{n\neq 0}|\langle\Phi_n|j_x^p|\Phi_0\rangle|^2\,\delta(\omega-E_n+E_0) \quad , \tag{15.170}$$

where the E_n are the energies of the excited states $|\Phi_n\rangle$ of the Hubbard system.

Numerical results for a 3×3 Hubbard cluster are shown in Fig. 15.28 for different doping concentrations. One notices an appreciable mid-infrared absorption below the Hubbard gap $\omega_g=5t$. The pseudogap below $\omega/t\simeq 2$ in the presence of 20 % doping is probably due to finite-size effects. The increase

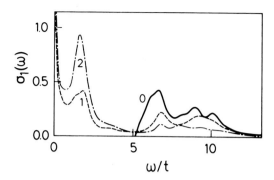

Fig. 15.28. Optical conductivity $\sigma_1(\omega)$ calculated for a 3×3 sites cluster within the Hubbard model. Curves 0, 1, 2 refer to half filling and to 10 % and 20 % hole doping, respectively, while t denotes the nearest-neighbor hopping matrix element. (From [425])

in absorption below the Hubbard gap goes hand in hand with a decrease of absorption in the upper Hubbard band. This agrees with the findings for the one-particle spectral density discussed in Sect. 8.2.

We obtain similar results when, instead of the Hubbard Hamiltonian, the $t - J$ Hamiltonian (10.101) is used. The upper Hubbard band is missing here, yet the mid-infrared absorption is similar in character as numerical diagonalizations of small clusters show [77]. When we calculate $\sigma_{xx}(\omega)$ with the help of Green's functions by using slave bosons, slave fermions, or other means of treating the strong correlations, we find that the mid-infrared absorption results predominantly from the incoherent part $G_{\mathrm{inc}}(\mathbf{k}, \omega)$ of $G(\mathbf{k}, \omega)$. It can be interpreted as a dipole transition from a s-like to a p-like spin bag or, alternatively, from the lowest quasiparticle band to an excited one [472, 493].

The absorption in the mid-infrared regime results from the coupling of the perturbing field (vector potential) to the small charge fluctuations associated with the spin-like excitations of the bag. This is a distinct mark of the strong, short-range antiferromagnetic correlations evidently present in the system even in the absence of long-range antiferromagnetic order.

We can also calculate the Drude weight D as function of the band-filling factor n_b, by using the sum rule for the conductivity. It relates the integrated optical conductivity $Re\{\sigma_{xx}(\omega)\} = \sigma_1(\omega)$ to the kinetic energy $2N\langle T_x(i)\rangle/e^2$ of the two dimensional system [307]. We do not prove this sum rule here but merely state the result

$$\int_0^\infty d\omega \sigma_1(\omega) = \frac{\pi}{2}\langle -T_x \rangle \quad . \tag{15.171}$$

One finds that D vanishes for $n = 1$ as well as for half filling ($n = 0$) and that it reaches a maximum for $n = 0.5$. A vanishing Drude weight at $n = 0$

indicates that the system is an insulator at half filling. More precisely, we find that $D \propto n/m^*$, where m^* is the effective mass of the holes near half filling.

15.5.5 Pairing Interactions

Next we consider the superconducting part of the phase diagram. The microscopic origin of the strong binding of electrons to Cooper pairs is an especially important point of consideration which will be discussed. Although phonons will certainly contribute to the pair formation, it is highly unlikely that they can cause the high T_c values which are observed. The nearness of antiferromagnetism and superconductivity together with the fact that electron correlations are strong in the high-T_c materials suggests an important contribution of spin fluctuations to the formation of Cooper pairs.

When studying the effects of spin fluctuations on the superconducting transition temperature, we have to distinguish between the overdoped and the underdoped regimes. In the overdoped regime the normal state shows Fermi-liquid behavior. The strong correlations are partially reduced here because of large hole concentrations. Then the probability becomes smaller that electrons scatter off each other by the Hubbard repulsion U. Therefore, an RPA approach, which is formulated in momentum- or **k** space is justified here. One should remember that the RPA has its strengths in treating long wavelength fluctuations, while it is inaccurate for short wavelengths. In underdoped samples short-range correlations are particularly important. They are the origin of the observed pseudogap. Therefore, we must think of methods which properly treat their effect on pairing. The $t-J$ model is a good starting point here. We also want to draw attention to the fact that in the calculation described before the coupling constant U has always been assumed to remain unchanged with varying hole concentration. That may be justified if at the end we adjust its value to fit experiments.

We begin with the overdoped regime. In order to determine the effect of antiferromagnetic spin fluctuations on pairing we have to solve Eliashberg's equations (15.129). For that we have to know the form of the corresponding boson propagator. There are two different approaches possible: one is to determine the propagator within a model, the other is to deduce it from experiments.

In the first case one starts from $G_0(\mathbf{p}, i\omega_n) = (i\omega_n - \epsilon_\mathbf{p})^{-1}$ where $\epsilon_\mathbf{p}$ is obtained from LDA or quantum chemical calculations. The energy dispersion is of the form (12.5). The unrenormalized values for t, t', t'' which have to be used in G_0 were obtained earlier by quantum chemical methods. They are listed in Sect. 12.1.1. The coupling of the conduction electrons to the spin fluctuations is given by

$$H_{\text{int}} = g \sum_{\mathbf{q}} \mathbf{s}(\mathbf{q}) \mathbf{S}(-\mathbf{q}) \tag{15.172}$$

where $\mathbf{S}(\mathbf{q})$ is the operator in terms of which spin fluctuations are expressed. For example, the retarded spin-susceptibility tensor is written as

$$\chi_{\alpha\beta}(\mathbf{q},t) = -i\theta(t)\left\langle [S_\alpha(\mathbf{q},t), S_\beta(-\mathbf{q},0)]_-\right\rangle \quad. \tag{15.173}$$

Since the spin fluctuations are set up by the same electrons which interact through them, the coupling constant g is in the Hubbard model given by the energy U. With the energy dispersion $\epsilon_\mathbf{p}$ we calculate Lindhard's function, i.e., the bare susceptibility (11.106). The temperature-dependent susceptibility $\chi(\mathbf{q}, i\omega_\nu)$ is determined by making use of the FLEX approximation (see Sect. 11.3.2). Next, the self-energy $\Sigma(\mathbf{p}, i\omega_n)$ is determined from the first of Eqs. (15.129). We write it here in the form

$$\Sigma(\mathbf{p}, i\omega_n) = g^2 T \sum_m \sum_{\mathbf{p}'} \chi\left(\mathbf{p}-\mathbf{p}', i(\omega_n - \omega_m)\right) G\left(\mathbf{p}', i\omega_m\right) \quad. \tag{15.174}$$

The role of the boson propagator $D(\mathbf{k}-\mathbf{k}', i\omega_\nu)$ is taken here by the magnetic susceptibility, i.e.,

$$D(\mathbf{k}, \omega_m) = \chi(\mathbf{k}, \omega_m) \quad. \tag{15.175}$$

This seems a simplification in view of the detailed discussion in Sect. 11.3.2 and of the self-energy diagrams shown in Fig. 11.14. However, we do not expect charge fluctuations to contribute much to Cooper pairing and, therefore neglecting them seems justified. Note that the prefactor g^2 is here $3U^2/2$ where the factor $3/2$ results from the three diagonal components of the susceptibility tensor. The computed $\Sigma(\mathbf{p}, i\omega_n)$ is used to determine the full Green's function $G(\mathbf{p}', i\omega_n)$ (see (7.16)). With its help, the modified electron-hole bubble and RPA susceptibility shown in Fig. 11.15 are redetermined. When set into (15.174) we obtain an improved self-energy $\Sigma(\mathbf{p}, i\omega_n)$. The process is continued until self-consistency is attained. The final quantities are set into the second of Eliashberg's equations (15.129), from which the superconducting transition temperature due to spin fluctuations is obtained.

This way of estimating T_c has been further substantiated by dynamical cluster Monte Carlo calculations based on the Hubbard Hamiltonian. They were performed for a 2×2 cluster embedded self-consistently, with 15 % hole concentration [304]. In agreement with the above it is found that the pairing interaction is well approximated by the form

$$V_{\text{eff}}(\mathbf{q}, \omega) = \frac{3}{2} U_{\text{eff}}^2 \chi(\mathbf{q}, \omega) \tag{15.176}$$

where U_{eff} is an effective Hubbard interaction which is adjusted to fit the numerical findings. The above discussion also provides for an explanation as to why the superconducting order parameter cannot be of the conventional s-wave type. For that purpose we consider the weak coupling or BCS limit for an interaction of the form of (15.176), which implies that the static susceptibility must be used. The self-consistency condition (15.53) here takes the form

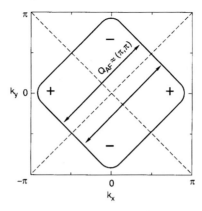

Fig. 15.29. Schematic plot of the large Fermi surface in overdoped La_2CuO_4 (compare with Fig. 15.26). For $\mathbf{q} = \mathbf{Q}_{AF}$ the spin susceptibility is strongly enhanced. It scatters the order parameter $\Delta_\mathbf{k}$ into $\Delta_{\mathbf{k}+\mathbf{q}}$. This is accompanied with a change of sign in case of d-wave pairing and, therefore, can lead to a solution for (15.177). Dashed lines indicate a vanishing gap $\Delta_\mathbf{k} = 0$. (From [109])

$$\Delta_\mathbf{k} = -\frac{3}{4} U_{\text{eff}}^2 \sum_{\mathbf{k}'} \frac{\chi(\mathbf{k}-\mathbf{k}')}{E(\mathbf{k}')} \Delta_{\mathbf{k}'} \quad . \quad (15.177)$$

In distinction to the attractive electron-phonon or BCS interaction (15.6) we are dealing here with a *repulsive* interaction which is strongly momentum dependent. In fact, $\chi(\mathbf{q})$ is strongly peaked near $\mathbf{q} = \mathbf{Q}$, i.e., the antiferromagnetic reciprocal lattice vector. As seen from Fig. 15.29 the vector $\mathbf{Q} = (\pi, \pi)$ connects nearly parallel parts of the Fermi surface. Therefore, in order to find a solution of (15.177) $\Delta_\mathbf{k}$ and $\Delta_{\mathbf{k}+\mathbf{Q}}$ must have different signs in order to overcome the minus sign on the right-hand side of (15.177). This excludes s-wave pairing as mentioned above, but favors d-wave pairing.

The second route, i.e., to deduce $\chi(\mathbf{q}, \omega)$ from available experiments and to calculate with it T_c, was pursued for $YBa_2Cu_3O_{6.6}$ which we consider in the following [80]. The reason that this substance was chosen is that detailed magnetic neutron scattering data as well as data from ARPES are available. In other materials, only one at most, of these data sets is known. However, both sets are required in order to derive accurate values for a parametrized function $\chi(\mathbf{q}, \omega)$ and also for U_{eff}. The latter is found to be $U_{\text{eff}} = 1.6$ eV. With the required input deduced from experiments, the solution of the linearized Eliashberg equation (15.129) yields a value of $T_c = 174$ K. This exceeds the measured value of 150 K. The calculations show that AF spin fluctuations can explain the high transition temperatures. At the same time, the ARPES data can also be quantitatively described by the theory [80, 309].

Next we consider the underdoped regime. Here we look for an alternative to the FLEX approximation. The limitation of the latter is apparent in view of the violation of Luttinger's theorem in the underdoped regime. That theorem

15.5 High-T$_c$ Superconductors

was proven by applying perturbation theory. It states that the volume in momentum space enclosed by the Fermi surface of a noninteracting system of electrons remains unchanged when electron interactions are switched on. This implies that a Mott insulator cannot be obtained from perturbation theory because it does not have a Fermi surface like the corresponding noninteracting system does.

On the other hand, the $t - J$ model seems an appropriate tool to treat the underdoped regime. The model itself was discussed at length in Sect. 10.5, and the motion of holes in an AF environment was also considered there. Here we want to extend the theory to the interaction of two holes. Intuitively it seems plausible that two holes should attract each other. The reasoning is as follows. A single hole breaks four bonds $J\mathbf{S}_i\mathbf{S}_j$ on an AF square lattice. When two holes are next to each other, only seven instead of eight bonds are broken (see Fig. 15.30a). This effect will be the more important, the larger the ratio J/t is. Moreover, when a single hole is moving through the system its bandwidth is of order $J(=4t^2/U)$, as previously discussed. However, two holes in close neighborhood can move with an effective hopping of order t, since a second hole can heal spin defects caused by the motion of the first one.

Therefore, the holes form a bound state with a binding energy which decreases as J/t decreases. This decrease is due to the loss of kinetic energy which is counterbalancing the gain in exchange energy. When two holes occupy neighboring sites, each of them can hop to only three sites instead of four sites. This limits the formation of a bound state to ratios J/t which exceed a critical value $(J/t)_{\text{crit}}$. Because $J = 4t^2/U$ we are in the limit $J \ll t$ when underdoped cuprates are considered. Numerical calculations for clusters up to 26 sites based on the Lanczos' method (see Appendix F) find a bound state for ratios $J/t \gtrsim 0.15$ [374]. However, the problem can be also studied by an analytic approach, which provides better insight into binding. For example, it is found that the pair wavefunction of two holes with total pairing momentum zero must be of a p- or d-wave type. This is an important result. The analytic method yields a generalization of previous results obtained, when the projection method discussed in Sect. 10.5 was applied to the motion of one hole [105, 491]. We sketch it here in form of a wavefunction approach.

The $t - J$ Hamiltonian H_{t-J} is divided into

$$H_{t-J} = H_0 + H_1 \tag{15.178}$$

where $H_0 = H_t + H_{\text{Ising}}$ contains the hopping and the Ising part (see (10.101,10.102)) while H_1 is given by (10.103). When a hole is created at site m, it remains confined to that site as long as only H_0 is acting on it. It merely generates a spin bag. This was explained in Sect. 10.6. Trugman paths are excluded here. For the eigenstates of H_{t-J} we make the ansatz

$$|\phi_m\rangle = \sum_\nu \alpha_\nu \sum_P |m, \nu, P\rangle \quad . \tag{15.179}$$

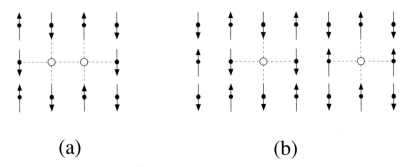

Fig. 15.30. Two holes on an AF square lattice (a) at neighboring sites and (b) farther apart.

Here ν denotes the number of hops of the hole which is starting from the state $c_{m\uparrow}|\Phi_{\text{Néel}}\rangle$ where $|\Phi_{\text{Néel}}\rangle$ is the Néel state. Furthermore, P labels the path taken by the hole. The α_ν are determined by minimizing $\langle\phi_m|H_0|\phi_m\rangle$. The state $|\phi_m\rangle$ can be thought of as a quasiparticle, i.e., a hole with its spin bag. As discussed in Sect. 10.6 it delocalizes due to the action of H_1. Note that when the quasiparticle is delocalized it remains on the sublattice on which it was created.

We are interested here in the two-hole wavefunction for which we make in analogy to (15.179) the ansatz

$$|\phi_{mn}\rangle = \sum_{\mu\nu} \alpha_{\mu\nu} \sum_{PP'} |mn, \mu\nu, PP'\rangle \quad . \tag{15.180}$$

Here, the starting point is the state $c_{m\uparrow}c_{n\downarrow}|\Phi_{\text{Néel}}\rangle$. It is important to restrict the states $|mn, \mu\nu, PP'\rangle$ to *irreducible* ones. A state with starting points m and n is called irreducible if it is not possible to generate that state with fewer hops by starting from a different pair of sites. An example of a reducible state, i.e., one which does not satisfy that criteria is shown in Fig. 15.31. Again, the $\alpha_{\mu\nu}$ are determined by minimization of $\langle\phi_{mn}|H_0|\phi_{mn}\rangle/\langle\phi_{mn}|\phi_{mn}\rangle$.

The Fourier transform of $|\phi_{mn}\rangle$ is

$$|\Phi(\mathbf{p},\mathbf{k})\rangle = \frac{2}{N}\sum_{m,n} e^{-i(\mathbf{k}-\frac{\mathbf{p}}{2})\mathbf{R}_n} e^{i(\mathbf{k}+\frac{\mathbf{p}}{2})\mathbf{R}_m} |\phi_{mn}\rangle \quad . \tag{15.181}$$

It corresponds to a pair of holes with total momentum $\mathbf{p}$ and relative momentum $\mathbf{k}$. Two holes connected by a string or irreducible path are called a spin bipolaron. It constitutes a compromise between a gain in kinetic energy and a reduced disturbance of the antiferromagnetic background. With the help of the $|\Phi(\mathbf{p},\mathbf{k})\rangle$ we construct the eigenstate of the two holes with total momentum $\mathbf{p}$ of the full Hamiltonian H_{t-J}. We make the ansatz

$$|\psi(\mathbf{p})\rangle = \sum_{\mathbf{k}} \gamma(\mathbf{k}) |\Phi(\mathbf{p},\mathbf{k})\rangle \tag{15.182}$$

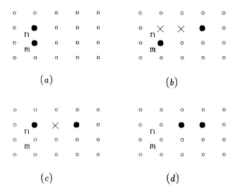

Fig. 15.31. Example of a state which is reducible. Starting from the state (a) the state (b) is obtained after two hops. Another hop leads to (c). This state, however, can be obtained by one hop only if one is starting from (d). Thus, (b) is an irreducible state with starting points m, n but (c) is not. (From [105])

where the sum is over all $\mathbf{k}$ states of the AF Brillouin zone. Minimizing the energy with respect to H_{t-J} results in an equation for the $\gamma(\mathbf{k})$ which has to be solved numerically [105]. From the forms of (15.181, 15.182) one can gain interesting insights. For that purpose we consider first the Fourier transform of $|\phi_m\rangle$, i.e.,

$$|\phi(\mathbf{k})\rangle = \sqrt{\frac{2}{N}} \sum_m e^{i\mathbf{k}\mathbf{R}_m} |\phi_m\rangle \quad . \tag{15.183}$$

The index m runs over the sites of only one sublattice. This is the reason for the prefactor $(2/N)^{1/2}$ rather than $(1/N)^{1/2}$. One notices that when $\mathbf{k}$ is changed to $\mathbf{k}+\mathbf{Q}_{1,2}$ where $\mathbf{Q}_1 = (\pi, \pi)$ and $\mathbf{Q}_2 = (\pi, -\pi)$ then $|\phi(\mathbf{k}+\mathbf{Q}_{1,2})\rangle = \pm|\phi(\mathbf{k})\rangle$. The plus sign applies only when m belongs to the same sublattice as site $(0,0)$ does. From that result, it follows that the two-hole wavefunction has the property

$$|\Phi(\mathbf{p}, \mathbf{k}+\mathbf{Q}_{1,2})\rangle = -|\Phi(\mathbf{p}, \mathbf{k})\rangle \quad . \tag{15.184}$$

Now consider $|\Phi(0, \mathbf{k})\rangle$, i.e., a pair function with zero pairing momentum. A symmetry transformation T with $T\mathbf{k} = \mathbf{k} + \mathbf{Q}_{1,2}$ leaves the pair function invariant and that requires

$$\gamma(\mathbf{k} + \mathbf{Q}_{1,2}) = -\gamma(\mathbf{k}) \quad . \tag{15.185}$$

Such a symmetry transformation may apply to $\mathbf{k}$ points on the surface of the AF Brillouin zone and in this case (15.185) is a boundary condition on the form of the two-hole wavefunction. Two examples of $|\Phi(0, \mathbf{k})\rangle$ which satisfy (15.185) are shown in Fig. 15.32.

The two situations correspond to p and d wave pairing. This shows in a transparent way why numerical calculations on clusters for the pair susceptibility within the $t - J$ model find peaks at low frequencies only in the p- and

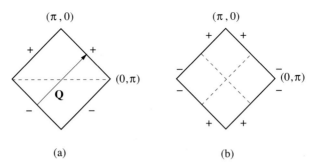

Fig. 15.32. Two examples of sign of $\gamma(\mathbf{k})$ on the boundary of the AF Brillouin zone which fulfill (15.185). Dashed lines separate regions with different signs of the $|\Phi(0, \mathbf{k})\rangle$. The two-particle wavefunctions correspond to p- and d-wave pairing. (From [105])

d-channels [41, 79]. The above theory together with further refinements [490] yields, for the binding energy, reasonable agreement with the results of exact diagonalizations of small clusters. Binding is obtained for $J/t \gtrsim 1/6$. The theory which here was presented in **k**-space can also be reformulated in **r**-space [492].

The above considerations provide an explanation for a pairing interaction in the underdoped regime. However, they are not able to describe superconductivity, i.e., pair condensation which takes place at low temperatures. What is gratifying is that they yield a pair function which is either of d or p type.

Let us now come back to the molecular-field description of the $t - J$ model and to the phase diagram, shown in Fig. 15.28. Recalling that an electron is decomposed as $\hat{a}_{i\sigma}^+ = f_{i\sigma}^+ b_i$, the superconducting order parameter $\langle \hat{a}_{i\sigma}^+ \hat{a}_{j-\sigma}^+ \rangle$ can be written in molecular-field approximation as

$$\langle \hat{a}_{i\sigma}^+ \hat{a}_{j-\sigma}^+ \rangle = \langle f_{i\sigma}^+ b_i f_{j-\sigma}^+ b_j \rangle$$
$$= \langle b \rangle^2 \langle f_{i\sigma}^+ f_{j-\sigma}^+ \rangle \quad . \qquad (15.186)$$

This implies that in the superconducting phase Δ_{ij} (defined by (10.146)) as well as $\langle b \rangle$ have to be nonzero [249]. In addition $\chi_{ij} \neq 0$ (see (15.153)) since holons must be able to move freely through the system. A finite $\langle b \rangle \neq 0$ can be interpreted as boson condensation. Looking at different self-consistent solutions of the mean-field Hamiltonian for a square lattice, one finds that the one where Δ_{ij} has a d-wave symmetry has lowest energy. It is also found that a superconducting solution does exist only below a critical doping concentration x_{crit}. Despite various reservations one might have against the molecular field approximation one should take notice that the simple model can describe a number of essential features of doped superconducting Mott-Hubbard insulators.

15.5.6 Stripe Formation

In the hole-doped part of the phase diagram shown in Fig. 15.24 one notices a dip in T_c around $x = 0.12$. It is related to the formation of stripes in some of the cuprates, notably in $La_{2-x}Ba_xCuO_4$ and $La_{2-x-y}Nd_ySr_xCuO_4$. Stripes are a general expression for one-dimensional density or spin-density waves which break the translational and rotational symmetry of the CuO_2 planes. We speak of density stripes when

$$\langle n(\mathbf{R}) \rangle = n_{av} + Re\left[e^{i\mathbf{QR}}\phi_n(\mathbf{R})\right] \tag{15.187}$$

and of spin-density stripes when

$$\langle m_\alpha(\mathbf{R}) \rangle = \mathbf{m}_{av} + Re\left[e^{i\mathbf{QR}}\phi_{m\alpha}(\mathbf{R})\right] \quad , \quad \alpha = 1, 2, 3 \tag{15.188}$$

Here n_{av} and $\mathbf{m}_{av}$ are the average density and magnetization, respectively. The prefactor $\exp(i\mathbf{QR})$ characterizes the wave length and direction of the charge-density wave (CDW) or spin-density wave (SDW). The functions $\phi_n(\mathbf{R})$, $\phi_{m\alpha}(\mathbf{R})$ are generally complex because the CDW or SDW can slide. They can be shifted continuously without any loss of energy, provided the $\mathbf{Q}$ vector is incommensurate with the reciprocal lattice vector. When $\mathbf{Q}$ is a reciprocal lattice vector the functions ϕ_n, $\phi_{m\alpha}$ are real. In case of CDW we may identify $\mathbf{R}$ as the center coordinate of an electron and a hole, i.e.,

$$\langle a_{i\sigma}^+ a_{j\sigma} \rangle = f_{av}(\mathbf{r}) + f_n(\mathbf{r}) Re\left[e^{i\mathbf{QR}}\phi_n(\mathbf{R})\right] \quad , \tag{15.189}$$

where $\mathbf{r} = \mathbf{R}_i - \mathbf{R}_j$ and $\mathbf{R} = \frac{1}{2}(\mathbf{R}_i + \mathbf{R}_j)$. The functions depending on $\mathbf{r}$ are short ranged. A CDW is classified according to the point symmetry of the order parameter, i.e., of $f_n(\mathbf{r})$. In the simplest case the Fourier transform of $f_n(\mathbf{r})$ is $\mathbf{k}$ independent; then, we speak of an s-CDW and the density changes within a unit cell remain symmetric. Similarly, we can define CDW of higher angular momenta. Special attention has been paid to d-CDWs [471].

Generally one can distinguish between charge modulations on lattice sites and on lattice bonds, which connect lattice sites. This can be visualized in Fig. 15.33 for charge- as well as spin-density waves. On a square lattice a unidirectional s- as well as d-CDW with vector $\mathbf{Q} = (0,1)$ or $(1,0)$ in (15.187) reduces the point symmetry from C_4 to C_2. Generally both types of CDW are superimposed.

As mentioned earlier, CDWs and SDWs have been observed in some of the hole-doped cuprates. Experimental verification of CDWs comes most directly from resonant soft X-ray scattering. In this way, stripes were unambiguously identified in $La_{2-x}Ba_xCuO_4$ for $x = 1/8$ [1]. For that system sublattice peaks were observed in the CuO_2 planes at $\mathbf{Q} = 2\pi(0.25 \pm 0.02, 0)$. The lattice constant has been set equal to unity here. Similar observations were made for $La_{1.8-x}Eu_{0.2}Sr_xCuO_4$, again at $x = 1/8$ [118]. In Sect. 10.9.1 we have seen that in the Hubbard model on a square lattice at a hole doping of $x \simeq 0.12$ the Fermi energy coincides with the (broadened) van Hove singularity in

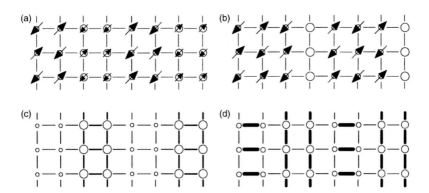

Fig. 15.33. Examples of bond (a) and site (b) centered SDWs as well as CDWs (c) and (d). The sizes of empty circles and of spins are in proportion to the hole density and magnetization vectors. The strength of a line between sites quantifies bond modulations. Note that the periods are eight and four for the SDW and CDW, respectively. (From [471])

the density of states. Therefore we expect that at this doping concentration the system is very sensitive to instabilities. This may serve as one possible explanation why strings are forming at that hole concentration.

When holes cluster in the form of a CDW there must be an effective attractive interaction between them. However, in the one-band (8.22) and three-band (12.7) Hubbard model only repulsive interactions do occur. So, how can we construct attractive interactions from the latter? The answer is that it is the kinetic energy in combination with the repulsive interactions which may result in an attraction. The simplest example is the $t - J$ model discussed in Sect. 10.6. A large Hubbard interaction U together with hopping matrix element $-t$ between nearest-neighbor sites transforms in the strong correlation limit into a Heisenberg spin-spin interaction with coupling constant $J = 4t^2/U$. Electrons at neighboring sites with antiparallel spin experience, therefore, an attractive interaction, although the Hubbard Hamiltonian only contains repulsive interactions. In the following we want to show that attractions of that type can lead to phase separations. Thereby we use the three-band Hubbard model as well as the $t - J$ model.

We begin with the three-band Hubbard model defined by the Hamiltonian (12.7). We neglect interactions between different oxygen sites, i.e., we set $U_{pp} = 0$. The binding energy of two holes is given by $E_b = (E_{2h} - E_0) - 2(E_{1h} - E_0)$, where E_0 is the ground-state energy of the undoped system, while E_{1h}, E_{2h} are the ones for a system with one and with two holes, respectively. In order to understand the origin of binding, we consider the limit $U_d \to \infty$. In that case, double occupancies of sites are excluded. If $U_{pd} \gg (\epsilon_p - \epsilon_d)$ the system tries to prevent configurations with a hole on a Cu and on a neighboring oxygen site. The resulting formation of hole droplets is easily

15.5 High-T_c Superconductors

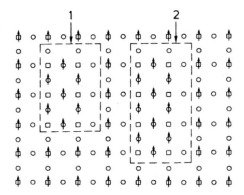

Fig. 15.34. Formation of hole droplets in the presence of hole doping when the interactions are strong, i.e., $U_d, U_p \to \infty$, $U_{pd} \gg (\epsilon_p - \epsilon_d)$. In the vicinity of doped holes, occupation of nearest neighbor sites is avoided by holes moving from Cu to O sites. Shown are (1) six unit cells with seven holes and (2) eight unit cells with ten holes. (From [426])

seen from Fig. 15.34. By moving holes from Cu sites to O sites, occupations of nearest neighbor sites can be avoided even when additional holes are doped into the system. Detailed numerical studies find hole binding for a certain parameter range in U_d and U_p. What seems to be important is that U_{pd} is sufficiently large compared with $(\epsilon_p - \epsilon_d)$ and with t [426].

Phase separation is also obtained in the $t - J$ model described by the Hamiltonian (10.101). For small hole doping and large ratios J/t this is seen as follows. When one hole is injected into a Heisenberg AF, four bonds $J\mathbf{S}_i\mathbf{S}_j$ are broken. When two holes are introduced and they are next to each other, only seven bonds are broken instead of eight (see Fig. 15.30). This favors clusters of holes. However, in the cuprates $J/t < 1$. Therefore, it is important to find out down to which ratios J/t the formation of stripes can be expected. Different numerical techniques have been applied to answer that question. Although different methods differ in their results, mainly due to small cluster sizes which can only be treated, there is evidence that for $J/t \simeq 0.5$ and hole doping $x \simeq 0.1$ the ground state of the $t - J$ model is striped. A critical evaluation of the different works is found in Ref. [483].

An interesting topic is whether or not an inhomogeneous ground state with stripes favors or disfavors Cooper-pair formation. From the phase diagram in Fig. 15.24 it seems that the latter is the case. The transition temperature of $\text{La}_{2-x}\text{Ba}_x\text{CuO}_4$ shows a distinct drop at $x = 1/8$. This hole concentration gives rise to stripes in the form of a SDW and CDW as shown in Fig. 15.33.

This completes the discussion of the different regions of the phase diagram in Fig. 15.28.

Acknowledgements

I have profited greatly from many discussions which I had with numerous colleagues. During the last years these were particularly J. Betouras, I. Eremin, L. Hozoi, Y. Kakehashi, S. Nishimoto, K. Penc, F. Pollmann, E. Runge, N. Shannon, K. Shtengel, O. Sikora, H. Stoll, A. Stoyanova, P. Thalmeier, P. Wrobel, Y. Yushankhai and G. Zwicknagl. My sincere thanks go to them.

Some colleagues were so kind as to look over different chapters of the book. Their criticism helped to improve at several places the clarity of writing. Special thanks go to R. Eder, Y. Kakehashi, E. Runge, H. Stoll, T. Takimoto and P. Thalmeier. Dr. C. Benassi helped by editorial work to improve considerably the quality of the English and so did Ms. S. Kang. I render sincere thanks to them.

Last but not least, I thank whole heartedly Mrs. Regine Schuppe who prepared the manuscript in a masterly manner. I seriously doubt that without her help the book would have been completed in the foreseeable future.

A

Some Relations for Cumulants

In the following we derive some relations for cumulants which are helpful for becoming better acquainted with them.

We start out showing how a cumulant $\langle \Phi_1 | A | \Phi_2 \rangle^c$ changes when the state Φ_2 is transformed into a state Φ_3. For that purpose we consider an infinitesimal transformation of Φ_2 into Φ_3, i.e.,

$$\Phi_3' = e^{\delta S} \Phi_2 = (1 + \delta S) \Phi_2 \; ; \; \delta S = \epsilon S \text{ with } \epsilon \ll 1 \; . \tag{A.1}$$

By using the form (5.19) we find that

$$\langle \Phi_1 | A | \Phi_3' \rangle^c = \langle \Phi_1 | A(1 + \delta S) | \Phi_2 \rangle^c \; . \tag{A.2}$$

Assume that we go from Φ_2 to Φ_3 through a sequence of N infinitesimal transformations δS_i. In the limit $N \to \infty$ we find

$$\langle \Phi_1 | A | \Phi_3 \rangle^c = \langle \Phi_1 | A \prod_{i=1}^{N}(1 + \delta S_i) | \Phi_2 \rangle^c$$
$$= \langle \Phi_1 | A \Omega_{2 \to 3} | \Phi_2 \rangle^c \; . \tag{A.3}$$

Here

$$\Omega_{2 \to 3} = \lim_{N \to \infty} \prod_{i=1}^{N}(1 + \delta S_i) \tag{A.4}$$

is called a *cumulant wave operator* because it appears in a cumulant and resembles the wave operator in quantum mechanics which transforms the ground state of H_0 into the one of H. Note that $\Omega_{2 \to 3}$ always begins with the number 1. It is not uniquely defined though, because many different paths in Hilbert space may connect Φ_2 with Φ_3. However, that has no consequences because different representations of $\Omega_{2 \to 3}$ lead to the same cumulant, i.e., the difference vanishes. Although these considerations seem a bit formal, they are extremely useful.

As pointed out before, the cumulant wave operator $|\Omega\rangle$ characterizes the exact ground state. In that respect it resembles the wave operator $\tilde{\Omega}$ (5.34) which transforms the unperturbed ground state $|\Phi_0\rangle$ into the exact one $|\psi_0\rangle$. A formal relationship between the two can be established by using that

$$\nabla_{\Phi_0}\langle\Phi_0 \mid 1 \mid \psi_0\rangle^c = \nabla_{\Phi_0} ln\langle\Phi_0 \mid \psi_0\rangle$$
$$= \frac{|\psi_0\rangle}{\langle\Phi_0 \mid \psi_0\rangle} \quad . \tag{A.5}$$

Reexpressing the left hand side in terms of Ω and the right hand side in terms of $\tilde{\Omega}$ we obtain

$$\nabla_{left\Phi_0}\langle\Phi_0 \mid \Omega \mid \Phi_0\rangle^c = \frac{|\psi_0\rangle}{\langle\Phi_0 \mid \psi_0\rangle}$$
$$= \frac{\tilde{\Omega} \mid \Phi_0\rangle}{\langle\Phi_0 \mid \tilde{\Omega} \mid \Phi_0\rangle} \quad . \tag{A.6}$$

This provides for the link between $|\Omega\rangle$ and $\tilde{\Omega}$.

B
Scattering Matrix in Single-Centre and Two-Centre Approximation

Here we want to present some more details as regard the decomposition of the scattering matrix in terms of increments. We start from

$$|S) = |\Omega - 1) = \lim_{z \to 0} \sum_{n=1}^{\infty} \left| \left(\frac{1}{z - H_{\text{SCF}}} H_{\text{res}} \right)^n \right) \quad (B.1)$$

and assume that SCF ground state has been expressed in terms of Wannier orbitals, see (5.59).

In order to include H_{res} by means of quantum-chemical program packages, we must slightly reformulate it. The creation and annihilation operators in H_{res} should refer to Wannier orbitals (occupied space) and to orbitals in unoccupied or virtual space. Therefore we express H_{res} not in terms of the $a_{i\sigma}^+$, $a_{i\sigma}$ operators, but instead in $c_{\nu\sigma}^+(I)$, $c_{\nu\sigma}(I)$, i.e., creation and annihilation operators of Wannier orbitals, and $\tilde{a}_{i\sigma}^+(I)$, $\tilde{a}_{i\sigma}(I)$ operators. The $\tilde{a}_{i\sigma}^+(I)$, $\tilde{a}_{i\sigma}(I)$ refer to modified basis functions $\tilde{f}_i(\mathbf{r})$. They are obtained by orthogonalizing the $f_i(\mathbf{r})$ to the occupied space, i.e., to the Wannier orbitals. The index I labels the site or bond they are centered on. This enables us to decompose the residual interactions in the form

$$H_{\text{res}} = \sum_I H_I + \sum_{\langle IJ \rangle} H_{IJ} + \sum_{\langle IJK \rangle} H_{IJK} + \sum_{\langle IJKL \rangle} H_{IJKL} \quad . \quad (B.2)$$

The different parts indicate to which centres the creation and annihilation operators $c_{\nu\sigma}^+(I)$, $c_{\nu\sigma}(I)$, $\tilde{a}_{i\sigma}^+(I)$, $\tilde{a}_{i\sigma}(I)$ are attached. The brackets $\langle \ldots \rangle$ refer to pairs, triples and quadruples. There are at most products of four creation and annihilation operators appearing in H_{SCF} and therefore the decomposition terminates with H_{IJKL}.

Keeping in mind that we want to reduce the correlation problem of N electrons to that of a few electrons, we introduce operators A_α of the form

$$A_\alpha = \lim_{z \to 0} \frac{1}{z - H_{\text{SCF}}} H_\alpha \quad , \quad (B.3)$$

with $\alpha = I$, $\langle IJ \rangle$, $\langle IJK \rangle$ and $\langle IJKL \rangle$. This enables us to rewrite

$$|S) = \sum_n \left| \left(\sum_\alpha A_\alpha \right)^n \right) \quad . \tag{B.4}$$

This form is particularly amenable to appropriate approximations. The method of increments is based on a decomposition of the right-hand side into terms with one Greek index only plus the remaining ones, i.e.,

$$|S) = \sum_\alpha \left| \left(\sum_{n=1}^\infty A_\alpha^n \right) \right) + \sum_{\alpha \neq \beta} |T_{\alpha\beta})$$
$$= \sum_\alpha |S_\alpha) + \sum_{\alpha \neq \beta} |T_{\alpha\beta}) \quad . \tag{B.5}$$

The operators $T_{\alpha\beta}$ begin from the left with products of A_α until A_β follows as the first operator with an index $\beta \neq \alpha$. Examples are $A_\alpha A_\alpha A_\beta A_\gamma$ or $A_\alpha A_\alpha A_\alpha A_\beta A_\alpha \ldots$. One notices hat S_α is the scattering operator of a Hamiltonian $H_{\text{SCF}} + H_\alpha$. The matrix $T_{\alpha\beta}$ can be written by factorizing out the part which depends on products of the two operators A_α and A_β only, i.e.,

$$T_{\alpha\beta} = (A_\alpha A_\beta + A_\alpha A_\beta A_\alpha + \ldots) \left(1 + \sum_{\gamma \neq \alpha, \beta} S_\gamma + \sum_{\gamma \neq \alpha, \beta, \delta} T_{\gamma\delta} \right) \quad . \tag{B.6}$$

Adding $T_{\alpha\beta} + T_{\beta\alpha}$ one notices that the terms containing A_α and A_β only, constitute the scattering matrix $\tilde{S}_{\alpha\beta}$ of a Hamiltonian $H_{\text{SCF}} + H_\alpha + H_\beta$, except that the contributions $S_\alpha + S_\beta$ are missing because they depend on one index only. Therefore we max write

$$T_{\alpha\beta} + T_{\beta\alpha} = \left(\tilde{S}_{\alpha\beta} - S_\alpha - S_\beta \right) \left(1 + \sum_{\gamma \neq \alpha, \beta} S_\gamma + \sum_{\gamma \neq \alpha, \beta, \delta} T_{\gamma\delta} \right) \quad . \tag{B.7}$$

The procedure may be continued by factorizing $T_{\gamma\delta}$ as done before for $T_{\alpha\beta}$ [139].

The largest contribution to $|S)$ comes certainly from the part H_I in (B.2). When only $|S)$ is kept, this is called the *single-centre approximation*. In that case all electrons in $|\Phi_{\text{SCF}}\rangle$ are kept frozen, except those in Wannier orbitals centred at I, i.e., only $\prod_{\nu\sigma} c_{\nu\sigma}^+(I)|0\rangle$ has to be correlated. The operator $S_{\alpha=1}$ is the scattering matrix of a Hamiltonian $H_{\text{SCF}} + H_I$. It describes the excitations of electrons from orthogonal localized occupied orbitals ν at centre I into virtual orbitals at the same centre. The N electron problem has hence been reduced to a problem involving few electrons only with a strongly reduced virtual space. Within the single-centre approximation we can write the cumulant scattering operator as

B Scattering Matrix in Single-Centre and Two-Centre Approximation

$$|S) = \sum_I |S_I) \qquad (B.8)$$

and

$$E_{\text{corr}} = \sum_I (H_I|S_I)$$

$$= \sum_I \epsilon_I \qquad (B.9)$$

because all other cumulants vanish.

An improved level of approximation consists in freezing all electrons in $|\Phi_{\text{SCF}}\rangle$ except those at centres I and J and furthermore neglecting H_{IJK} and H_{IJKL}. Thus we include in the calculations additionally S_α with $\alpha = IJ$ and $T_{\alpha\beta}$ with α and β being I, J and IJ. We call this a *two-centre approximation*. It implies replacing the second bracket in (B.7) by 1. The scattering operator is

$$|S) = \sum_I |S_I) + \sum_{\langle IJ \rangle} |S_{IJ} - S_I - S_J) \quad . \qquad (B.10)$$

Here $|S_{IJ})$ is the scattering operator of a Hamiltonian $H_{\text{eff}}(I,J) = H_{\text{SCF}} + H_I + H_J + H_{IJ}$, i.e.

$$|S_{IJ}) = \sum_{n=1}^{\infty} |(A_I + A_J + A_{IJ})^n) \quad . \qquad (B.11)$$

Note the difference to the operator $|\tilde{S}_{IJ}\rangle$. Within the two-centre approximation the correlation energy is

$$E_{\text{corr}} = \sum_I (H|S_I) + \sum_{\langle IJ \rangle} (H|\delta S_{IJ})$$

$$= \sum_I \epsilon_I + \sum_{\langle IJ \rangle} \epsilon_{IJ} \quad , \qquad (B.12)$$

where $\delta S_{IJ} = S_{IJ} - S_I - S_J$. The expansion can be continued to include three-centre and higher contributions. Specific examples presented in Sect. 6.1 show that the series is rapidly convergent as the number of centres increases and also with increasing distances between different centres.

C
Intra-atomic Correlations in a C Atom

For an estimate of the intra-atomic correlation energy of diamond one needs to know the correlation energy for different numbers ν of valence electrons on a carbon atom. A list of that energy as function of ν is given in Table C.1. Note that as discussed in Sect. 6.1.2 contributions of the $s^2 \to p^2$ excitation have been excluded because they are contained in the interatomic correlation contribution.

As far as negative ions are concerned, calculations for a free atom prove less useful given that ionic radii become much larger than the available space in a solid. Since we are also lacking experimental information on this point, we had to obtain the listed data by extrapolation of the data for a corresponding state in heavier elements. An estimation of the error introduced by such a procedure is hardly possible.

C Intra-atomic Correlations in a C Atom

Table C.1. List of the different $\epsilon_\nu^{at}(i)$ for the C atom [eV]. They do *not* contain the $s^2 \to p^2$ excitation, which is listed in the last column. Values for the negative ions are by extrapolation (From [86, 285, 467]).

ν	Configuration i	$\epsilon_\nu^{at}(i)$	$-\epsilon_{\text{corr}}(s^2 \to p^2)$
8	$s^2 p^6$	11.24	
7	$s^2 p^5$	8.65	
	$s^1 p^6$	12.22	
6	$s^2 p^4$	6.99	0.05
	$s^1 p^5$	8.57	
	$s^0 p^6$	6.18	
5	$s^2 p^3$	5.31	0.24
	$s^1 p^4$	6.64	
	$s^0 p^5$	8.98	
4	$s^2 p^2$	3.89	0.54
	$s^1 p^3$	4.63	
	$s^0 p^4$	6.18	
3	$s^2 p^1$	2.67	1.12
	$s^1 p^2$	3.07	
	$s^0 p^3$	3.97	
2	$s^2 p^0$	1.47	1.99
	$s^1 p^1$	1.99	
	$s^0 p^2$	2.31	
1	$s^1 p^0$	1.36	
	$s^0 p^1$	1.41	
0	$s^0 p^0$	1.22	

D

Landau Parameter: Quasiparticle Mass

In Sect. 7.2 we have given examples how experimental quantities relate to the quasiparticle interaction parameters F_ℓ^λ. Here we want to demonstrate this relation explicitly for the effective mass m^* of the quasiparticles (see (7.97)). This relation follows from Galilean invariance and therefore applies to homogeneous systems only.

Consider the ground state of a system of $(N-1)$ electrons with energy E_0 to which we add a quasiparticle with momentum $\mathbf{p}$. In the rest frame of the system the energy is $H = E_0 + \epsilon_{\mathbf{p}\sigma}^{(0)}$. By going over to a moving frame, each electron obtains an additional momentum $\mathbf{q}$. The energy in that frame is therefore

$$H_\mathbf{q} = \sum_i^N \frac{1}{2m}(\mathbf{k}_i + \mathbf{q})^2 + V$$

$$= H + \mathbf{q}\sum_i^N \frac{\mathbf{k}_i}{m} + \frac{Nq^2}{2m}, \qquad (D.1)$$

where V denotes the electron interaction energy. In the moving frame the Fermi sphere shifts by $+\mathbf{q}$, permitting the presence of a number of quasiparticles. They are indicated in Fig. D.1 by the shaded area and represent the deviations of the distribution function $n_{\mathbf{k}\sigma}^{(0)}$ from (7.84) when $n_{\mathbf{k}\sigma}^{(0)}$ refers to the moving frame. According to (7.90), the energy of the system is

$$H_\mathbf{q} = \frac{(N-1)}{2m}q^2 + E_0 + \epsilon_{\mathbf{p}+\mathbf{q}\sigma}^{(0)} + 2\sum_{\mathbf{p}'} f^s(\mathbf{p}+\mathbf{q},\mathbf{p}')\,\delta n_{\mathbf{p}'}, \qquad (D.2)$$

where we have used that $\delta n_{\mathbf{p}'\sigma'} = \delta n_{\mathbf{k}'\sigma} = \delta n_{\mathbf{p}'}\mathbb{1}$. The first two terms give the energy of the interacting $(N-1)$-electron system in the moving frame. The last term describes the interaction of the original quasiparticle with the quasiparticles created by going into the moving frame. Furthermore, $\epsilon_{\mathbf{p}+\mathbf{q}\sigma}^{(0)} =$

D Landau Parameter: Quasiparticle Mass

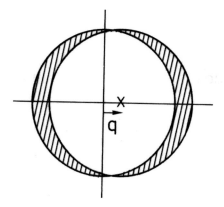

Fig. D.1. Momentum distribution of the electrons within the ground state in the rest frame and in a frame moving with velocity $v = \mathbf{q}/m$. The quasiparticles in the moving frame are indicated by the shaded area.

$\epsilon_{\mathbf{p}\sigma}^{(0)} + \mathbf{p}\cdot\mathbf{q}/m^* + q^2/2m^*$. By comparing (D.1) with (D.2) and using (7.90) one obtains in the limit $\mathbf{q} \to 0$

$$\frac{\mathbf{q}\cdot\mathbf{p}}{m} = \frac{\mathbf{q}\cdot\mathbf{p}}{m^*} + 2\sum_{\mathbf{p}'} f^s(\mathbf{p},\mathbf{p}')\,\delta n_{\mathbf{p}'} \quad . \tag{D.3}$$

This equation can be written as

$$\frac{\mathbf{q}\cdot\mathbf{p}}{m} = \frac{\mathbf{q}\cdot\mathbf{p}}{m^*} + 2\sum_{\mathbf{p}'} f^s(\mathbf{p},\mathbf{p}')\,\mathbf{q}\cdot\mathbf{v}'_{\mathbf{p}}\,\frac{\partial n_{\mathbf{p}'}^{(0)}}{\partial \epsilon_{\mathbf{p}'}} \quad , \tag{D.4}$$

where we have used the relation $\partial \epsilon_{\mathbf{p}'}/\partial \mathbf{p}' = \mathbf{v}'_{\mathbf{p}}$. Expanding $f^s(\theta)$ according to (7.96), we notice that only the term with $l = 1$ contributes when the sum over $\mathbf{p}'$ is taken. Note that $\partial n_{\mathbf{p}'}^{(0)}/\partial \epsilon_{\mathbf{p}'} = -\delta(\epsilon_{\mathbf{p}'} - \mu)$. This leads immediately to (7.97).

E

Kondo Lattices: Quasiparticle Interactions

We want to show how the quasiparticle interactions in Kondo lattice systems affect their low temperature thermodynamic properties, in particular the specific heat and the spin susceptibility. For that purpose we start from (13.3) and assume that the quasiparticle interactions are hard-core or δ-function like, which would mean that $\eta_\tau(\epsilon)$ depends only on $\delta n_{-\tau}(\epsilon)$, i.e., $\phi_{\tau\tau} = 0$. The f orbital occupancy n_f is independent of small changes in the Fermi energy. Therefore the phase shift $\eta_\tau(\epsilon_F) \simeq \pi/2$ must follow the Fermi energy when the latter shifts by an amount $\Delta\epsilon$. When this shift takes place, a number $\delta n_{-\tau} = [N(0) + \delta N(0)]\Delta\epsilon \simeq N(0)\Delta\epsilon$ of quasiparticles is generated. Note that $\delta N(0)/N(0)$ is on the order of the inverse electron number N^{-1} because we are considering the case of one impurity. In order for the phase shift to remain unchanged, the following relation must hold:

$$\frac{1}{k_B T_0} + N(0)\phi_{\tau,-\tau} = 0 \qquad (E.1)$$

or

$$\phi_{\tau,-\tau} = -\frac{\pi \delta N(0)}{N(0)} \ . \qquad (E.2)$$

Therefore (13.3) simplifies to

$$\eta_\tau(\epsilon) = \eta(\mu) + \frac{1}{k_B T_0}(\epsilon - \mu) - \frac{1}{N(0)k_B T_0}\sum_{\epsilon'}\delta n_{-\tau}(\epsilon') \ . \qquad (E.3)$$

Using a grand canonical ensemble, we have replaced the Fermi energy by the chemical potential μ. The last equation contains in a condensed form the influence of a magnetic impurity on the electronic system in the regime $T \ll T_0$ where the magnetic moment is quenched[1]. It can be used to derive an expression for the excess specific heat δC and magnetic susceptibility $\delta\chi_m$. The former follows directly from the excess density of states $\delta N(0)$, i.e.,

---

[1] see [346]

$$\frac{\delta C}{C} = \frac{\delta N(0)}{N(0)} \quad , \tag{E.4}$$

with $\delta N(0)$ given by (13.4). Notice that the quasiparticle interactions characterized by $\phi_{\tau,-\tau}$ do not enter δC. Therefore the specific heat remains unaffected by them.

If we calculate $\delta \chi_m$, we find this point to be different. In the presence of an applied magnetic field h.

$$\eta_\tau(\mu) - \eta_{-\tau}(\mu) = 2\mu_{\text{eff}} h \pi \delta N(0) - m\phi_{\tau,-\tau}$$
$$= \pi \delta m \tag{E.5}$$

Here we have μ_{eff} as the effective magnetic moment, for example that of the Ce^{3+} ground-state doublet in the presence of a crystalline electric field. Furthermore, m is the difference in the spin-up and spin-down population of the electronic system and δm corresponds to the increment due to the presence of the impurity. We have $m = m_0 + \delta m$ and $m_0 = 2\mu_B h N(0)$. Usually the difference between μ_{eff} and the Bohr magneton μ_B is neglected; we will follow suit here. From (E.5) we then obtain

$$m = 2\mu_B h N(0) \frac{1 + \delta N(0)/N(0)}{1 + \phi_{\tau,-\tau}/\pi}$$
$$\simeq 2\mu_B h N(0) \left[1 + \delta N(0)/N(0)\right] \left[1 - \phi_{\tau,-\tau}/\pi\right] \quad . \tag{E.6}$$

We have made use of the fact that $\phi_{\tau,-\tau}$, like $\delta N(0)/N(0)$, is small. The magnetic susceptibility is defined by

$$\chi_m = \frac{\mu_B m}{h} \quad . \tag{E.7}$$

Let $\delta \chi_m$ denote the increment in the susceptibility due to the presence of the impurity. It is noticed from (E.6) that quasiparticle interactions may have a sizable effect on that increment. By means of (E.2-E.7), we obtain for the ratio

$$\frac{\delta \chi_m/\chi_m}{\delta C/C} = 1 - \frac{\phi_{\tau,-\tau}}{\pi \delta C/C}$$
$$= 2 \quad . \tag{E.8}$$

That this ratio differs from unity is solely due to the quasiparticle interactions represented by $\phi_{\tau,-\tau}$. Equation (E.8) was first obtained by renormalization-group techniques [489].

F
Lanczos Method

The Lanczos algorithm is an efficient method for numerically calculating physical quantities of finite systems, such as the ground-state energy or correlation functions. The method was originally used by mathematicians for the diagonalization of large, sparse matrices[1], but has found wide applications in solid-state theory[2].

One way of solving for a finite system is by diagonalization of the full Hamiltonian matrix. But often the system are too large for such a diagonalization to be done. With the help of the Lanczos method we can find iterative solutions of the Hamiltonian even in that case. We consider the resolvent operator $(z - H)^{-1}$ which contains information about the energy spectrum of a system. In order to determine its matrix elements we begin with a normalized trial state $|\Phi_0\rangle$ of the system that has a finite component on the subspace we want to limit ourselves and construct the following series of states:

$$\begin{aligned}
|\Phi_1\rangle &= H\,|\Phi_0\rangle - |\Phi_0\rangle\langle\Phi_0\,|\,H\,|\,\Phi_0\rangle \\
|\Phi_2\rangle &= H\,|\Phi_1\rangle - |\Phi_0\rangle\langle\Phi_0\,|\,H\,|\,\Phi_1\rangle - |\Phi_1\rangle\langle\Phi_1\,|\,H\,|\,\Phi_1\rangle\langle\Phi_1\,|\,\Phi_1\rangle^{-1} \\
&\vdots
\end{aligned} \tag{F.1}$$

We notice that the $|\Phi_i\rangle$ are mutually orthogonal. The only nonvanishing matrix elements of H in the basis of the $|\Phi_i\rangle$ are

$$\begin{aligned}
a_i &= \langle\Phi_i\,|\,H\,|\,\Phi_i\rangle\langle\Phi_i\,|\,\Phi_i\rangle^{-1}\,, \\
b_i &= \langle\Phi_i\,|\,H\,|\,\Phi_{i+1}\rangle\langle\Phi_i\,|\,\Phi_i\rangle^{-1}
\end{aligned} \tag{F.2}$$

implying that the Hermitian matrix H_{ij} is tridiagonal in this representation.

We are interested in the diagonal matrix element $\langle\Phi_0|(z - H)^{-1}|\Phi_0\rangle$, the poles of which yield the excitation energies. From the identity

---

[1] see [264, 488]
[2] see, e.g., [366]

$$\sum_\beta (z-H)_{\alpha\beta}(z-H)^{-1}_{\beta\gamma} = \delta_{\alpha\gamma} \qquad (F.3)$$

it follows that the vector $x_\beta = (z-H)^{-1}_{\beta 1}$ satisfies an equation of the form

$$\sum_\beta (z-H)_{\alpha\beta} x_\beta = e_\alpha \qquad (F.4)$$

with $e_\alpha = \delta_{\alpha 1}$. By definition, $\langle \Phi_0 | (z-H)^{-1} | \Phi_0 \rangle = x_1$.

The inhomogeneous system of linear equations (F.4) is solved by applying Cramer's rule. In order to compute x_1 a determinant A is defined with elements

$$A_{\alpha\beta} = (z-H)_{\alpha\beta}(1-\delta_{\beta 1}) + e_\alpha \delta_{\beta 1} \quad , \qquad (F.5)$$

i.e., the first column of the matrix $(z-H)$ has been replaced by the vector **e**. From Cramer's rule we obtain

$$x_1 = \frac{\det A}{\det(z-H)} \quad . \qquad (F.6)$$

By expanding the two determinants, making use of the tridiagonal form of their elements, we find

$$x_1 = \frac{1}{z - a_0 + |b_1|^2 \dfrac{\det D_2}{\det D_1}} \qquad (F.7)$$

where the matrix D_ν is obtained from $(z-H)_{\alpha\beta}$ by discarding the first ν rows and columns. By continuing the expansion we obtain

$$x_1 = \cfrac{1}{z - a_0 + \cfrac{|b_0|^2}{z - a_1 + \cfrac{|b_1|^2}{z - a_2 + \ldots}}} \qquad (F.8)$$

The form of a continued fraction suggests a relation between the Lanczos algorithm and the projection method presented in Sect. 5.4. Indeed, we could have derived (F.8) as well by using the projection method.

In application of the method the values for the lowest eigenvalues of the denominator of (F.8) are usually rapidly convergent with increasing dimension of the matrix $(z-H)$, i.e., with an increasing number of states $|\Phi_i\rangle$ used in the calculation. Therefore the algorithm is suitable for determining, for example, the ground-state energy of a finite electron system or the spectrum of its low-energy excitations. We can compute correlation functions of the form

$$C_{AA}(z) = \left\langle \psi_0 \left| A^+ \frac{1}{z-H} A \right| \psi_0 \right\rangle \quad , \qquad (F.9)$$

where $|\psi_0\rangle$ is the ground state of the system, by starting from the product

$$|\tilde{\Phi}_0\rangle = A\,|\,\Phi_0\rangle \quad . \tag{F.10}$$

The ground state $|\psi_0\rangle$ is determined numerically by finding the eigenstate of the lowest eigenvalue of the matrix $\langle\tilde{\Phi}_i|H|\tilde{\Phi}_j\rangle$. Again, good convergence is found in most applications, i.e., $|\psi_0\rangle$ does not change by any appreciable amount any more when the dimension of the matrix exceeds a certain size.

G

Density Matrix Renormalization Group

The Density Matrix Renormalization Group (DMRG) technique has developed into a powerful tool for studying one-dimensional quantum lattice problems. It is a numerical method with the help of which one can approximately diagonalize chain systems which are too large for exact diagonalization. The numerical accuracies which are thereby achieved are remarkable. For example, the ground-state energy of a chain of several hundreds of sites with one orbital per site and any reasonable interaction can be calculated with a relative error of 10^{-10}. DMRG is ideally suited for investigating spin systems, e.g., Heisenberg chains as well as one-dimensional systems described by the Hubbard Hamiltonian.

The idea of using renormalization groups for a numerical treatment of many-body problems goes back to *Wilson*. But when, after its successful application to the Kondo impurity problem, the method was applied to one-dimensional lattice systems it failed badly. The understanding of the origin of this failure was the starting point of the development of DMRG by *White* [482]. This is achieved best by considering a simple model, i.e., a particle on an chain. Consider a chain of sites i with a Hamiltonian in matrix form

$$H_{ij} = \begin{cases} 2, & i = j \\ -1, & |i-j| = 1 \\ 0, & \text{otherwise} \end{cases} \quad . \tag{G.1}$$

In the continuum limit this Hamiltonian goes over onto $H = -\partial^2/\partial x^2$. This is seen by writing the equivalent of the second derivative for a discrete system in terms of the displacement δr_i as $(\delta r_{i+1} - \delta r_i) - (\delta r_i - \delta r_{i-1}) = \delta r_{i+1} - 2\delta r_i + \delta r_{i-1}$.

The standard renormalization group approach consists in forming a block of a number of adjacent sites, to diagonalize the Hamiltonian for that block and to determine its eigenstates. The sequence of these states is truncated by keeping only those, say m, states with the lowest eigenvalues. They are used to construct an effective Hamiltonian for a new, i.e., larger block obtained by adding one site to the original block (see Fig. G.1). For a noninteracting

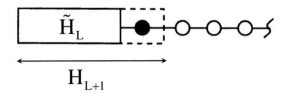

Fig. G.1. Block of L sites with an effective Hamiltonian $\tilde{H}_L$ to which an additional site is added resulting in H_{L+1}. (From [345])

system the dimension of the Hilbert space for a chain consisting of L sites is simply L, instead of an exponential of L as for an interacting system. In that case we may choose for the larger block the addition of two blocks of the same size, rather than adding a single site only. For example, when the Hamiltonian (G.1) is used, the first step consists in breaking up the Hamiltonian matrix into blocks containing L sites.

$$H = \begin{pmatrix} H_L & T_L & 0 & 0 & \cdots \\ T_L^+ & H_L & T_L & 0 & \cdots \\ 0 & T_L^+ & H_L & T_L & \cdots \\ \vdots & & & \ddots & \end{pmatrix} . \tag{G.2}$$

Each matrix H_L is of the trigonal $L \times L$ form

$$H_L = \begin{pmatrix} 2 & -1 & 0 & 0 & \cdots \\ -1 & 2 & -1 & 0 & \cdots \\ 0 & -1 & 2 & -1 & \cdots \\ \cdots & & & & \end{pmatrix} . \tag{G.3}$$

while the matrix T connects only sites at the ends of the block. After H_L has been diagonalized and the lowest m eigenvalues and eigenvectors $\psi_L^{(\nu)}$ are kept the new $m \times m$ effective Hamiltonian matrix $\tilde{H}_L$ is constructed as

$$\left(\tilde{H}_L\right)_{\nu\nu'} = \sum_{ij}^{L} \psi_L^{(\nu)}(i)\, (H_L)_{ij}\, \psi_L^{(\nu')}(j) \ . \tag{G.4}$$

In matrix notation this equation and a similar one for $\tilde{T}$ are

$$\begin{aligned} \tilde{H}_L &= O_L^+ H_L O \\ \tilde{T}_L &= O_L^+ T O \end{aligned} \tag{G.5}$$

where the $m \times L$ matrix $O_{\nu\mu}$ has the m eigenvectors $\psi_L^{(\nu)}$ as columns.

In a next step two blocks of size L are merged to form a block of size $2L$. We form the matrix

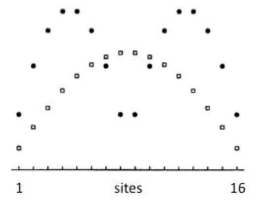

Fig. G.2. Lowest eigenstate of a 16 site block (open squares) and of two disconnected 8 site blocks (solid dots) with fixed boundary conditions. (From [345])

$$H_{2L} = \begin{pmatrix} \tilde{H}_L & \tilde{T}_L \\ \tilde{T}_L^+ & \tilde{H}_L \end{pmatrix} \quad (G.6)$$

and diagonalize H_{2L}. This procedure is repeated, but the results remain poor.

The failure of such an iterative procedure for the construction of the ground state of the system is seen as follows. The eigenstates of a block are just the ones of a particle in a box. Those eigenfunctions vanish at the ends of the block (fixed boundary condition). Therefore, when in the next iteration step two neighboring blocks are joined, a state constructed only from low-energy states of the previous iteration must have a dent in the middle. In particular it is not possible to construct the lowest energy state of the joined blocks, which has its maximum at the link of the two (see Fig. G.2). In order to construct the ground-state wavefunction of the enlarged system in terms of the eigenfunctions of the two smaller systems we must include not only low-energy eigenstates of the latter but also high-energy states. The DMRG solves this problem of the boundaries by not selecting the states with the lowest energies but rather choosing the eigenstates of the density matrix with the largest weight. This is explained in the following.

We start with the introduction of a superblock. It contains the original block under consideration plus an *environment* (see Fig. G.3). The superblock is diagonalized and the eigenfunctions are expressed in terms of the (in general many-body) states $|i\rangle$ of the block and the states $|\alpha\rangle$ of the environment.

Let us denote with $|i\rangle$ the many-body states of the block and with $|\alpha\rangle$ those of the environment. The ground state $|\psi\rangle$ of the superblock can then be expanded in the form

$$|\psi\rangle = \sum_{i\alpha} \psi_{i\alpha} |i\rangle |\alpha\rangle \quad . \quad (G.7)$$

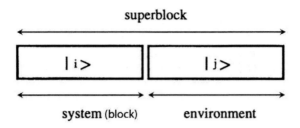

Fig. G.3. A block with states $|i\rangle$ and the environment with states $|j\rangle$ forming a superblock. (From [345])

The reduced density matrix is then defined by

$$\rho_{ii'} = \sum_\alpha \psi^*_{i\alpha} \psi_{i'\alpha} \quad . \tag{G.8}$$

The sum is taken over the states of the environment. It contains all required information on $|\psi\rangle$ when we calculate expectation values of operators acting on the superblock. The expection value of any operator O with respect to the ground state of the superblock can be expressed in terms of the reduced density matrix as

$$\langle O \rangle = Tr\rho\, O \quad . \tag{G.9}$$

We assume that we can diagonalize the density matrix and we denote its eigenvectors and eigenvalues by $|d_n\rangle$ and ω_n. Since $Tr\rho = 1$ we find that $\sum_n \omega_n = 1$. Equation (G.9) is then rewritten as

$$\langle O \rangle = \sum_n \omega_n \langle d_n | O | d_n \rangle \quad . \tag{G.10}$$

The main point of DMRG is the realization that we may discard the states $|d_n\rangle$ with the smallest eigenvalues ω_n without noticeable effect on expectation values. The ω_n decrease nearly exponentially with n. Thus not the eigenvalues of the Hamiltonian tell us which states may be discarded but instead the eigenvalues of the reduced density matrix do that.

We want to discuss this point in some more detail. At this stage the *singular value decomposition* becomes very useful. We digress briefly in order to explain it. According to that decomposition it is always possible to rewrite any rectangular matrix M of dimension $N_A \times N_B$ in the form

$$M = UDV^+ \quad . \tag{G.11}$$

Here U is a matrix of dimension $N_A \times N_A$ with the property that $U^+U = 1$. The matrix D has dimension $N_A \times N_B$ and is diagonal with non-negative matrix elements d_n, referred to as the singular values. The number of non-zero singular values defines the rank of the matrix M. Finally, the matrix V^+ has dimension $N_B \times N_B$ and it holds that $V^+V = 1$.

Equation (G.11) can be used as follows. Assume that we want to approximate $|\psi\rangle$ given by (G.7) by m product states (Schmidt decomposition)

$$|\psi'\rangle = \sum_{I=1}^{m} a_I |u^I\rangle |v^I\rangle \qquad (G.12)$$

where the $|u^I\rangle$ are functions of the block and $|v^I\rangle$ of the symmetric environment. We want to choose the a_I and $|u^I\rangle|v^I\rangle$ so that

$$S = (|\psi\rangle - |\psi'\rangle)^2 = \text{minimum}$$

$$= \sum_{i\alpha} \left(\psi_{i\alpha} - \sum_{I=1}^{m} a_I u_i^I v_\alpha^I \right)^2 \quad . \qquad (G.13)$$

The $u_i^I v_\alpha^I$ are obtained by decomposing

$$|u^I\rangle = \sum_i U_i^I |i\rangle \quad , \quad |v^I\rangle = \sum_\alpha V_\alpha^I |\alpha\rangle \quad . \qquad (G.14)$$

If we choose for the a_I the largest eigenvalues of D and for $|u^I\rangle$ and $|v^I\rangle$ the corresponding eigenvectors then S is minimized. Note that the $|u^I\rangle$ are also eigenvectors of the reduced density matrix $\rho_{ii'}$ of the block. Indeed, by setting (G.12) and (G.14) into (G.8) we obtain after summation over the states of the environment

$$\rho = U D^2 U^T \quad . \qquad (G.15)$$

This shows that the $|u^I\rangle$ diagonalize also the reduced density matrix ρ of the superblock. The conclusion is that the most significant eigenstates of ρ are sufficient for a good approximate representation of a wavefunction $|\psi\rangle$ of the whole system.

When the reduced density matrix is known and only the eigenstates $|d_n\rangle$ with the largest eigenvalues ω_n are used, we have solved the problem of truncation for that superblock. The truncation gives rise to an effective Hamiltonian $\tilde{H}_L$ for the new block as well as for the symmetric environment. In a next step the block consisting, e.g., of L sites is enlarged by adding one site and both are mirrored for the environment, i.e., for constructing the superblock. That is shown in Fig. G.4.

The algorithm of the DMRG is then the following. A superblock is formed consisting of a block and the environment. The latter is often chosen symmetric to the block. The system is diagonalized and the reduced density matrix is calculated according to (G.8) by using for $|\psi\rangle$ the ground state. A state which is used for the construction of the reduced density matrix is called a target state.

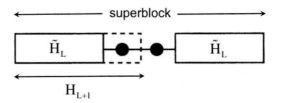

Fig. G.4. Superblock configuration for the algorithm in case of an infinite chain.

One way of starting is from a four-site chain, i.e., from a block consisting of two sites and the symmetric environment. The Hamiltonian of that system is diagonalized, e.g., by making use of the Lanczos method and the density matrix for the block is calculated. By using the eigenstates of the density matrix with the largest eigenvalues as target states, the effective Hamiltonian for the block is determined.

Next two sites one added to the system as indicated in Fig. G.4. The Hamiltonian of the six-sites system consists of the effective one for the original 2-site block and environment, the transfer to the added sites determined according to (G.5) and the one for the added sites. The new system is again diagonalized and the procedure described above is repeated step by step.

This iteration process continues until a required length of the chain is obtained. The procedure just described is called Infinite-System DMRG. The accuracy which is obtained this way is limited, though. One of the reasons is the small size of the starting block. The states chosen from it are different from the ones we would choose if the block is embedded in a final system of given length L. As a consequence the system may end up in a state with a relative but not absolute minimum in energy.

The situation is much improved when an Infinite-System DMRG is followed by a finite-system DMRG. It corrects the selection of the reduced number of states made for a superblock smaller than the final one. The finite-system DMRG proceeds like the infinite-system DMRG. However, starting point is here the last step of the infinite-system DMRG with two blocks A and B of length $L/2$. Yet now, in distinction to the infinite DMRG the size of block A is increased at the expense of the length of block B when a superblock is formed. Thus block B shrinks in each consecutive step by one site until only one site remains. Now the direction is reversed and block B is growing again at the expense of block A. These shifts between the two blocks are continued until complete convergence has been attained. The accuracy, which can be obtained, is extremely high and of the order mentioned before. This can be understood by returning to (G.10) and investigating how fast the eigenvalues ω_n of the reduced density may fall off as function of n. Analyses which have been performed show that the ω_n decay exponentially when the excitations of the system are gapped. In these cases the accuracy of the results increases very fast with increasing number of n.

G Density Matrix Renormalization Group

Recently it has been realized that the DMRG is closely related to the problem of entanglement and more over to so called area law. By connecting two parts of a chain, the entanglement of the two parts is limited to a small local regime in the vicinity of the interface. This limitation is called Area Law because the entanglement increases in proportion to the size of the interface rather than the volume. The same reasoning holds also for higher dimension. The DMRG is based on this law. It enables us to construct the ground state of a long chain by starting from a small chain. For more details we refer to the review [402].

H

Monte Carlo Methods

Originally a way of numerically evaluating multidimensional integrals which otherwise cannot be computed, the Monte Carlo technique has become a fruitful tool not only in statistical mechanics, but also in many-body theory[1].

In physics one is often faced with integrals of the form

$$\langle f \rangle_p = \int d\mathbf{x} f(\mathbf{x}) p(\mathbf{x}) \ , \tag{H.1}$$

where $\mathbf{x}$ is a multidimensional vector and $p(\mathbf{x})$ is a probability distribution function. Consider, for example, a system of N classical, free particles. Then

$$\langle A \rangle = \frac{1}{Z} \int d\mathbf{x} \, e^{-\beta E(\mathbf{x})} A(\mathbf{x}) \tag{H.2}$$

describes the thermal average of a quantity A which depends on the momenta $\mathbf{p}_n$ of the particles. The vector $\mathbf{x}$ in this case is $\mathbf{x} = (\mathbf{p}_1, \mathbf{p}_2, \ldots, \mathbf{p}_N)$. The energy of the system is $E(\mathbf{x})$, Z is the partition function, and $\beta = (k_\mathbf{B} T)^{-1}$. Equation (H.2) is a $3N$-dimensional integral. The function

$$p(\mathbf{x}) = \frac{1}{Z} e^{-\beta E(\mathbf{x})} \tag{H.3}$$

is a probability distribution function in the $3N$-dimensional phase space. It describes the statistical weight of a state $\mathbf{x}$ of the system in thermal equilibrium. When the particles are free, integrals of the form of (H.2) can often be evaluated analytically. However, when interactions among the particles are included, we can compute integrals of the form (H.2), or more generally of the form (H.1), only by Monte Carlo techniques.

How can the Monte Carlo technique be made applicable to ground-state energy calculations of a many-electron system, for example? In order to demonstrate this, consider the ground state of a system of N electrons. We assume

---

[1] see, e.g., [38, 58, 343, 442]

for its wavefunction a trial function $\psi_\eta(\mathbf{R})$, where η stands for the variational parameters contained in it and $\mathbf{R} = (\mathbf{r}_1, \mathbf{r}_2, \ldots, \mathbf{r}_N)$. The corresponding energy $E(\eta)$ is

$$E(\eta) = \int d\mathbf{R} \frac{\langle \psi_\eta(\mathbf{R}) | H | \psi_\eta(\mathbf{R}) \rangle}{\langle \psi_\eta(\mathbf{R}) | \psi_\eta(\mathbf{R}) \rangle} \quad . \tag{H.4}$$

This expression can be written in the form

$$E(\eta) = \int d\mathbf{R}\, p_\eta(\mathbf{R}) \frac{(H \psi_\eta(\mathbf{R}))}{\psi_\eta(\mathbf{R})} \tag{H.5}$$

with

$$p_\eta(\mathbf{R}) = \frac{|\psi_\eta(\mathbf{R})|^2}{\int d\mathbf{R}'\, |\psi_\eta(\mathbf{R}')|^2} \quad . \tag{H.6}$$

We see here that (H.5) resembles (H.1) and is therefore amenable to Monte Carlo calculations (variational Monte Carlo method). Our task is then to calculate integrals like (H.1) in the most economical way.

H.1 Sampling Techniques

A simple, if impractical, way of sampling would consist of introducing a grid with equal spacing in the $3N$-dimensional space (hypercube). If n_0 is the number of points in a given direction of the cube, then n_0^{3N} is the total number of points in the cube, a number usually much too large for practical computations. Systematic errors may arise when important contributions to the integrals come from regions between points on the grid. In addition, for large N, practically all points appear on the surface of the hypercube, not in its interior, a somewhat surprising aspect which can be easily demonstrated. Of the n_0 points in one direction, two (i.e., the end points) are on the surface, while $n_0 - 2$ are in the interior. The ratio of points in the interior of the hypercube to the total number of points is

$$\left(\frac{n_0 - 2}{n_0}\right)^{3N} = \exp\left[3N \log\left(1 - \frac{2}{n_0}\right)\right] \simeq e^{-6N/n_0} \xrightarrow[N \to \infty]{} 0 \quad . \tag{H.7}$$

A better sampling method, called *simple sampling*, is to produce M random values $\mathbf{x}_n$ by a random number generator. The integral $\langle f \rangle_p$ is then approximated by

$$\langle f \rangle_p = \sum_{n=1}^{M} f(\mathbf{x}_n) p(\mathbf{x}_n) \quad \text{with} \quad \sum_{n=1}^{M} p(\mathbf{x}_n) = 1 \quad . \tag{H.8}$$

The partition function which appears in $p(\mathbf{x})$ is similarly evaluated. This method has the advantage that systematic errors can be avoided even if n_0 is small. As before, most points are on the surface of the hypercube. When we generate several different sets of M random values $\mathbf{x}_n$, the right-hand side of (H.8) yields slightly different values for each set. We can show that these values form a normal distribution around $\langle f \rangle_p$ and write it as

$$\langle f \rangle_p = \sum_{n=1}^{M} f(\mathbf{x}_n) p(\mathbf{x}_n) \pm \frac{1}{\sqrt{M}} \left(\langle f^2 \rangle_p - \langle f \rangle_p^2 \right)^{1/2} \quad , \tag{H.9}$$

noticing that the statistical error decreases as $M^{-1/2}$.

The convergence can be improved if the $\mathbf{x}_n$ are not chosen at random, but rather such that those configurations $\mathbf{x}$ appear most often in the sampling, which make particularly large contributions to $\langle f \rangle_p$. We call this form of sampling *importance sampling*. Assume that the $\mathbf{x}_n$ are chosen according to a probability distribution $P(\mathbf{x}_n)$; then we replace (H.8) with

$$\langle f \rangle_p = \frac{1}{M} \sum_{n=1}^{M} f(\mathbf{x}_n) \frac{p(\mathbf{x}_n)}{P(\mathbf{x}_n)} \quad . \tag{H.10}$$

If we choose for $P(\mathbf{x}_n)$ the function $p(\mathbf{x}_n)$, then

$$\langle f \rangle_p = \frac{1}{M} \sum_{n=1}^{M} {}_{(p)} f(\mathbf{x}_n) \quad . \tag{H.11}$$

The subscript (p) is a reminder that the $\mathbf{x}_n$ are sampled according to $p(\mathbf{x}_n)$.

A popular way of achieving importance sampling in practice is the *Metropolis method*, named after its inventors, *Metropolis* et al. [321]. The $\mathbf{x}_n$ are chosen in the form of a Markov chain, i.e., the $\mathbf{x}_n$ are not independent of each other but rather $\mathbf{x}_{n+1}$ depends on $\mathbf{x}_n$. The former is constructed from the latter via a properly chosen transition probability $P(\mathbf{x}_n \to \mathbf{x}_{n+1})$, which ought to be such that in the limit of large M the distribution of $\mathbf{x}_n$ values converges towards the equilibrium distribution $p(\mathbf{x}_n)$. We achieve this by requiring that a microreversibility condition be satisfied, i.e.,

$$p(\mathbf{x}_n) P(\mathbf{x}_n \to \mathbf{x}_{n+1}) = p(\mathbf{x}_{n+1}) P(\mathbf{x}_{n+1} \to \mathbf{x}_n) \quad . \tag{H.12}$$

The proof that this is indeed a sufficient condition can be found in books on Monte Carlo techniques, e.g., in [37]. When $p(\mathbf{x})$ takes the form of (H.3), we can write (H.12) as

$$\frac{P(\mathbf{x}_n \to \mathbf{x}_{n+1})}{P(\mathbf{x}_{n+1} \to \mathbf{x}_n)} = e^{-\beta [E(\mathbf{x}_{n+1}) - E(\mathbf{x}_n)]} \quad . \tag{H.13}$$

We often use the following form of $P(\mathbf{x}_n \to \mathbf{x}_{n+1})$:

$$P(\mathbf{x}_n \to \mathbf{x}_{n+1}) = \begin{cases} e^{-\beta [E(\mathbf{x}_{n+1}) - E(\mathbf{x}_n)]} & , \quad \text{if } E(\mathbf{x}_{n+1}) > E(\mathbf{x}_n) \\ 1 & , \quad \text{otherwise} \end{cases} \tag{H.14}$$

We see here that this form satisfies (H.13). Thus, when $\mathbf{x}_{n+1}$ has a lower energy than $\mathbf{x}_n$, i.e., when $\Delta E(n \to n+1) = E(\mathbf{x}_{n+1}) - E(\mathbf{x}_n)$ is negative, the new value is always accepted. However, when $\Delta E(n \to n+1) > 0$, this new value is accepted only with a probability $p(\mathbf{x}_{n+1})/p(\mathbf{x}_n)$. In practice, when we move from $\mathbf{x}_n$ to $\mathbf{x}_{n+1}$ we change only one or a few components of $\mathbf{x}$. Otherwise, with $3N \gg 1$, we would expect an abrupt decrease of the ratio $p(\mathbf{x}_{n+1})/p(\mathbf{x}_n)$ when $\Delta E(n \to n+1) > 0$, because each of the $3N$ dimensions is expected to contribute an amount of order, say, $k_B T$ to the energy change, depending on the chosen interval size. The transition region to lower energies becomes then very narrow. Such a decrease would imply that most of the attempted sampling moves are not executed and that the system gets stuck in its original configuration.

H.2 Ground-State Energy

We can apply the variational Monte Carlo method to calculate the ground-state energy of a solid. A simple ansatz for the trial function $\psi_\eta(\mathbf{r}_1, \ldots, \mathbf{r}_N)$ in (H.5) is the Jastrow wavefunction

$$\psi_\eta(\mathbf{R}) = \exp\left(\sum_i d(\mathbf{r}_i) + \sum_{ij} f(\mathbf{r}_i - \mathbf{r}_j)\right) \Phi(\mathbf{r}_1, \ldots, \mathbf{r}_N) \quad , \tag{H.15}$$

where $\Phi(\mathbf{r}_1, \ldots, \mathbf{r}_N)$ is a Slater determinant. For example, when studying a semiconductor like diamond or silicon, $\Phi(\mathbf{R})$ can be constructed in form of a Slater determinant from the solutions of the Kohn-Sham equation within the local density approximation to density functional theory [115]. The pair function $f(\mathbf{r})$ introduces electron correlations into the ground-state wavefunction $\psi_\eta(\mathbf{R})$ and contains adjustable parameters η. The function $f(\mathbf{r})$ can be chosen so that the correlation cusp in the pair distribution function $g(\mathbf{r}, \mathbf{r}')$ is properly accounted for (compare with Sect. 5.1). The function $d(\mathbf{r}_i)$ aims at ensuring that the electron charge distribution $\rho(\mathbf{r})$ is properly adjusted. A Slater determinant constructed from solutions of the Kohn-Sham equation yields a slightly different density distribution than a SCF calculation does. Correlations modify the optimal $\rho(\mathbf{r})$ slightly. We refer the reader here to the discussion in Sect. 5.4.4.

We want to obtain the ground-state energy by solving the N-particle Schrödinger equation with Monte Carlo techniques. Replacing t by $-i\tau$ in the time-dependent Schrödinger equation

$$i\frac{\partial \psi(\mathbf{R}, \tau)}{\partial t} = H\psi(\mathbf{R}, t) \quad , \tag{H.16}$$

we end up with a diffusion equation, i.e.,

$$-\frac{\partial}{\partial \tau}\psi(\mathbf{R}, \tau) = H\psi(\mathbf{R}, \tau) \quad . \tag{H.17}$$

Explicitly, it is of the form

$$-\frac{\partial}{\partial \tau}\psi(\mathbf{R},\tau) = -\sum_{i=1}^{N}\frac{\nabla_i^2}{2m}\psi(\mathbf{R},\tau) + [V(\mathbf{R}) - E_0]\psi(\mathbf{R},\tau) \quad . \tag{H.18}$$

The potential $V(\mathbf{R})$ contains the external potential as well as the electron-electron Coulomb repulsions. Here we have subtracted an energy E_0 for convenience. Finally E_0 is adjusted so that it agrees with the ground-state energy. Provided that the real function $\psi(\mathbf{R},\tau)$ does not change sign, (H.17) can be interpreted as a classical diffusion equation, which would make $|\psi(\mathbf{R},\tau)|$ a probability density. In the discussion that follows, we assume that $\psi(\mathbf{R},\tau) \geq 0$. The necessary generalization to fermions, i.e., electrons, will be brought in later.

It is easy to see that $\psi(\mathbf{R},\tau)$ relaxes exponentially fast towards the ground state, the decay time being given by the excitation energies of the system. Thus we replace (H.17) with the modified form

$$-\frac{\partial}{\partial \tau}\psi = (H - E_0)\psi \tag{H.19}$$

and imagine that ψ is decomposed in terms of eigenfunctions of H. The above statement about relaxation then follows immediately, provided E_0 is the ground-state energy; otherwise $\psi(\mathbf{R},\tau)$ goes exponentially to zero or grows exponentially.

We treat (H.18) numerically as follows. We start with an ensemble of M different configurations $\mathbf{R}_1^{(0)},\ldots,\mathbf{R}_M^{(0)}$, assuming their distribution to be such that the density of selected configurations is $\psi_0(\mathbf{R})$, a convenient starting function, e.g., a wavefunction of the Jastrow type. The number M is typically of order 100 - 1000. In the next step each configuration is modified slightly by displacing the particles in a random fashion with a mean-square displacement given by $\Delta\tau/2m$. The quantity $\Delta\tau$ denotes the time interval into which the diffusion process is discretized. The new configurations are denoted by $\tilde{\mathbf{R}}_1^{(0)},\ldots,\tilde{\mathbf{R}}_M^{(0)}$. They are replicated or deleted according to $[W(\tilde{\mathbf{R}}_n^{(0)})]$. The function $W(\mathbf{R})$ is given by

$$W(\mathbf{R}) = e^{-\Delta\tau[V(\mathbf{R}) - E_\tau]} + z \quad , \tag{H.20}$$

where the unknown energy E_0 has been replaced by a trial energy E_τ and z is a random number taken from the interval $[0, 1]$. It describes the exponential relaxation due to diffusion. The square brackets around $W(\tilde{\mathbf{R}}_n^{(0)})$ imply taking the largest integer which is less than $W(\tilde{\mathbf{R}}_n^{(0)})$. If $V(\mathbf{R}) < E_\tau$, the function $\exp\{-\Delta\tau[V(\mathbf{R}) - E_\tau]\}$ can become larger than 1, and if z is added, e.g., between 2 and 3. In this case $[W(\tilde{\mathbf{R}}_n)] = 2$ and the configuration $\tilde{\mathbf{R}}_i^{(0)}$ is doubled. According to the above procedure, we have gone over from the ensemble $\mathbf{R}_1^{(0)}, \mathbf{R}_2^{(0)},\ldots,\mathbf{R}_M^{(0)}$ to a new ensemble $\mathbf{R}_1^{(1)}, \mathbf{R}_2^{(1)},\ldots,\mathbf{R}_M^{(1)}$. In order

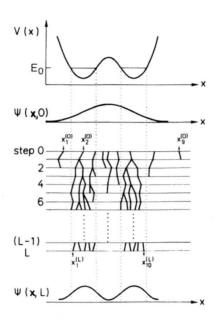

Fig. H.1. Calculation of the ground-state wavefunction $\psi_0(x)$ of a particle in an external double-well potential $V(x)$. The ensemble $\mathbf{R}_1^{(0)},\ldots\mathbf{R}_M^{(0)}$ consists of nine coordinates $x_1^{(0)},\ldots x_9^{(0)}$. For the purpose of better illustration, a poor starting distribution has been chosen, corresponding to a wavefunction $\psi(x,0)$. As the system evolves, branchings occur in regions where $V(x)$ is small, and depletions in regions where $V(x)$ is large. After L steps the ensemble $x_1^{(L)},\ldots,x_{10}^{(L)}$ is obtained, which corresponds to a wavefunction $\psi(x,L)$ that is close to $\psi_0(x)$.

to ensure that the total population remains unchanged, i.e., that $M' \simeq M$, the energy E_T has to be properly adjusted after taking a number of time steps. Eventually E_T will coincide with E_0. The new ensemble corresponds to an improved wavefunction $\psi_1(\mathbf{R})$. If we repeat these steps a number of times, the ground-state wavefunction $\psi(\mathbf{R})$ will evolve. We show this schematically in Fig. H.1.

The described numerical calculation of the ground-state wavefunction and energy can be considerably improved by introducing a function

$$f(\mathbf{R},\tau) = \Phi_t(\mathbf{R})\psi(\mathbf{R},\tau) \quad , \tag{H.21}$$

where $\Phi_t(\mathbf{R})$ is a trial wavefunction which is kept fixed and comes as close as possible to the exact wavefunction. If $\psi(\mathbf{R},\tau)$ satisfies (H.18), then $f(\mathbf{R},\tau)$ satisfies

$$-\frac{\partial f}{\partial \tau} = -\sum_{i=1}^{N}\frac{1}{2m}\nabla_i\left[\nabla_i f - f\nabla_i\left(\ln\Phi_t^2\right)\right] + \left[(H-E_0)\Phi_t\right]\frac{f}{\Phi_t} \quad . \tag{H.22}$$

In contrast to the imaginary-time Schrödinger equation (H.18) this equation contains an additional drift term $-\nabla_i f \nabla_i (\ln \Phi_t^2)/2m$ superimposed on the diffusion. Furthermore, the source-sink term is modified and has the form $[(H - E_0)\Phi_t](f/\Phi_t)$. We notice that in the limit where Φ_t coincides with the exact ground state, the source-sink term vanishes. Therefore, if the function $\Phi_t(\mathbf{R}, \tau)$ is well chosen, replication and deletion of configurations take place to a much lesser extent as the system evolves in τ than they do, when we treat (H.18). The presence of the drift term implies that, when considering the evolution of the system during a time interval $\Delta\tau$, we have to add a drift displacement $\Delta\tau \nabla_i (\ln \Phi_t)/m$ to each particle i in a configuration $\mathbf{R}$.

A slightly more sophisticated method of solving the imaginary-time Schrödinger equation (H.18) is the Green's function Monte Carlo technique [228], in which the differential equation is converted into an integral equation. In order to explain it, we introduce an energy scale with a fixed value given by the ground-state energy E_0. Instead of the electron coordinates $\mathbf{r}_i$, we use renormalized coordinates $\mathbf{x}_i = (-2mE_0)^{1/2}\mathbf{r}_i$. The Schrödinger equation for the ground state is then of the simple form

$$\left(-\nabla_\mathbf{x}^2 + 1\right)\psi(\mathbf{X}) = \lambda W(\mathbf{X})\psi(\mathbf{X}) \quad , \tag{H.23}$$

where $\mathbf{X} = (\mathbf{x}_1, \mathbf{x}_2, \ldots, \mathbf{x}_N)$ and $\nabla_\mathbf{x}^2 = \sum_{i=1}^{N} \nabla_i^2$. Furthermore, we have written $V(\mathbf{R})$ as

$$V(\mathbf{R}) = \lambda E_0 W(\mathbf{X}) \quad , \tag{H.24}$$

where the renormalization factor λ depends on E_0. It determines the strength of the interaction potential required to produce a binding energy $-E_0$ for the ground state. The Green's function $G_0(\mathbf{X}_0, \mathbf{X})$ is defined as the solution of the equation

$$(-\nabla_\mathbf{x}^2 + 1)G_0(\mathbf{X}_0, \mathbf{X}) = \delta(\mathbf{X} - \mathbf{X}_0) \quad . \tag{H.25}$$

Because this is an equation for noninteracting particles, one can solve it for given boundary conditions. The solution $\psi(\mathbf{X})$ of (H.23) can be expressed in terms of $G_0(\mathbf{X}', \mathbf{X})$ in the form of an integral equation

$$\psi(\mathbf{X}) = \lambda \int d\mathbf{X}' G_0(\mathbf{X}', \mathbf{X}) W(\mathbf{X}') \psi(\mathbf{X}') \quad . \tag{H.26}$$

This equation can be solved by iteration, i.e., by computing

$$\psi_{n+1}(\mathbf{X}) = \lambda \int d\mathbf{X}' G_0(\mathbf{X}', \mathbf{X}) W(\mathbf{X}') \psi_n(\mathbf{X}') \tag{H.27}$$

starting from a properly chosen $\psi_0(\mathbf{X})$. A self-consistent solution exists only for the right value of λ. A population of points $\mathbf{X}'$ is chosen in correspondence to $\psi_n(\mathbf{X}')$. For each point $\mathbf{X}'$ the function $\lambda W(\mathbf{X}')$ is calculated and $p(\mathbf{X}') = [\lambda W(\mathbf{X}')]$ new points $\mathbf{X}$ are generated according to the probability distribution $G_0(\mathbf{X}', \mathbf{X})$. The meaning of the brackets $[\ldots]$ is the same as in connection with (H.20). When we take an average over the distributions $\mathbf{X}'$ and $p(\mathbf{X}')$ the distribution of points $\mathbf{X}$ is according to $\psi_{n+1}(\mathbf{X})$. Only when λ has the proper value does the population remain constant. Otherwise it decays or grows exponentially. For more details the reader should consult the original literature [59, 228].

Until now it has been assumed that $\psi(\mathbf{R}, \tau) \geq 0$, which excludes the discussion of fermions. For these systems, the Pauli principle requires the ground-state wavefunction to be antisymmetric with respect to particle exchange. Since we are interested in electrons, it is important to know how the previous considerations can be extended to Fermi systems. Consider the wavefunction (H.15). Antisymmetry is built into the wavefunction by means of the Slater determinant $\Phi(\mathbf{R})$; the single-particle orbitals which it contains determine the node structure. The Jastrow prefactor is always positive. We can extend the method to fermions if we apply the fixed-node approximation. The trial wavefunction is considered in different regions of configuration space which are bounded by nodal surfaces. Within each of the regions, the sign of the wavefunction is fixed, i.e., either plus or minus. We evaluate (H.22) under the supplementary condition that whenever a nodal surface of $\Phi_t(\mathbf{R}, \tau)$ is crossed in a walk $\mathbf{R}_\nu \to \tilde{\mathbf{R}}_\nu$, i.e., whenever $\Phi_t(\mathbf{R}, \tau)\Phi_t(\tilde{\mathbf{R}}, \tau + \Delta\tau) < 0$, the trial step is eliminated. There exist improvements of the fixed node approximation, but they go beyond the scope of the present study.

I
Computing the Memory Function by Increments

In the following we want to show in more detail how the memory function (9.21)

$$M_{ij}(z) = \left(A_i \left| \frac{1}{z - QLQ} \right| A_j \right)_+ \tag{I.1}$$

can be calculated by applying the method of increments. The notation is the same as in Sect. 9.2. We decompose L into a coherent part $\tilde{L}(z)$ and an interaction part L_I, i.e.,

$$L = \tilde{L}(z) + L_I(z) \tag{I.2}$$

where $\tilde{L}(z)$ corresponds to $\tilde{H}(z)$ (see (9.23)) while L_I corresponds to

$$\begin{aligned} H_I(z) &= \sum_i \left(U \delta n_{i\uparrow} \delta n_{i\downarrow} - \tilde{\Sigma}_\sigma(z) n_{i\sigma} \right) \\ &= \sum_i H_I^{(i)} \end{aligned} \tag{I.3}$$

With this notation and using (I.2) we can introduce a scattering operator T by writing

$$\frac{1}{z - QLQ} = g_0 + g_0 T g_0 \tag{I.4}$$

where g_0 is defined according to (9.34). The T operator is

$$\begin{aligned} T &= QL_IQ \frac{1}{1 - g_0 QL_IQ} \\ &= QL_IQ + (QL_IQ)g_0(QL_IQ) + \ldots \end{aligned} \tag{I.5}$$

It can be compared with (9.11).

Like in Appendix B the scattering operator can be expanded in terms of cluster scattering operators

$$T = \sum_i T_i + \sum_{\langle ij \rangle} \delta T_{ij} + \dots \tag{I.6}$$

where

$$\delta T_{ij} = T_{ij} - T_i - T_j \quad \text{etc.} \tag{I.7}$$

For higher order terms see [225]. The operators T_i, T_{ij}, ... are scattering operators for a cluster consisting of a single site i, of two sites i,j etc. We denote these sets of sites by an index $c = i, ij, \dots$

Consider a cluster c with the cluster memory function $M_{ij}^{(c)}(z)$ defined by (9.30) together with (9.31) and (9.32). The simplest approximation is restricting the cluster to a single site, i.e., $c = i$. This is the case treated in DMFT and the only matrix element required is

$$M_{ii}^{(i)}(z) = \left(A_{i\sigma} | g_0 + g_0 T_i g_0 | A_{i\sigma}^+ \right)_+ \quad . \tag{I.8}$$

According to the above the cluster scattering operator is here $T = T_i$. It is equivalent to an impurity problem with the interaction given by $H_I^{(i)}$ (see I.3). When we include also two-site clusters this matrix element changes into

$$M_{ii}^{(il)}(z) = M_{ii}^{(i)}(z) + \left(A_{i\sigma}^+ | g_0 \left(T_i + \delta T_{il} \right) A_{i\sigma}^+ \right)_+ \tag{I.9}$$

so that the change is

$$\delta M_{ii}^{(il)}(z) = \left(A_{i\sigma}^+ | g_0 \left(T_i + \delta T_{il} \right) A_{i\sigma}^+ \right)_+ \quad . \tag{I.10}$$

Here the cluster scattering operator is $T = T_i + T_j + \delta T_{ij}$. Similarly we find for the off-diagonal matrix elements

$$M_{ij}^{(ij)}(z) = \left(A_{i\sigma}^+ | (g_0 + g_0 \left(T_i + T_j \right) g_0) A_{j\sigma}^+ \right)_+ \tag{I.11}$$

and (see (9.32))

$$\delta M_{ij}^{(ijl)}(z) = M_{ij}^{(ijl)}(z) - M_{ij}^{(ij)}(z) \quad . \tag{I.12}$$

We want to draw attention that only matrix elements $M_{ij}^{(c)}(z)$ contribute where sites i and j are parts of the cluster. Further details are found in [225].

J

Kagome Lattice at 1/3 Filling

We want to show that the narrow peak in the hole spectral density seen in Fig. 14.23 appears because

$$\left|\psi_{\mathbf{k}=0}^{N-1}\right\rangle = c_{\mathbf{k}=0}\left|\psi_0\right\rangle$$
$$= \frac{1}{\sqrt{3N}} \sum_i c_i \left|\psi_0\right\rangle \quad (J.1)$$

is an eigenstate of H defined by (14.30) in the limit $t/V \to 0$. An equivalent statement holds for the particle spectral density. Because we consider a 1/3 filling of the kagome lattice, the operator $c_{\mathbf{k}=0}$ corresponds in the noninteracting case to the lowest band in Fig. 14.11. The ground state $|\psi_0\rangle$ of H refers to 1/3 filling. We divide H into a kinetic energy part H_0 and an interacting part H_{int} and compute $[H, c_{\mathbf{k}=0}]_-$. It is

$$[H_0, c_{\mathbf{k}=0}]_- = -4t\, c_{\mathbf{k}=0} \quad (J.2)$$

because each site has four nearest neighbors. We find also that

$$[H_{\text{int}}, c_{\mathbf{k}=0}]_- = \frac{V}{\sqrt{3N}} \sum_{\langle ij \rangle} (c_j n_i + c_i n_j)$$
$$= \frac{V}{\sqrt{3N}} \sum_i \frac{4t}{V} c_i |\psi_0\rangle$$
$$= 4t \quad . \quad (J.3)$$

Here we have used that the four nearest neighbors of a site are occupied each with probability t/V due to quantum fluctuations contained in $|\psi_0\rangle$. This holds true in the strong correlation limit $t/V \ll 1$. Charge order does not destroy that result provided we average over the degenerate ground states. When summed up we obtain

$$[H, c_{\mathbf{k}=0}]_- = 0 \quad (J.4)$$

J Kagome Lattice at 1/3 Filling

and hence a δ-function peak in the hole spectral density at $\omega = 0$. Remember that the commutator gives us the excitation energy of an eigenstate.

A similar calculation for $c^+_{\mathbf{k}=0}|\psi_0\rangle$ gives us

$$\left[H, c^+_{\mathbf{k}=0}\right]_- = 2(V-t)c^+_{\mathbf{k}=0} \qquad (\text{J.5})$$

and therefore a peak in the particle spectral density near $2(V-t)$ (see Fig. 14.23).

References

1. Abbamonte, P., Rusydi, A., Smadici, S., Gu, G.D., Sawatzky, G.A., Feng, D.L.: Nature Phys. **1**, 155 (2005)
2. Abramowitz, M., Stegun, I.A.: Handbook of Mathematical Functions. Dover Publ., New York (1965)
3. Abrikosov, A.A., Gorkov, L.P.: Zh. Eksp. Teor. Fiz. **39**, 1781 (1960). Engl. transl.: Sov. Phys.-JETP **12**, 1243 (1961)
4. Abrikosov, A.A., Gorkov, L.P., Dzyaloshinski, I.E.: *Methods of Quantum Field Theory in Statistical Physics*. Prentice-Hall, Englewood Cliffs, New York (1963)
5. Amici, A., Thalmeier, P., Fulde, P.: Phys. Rev. Lett. **84**, 1800 (2000)
6. Anderson, P.W.: Science **235**, 1196 (1987)
7. Anderson, P.W.: *The Theory of Superconductivity in the High-T_c Cuprates*. Princeton Univ. Press, Princeton (1997)
8. Anisimov, V.I., Aryasetiawan, F., Lichtenstein, A.I.: J. Phys.: Condens. Matter **9**, 767 (1997)
9. Anisimov, V.I., Izyumov, Y.A.: *Electronic Structure of Strongly Correlated Materials*. Springer Series in Solid State Sciences. Springer, Heidelberg (2009)
10. Anisimov, V.I., Nekrasov, I.A., Kondakov, D.E., Rice, T.M., Sigrist, M.: Eur. Phys. J. B **25**, 191 (2002)
11. Anisimov, V.I., Zaanen, J., Andersen, O.K.: Phys. Rev. B **44**, 943 (1991)
12. Aoki, H., Uji, S., Albessard, A.K., Onuki, Y.: Phys. Rev. Lett. **71**, 2110 (1993)
13. Appelquist, T.W., Politzer, H.D.: Phys. Rev. Lett. **34**, 43 (1975)
14. Arovas, D.P., Auerbach, A.: Phys. Rev. B **38**, 316 (1988)
15. Aryasetiawan, F.: Phys. Rev. B **46**, 13051 (1992)
16. Auerbach, A.: *Interacting electrons and quantum magnetism*. Springer-Verlag, Berlin, Heidelberg, New York (1994)
17. Auerbach, A., Arovas, D.P.: Phys. Rev. Lett. **61**, 617 (1988)
18. Aulbur, W.G., Jonsson, L., Wilkins, J.W.: In: H. Ehrenreich, D. Turnbull (eds.) *Solid State Physics*, vol. 54. Academic Press, New York (2000)
19. Avella, A., Mancini, F., (eds.): *Strongly correlated Systems, in Springer Series in Solid-State Sciences*, vol. 171. Springer-Verlag, Heidelberg (2012)
20. Baltensperger, W., Straessler, S.: Phys. Kondens. Materie **1**, 20 (1963)
21. Bardeen, J., Cooper, L.N., Schrieffer, J.R.: Phys. Rev. **108**, 1175 (1957)
22. Bardeen, J., Pines, D.: Phys. Rev. **99**, 1140 (1955)
23. Baym, G.: Phys. Rev. **127**, 1391 (1962)

24. Becke, A.D.: J. Chem. Phys. **84**, 4524 (1986)
25. Becke, A.D.: Phys. Rev. A **38**, 3038 (1988)
26. Becker, K.W., Won, H., Fulde, P.: Z. Phys. B **75**, 335 (1989)
27. Bednorz, J.G., Müller, K.A.: Z. Phys. B **64**, 189 (1986)
28. Berk, N.F., Schrieffer, J.R.: Phys. Rev. Lett. **17**, 433 (1966)
29. Bernert, A., Thalmeier, P., Fulde, P.: Phys. Rev. B **66**, 165108 (2002)
30. Bernhoeft, N., Sato, N., Roessli, B., Aso, N., Hiess, A., Lander, G.H., Endoh, Y., Komatsubara, T.: Phys. Rev. Lett. **81**, 4244 (1998)
31. Berry, M.V.: Proc. Royal Soc. London Sev. A **392**, 45 (1984)
32. Bethe, H.A.: Z. Phys. **71**, 205 (1931)
33. Bethe, H.A., Goldstone, J.: Proc. R. Soc. A (London) **238**, 551 (1957)
34. Bianchi, A., Movshovich, R., Capan, C., Pagliuso, P.G., Sarrao, J.L.: Phys. Rev. Lett. **91**, 187004 (2003)
35. Bickers, N.E., Cox, D.L., Wilkins, J.W.: Phys. Rev. B **36**, 2036 (1987)
36. Bickers, N.E., Scalapino, D.J., White, S.R.: Phys. Rev. Lett. **62**, 961 (1989)
37. Binder, K. (ed.): *Monte Carlo Methods in Statistical Physics, Topics Curr. Phys.*, vol. 7, 2nd edn. Springer, Berlin, Heidelberg (1986)
38. Binder, K. (ed.): *Applications of the Monte Calo Method in Statistical Physics, Topics Curr. Phys.*, vol. 36, 2nd edn. Springer, Berlin, Heidelberg (1987)
39. Blankenbecker, R., Scalapino, D.J., Sugar, R.L.: Phys. Rev. D **24**, 2278 (1981)
40. Bonner, J.C., Fisher, M.E.: Phys. Rev. A **135**, 640 (1964)
41. Bonča, J., Prelovšek, P., Sega, I.: Phys. Rev. B **42**, 10706 (1990)
42. Borrmann, W., Fulde, P.: Phys. Rev. B **31**, 7800 (1985)
43. Bouwers, J.A., Rajagopal, K.: Phys. Rev. D **66**, 065002 (2002)
44. Bramwell, S.T., Giblin, S.R., Calder, S., Aldus, R., Fennell, D.P.T.: Nature **461**, 956 (2009)
45. Brinkmann, W.F., Rice, T.M.: Phys. Rev. B **2**, 4302 (1970)
46. Bruls, G., Weber, D., Wolf, B., Thalmeier, P., Lüthi, B., de Visser, A., Menovsky, A.: Phys. Rev. Lett. **65**, 2294 (1990)
47. Bulaevskii, L.N.: Sov. Phys. JETP **38**, 634 (1974)
48. Bulaevskii, L.N., Kuzii, V.V., Sobyanin, A.A.: Pisma Zh. Eksp. Teor. Fiz **25**, 314 (1977). (JETP Lett. 25, 290 (1977))
49. Buzdin, A.I.: Rev. Mod. Phys. **77**, 935 (2005)
50. Buzdin, A.I., Bulaevskii, L.N., Panjukov, S.V.: Pisma Zh. Eksp. Teor. Fiz **35**, 147 (1982). (JETP Lett. 35, 178 (1982))
51. Callen, H.B., Welton, R.F.: Phys. Rev. **86**, 702 (1952)
52. Cane, C.L., Lee, P.A., Read, N.: Phys. Rev. B **39**, 6880 (1989)
53. Care, C.M., March, N.H.: Adv. Phys. **24**, 101 (1975)
54. Carr, J.W.: Phys. Rev. **122**, 1437 (1961)
55. Casalbuoni, R., Nardulli, G.: Rev. Mod. Phys. **76**, 263 (2004)
56. Castelnovo, C., Moessner, R., Sondhi, S.L.: Nature **451**, 42 (2008)
57. Cava, R.J., Takagi, H., Zandbergen, H.W., Krajewski, J.J., Peek, W.F., Siegrist, T., Batlogg, B., v. Dover, R.B., Felder, R.J., Mizuhashi, K., Lee, J.O., Elsaki, H., Uchida, S.: Nature **367**, 252 (1994)
58. Ceperley, D.M., Alder, B.: Science **231**, 555 (1986)
59. Ceperley, D.M., Kalos, M.H.: In: K. Binder (ed.) *Monte Carlo Methods in Statistical Physics*, Topics Curr. Phys., 2nd edn., p. 145. Springer, Berlin, Heidelberg (1986)

60. Chakraborty, T., Pietiläinen, P.: *The Quantum Hall Effects: Fractional and Integral*, in Springer Series in Solid-State Sciences, vol. 85, 2nd edn. Springer-Verlag, Berlin, Heidelberg (1995)
61. Chang, J., Eremin, I., Thalmeier, P., Fulde, P.: Phys. Rev, B **75**, 024503 (2007)
62. Chatterji, T., Regnault, L.P., Thalmeier, P., van de Kamp, R., Schmidt, W., Hiess, A., Vorderwisch, P., Suryanarayanan, R., Dahalenne, G., Revcolevschi, A.: J. Alloys and Compounds **326**, 15 (2001)
63. Chatterji, T., Regnault, L.P., Thalmeier, P., Suryanarayanan, R., Dhalenne, G., Revcolevschi, A.: Phys. Rev. B **60**, R6965 (1999)
64. Chen, Q., Stajic, J., Levin, K.: Fiz. Nizkikh. Temp. **32**, 538 (2006)
65. Chen, X.H., Wu, T., Liu, R.H., Chen, H., Fang, D.F.: Nature **453**, 761 (2008)
66. Christianson, A.D., Goremychkin, E.A., Osborn, R., Rosenkranz, S., Lumsden, M.D., Malliakas, C.D., Todorov, I.S., Claus, H., Chung, D.Y., Kanatzikis, M.G., Bewley, R.I., Guidi, T.: Nature **456**, 930 (2008)
67. Chu, C.W.: In: L.N. Cooper, D. Feldmann (eds.) *BCS: 50 years*. World Scientific, Singapore (2011)
68. des Cloizeaux, J., Pearson, J.J.: Phys. Rev. **128**, 2131 (1962)
69. Combescot, R., Mora, C.: Phys. Rev. B **71**, 144517 (2005)
70. Continentino, M.A.: *Quantum Scaling in Many-Body Systems*, World Scientific Lecture Notes in Physics, vol. 67. World Scientific, Singapore (2001)
71. Cooper, L.N.: Phys. Rev. **104**, 1189 (1956)
72. Coqblin, B., Schrieffer, R.: Phys. Rev. **185**, 847 (1969)
73. CRC: In: D.R. Lide (ed.) *Handbook of Chemistry and Physics*, 75nd edn. CRC Press, Boca Raton (1994/95)
74. Cross, M.C., Fisher, D.S.: Phys. Rev. B. **19**, 402 (1979)
75. Cyrot, M.: Phys. Lett. **37**, 189 (1971)
76. Cyrot, M.: J. Phys. (Paris) **33**, 25 (1972)
77. Dagotto, E.: Rev. Mod. Phys. **66**, 763 (1984)
78. Dagotto, E.: *Nanoscale Phase Separation and Colossal Magnetoresistance: The Physics of Manganites and Related Compounds*, Springer Series in Solid State Sciences, vol. 136. Springer, Berlin (2003)
79. Dagotto, E., Riera, J., Young, A.P.: Phys. Rev. B **42**, 2347 (1990)
80. Dahm, T., Hinkov, V., Borisenko, S.V., Kordyuk, A.A., Zabolotnyy, V.B., Fink, J., Büchner, B., Scalapino, D.J., Hanke, W., Keimer, B.: Nature Phys. **5**, 217 (2009)
81. de Gennes, P.G.: Phys. Kondens. Materie **3**, 79 (1964)
82. de Gennes, P.G.: *Superconductivity of Metals and Alloys*. W. A. Benjamin Inc., New York (1966)
83. de Gennes, P.G., Tinkham, M.: Phys. Kondens. Materie **1**, 107 (1964)
84. de Graaf, C.: private communication (2008)
85. de Lara-Castells, M., Mitrushenkov, A.: J. Phys. Chem. C **115**, 17540 (2011)
86. Declaux, P.G., Moser, C.M., Verhaegen, G.: J. Phys. B **4**, 296 (1971)
87. Demler, E.A., Arnold, G.B., Beasley, M.R.: Phys. Rev. B **55**, 15174 (1997)
88. Denlinger, J.D., Gweon, G.H., Allen, J.W., Olson, C.G., Daliachaouch, Y., Lee, B.W., Maple, M.B., Fisk, Z., Canfield, P.C., Armstrong, P.E.: Physica B **281 & 282**, 716 (2000)
89. Denlinger, J.D., Gweon, G.H., Allen, J.W., Olson, C.G., Maple, M.B., Sarrao, J.L., Armstrong, P.E., Fisk, Z., Yamagami, H.: J. Electron Spectrosc. Relat. Phenom. **117 & 118**, 347 (2001)

90. Dieterich, W.: Adv. Phys. **25**, 615 (1976)
91. Dolg, M.: In: P. Schwerdtfeger (ed.) *Relativistic Electronic Structure Theory – Fundamentals, Theoret. and Comput. Chemistry*, vol. 11. Elsevier, Amsterdam (2003)
92. Dolg, M., Fulde, P., Kuechle, W., Neumann, C.S., Stoll, H.: J. Chem. Phys. **94**, 1360 (1991)
93. Doll, K., Dolg, M., Fulde, P., Stoll, H.: Phys. Rev. B **52**, 4842 (1995). Ibid: Phys. Rev. B **55**, 10282 (1997)
94. Doll, K., Dolg, M., Stoll, H.: Phys. Rev. B **54**, 13529 (1996)
95. Doniach, S., Sondheimer, E.H.: *Green's Functions for Solid State Physicists.* Benjamin/Cummings, London (1974)
96. Dopf, G., Wagner, J., Dietrich, P., Muramatsu, A., Hanke, W.: Phys. Rev. Lett. **68**, 2082 (1992)
97. Doucot, B., Duplantier, B., Pasquier, V., Rivasseau, V., eds.: *The Quantum Hall Effect.* Birkhäuser, Basel (2005)
98. Dreizler, R.M., Gross, E.K.U.: *Density Functional Theory.* Springer-Verlag, Berlin (1990)
99. Dunning, T.M.J.: J. Chem. Phys. **90**, 1007 (1989)
100. Dupuis, N.: Phys. Rev. B **51**, 9074 (1995)
101. Eberhardt, W., Plummer, E.W.: Phys. Rev. B **21**, 3245 (1980)
102. Ebert, D.: *Eichtheorien.* Akademie Verlag, Berlin (1989)
103. Eder, R.: private communication
104. Eder, R.: Phys. Rev. B **43**, 10706 (1991)
105. Eder, R.: Phys. Rev. B **45**, 319 (1992)
106. Eder, R.: private communication (2011)
107. Efremov, D., Hasselmann, N., Runge, E., Fulde, P., Zwicknagl, G.: Phys. Rev. B **69**, 115114 (2004)
108. Eremin, I.: Phys. Rev. Lett. **101**, 187001 (2008)
109. Eremin, I., Manske, D.: Low Temp. Phys. **32**, 519 (2006)
110. Eschrig, H.: *The Fundamentals of Density Functional Theory.* B. C. Teubner, Stuttgart (1996)
111. Eschrig, M.: Phys. Today **64**, 43 (2011)
112. Eyert, V., Höck, K.H., Horn, S., Loidl, A., Riseborough, P.S.: Europhys. Lett. **46**, 762 (1999)
113. Ezawa, Z.F.: *Quantum Hall Effect.* World Scientific, Singapore (2000)
114. Faddeev, L.D.: Zh. Eksp. Teor. Fiz. **39**, 1459 (1960). Engl. Transl.: Sov. Phys.-JETP **12**, 1014 (1961)
115. Fahy, S., Wang, X.W., Louie, S.G.: Phys. Rev. B **42**, 3503 (1990)
116. Feofanov, A.K., Oboznov, V.A., Bolginov, V.V., Lisenfeld, J., Poletto, S., Ryazanov, V.V., Rossolenko, A.N., Khabipov, M., Balashov, D., Zorin, A.B., Dmitriev, P.N., Koshelets, V.P., Ustinov, A.V.: Nature Physics **6**, 593 (2010)
117. Fetter, A.L., Walecka, J.D.: *Quantum Theory of Many-Particle Systems.* Mc Graw-Hill, New York (1971)
118. Fink, J., Schierle, E., Weschke, E., Geck, J., Hawthorn, D., Wadati, H., Hu, H.H., Dürr, H.A., Wizent, N., Büchner, B., Sawatzky, G.A.: Phys. Rev. B **79**, 100502(R) (2009)
119. Fiolhais, C., Nogueira, F., Marques, M.A.L., eds.: In: *A Primer in Density Functional Theory, in Lect. Notes Phys.*, vol. 620. Springer, Berlin (2003)
120. Fisher, R.A., Ballon, R., Emerson, J.P., Lelivre-Berna, E., Phillips, N.E.: Int. J. Mod. Phys. B **7**, 830 (1992)

121. Foster, J.M., Boys, S.F.: Rev. Mod. Phys. **32**, 300 (1960)
122. Fradkin, E.: *Field Theories of Condensed Matter Systems*. Addison-Wesley Publ., Redwood City (1991)
123. Frahm, H.: Phys. Rev. B **42**, 10553 (1990). Ibid. B **43**, 5653 (1991)
124. Friedel, J.: Phil. Mag. **43**, 153 (1952)
125. Friedel, J.: *The Physics of Metals: 1. Electrons*. Cambridge Univ. Press, Cambridge (1969)
126. Friedel, J., Sayers, C.M.: J. de Phys. **38**, 697 (1977)
127. Fujikawa, K., Miyoshi, M., Takenchi, J., Shimaoka, Y., Kobayashi, T.: J. Phys.: Cond. Matt. **16**, S615 (2004)
128. Fujimori, A., Minami, F., Sugano, S.: Phys. Rev. B **29**, 5225 (1984)
129. Fujimori, A., Minami, F., Sugano, S.: Phys. Rev. B **30**, 957 (1984)
130. Fulde, P.: In: K.A. Gschneidner, L. Eyring (eds.) *Handbook of the Physics and Chemistry of Rare Earths*, vol. 2, chap. 17, p. 295. North-Holland, Amsterdam (1979)
131. Fulde, P.: *Electron Correlations in Molecules and Solids, in Springer Series in Solid-State Sciences*, vol. 100, 3rd edn. Springer-Verlag, Berlin, Heidelberg (1995)
132. Fulde, P., Ferrell, R.A.: Phys. Rev. **135**, A550 (1964)
133. Fulde, P., Hirst, L.L., Luther, A.: Z. Phys. **238**, 99 (1970)
134. Fulde, P., Jensen, J.: Phys. Rev. B **27**, 4085 (1983)
135. Fulde, P., Kakehashi, Y., Stollhoff, G.: In: H. Capellmann (ed.) *Metallic Magnetism, in Topics Curr. Phys.*, vol. 42. Springer, Berlin, Heidelberg (1987)
136. Fulde, P., Keller, J., Zwicknagl, G.: In: H. Ehrenreich, D. Turnbull (eds.) *Solid State Physics*, vol. 41, p.1. Academic Press, New York (1988)
137. Fulde, P., Penc, K., Shannon, N.: Ann. Phys. (Leipzig) **11**, 892 (2002)
138. Fulde, P., Schmidt, B., Thalmeier, P.: Europhys. Lett. **31**, 323 (1995)
139. Fulde, P., Stoll, H., Kladko, K.: Chem. Phys. Lett. **299**, 481 (1999)
140. Furukawa, N., Imada, M.: J. Phys. Soc. Jpn. **61**, 3331 (1992)
141. Galitzkii, V.M.: Zh. Eksp. Teor. Fiz. **34**, 151 (1958). Engl. transl.: Sov. Phys. - JETP 7, 104 (1958)
142. Gaskell, T.: Proc. Phys. Soc. **77**, 1182 (1961)
143. Geck, J., Wochner, P., Kiele, S., Klingeler, R., Revcolevschi, A., v. Zimmermann, M., Büchner, B., Reutler, P.: New J. of Phys. **6**, 152 (2004)
144. Geldart, D.J.W., Rasolt, M.: In: N.H. March, B.M. Debb (eds.) *The Single-Particle Density in Physics and Chemistry*. Acadamic, New York (1987)
145. Gell-Mann, M., Bruckner, K.: Phys. Rev. **106**, 364 (1957)
146. Georges, A., Kotliar, G.: Phys. Rev. B **45**, 6479 (1992)
147. Georges, A., Kotliar, G., Krauth, W., Rosenberg, M.J.: Rev. Mod. Phys. **68**, 13 (1996)
148. Goldstone, J.: Proc. R. Soc. A (London) **239**, 267 (1957)
149. Goodenough, J.B.: *Magnetism and the Chemical Bond*. Interscience–Wiley, New York (1963)
150. Goremychkin, E.A., Osborn, R.: Phys. Rev. B **47**, 14280 (1993)
151. Goremychkin, E.A., Osborn, R., Bauer, E.D., Maple, M.B., Frederick, N.A., Yuhasz, W.M., Woodward, F.M., Lynn, J.W.: Phys. Rev. Lett **93**, 157003 (2004)
152. Goto, T., Lüthi, B.: Adv. Phys. **52**, 67 (2003)
153. Goto, T., Nemoto, Y., Ochiai, A.: Phys. Rev. B **59**, 269 (1999)

154. Götzel, D., Segall, B., Andersen, O.K.: Solid State Comm. **36**, 403 (1980)
155. Gräfenstein, J., Stoll, H., Fulde, P.: Phys. Rev. B **55**, 13588 (1997)
156. Grewe, N., Steglich, F.: In: K.A. Gschneidner, L. Eyring (eds.) *Handbook on the Physics and Chemistry of Rare Earths*, vol. 17. North-Holland, Amsterdam (1991)
157. Gros, C.: Phys. Rev. B **38**, 931 (1988)
158. Gross, E.K., Runge, E., Heinonen, O.: *Many-Particle Theory*. A. Hilger, Bristol (1991)
159. Gschneidner, K.A., Eyring, L., eds.: *Handbook on the Physics and Chemistry of Rare Earth*. North-Holland Publ. Comp. (1978)
160. Gunnarsson, O., Jonson, M., Lundqvist, B.I.: Phys. Rev. B **20**, 3136 (1979)
161. Gunnarsson, O., Lundqvist, B.I.: Phys. Rev. B **13**, 4274 (1976)
162. Gunnarsson, O., Schönhammer, K.: Phys. Rev. B **28**, 4315 (1983)
163. Gunnarsson, O., Schönhammer, K.: Phys. Rev. B **31**, 4815 (1985)
164. Gutzwiller, M.C.: Phys. Rev. Lett. **10**, 159 (1963)
165. Gutzwiller, M.C.: Phys. Rev. A **134**, 923 (1964)
166. Gutzwiller, M.C.: Phys. Rev. A **137**, 1726 (1965)
167. Haldane, F.D.M.: J. Phys. C **14**, 2585 (1981)
168. Haldane, F.D.M.: Phys. Lett. A **93**, 464 (1983)
169. Hamada, N., Sawada, H., Solovyev, I., Terakura, K.: Physica B **237-238**, 11 (1997)
170. Harris, A.B., Lange, R.V.: Phys. Rev. **157**, 295 (1967)
171. Harrison: J. Phys.: Cond. Matt. **16**, S553 (2004)
172. Hase, M., Terasaki, I., Uchinokura, K.: Phys. Rev. Lett. **70**, 3651 (1993)
173. Hasegawa, H.: J. Phys. Soc. Jpn. **46**, 1504 (1979)
174. Hasegawa, H.: J. Phys. Soc. Jpn. **49**, 178 (1980)
175. Hedin, L.: Phys. Rev. **139**, A796 (1965)
176. Hedin, L., Lundqvist, S.: In: F. Seitz, D. Turnbull, H. Ehrenreich (eds.) *Solid State Physics*, vol. 23. Academic Press, New York (1969)
177. Heinonen, O., ed.: *Composite Fermions*. World Scientific, Singapore (1998)
178. Henley, C.L.: Ann. Rev. Condens. Matter Phys. **1**, 179 (2010)
179. Hertz, J.A.: Phys. Rev. B **14**, 1165 (1976)
180. Hertz, J.A., Klenin, M.A.: Phys. Rev. B **10**, 1984 (1974)
181. Hewat, A.W.: In: E. Kaldis (ed.) *Materials and Crystallographic Aspects of High-Tc Superconductivity, in NATO ASI Series, Series E: Applied Sciences*, vol. 263. Kluwer, Dordrecht (1994)
182. Hiess, A., Bernhoeft, N., Metoki, N., Lander, G.H., Roessli, B., Sato, N.K., Aso, N., Haga, Y., Koike, Y., Komatsubara, T., Onuki, Y.: J. Phys.: Condens. Matt. **18**, R437 (2006)
183. Hirota, K., Ishihara, S., Fujioka, H., Kubota, M., Yoshizawa, H., Moritomo, Y., Endoh, Y., Maekawa, S.: Phys. Rev. B **65**, 64414 (1998)
184. Hirsch, J.: Phys. Rev. **31**, 4403 (1985)
185. Hoang, A.T., Thalmeier, P.: J. Phys. Condens. Matter **14**, 6639 (2002)
186. Hohenberg, P., Kohn, W.: Phys. Rev. B **136**, 864 (1964)
187. Horsch, S., Horsch, P., Fulde, P.: Phys. Rev. B **29**, 1870 (1984)
188. Hotta, T., Dagotto, E.: Phys. Rev. Lett. **88**, 017201 (2002)
189. Houzet, M., Buzdin, A.: Phys. Rev. B **63**, 184521 (2001)
190. Hozoi, L., Birkenheuer, U., Fulde, P., Mitrushchenko, A., Stoll, H.: Phys. Rev. B **76**, 085109 (2007)

191. Hozoi, L., Laad, M.S., Fulde, P.: Phys. Rev. B **78**, 165107 (2008)
192. Hozoi, L., Laad, M.S., Fulde, P.: Erratum Phys. Rev. B **81**, 159904 (2010)
193. Hubbard, J.: Phys. Rev. Lett. **3**, 77 (1959)
194. Hubbard, J.: Proc. R. Soc. London A **276**, 238 (1963)
195. Hubbard, J.: Proc. R. Soc. London A **281**, 401 (1964)
196. Hüfner, S., Wertheim, G.K.: Phys. Lett. A **47**, 349 (1974)
197. Huzinaga, S.: J. Chem. Phys. **42**, 1293 (1965)
198. Hybertsen, M.S., Louie, S.: Phys. Rev. B **34**, 5390 (1986)
199. Hybertsen, M.S., Louie, S.G.: Phys. Rev. Lett. **55**, 1418 (1985)
200. Hybertsen, M.S., Schlüter, M., Christensen, N.E.: Phys. Rev. B **39**, 9028 (1989)
201. Igarashi, J.: J. Phys. Soc. Jpn. **54**, 260 (1985)
202. Igarashi, J., Fulde, P.: Phys. Rev. B **45**, 10419 (1992). Ibid 12357; ibid B **48**, 998 (1993)
203. Igarashi, J., Unger, P., Hirai, K., Fulde, P.: Phys. Rev. B **49**, 16181 (1994)
204. Inada, Y., Yamagani, H., Haga, Y., Sakurai, K., Tokiwa, Y., Hinma, T., Yamamoto, E., Onuki, Y., Yanagisawa, T.: J. Phys. Soc. Jpn. **68**, 3643 (1999)
205. Ishida, K., Nakai, Y., Hosono, H.: J. Phys. Soc. Jpn. **78**, 062001 (2009)
206. Isobe, M., Ueda, Y.: Phys. Soc. Jpn. **65**, 1178 (1996)
207. Isoda, M., Mori, S.: J. Phys. Soc. Jpn. **69**, 1509 (2000)
208. Izuyama, T., Kim, D.J., Kubo, R.: J. Phys. Soc. Jpn. **18**, 1025 (1963)
209. Izyumov, Y.A., Kurmaev, E.: *High-Tc Superconductivity in FeAs Intermetallic Systems*. R and C Dynamics, Moscow (2008). (in Russian)
210. Izyumov, Y.A., Letfulov, B.M.: J. Phys.: Cond. Matt. **2**, 8905 (1990)
211. Izyumov, Y.A., Proshin, Y.N., Khusainov, M.G.: Physics-Uspekhi **45**, 109 (2002)
212. Jackiw, Rebbi, C.: Phys. Rev. D **13**, 3398 (1976)
213. Jaclič, J., Prelovšek, P.: Phys. Rev. B **49**, 5065 (1994)
214. Jarrell, M.: Phys. Rev. Lett. **69**, 168 (1992)
215. Jarrell, M., Krishnamurthy, H.R.: Phys. Rev. B **63**, 125102 (2001)
216. Jastrow, R.: Phys. Rev. **98**, 1479 (1955)
217. Jayprakash, C., Krishnamurthy, H.R., Sarker, S.: Phys. Rev. B **40**, 2610 (1989)
218. Jayprakash, C., Krishnamurthy, H.R., Wilkins, J.W.: Phys. Rev. Lett. **47**, 737 (1981)
219. Jerome, D.: *Organic Conductors*. Dekker, New York (1994)
220. Johnston, D.C.: Adv. Phys. **59**, 803 (2010)
221. Jonker, G.H., van Santen, J.H.: Physica **16**, 377 (1950)
222. Kajzar, F., Friedel, J.: J. de Phys. **39**, 379 (1978)
223. Kakehashi, Y.: J. Magn. Magn. Mater. **104-107**, 677 (1992)
224. Kakehashi, Y.: Phys. Rev. B **65**, 184420 (2002)
225. Kakehashi, Y.: Adv. Phys. **53**, 497 (2004)
226. Kakehashi, Y., Miyagi, K.: private communication (2012)
227. Kakehashi, Y., Nakamura, T., Fulde, P.: J. Phys. Soc. Jpn. **78**, 124710 (2009)
228. Kalos, M.H.: Phys. Rev. **128**, 1791 (1962)
229. Kampf, A.P.: Phys. Reports **249**, 219 (1994)
230. Kanamori, J.: Progr. Theor. Phys. **30**, 275 (1963)
231. Kawabate, A.: J. Phys. F **4**, 1477 (1974)
232. Kawakami, N., Okiji, A.: In: H. Fukuyama, S. Maekawa, A.P. Malozemoff (eds.) *Strong Correlation and Superconductivity*, in Springer Ser. Solid-State Sci., vol. 84. Springer, Berlin, Heidelberg (1989)

233. Keimer, B., Casa, D., Ivanov, A., Lynn, J.W., v. Zimmermann, M., Hill, J.P., Gibbs, D., Taguchi, Y., Tokura, Y.: Phys. Rev. Lett. **85**, 3946 (2000)
234. Keldysh, L.V.: Zh. Eksp. Teor. Fiz. **47**, 1515 (1964). Engl. transl.: Sov. Phys. JETP **20**, 1018 (1965)
235. Kenzelmann, M., Strässle, T., Niedermayer, C., Sigrist, M., Padmanabham, B., Zolliker, M., Bianchi, A.D., Movshovichi, R., Bauer, E.D., Sarrao, J.L., Thompson, J.D.: Science **321**, 1652 (2008)
236. Ketterle, W., Shin, Y., Schirotzek, A., Schuuk, C.H.: In: L.N. Cooper, D. Feldman (eds.) *BCS: 50 years*, p. 491. World Scientific Publ., Singapore (2011)
237. Khaliullin, G.G.: Pis'ma Zh. Eksp. Teor. Fiz. **52**, 999 (1990). Engl. transl.: Sov. Phys.-JETP Lett. **52**, 389 (1990)
238. Khusainov, M.G., Khusainov, M.M., Proshin, Y.: In: O.A. Chang (ed.) *Progress in Superconductivity Research*, p. 79. Nova Science Publ. (2008)
239. Kiel, B., Stollhoff, G., Weigel, C., Fulde, P., Stoll, H.: Z. Phys. B **46**, 1 (1982)
240. Kimura, T., Kumai, R., Tokura, Y., Li, J.Q., Matsui, Y.: Phys. Rev. B **58**, 11081 (1998)
241. King, C.A., Lonzarich, G.G.: Physica B **171**, 161 (1991)
242. Kittel, C.: *Quantum Theory of Solids*. Wiley, New York (1963)
243. Kladko, K., Fulde, P.: Int. J. of Quant. Chem. **66**, 377 (1998)
244. Kohgi, M., Iwasa, K., Mignot, J.M., Ochiai, A., Suzuki, T.: Phys. Rev. B **56**, R11388 (1997)
245. Kohgi, M., Iwasa, K., Mignot, J.M., Pyka, N., Ochiai, A., Aoki, H., Suzuki, T.: Physica B **259-261**, 269 (1999)
246. Kohn, W., Sham, L.: Phys. Rev. A **140**, 1133 (1965)
247. Koitzsch, A.: Priv. comm.
248. Kondo, S., Johnston, D.C., Swenson, C.A., Borsa, F., Mahajan, A.V., Miller, L.L., Gu, T., Goldman, A.I., Maple, M.B., Gajewski, D.A., Freeman, E.J., Dilley, N.R., Dickey, R.P., Merrin, J., Kojima, K., Luke, G.M., Uemura, Y.J., Chmaissem, O., Jorgensen, J.D.: Phys. Rev. Lett. **78**, 3729 (1997)
249. Kotliar, G., Liu, J.: Phys. Rev. B **38**, 5142 (1988)
250. Kotliar, G., Ruckenstein, A.E.: Phys. Rev. Lett. **57**, 1362 (1986)
251. Krimmel, A., Loidl, A., Klemm, M., Horn, S., Schrober, H.: Phys. Rev. Lett. **82**, 2919 (1999)
252. Kristyan, S., Pulay, P.: Phys. Lett. **229**, 175 (1994)
253. Kubo, R.: J. Phys. Soc. Jpn. **17**, 1100 (1962)
254. Kubo, R.: Rep. Prog. Phys. **29(1)**, 255 (1966)
255. Kudasov, Y.B.: Physics-Uspekhi **46(2)**, 117 (2003)
256. Kugel, K.I., Khomskii, D.I.: Sov. Phys. Usp. **25**, 231 (1982)
257. Kuhn, H.: Helv. Chim. Acta **31**, 1441 (1948)
258. Kumagai, K., Shishido, H., Shibanchi, T., Matsuda, T.: Phys. Rev. Lett **106**, 137004 (2011)
259. Kümmel, H., Lührmann, K.H., Zabolitzky, J.G.: Phys. Lett. C **36**, 1 (1978)
260. Kuramoto, Y., Kusunose, H., Kiss, A.: J. Phys. Soc. Jpn. **78**, 072001 (2009)
261. Kutzelnigg, W.: In: H.F. SchaeferIII (ed.) *Modern Theoretical Chemistry*, vol. 3. Plenum, New York (1977)
262. Kuwahara, K., Iwasa, K., Kohgi, M., Kaneko, K., Araki, S., Metoki, N., Sugawara, H., Aoki, Y., Sato, H.: J. Phys. Soc. Jpn. **73**, 1438 (2004)
263. Labbé, J., Friedel, J.: Journ. Phys (Paris) **27**, 153, 303 (1966)
264. Lanczos, C.: J. Res. Natl. Bur. Stand. **45**, 255 (1950)

265. Landau, L.D.: Zh. Eksp. Teor. Fiz. **30**, 1058 (1956)
266. Landau, L.D.: Zh. Eksp. Teor. Fiz. **32**, 59 (1957)
267. Landau, L.D., Pitajewski, L.P.: *Physical Kinetics, in Course of Theor. Physics*, vol. 10. Pergamon, Oxford (1981)
268. Langreth, D.C., Mehl, M.J.: Phys. Rev. B **28**, 1809 (1983)
269. Langreth, D.C., Perdew, J.P.: Solid State Comm. **17**, 1425 (1975)
270. Larkin, A.I., Ovchinnikov, Y.N.: Zh. Eksp. Teor. Fiz. **47**, 1136 (1964). Engl. transl.: Sov. Phys. JETP 20, 762 (1965)
271. Larkin, W.: J. Chem. Phys. **43**, 2954 (1965)
272. Laughlin, R.B.: Phys. Rev. Lett. **52**, 2304 (1984)
273. Lavagna, M., Stemmann, G.: Phys. Rev. B **49**, 4235 (1994)
274. Lee, P., Nagaosa, N., Wen, X.G.: Rev. Mod. Phys. **78**, 17 (2006)
275. Lee, S., Qin, Y., Bornholm, C., Ueda, Y., Rush, J.J.: Phys. Rev. Lett. **86**, 5554 (2001)
276. Leggett, A.J.: Rev. Mod. Phys. **47**, 331 (1975)
277. Leggett, A.J.: *Quantum Liquids*. Oxford Univ. Press, Oxford, UK (2006)
278. LeHur, K., Rice, T.M.: Annals of Physics **324**, 1452 (2009)
279. Leinaas, J.M., Myrheim, J.: Nuovo Cimento B **37**, 1 (1977)
280. Levy, M.: Proc. Natl. Acad. Sci. USA **76**, 6062 (1979)
281. Liao, Y., Rittner, A.S.C., Paprotta, T., Li, W., Partridge, G.B., Hulet, R.G., Baur, S.K., Mueller, E.J.: Nature **467**, 567 (2010)
282. Lichtenstein, A.I., Anisimov, V., Zaanen, J.: Phys. Rev. B **52**, R5467 (1995)
283. Lieb, L.H., Wu, F.Y.: Phys. Rev. Lett. **20**, 1445 (1968)
284. Liebsch, A.: Phys. Rev. Lett. **43**, 1431 (1979)
285. Lievin, J., Breulet, J., Verhaegen, G.: Theor. Chim. Acta **60**, 339 (1981)
286. Lifshitz, E.M., Pitajewski, L.P.: *Statistical Physics, in Course of Theoret. Physics*, vol. 9. Pergamon, Oxford (1981)
287. v. Löhneysen, H., Pietrus, T., Portisch, G., Schlager, H.G., Schröder, A., Sieck, M., Trappmann, T.: Phys. Rev. Lett. **72**, 3262 (1994)
288. Loktev, V.M., Quick, R.M., Sharapov, S.G.: Phys. Reports **349**, 1 (2001)
289. Longuet-Higgins, H.C., Salem, L.: Proc. R. Soc. London A **251**, 172 (1959)
290. Lonzarich, G.G.: J. Magn. Magn. Mat. **76-77**, 1 (1988)
291. Lortz, R., Wang, Y., Demner, A., Böttger, P.H.M., Bergk, B., Zwicknagl, G., Nakazawa, Y., Wosnitza, J.: Phys. Rev. Lett. **99**, 187002 (2007)
292. Löwdin, P.O.: J. Mol. Spectrosc. **10**, 12 (1963)
293. Löwdin, P.O.: J. Mol. Spectrosc. **13**, 326 (1964)
294. Löwdin, P.O.: J. Mol. Spectrosc. **14**, 112 (1964)
295. Lüdecke, J., Jobst, A., van Smaalen, S., Morré, E., Geibel, C., Krane, H.G.: Phys. Rev. Lett. **82**, 3633 (1999)
296. Luttinger, J.M.: J. Math. Phys. **4**, 1154 (1963)
297. Lynn, J.W., Skanthakumar, S., Huang, Q., Sinha, S.K., Hossain, Z., Gupta, L.C., Nagarajan, R., Godart, C.: Phys. Rev. B **55**, 6584 (1997)
298. Machida, K., Nakanishi, H.: Phys. Rev. B **30**, 122 (1984)
299. Macke, W.: Z. Naturforsch. **5a**, 192 (1950)
300. Mackenzie, A.P., Maeno, Y.: Rev. Mod. Phys. **75**, 657 (2003)
301. Mahan, G.D.: *Many Particle Physics*. Plenum, New York (1981)
302. Maier, T.A., Jarrell, M., Pruschke, T., Hettler, M.H.: Rev. Mod. Phys. **77**, 1027 (2005)
303. Maier, T.A., Jarrell, M., Schulthess, T.C., Kent, P.R.C., White, J.B.: Phys. Rev. Lett. **95**, 237001 (2005)

518   References

304. Maier, T.A., Macridin, A., Jarrell, M., Scalapino, D.J.: Phys. Rev. B **76**, 144516 (2007)
305. Maki, K.: Phys. kond. Materie **1**, 21 (1964)
306. Maki, K.: In: R.D. Parks (ed.) *Superconductivity*, vol. 2. Dekker, New York (1969)
307. Maldague, P.F.: Phys. Rev. B **16**, 2437 (1977)
308. Mancini, F., Avella, A.: Adv. Phys. **53**, 537 (2004)
309. Manske, D., Eremin, I., Bennemann, K.H.: Phys. Rev. B **67**, 134520 (2003)
310. Maple, M.B., Fertig, W.A., Mota, A.C., DeLong, L.E., Wohlleben, D., Fitzgerald, R.: Sol. State Comm. **11**, 829 (1972)
311. Maple, M.B., Fischer, Ø., eds.: *Superconductivity in Ternary Compounds*, vol. I+II. Springer, Berlin (1982). In Lect. Notes Phys.
312. March, N.H., Sampanthas, S.: Acta Phys. Hung. **14**, 67 (1962)
313. van der Marel, D., Sawatzky, G.: Phys. Rev. B **37**, 10674 (1988)
314. Martinez, G., Horsch, P.: Phys. Rev. B **44**, 317 (1991)
315. Mason, T.E., Aeppli, G.: Matematisk-fysiske Meddelelser. **45**, 231 (1997)
316. Mattheiss, L.F.: Phys. Rev. Lett. **58**, 1026 (1987)
317. McHale, P., Thalmeier, P., Fulde, P.: Phys. Rev. B **70**, 014513 (2004)
318. Meng, J., Liu, G., Zhang, W., Zhao, L., Liu, H., Jia, X., Mu, D., Liu, S., Dong, X., Zhang, J., Lu, W., Wang, G., Zhou, Y., Zhu, Y., Wang, X., Xu, Z., Chen, C., Zhou, X.J.: Nature **462**, 335 (2009)
319. Meservey, R., Tedrow, P.M., Fulde, P.: Phys. Rev. Lett. **25**, 1270 (1970)
320. Messiah, A.: *Quantum Mechanics*. North Holland, Amsterdam (1965)
321. Metropolis, N., Rosenbluth, A.W., Teller, A., Teller, E.: J. Chem. Phys. **21**, 1087 (1953)
322. Metzner, W., Vollhardt, D.: Phys. Rev. Lett. **59**, 121 (1987)
323. Metzner, W., Vollhardt, D.: Phys. Rev. Lett. **62**, 324 (1989)
324. Meul, H.W., Rossel, C., Decroux, M., Fischer, Ø., Remenyi, G., Briggs, A.: Phys. Rev. Lett. **53**, 497 (1984)
325. Millis, A.J.: Phys. Rev. B **48**, 7183 (1993)
326. Miyake, K.: J. Phys. Soc. Jpn. **77**, 123703 (2008)
327. Mizokawa, T., Fujimori, A.: Phys. Rev. B **54**, 5368 (1996)
328. Moessner, R., Sondhi, S.L., Chandra, P.: Phys. Rev. B **64**, 144416 (2001)
329. MOLCAS6. Lund, Sweden (2004). Univ. of Lund, Dept. of Theoret. Chemistry
330. Mori, H.: Progr. Theor. Phys. **33**, 423 (1965)
331. Moriya, T., Kawabata, A.: J. Phys. Soc. Jpn. **35**, 669 (1973)
332. Moriya, T., Kawabata, A.: J. Phys. Soc. Jpn. **34**, 639 (1973)
333. Morosov, A.J.: Sov. Phys. Solid State **22**, 1974 (1980)
334. Morris, D.J.P., Tennant, D.A., Grigera, S.A., Klemke, B., Castelnovo, C., Moessner, R., Czternasty, C., Meissner, M., Rule, K.C., Hoffmann, J.U., Kiefer, K., Gerischer, S., Slobinsky, D., Perry, R.S.: Science **326**, 411 (2009)
335. Müller-Hartmann, E.: Z. Phys. B **74**, 507 (1989)
336. Müller-Hartmann, E., Zittartz, J.: Z. Phys. **234**, 58 (1970)
337. Murakami, Y., Hill, J.P., Gibbs, D., Blume, M., Koyama, I., Tanaka, M., Kawata, H., Arima, T., Tokura, Y., Hirota, K., Endoh, Y.: Phys. Rev. Lett. **81**, 582 (1998)
338. Muzikar, P.: Ph.D. thesis, Cornell University (1980)
339. N. Andrei, F.F., Loewenstein, J.H.: Rev. Mod. Phys. **55**, 331 (1983)
340. Nagamatsu, J., Nakagawa, N., Muranaka, T., Zenitani, Y., Akimitsu, J.: Nature **410**, 63 (2001)

341. Nagarajan, R., Mazumdar, C., Hossain, Z., Dhar, S.K., Gopalakrishnan, K.V., Gupta, L.C., Godart, C., Padalia, B., Vijayaraghavan, R.: Phys. Rev. Lett. **72**, 274 (1994)
342. Neef, M.: *Kooperative Phänomene in CeCu2Si2*. Master's thesis, Technische Universität Braunschweig (2004)
343. Negele, J.W., Orland, H.: *Quantum Many-Particle Systems*. Addison-Wesley, Redwood City, CA (1988)
344. Nesbet, R.K.: Rev. Mod. Phys. **33**, 28 (1961)
345. Noack, R., White, S.: In: I. Peschel, X. Wang, M. Kaulke, K. Hallberg (eds.) *Density Matrix Renormalization*. Springer, Heidelberg (1999)
346. Noziéres, P.: J. Low Temp. Phys. **17**, 31 (1974). and J. de Physique **39**, 1117 (1978)
347. Nücker, N., Fink, J., Renker, B., Ewert, D., Politis, C., Weijs, P.J.W., Fuggle, J.C.: Z. Phys. B **67**, 9 (1987)
348. O'Brien, A., Pollmann, F., Fulde, P.: Phys. Rev. B **81**, 235115 (2010)
349. Ochiai, Suzuki, T., Kasuya, T.: J. Phys. Soc. Jpn. **59**, 4129 (1990)
350. Ogata, M., Fukuyama, H.: Rep. Prog. Phys. **71**, 036501 (2008)
351. Ogawa, T., Kanda, K., Matsubara, T.: Progr. Theor. Phys. **53**, 614 (1975)
352. Oleś, A., Pfirsch, F., Fulde, P.: Z. Phys. B **66**, 359 (1987)
353. Onsager, L.: J. Am. Chem. Soc. **58**, 1486 (1936)
354. Parks, R.D., ed.: *Superconductivity*. Dekker, New York (1969)
355. Pauling, L.: *The Nature of the Chemical Bond*. Cornell Univ. Press, Ithaca, NY (1939)
356. Pauling, L., Corey, R.B.: Proc. Natl. Acad. Sci. **39**, 551 (1953)
357. Paulus, B.: Physics Reports **428**, 1 (2006)
358. Paulus, B., Fulde, P., Stoll, H.: Phys. Rev. B **51**, 10572 (1995)
359. Paulus, B., Fulde, P., Stoll, H.: Phys. Rev. B **54**, 2556 (1996)
360. Penn, D.R.: Phys. Rev. Lett. **42**, 921 (1979)
361. Perdew, J.P., Burke, K., Wang, Y.: Phys. Rev. B **54**, 16533 (1996)
362. Perdew, J.P., Wang, Y.: Phys. Rev. B **33**, 8800 (1986)
363. Perdew, J.P., Zunger, A.: Phys. Rev. B **23**, 5048 (1981)
364. Perez-Jorda, J.M., Becke, A.D.: Chem. Phys. Lett. **233**, 134 (1995)
365. Perring, T.G., Adroja, D.T., Chaboussant, G., Aeppli, G., Kimura, T., Tokura, Y.: Phys. Rev. Lett. **87**, 217201 (2001)
366. Pettifor, D.G., Weare, D.L.: *The Recursion Method and its Applications*, in Springer Ser. Solid State Sci., vol. 58. Springer, Berlin, Heidelberg (1985)
367. Pfirsch, F., Böhm, M.C., Fulde, P.: Z. Phys. B **60**, 171 (1985)
368. Pimentel, I.R., Orbach, R.: Phys. Rev. B **46**, 2920 (1992)
369. Pines, D., Nozieres, P.: *The Theory of Quantum Liquids*, vol. 1. W. A. Benjamin, New York (1966)
370. Pipek, J., Mezey, P.G.: Chem. Phys. **90**, 4916 (1989)
371. Pisani, C., Dovesi, R., Roetti, C.: *Program package CRYSTAL in: Hartree-Fock Ab Initio Treatment of Crystalline Solids*, in Lect. Notes Chem., vol. 48. Springer-Verlag, Berlin (1988)
372. Pisani, C., Dovesi, R., Roetti, C.: *Program package CRYSTAL in: Hartree-Fock Ab Initio Treatment of Crystalline Systems*, in Topics in Current Physics, vol. 7, 2nd edn. Springer-Verlag, Berlin (1988)
373. Plakida, N.: *High-Temperature Cuprate Superconductors*. in Springer Series in Solid State Sciences. Springer, Berlin (2010)

374. Poilblanc, D., Riera, J., Dagotto, E.: Phys. Rev. B **49**, 12318 (1994)
375. Pollmann, F., Betouras, J., Shtengel, K., Fulde, P.: Phys. Rev. Lett **97**, 170407 (2006)
376. Pollmann, F., Fulde, P.: Europhys. Lett. **75**, 133 (2006)
377. Polyakov, A.M.: *Gauge Fields and Strings, in Contemporary Concepts in Physics*, vol. 3. Harwood Academic Publ. (1993)
378. Prange, R.E., Girvin, S.M., eds.: *The Quantum Hall Effect*. Springer-Verlag, Berlin, Heidelberg (1987)
379. Preuss, R., Hanke, W., Gröber, C., Evertz, H.G.: Phys. Rev. Lett. **79**, 1122 (1997)
380. Pulay, P.: Chem. Phys. Lett. **100**, 151 (1983)
381. Pytte, E.: Phys. Rev. B **10**, 4637 (1974)
382. Pyykkoe, P., Stoll, H.: In: A. Hinchliffe (ed.) *R. S. C. Specialist Periodical Reports: Chemical Modelling, Applications and Theory*, vol. 1. RSC, London (2000)
383. Qiu, X., Billinge, S.J.L., Kmety, C.R., Mitchell, J.F.: J. Phys. and Chem. of Solids **65**, 1423 (2004)
384. Radovan, H.A., Fortune, N.A., Murphy, T.P., Hannahs, S.T., Palm, E.C., Tozer, S.W., Hall, D.: Nature **425**, 51 (2003)
385. Ren, Y., a. Nugroho, A., Menovsky, A.A., Strempfer, J., Rütt, U., Iga, F., Takabatake, T., Kimball, C.W.: Phys. Rev. B **67**, 014107 (2003)
386. Renk, K.F.: In: A. Narlikar (ed.) *Studies of High-Temperature Superconductors*. Nova Science, New York (1992)
387. Roothan, C.C.J.: Rev. Mod. Phys. **23**, 69 (1951)
388. Rosch, A.: Phys. Rev. Lett. **82**, 4280 (1999)
389. Rosciszewski, K., Paulus, B., Fulde, P., Stoll, H.: Phys. Rev. B **60**, 7905 (1999)
390. Ross: Information Service, Research Center Dresden-Rossendorf (2010)
391. Rossat-Mignot, J., Regnault, L.P., Vettier, C., Bourges, P., Burlet, P., Bossy, J.: Physica C **185-189**, 86 (1991)
392. Rújula, A.D., Glashow, S.L.: Phys. Rev. Lett. **34**, 46 (1975)
393. Runge, E., Fulde, P.: Phys. Rev. B **70**, 245113 (2004)
394. Runge, E., Gross, E.K.U.: Phys. Rev. Lett. **52**, 997 (1984)
395. Runge, E., Zwicknagl, G.: Ann. Physik **5**, 333 (1996)
396. Sachdev, S.: *Quantum Phase Transitions*. Cambridge University Press, Cambridge (1999)
397. Sato, N., Aso, N., Miyake, K., Shiina, R., Thalmeier, P., Varelogiannis, G., Geibel, C., Steglich, F., Fulde, P., Komatsubara, T.: Nature **410**, 340 (2001)
398. Saxena, S.S., Agarwal, P., Ahilan, K., Grosche, F.M., Haselwimmer, R.K.W., Steiner, M.J., Pugh, E., Walker, I.R., Julian, S.R., Monthoux, P., Lonzarich, G.G., Huxley, A., Sheikia, I., Braithweite, D., Flouquet, J.: J. Magn. Magn. Mater. **226**, 45 (2001)
399. Scalapino, D.J.: In: M.P. Das, J. Mohanty (eds.) *Modern Perspectives in Many-Body Physics*, p. 199. World Scientific (1994)
400. Scalapino, D.J.: In: J.R. Schrieffer, J.S. Brooks (eds.) *Handbook of High-Temperature Superconductivity*. Springer, New York (2007)
401. Schlottmann, P., Sacramento, P.D.: Adv. Phys. **42**, 641 (1993)
402. Schollwöck, U.: Annals of Physics **326**, 96 (2011)
403. Schrieffer, J.R.: *Theory of Superconductivity*. W. A. Benjamin Inc., New York (1964)

404. Schrieffer, J.R., Wen, X.G., Zhang, S.C.: Phys. Rev. Lett. **60**, 944 (1988)
405. Schrieffer, J.R., Wolff, P.A.: Phys. Rev. **149**, 491 (1966)
406. Schulz, H.J.: Phys. Rev. Lett. **64**, 2831 (1990)
407. Shannon, N., Chatterji, T., Ouchni, F., Thalmeier, P.: Eur. Phys. J. B **27**, 287 (2002)
408. Shannon, N., Sikora, O., Pollmann, F., Penc, K., Fulde, P.: Phys. Rev. Lett. **108**, 067204 (2012)
409. Shiba, H., Ogata, M.: In: G. Baskaran, A.E. Ruckenstein, E. Tosatti, Y. Lu (eds.) *Strongly Correlated Electron Systems, in Progress in High Temperature Superconductivity*, vol. 29. World Scientific, Singapore (1991)
410. Shiba, H., Ueda, K., Sakai, O.: J. Phys. Soc. Jpn. **69**, 1493 (2000)
411. Shiina, R.: J. Phys. Soc. Jpn. **73**, 2257 (2004)
412. Shimahara, H.: J. Phys. Soc. Jpn. **67**, 736 (1998)
413. Shraiman, B.I., Siggia, E.D.: Phys. Rev. Lett. **60**, 740 (1988)
414. Shukla, A., Dolg, M., Stoll, H., Fulde, P.: Chem. Phys. Lett. **262**, 213 (1996)
415. Sigrist, M., Ueda, K.: Rev. Mod. Phys. **63**, 239 (1991)
416. Sikora, O., Pollmann, F., Shannon, N., Penc, K., Fulde, P.: Phys. Rev. Lett. **103**, 247001 (2009)
417. Singleton, J., Symington, J., Nam, M., Ardavan, A., Kurmoo, K., Day, P.: J. Phys.: Cond. Matter **12**, L641 (2000)
418. Slater, J.C.: Phys. Rev. **49**, 537 (1936)
419. Solyom, J.: Adv. Phys. **28**, 201 (1979)
420. Sommerfeld, A., Bethe, H.: *Elektronentheorie der Metalle, in Handbuch der Physik*, vol. 24/2, 2nd edn. Springer, Berlin, Heidelberg (1933)
421. Sorella, S., Parola, A., Parinello, M., Tosatti, E.: Europhys. Lett. **12**, 721 (1990)
422. Soven, P.: Phys. Rev. **156**, 809 (1967)
423. Springford, M.: Physica B **171**, 151 (1991)
424. Stemmann, G., Pepin, C., Lavagna, M.: Phys. Rev. B **50**, 4075 (1994)
425. Stephan, W.H., Horsch, P.: Phys. Rev. B **42**, 8736 (1990)
426. Stephan, W.H., v. d. Linden, W., Horsch, P.: Phys. Rev. B **39**, 2924 (1989)
427. Stewart, G.R.: Rev. Mod. Phys. **56**, 755 (1984)
428. Stock, C., Broholm, C., Hudis, J., Kang, H.J., Petrovic, C.: Phys. Rev. Lett. **100**, 087001 (2008)
429. Stockert, O., Arndta, J., Schneidewind, A., Schneider, H., Jeevan, H.S., Geibel, C., Steglich, F., Loewenhaupt, M.: Physica **403B**, 973 (2008)
430. Stockert, O., Faulhaber, E., Zwicknagl, G., Stuesser, N., Jeevan, H.S., Cichorek, T., Loewenhaupt, M., Geibel, C., Steglich, F.: Phys. Rev. Lett. **92**, 136401 (2004)
431. Stoermer, H.: Rev. Mod. Phys. **71**, 875 (1999)
432. Stoll, H.: Phys. Rev. B **46**, 6700 (1992)
433. Stoll, H., Paulus, B., Fulde, P.: Chem. Phys. Lett. **469**, 90 (2009)
434. Stollhoff, G.: J. Chem. Phys. **105**, 227 (1996)
435. Stollhoff, G., Fulde, P.: J. Chem. Phys. **73**, 4548 (1980)
436. Stone, M., ed.: World Scientific, Singapore (1992)
437. Stoyanova, A., Hozoi, L., Fulde, P., Stoll, H.: Phys. Rev. B **83**, 205119 (2011)
438. Stratonovich, R.L.: Dokl. Akad. Nauk SSSR **115**, 1907 (1957). (Engl. transl.: Sov. Phys. Dokl. 2, 416, (1958))
439. Su, W.P., Schrieffer, J.R.: Phys. Rev. Lett. **46**, 738 (1981)
440. Su, W.P., Schrieffer, J.R., Heeger, A.J.: Phys. Rev. Lett. **42**, 1698 (1979)

441. Sushkov, O.P., Sawatzky, G.A., Eder, R., Eskes, H.: Phys. Rev. B **56**, 11769 (1997)
442. Suzuki, M.: *Quantum Monte Carlo Methods, in Springer Ser. Solid-State Sc.*, vol. 74. Springer, Berlin, Heidelberg (1987)
443. Svane, A., Gunnarsson, O.: Phys. Rev. B **37**, 9919 (1988)
444. Takahashi, M., Igarashi, J., Fulde, P.: J. Phys. Soc. Jpn. **68**, 2530 (1999)
445. Tanatar, B., Ceperley, D.M.: Phys. Rev. B **39**, 5005 (1989)
446. Tanner, D.B., Timusk, T.: Optical properties of high-temperature superconductors. In: D.M. Ginsberg (ed.) *Physical Properties of High-Temperature Superconductors III*. World Scientific, Singapore (1992)
447. Taylor, D.W.: Phys. Rev. **156**, 1017 (1967)
448. Thalmeier, P.: Europhys. J. B **27**, 29 (2002)
449. Tinkham, M.: *Introduction to Superconductivity*. McGraw-Hill, New York (1975)
450. Tohyama, T., Maekawa, S.: Physica C **191**, 193 (1992)
451. Tokura, Y.: Physics Today **56(7)**, 50 (2003)
452. Tomioka, Y., Asamitsu, A., Kuwahara, H., Moritomo, Y., Tokura, Y.: Phys. Rev. B. **53**, R1689 (1996)
453. Tomonaga, S.: Progr. Theor. Phys. **5**, 544 (1950)
454. Treglia, G., Ducastelle, F., Spanjaard, D.: J. Phys. (Paris) **41**, 281 (1980)
455. Tsui, D.C., Störmer, H.L., Gossard, A.C.: Phys. Rev. Lett **48**, 1559 (1982)
456. Tsvelick, A.M., Wiegmann, P.B.: Adv. Phys. **32**, 453 (1983)
457. Uchida, S.: Mod. Phys. Lett. B **4**, 513 (1990)
458. Uimin, G., Kudasov, Y., Fulde, P., Ovchinnikov, A.: Eur. Phys. J. B **16**, 241 (2000)
459. Uji, S., Shinagawa, H., Terashima, T., Yakabe, T., Terai, Y., Tokumoto, M., Kobayashi, A., Tanaka, H., Kobayashi, H.: Nature **410**, 908 (2001)
460. Uji, S., Terashima, T., Nishimura, M., Takahide, Y., Konoike, T., Enomoto, K., Cui, H., Kobayashi, H., Kobayashi, A., Tanaka, H., Tokumoto, M., Choi, E.S., Tokumoto, T., Graf, D., Brooks, J.S.: Phys. Rev. Lett. **97**, 157001 (2006)
461. Unger, P., Fulde, P.: Phys. Rev. B **48**, 16607 (1993). Ibid B **47**, 8947 (1993)
462. Unger, P., Igarashi, J., Fulde, P.: Phys. Rev. B **50**, 10485 (1994)
463. Varma, C.M.: Phys. Rev. B **73**, 155113 (2006)
464. Varma, C.M., Littlewood, P.B., Schmitt-Rink, S., Abrahams, E., Ruckenstein, A.: Phys. Rev. Lett. **63**, 1996 (1989)
465. Varma, C.M., Yafet, Y.: Phys. Rev. B **13**, 2950 (1976)
466. Velicky, B., Kirkpatrick, S., Ehrenreich, H.: Phys. Rev. **175**, 747 (1968)
467. Verhaegen, G., Moser, C.M.: J. Phys. B **3**, 478 (1970)
468. Verwey, E.J.W., Haayman, P.W.: Physica **8**, 979 (1941)
469. Vidhyadhiraja, N.S., Macridin, A., Sen, S., Jarrell, M., Ma, M.: Phys. Rev. Lett. **102**, 206407 (2009)
470. Vojta, M.: Reports on Progr. Phys. **66**, 2069 (2003)
471. Vojta, M.: Adv. Phys. **58**, 699 (2009)
472. Vojta, M., Becker, K.: Europhys. Lett. **38**, 607 (1997)
473. Volovik, G.E., Gorkov, L.P.: Zh. Eksp. Teor. Fiz. **88**, 1412 (1985). Engl. transl.: Sov. Phys.-JETP **61**, 843 (1985)
474. Vonsovsky, S.V.: Sov. Phys.-JETP **16**, 980 (1946)
475. Wada, H., Nakamura, H., Fukami, E., Yoshimura, K., Shiga, M., Nakamura, Y.: J. Magn. Magn. Mater. **70**, 17 (1987)

476. Wang, C.S., Callaway, J.: Phys. Rev. B **15**, 298 (1977)
477. Wang, Y., Yu, Z., Kakeshita, T., Uchida, S., Ono, S., Ando, Y., Ong, N.P.: Phys. Rev. B **64**, 224519 (2001)
478. Watanabe, T., Izawa, K., Kasahara, Y., Haga, Y., Onuki, Y., Thalmeier, P., Maki, K., Matsuda, Y.: Phys. Rev. B **70**, 184502 (2004)
479. Weiden, M., Hauptmann, R., Geibel, C., Steglich, F., Fischer, M., Lemmens, P., Güntherod, G.: Z. Phys. B **103**, 1 (1997)
480. Wen, H.H., Li, S.: Ann. Review Cond. Matt. Phys. **2**, 121 (2011)
481. Werner, H.J., Knowles, P.J.: *MOLPRO - program package.* with contributions from J. Almlöf and R. Amos and M. J. O. Deegan and S. T. Elbert and C. Hampel and W. Meyer and K. Peterson and R. Pitzer and A. J. Stone and P. R. Taylor
482. White, S.R.: Phys. Rev. Lett. **69**, 2863 (1992)
483. White, S.R., Scalapino, D.J.: Phys. Rev. B **61**, 6320 (2000)
484. Whitman, D.R., Hornback, C.J.: J. Chem. Phys. **51**, 398 (1968)
485. Wick, G.C.: Phys. Rev. **80**, 268 (1950)
486. Wigner, E.: Phys. Rev. **46**, 1002 (1934)
487. Wilczek, F.: Phys. Rev. Lett. **49**, 957 (1982)
488. Wilkonson, J.H.: *The Algebraic Eigenvalue Problem.* Clarendon, Oxford (1965)
489. Wilson, K.G.: Rev. Mod. Phys. **47**, 773 (1975)
490. Wrobel, P., Eder, R.: Phys. Rev. B **49**, 1233 (1994)
491. Wrobel, P., Eder, R.: Phys. Rev. B **58**, 15160 (1998)
492. Wrobel, P., Eder, R., Fulde, P.: J. Phys.: Cond. Mat. **15**, 6599 (2003)
493. Wrobel, P., Eder, R., Ohta, Y.: Phys. Rev. B **54**, 11034 (1996)
494. Yamashita, Y., Ueda, K.: Phys. Rev. B **67**, 195107 (2003)
495. Yanase, Y., Sigrist, M.: J. Phys. Soc. Jpn. **78**, 114715 (2009)
496. Yang, C.N.: Rev. Mod. Phys. **34**, 694 (1962)
497. Yaresko, A.N., Antonov, V.N., Eschrig, H., Thalmeier, P., Fulde, P.: Phys. Rev. B **62**, 15538 (2000)
498. Yin, M.T., Cohen, M.L.: Phys. Rev. B **24**, 6121 (1981)
499. Yokoyama, H., Shiba, H.: J. Phys. Soc. Jpn. **56**, 3570 (1987)
500. Yonezawa, F.: Progr. Theor. Phys. **40**, 734 (1968)
501. Yonezawa, S., Kusuba, S., Maeno, Y., Auban-Senzier, P., Pasquier, C., Jerome, D.: J. Phys. Soc. Jpn. **77**, 054712 (2008)
502. Yoshioka, D.: *The Quantum Hall Effect, in Springer Series in Solid State Sciences*, vol. 133. Springer-Verlag, Berlin, Heidelberg (2002)
503. Yuan, Y.M., Kaxiras, E., Gordon, R.G.: Phys. Rev. B **48**, 14944 (1993)
504. Yushankhai, V., Thalmeier, P., Takimoto, T.: Phys. Rev. B **77**, 125126 (2008)
505. Yushankhai, V., Yaresko, A., Fulde, P., Thalmeier, P.: Phys. Rev. B **76**, 085111 (2007)
506. Zaanen, J., Sawatzky, G.A., Allen, J.: Phys. Rev. Lett. **55**, 418 (1985)
507. Zwanzig, R.: *Lectures in Theoretical Physics*, vol. 3. Interscience, New York (1961)
508. Zwicknagl, G.: Adv. Phys. **41**, 203 (1992)
509. Zwicknagl, G., Fulde, P.: Z. Phys. B **43**, 23 (1981)
510. Zwicknagl, G., Fulde, P.: J. Phys.: Condens. Matter **15**, S1911 (2003)
511. Zwicknagl, G., Pulst, U.: Physica B **186**, 895 (1993)
512. Zwicknagl, G., Runge, E., Christensen, N.E.: Physica B **163**, 97 (1990)
513. Zwicknagl, G., Wosnitza, J.: In: L.N. Cooper, D. Feldman (eds.) *BCS: 50 years*, p. 337. World Scientific Publ., Singapore (2011)

514. Zwicknagl, G., Yaresko, A.N., Fulde, P.: Phys. Rev. **65**, 081103(R) (2002)
515. Zwierlein, M.W., Schirotzek, A., Schnuk, C.H., Ketterle, W.: Science **311**, 492 (2006)

Index

Abrikosov flux lattice, 424
Aharonov-Bohm phase, 375
almost ferromagnetic materials, 268
Anderson
    Hamiltonian, 170, 326
    impurity, 167
    lattice, 336
angular
    correlations, 24
    resolved photo electron spectroscopy, 454
anti-Th_3P_4 structure, 341
antibonding, 285
    orbitals, 85
antiferromagnetic
    background, 464
    correlations, 144, 459
    length, 291
    coupling, 305
    ground state, 207
    interaction, 300
    order, 230
    spin
        fluctuations, 279
        waves, 450
anyonic statistics, 363
ARPES, 291, 335, 462
atomic units, 28
autoionization, 397

κ-$(BEDT-TTF)_2Cu(NCS)_2$, 430
λ-$(BETS)_2FeCl_4$, 429
λ-$(BETS)_2GaCl_4$, 429

B atom, 17
backflow, 380
band gap, 129
basis functions, 11
BCS, 399
    ground-state, 425
    wavefunction, 211
    Hamiltonian, 210
Berezinskii-Kosterlitz-Thouless
    transition, 385, 452
Berry phase, 374
Bethe
    ansatz, 238
    methods, 177
    lattice approximation, 196
Bethe-Goldstone expansion, 69
Bloch orbitals, 16
blocking effect, 427
Bogoliubov transformation, 200
Bogoliubov-de Gennes equations, 428
bond
    length, 2
    orbital approximation, 84, 126
bonding, 285
    orbital, 2
Born approximation, 450
Bose condensation, 372
boson propagator, 439
bound states, 389
Brillouin
    theorem, 60
    zone, 180, 293
        reduced, 220

Brillouin-Wigner perturbation theory, 58
Brinkmann-Rice transition, 186
BSCCO, 283

CH_4, 61
C_2H_6, 61
III-V compounds, 91
C_6H_6, 20
C-C bonds, 22
Ca_2RuO_4, 319
CASSCF, 3, 62, 99, 284, 288, 293
causality, 266
CDMF theory, 234
CDW, 220, 315
  d-CDW, 467
  s-CDW, 467
Ce, 322
Ce^{3+}, 157
CeB_6, 281
$CeCoIn_5$, 335, 430, 431, 445, 447
$CeCu_2Si_2$, 332, 335, 445
CEF, 299, 352, 438
  levels, 438
  singlet ground state, 438
  splitting, 349
cerocene, 166
$CeRu_2Ge_2$, 333, 334
$CeRu_2Si_2$, 324, 325, 328, 331, 333
chains, 343
charge
  fluctuations, 157, 245, 461
  order, 311, 313, 507
  ordering, 317, 340
  transfer insulator, 400, 448
checkerboard lattice, 375, 379, 380, 391
chemical potential, 112
Clebsch-Gordan coefficients, 350
Clogston limit, 426
cluster
  expansion, 153
  memory function, 506
Co, 242
CO molecule, 18
coherence length, 424, 429
coherent
  hole motion, 196
  potential approximations (CPA), 145
  states, 410, 411

colossal magnetoresistance, 301
composite operators, 199, 453
condensation energy, 414
conduction bands, 134
configuration-interaction
  method, 58
conserving approximations, 272
constrained LDA, 294
continued fraction, 486
CoO, 46
Cooper pair, 6, 280, 373, 399
  formation, 435
correlation
  cusp, 61
  hole, 137
  strength, 160
Coulomb
  integrals, 438
  interaction
    screened, 52
  law, 395
  liquid, 397
  phase, 393
  repulsions, 501
coupled cluster method, 72
coupled electron pair approximation
  (CEPA), 72
  CEPA-0, 72
  CEPA-2, 72
CPA, 311
  dynamical, 149
  equation, 260
  many-body, 149
  single-site theory, 147
Cramer's rule, 486
critical doping, 451
CRYSTAL, 84, 90, 132
crystalline electric field, 158
Cu, 45
Cu-Ni alloy, 432
Cu-O
  planes, 160, 232, 282
cumulants, 3, 63, 114, 473, 477
  wave operator, 65, 473
CuO_2, 448
  planes, 436, 467
Curie
  like behavior, 322
  temperature, 263, 269, 333

Curie-Weiss behavior, 256, 262, 268, 355
cyclotron
  motion, 424
  orbit, 370
  resonance, 235

d electron, 247, 314
  bandwidth, 262
  number, 49
d holes, 297
d wave
  pairing, 401, 420
  superconductor, 430
  symmetry, 466
$3d$ systems, 241, 299, 348
$4d$ systems, 299
DCPA, 263
de Haas-van Alphen effect, 332
  experiments, 324, 333
Debye frequency, 416
deformation potential, 345
density correlation function, 238
density fluctuations, 81
density functional theory (DFT), 2, 213
  gradient corrections, 45
density matrix renormalization group, 5, 489
density of states, 322, 330, 415, 425, 453
dielectric constant, 130
diffusion equation, 501
dimer models, 386
dimerization, 317, 319, 365
dipolar interactions, 394
Dirac string, 395
disconnected diagrams, 67
disorder, 311
  dynamic, 145, 153
  static, 145
DMFT, 222, 263
domain walls, 237, 366
Doniach criterion, 324
double
  exchange, 301
  layer, 313
  zeta basis set, 12
Drude
  peak, 454
  term, 458

weight, 459
dual model, 224, 440, 442
$Dy_2Ti_2O_7$, 394
dynamic variables, 138
Dynamical Mean-Field Theory (DMFT), 145
  cluster, 150

e_g doublet, 299
  multiplet, 348
effective
  Hamiltonian, 167, 306, 493
  hybridization, 332
  interactions, 302
  mass, 48
elastic
  constant, 345
  energy, 303, 344
  strain, 343
electrical resistivity, 347
electron
  crystallization, 4
  energy loss spectroscopy, 290
  gas
    two-dimensional, 25
  hole pair, 271
  phonon
    coupling, 366
    interaction, 435
Eliashberg's equation, 435, 439, 460
energy
  cohesive, 92
  dispersion, 3
  gap, 3
  problem, 123
entropy, 119, 356
environment, 492, 493
equation of motion, 104
ergodic behavior, 419, 421, 422
exact diagonalization, 193, 466
exchange, 129
  double, 313
  field, 433
  indirect, 300
exchange-correlation, 215
  energy, 43
  potential, 42, 359
excitation
  double, 60

energy, 4
single, 60

f electrons
  itinerant, 442
f resonance, 330
f-like band, 329
$4f$, 321
  electrons, 5, 322, 409, 422
  hole, 342
  shell, 299, 394, 437
  systems, 299
$5f$, 321
  electrons, 6, 409, 440, 445
  shell, 437
  systems, 348
$5f^2$ subsystem, 352
Fe, 242
  pnictides, 7, 448
Fe_3O_4, 356, 376
FeO, 46
Fermi liquid, 121, 293, 324
  marginal, 233
  theory, 5, 101, 231, 347
Fermi surface, 293, 315, 328, 331, 405, 447
  UPd_2Al_3, 354
ferromagnet, 19, 320
  strong, 241
ferromagnetic
  ground state, 46
  layer, 431, 433
  order, 314
  superconductors, 401
  system, 260
ferromagnetism
  weak, 262
Feshbach resonance, 434
Feynman diagrams, 114
FFLO, 425, 428, 429
  phase, 430
  state, 430, 434
filling factor, 371, 372
flat band, 357
FLEX approximation, 274, 275, 461
fluctuation-dissipation theorem, 266, 297
flux
  quanta, 373
  quantum, 373
Fock matrix, 128
Foster-Boys
  localized orbitals, 16
  orbitals, 125
fractional
  charge, 25, 363, 382, 384, 388
  quantum Hall effect (FQHE), 6, 25, 363, 370, 375
free energy, 345
frequency
  binary dividers, 434
frequency matrix, 140
Friedel
  oscillations, 238, 271
  sum rule, 326
frustration, 318
FSCP operator technique, 233
fullerenes, 409
functional integrals, 149
  method, 251

GaAs, 92
gauge
  theory, 210
  transformation, 387
Gauss' law, 393
Gaussian
  average, 252
  type orbitals, 12
geometrical frustration, 356, 361
geometrically frustrated lattices, 375
Ginzburg-Landau
  expansion, 344
  regime, 428
Goodenough-Kanamori, 300
Green's function, 5, 102, 312, 450
  advanced, 103
  Matsubara, 115
  Monte Carlo technique, 503
  retarded, 103, 141, 149, 293
  temperature, 111
ground-state
  J multiplet, 176
  charge-ordered, 382
  configurations, 387
  crystal-field, 327
  energy, 67, 500
  multiplet, 437

Index   529

wavefunction, 2, 35, 306, 491, 502
Gutzwiller
　ansatz, 219
　approximation, 183
　projector, 211
　wavefunction, 79, 183
GW approximation, 52, 130, 249

H_2, 1, 18, 19, 159
Hall coefficient, 342
Hamiltonian
　effective, 340, 387
　reduced, 412
　single-particle, 311
Hartree-Fock, 4, 257
　approximation, 28
　equations, 9, 15, 39
　limit, 9
heavy
　electrons, 166
　fermion
　　superconductors, 409
　　systems, 323
　quasiparticles, 48, 321, 323, 332, 356
　　charge-neutral, 235
Heisenberg
　AF, 211
　chain, 236, 347, 390, 489
　exchange, 208
　ferromagnet, 261
　Hamiltonian, 199
　representation, 107
　spin chains, 347
Heitler-London
　limit, 2, 160
　wavefunction, 1, 2
heterojunctions, 36
hexagons, 381
　flipable, 382
high-T_c
　cuprates, 437
　materials, 400
　superconductivity, 448
hight
　field, 386
　representation, 385
Hilbert space, 58, 66, 389
　reduced, 167
Hohenberg, Kohn, Sham theory, 39

hole
　doping, 154, 310
　pockets, 202, 291, 293
hole doped
　cuprates, 448
　system, 308
holes, 290, 297
　doped, 294
holon, 206, 208, 453
Holstein-Primakoff transformation, 200, 212
Hubbard
　band, 143, 187
　　lower, 144, 286
　　splitting, 283
　　upper, 144, 286, 298
　chain, 237
　Hamiltonian, 79, 142, 178, 222
　I approximation, 142, 186
　III approximation, 188, 231
　interaction, 265
　model, 5, 155, 158, 377
　　1D, 236
　operator, 171
Hubbard-Stratonovich
　field, 226
　transformation, 149, 252
Hund's rule, 6, 299, 314, 349, 394
　correlations, 161, 246, 440
　coupling, 256
hypercube, 498

imaginary time, 113, 225
importance sampling, 227
impurity
　magnetic, 168
　solver, 145
in-out correlations, 24
incoherent
　excitation, 137
　structure, 206
incremental decomposition, 69
increments
　many-body, 92
　one-body, 94
　three-body, 95
　two-body, 95
independent
　electron approximation, 20

mode approximation, 80
induced moment system, 352
inelastic neutron scattering, 347, 446
Infinite-System DMRG, 494
inhomogeneous phase, 429
INS, 355
insulator-metal transition, 385
inter-pair correlations, 98
interaction representation, 116
interatomic
  correlations, 23, 84, 88
intra-atomic
  correlations, 23, 89, 162, 349, 351, 479
  excitations, 441
inversion center, 407
ionic
  bonding, 83
  configurations, 2, 20
isospin, 303, 307
isotope mass, 435

jj coupling, 352
Jahn-Teller effect, 299, 303, 346
Jastrow
  prefactor, 81, 504
  type ansatz, 371
  wavefunction, 32, 500

Kadowaki-Woods ratio, 342
kagome lattice, 375, 378
Kanamori limit, 189, 249
kink, 366
Kohn-Sham
  eigenvalues, 216
  equations, 39, 43, 500
  orbitals, 43
Kondo
  effect, 6, 158, 163, 323
  Hamiltonian, 167, 169
  impurities, 425
  lattice, 483
    model, 323
    system, 323, 324, 336
  problem, 327
  regime, 175
  resonance, 177
  temperature, 175, 321, 327
Kramers'
  degeneracy, 299

doublets, 326, 394
Kramers-Kronig relations, 122, 232, 444
Kubo relation, 457

Li_2, 61, 97
La_2CuO_4, 160, 221, 282
$La_2Mn_2O_7$, 307
$La_{1.8-x}Eu_{0.2}Sr_xCuO_4$, 467
$La_{2-x}Ba_xCuO_4$, 469
$La_{2-x}Sr_xCuO_4$, 449
Lagrange parameters, 15
$LaMnO_3$, 301, 303, 307
Lanczos
  algorithm, 225, 458, 485, 486
  method, 463, 494
Landé factor, 425
Landau
  damping, 443
  level, 25, 369, 371, 374
  orbital, 6
  parameter, 122, 266, 481
$LaRu_2Si_2$, 335
lattice degrees of freedom, 341
$LaVO_3$, 314
LDA, 150, 216, 281, 328, 345
  adiabatic, 56
LDA+U, 47, 281, 308, 317, 319
Li bulk, 97
Lindemann's criterion, 36
Lindhard's
  function, 271, 278, 461
  spin susceptibility, 447
linear response theory, 457
linked cluster theorem, 71
links, 376, 392
Liouville operator, 138
Liouvillean, 295
  cluster, 152
$LiTiO_3$, 320
LiV_2O_4, 321, 323, 355, 376
local
  ansatz, 77
  density approximation (LDA), 3, 39
  spin-density approximation (LSDA), 44
localization of electrons, 319
localized electrons, 437, 442
loop, 383
  models, 378

LS-coupling, 352
LSDA, 216
Luttinger
　liquid, 162
　theorem, 462

Madelung energy, 34
magnetic
　charges, 395, 397
　exciton, 353, 354, 442–444
　field, 422, 423
　impurity, 321, 418
　monopoles, 377, 394
　resonances, 441
　susceptibility, 483, 484
magnetization, 258
manganites, 301, 307
　double-layer, 309
many-electron wavefunctions, 3
marginal Fermi liquid, 159
Markov chain, 499
mass
　enhancement, 440
　renormalization, 277
Maxwell's theory, 394
mean free path, 118
mean-field
　approach, 454
　approximation, 159, 207, 261, 266, 337
　theory, 340
　transition, 345
medial lattice, 376, 387
Meissner effect, 373, 406
memory
　function, 505
　matrix, 140
Mermin-Wagner theorem, 150, 204, 239
metal-insulator transition, 224, 319
method of
　increments, 476
　steepest descent, 254
Metropolis method, 499
MgB_2, 401
MgO, 132
mid-infrared absorption, 458
midgap, 455
　state, 367
MnO_6 octahedra, 308, 309

model Hamiltonian, 366
MOLCAS, 289
molecular-orbital
　theory, 1
MOLPRO, 74, 84, 94
momentum distribution, 106
Monte Carlo
　calculations, 4
　methods, 193
　technique, 497, 499
　treatment, 382
Mott insulator, 207
Mott-Hubbard
　insulators, 6, 158, 192, 314, 452, 466
　transitions, 281
Møller-Plesset
　expansion, 70
　perturbation, 96

α'-NaV_2O_5, 315, 319
N_2, 18
Néel
　order, 320
　state, 179, 212
　temperature, 150
Nb, 432, 433
Nd_2CuO_4, 282, 455
$Nd_{2-x}Ce_xCuO_4$, 323
negative ions, 479
Nernst effect, 449, 453
neutron
　scattering, 395
　experiments, 440
　stars, 434
Ni, 242, 248
Ni_3Al, 264
Ni_3Ga, 264
NiO, 93
node lines, 420
non-Fermi liquid, 240
Noncrossing Approximation (NCA), 325
nonergodic
　behavior, 421
　processes, 419
noninteracting electrons, 31

off-diagonal long-range order, 406
one-particle Hamiltonian, 258

Index

Onsager's
  reaction field, 267
  theory, 397
open-shell systems, 17
optical conductivity, 454, 457
optimal hole-doping, 449
orbital
  canonical molecular, 59
  local, 59
  ordering, 301, 308
  selection, 350
  selective localization, 348
overdoped systems, 449

π band, 364
π junction, 431, 434
p holes, 297
p-wave pairing, 409, 420
$4p$ holes, 344, 347
pair
  breaking, 402, 418
    parameter, 419, 423
  condensate, 408
  correlations, 7, 399
  distribution function, 20, 29, 51
  excitation, 415
  function, 405
  wavefunction
    even-parity, 433
    odd-parity, 433
  weakening, 419
pairing
  d-wave, 462
  s-wave, 462
  conventional, 409
  states, 400
  unconventional, 400, 405, 409
paramagnetic impurities, 419
paramagnons, 271, 276
particle-hole excitation, 210
partition function, 112, 251
Pauli paramagnetic susceptibility, 321
Pd, 264
Peierls
  distortion, 220
  systems, 365
perovskite, 314
  structure, 282, 301, 304
phase
  diagram, 449, 469
  shifts, 325, 328, 483
  stiffness constant, 452
  transition, 343
    structural, 343
  unlocking, 452
phonons, 7, 422, 439
photoemission, 162, 177
  experiment, 322, 335
polarization
  cloud, 129, 133
  functions, 12
polarons, 314
population imbalance, 423
Pr metal, 353
probability
  density, 501
  distribution, 497, 503
projection
  method, 71, 294, 463
    fully self-consistent, 154
  technique, 58, 137, 193
PrOs$_4$Sb$_{12}$, 401, 437
pseudogap, 449, 451, 454
pseudopotential, 94
pyrochlore
  lattice, 375, 379
  structure, 356, 359

quadrupole moment, 438
quantum
  chemistry, 3
  critical point, 239
  fluctuations, 199, 239
  Hall effect
    fractional (FQHE), 368
    integer, 368
quark
  antiquark pairs, 384
  confinement, 384
quasihole, 101, 373, 375
quasiparticle, 5, 101, 120, 373
  approximation, 124
  bands, 338
  interaction, 483, 484
  mass, 121

(RE)Mo$_6$S$_8$, 421
(RE)Ni$_2$B$_2$C, 421

(RE)Rh$_4$B$_4$, 421
random phase approximation, 25
rare-earth compounds, 299
rare-gas
  crystals, 96
  dimers, 96
  solids, 93
Rayleigh-Schrödinger
  expansion, 71
  perturbation theory, 58
reduced density matrix, 492
reentrant behavior, 312
relaxation cloud, 101
renormalization constant, 106
renormalized
  band theory, 50, 336
  bandstructure calculations, 324, 330
  perturbation theory, 153
repulsive interactions, 462, 468
resistivity, 342, 355, 455
resolvent operator, 485
resonance peak, 441, 444, 446
resonating
  plaquettes, 388
  valence bonds, 207
ring-hopping processes, 380, 386, 392
Rokhsar-Kivelson point, 387
RPA, 269, 277, 357, 362, 446
Ruderman-Kittel-Kasuya-Yosida
  oscillations, 238, 271
Runge-Gross theorem, 53
ruthenate perovskites, 319

σ bonding, 364
sp^2 hybridization, 364
sp^3 hybrid, 21
saddle-point approximation, 254
sampling
  importance, 499
  simple, 498
satellite structure, 144, 248
Sc$_3$In, 264
scattering
  matrix, 68, 152, 475
  operator, 505
  potential, 325, 326
Schmidt decomposition, 493
Schrödinger equation, 39, 395, 403, 500

Schrieffer-Wolff transformation, 168, 302
Schwinger bosons, 450
SDW, 220, 315, 430, 431, 467
  d-wave, 211
  approximation, 182
  ground state, 180
  phase, 332
self
  biased phase qubits, 434
  energy, 105, 159
  interaction, 16
    correction, 46
self-consistent Born approximation, 199
self-consistent field (SCF), 3
  approximation, 126, 212, 312
    unrestricted, 179, 213
  ground-state, 9, 85
  localized orbitals, 16
  multiconfiguration, 62
  unrestricted wavefunctions, 9, 19
self-consistent renormalized RPA, 275
semimetal, 323
shadow band, 183
shake-up peaks, 242
shift operator, 391
single
  centre approximation , 476
  flux quantum cells, 434
  mode approximation, 79
  particle
    excitation, 415
    Hamiltonian effective, 217
    spectral weight, 229
  site approximation, 223, 259
singlet
  state, 305
  triplet excitation, 327
size-extensive quantity, 58
Slater
  determinant, 13, 236, 370, 372, 500
  functions, 11
slave
  boson, 459
    mean-field approximation, 218
  fermion, 459
solid-state theory, 3
soliton, 366
Sommerfeld-Wilson ratio, 322, 347

specific heat, 122, 321, 334, 361, 483
spectral
  density, 295, 507
    incoherent part, 141
  function, 388
  representation, 103
spin
  bag, 195, 198, 206, 450, 459
  bipolaron, 464
  chain, 361
  charge separation, 367
  defect, 194, 463
  density
    matrix, 41
    wave state, 178
  excitation energy, 205
  fluctuations, 201, 269, 276, 359, 460
    local, 251
  gap, 314, 319
  ice, 394
  interaction, 199
  orbit
    interaction, 299, 407
    scattering, 417, 426
  orbital, 14
  ordering, 308
  Peierls transition, 316, 319
  polaron, 201
  singlet order parameter, 408
  spin correlation function, 230, 238
  spin interaction, 300
  susceptibility, 122, 232, 361, 442
  triplet, 307
  triplet order parameter, 408
  wave velocity, 451
spinels, 356, 376
spinless fermions, 199, 236, 363, 367, 377
spinon, 207
spiral
  phase, 204
  states, 179
square lattice, 154
Sr_2RuO_4, 319, 401
Sr_2VO_4, 305
standard basis operator, 171
static approximation, 254
Stoner
  enhancement, 122

theory, 261
string, 382, 384, 464
  tension, 389
stripe formation, 467
strong coupling superconductivity, 416
$SU(2)$ symmetry, 225
superblock, 491–493
superconducting thin film, 424
superconductivity, 399
  and antiferromagnetism, 421
  gapless, 420
  reentrant, 425
superconductors
  type II, 424
supercurrent, 426, 452
superexchange, 300, 308, 313, 318
superoperator
  L, 66
  method, 138
surface roughness, 421
symmetry
  d-wave, 210
  group, 408

t-matrix, 146, 147, 271
$t-J$
  Hamiltonian, 189, 463
  model, 159, 450, 455, 460, 469
$t-J^z$ model, 199
t_{2g}
  multiplet, 348
  triplet, 299
$(TMTSF)_2ClO_4$, 429
$TbMo_6S_8$, 421
tetrahedron rule, 360, 378, 379
thermal
  broadening, 119
  conductivity, 347
thermodynamic potential, 114
time
  dependent density function theory, 40, 53
  ordering operator, 102
  reversal
    operator, 418
    symmetry, 417, 441
  reversed states, 401, 417
TiO_2, 135
titanates, 301

Tomonaga-Luttinger liquid, 238
topological structure, 382
trans-polyacetylene, 363, 364
transition metal, 242
    oxides, 299
Trellis lattice structure, 316
trial
    function, 498, 500
    wavefunction, 504
triplet state, 166, 306, 407
tunnelling measurements, 424
two
    centre approximation, 477
    particle density matrix, 406
    saddle-points approximation, 262

$U(1)$ gauge theory, 375, 390
UBe_{13}, 406
UGa_2, 401
ultracold
    atoms, 402
    fermionic atoms, 434
uncorrelated electrons, 26
underdoped
    cuprates, 463
    regime, 463
    systems, 449
UPd_2Al_3, 323, 349, 401, 442
upper critical magnetic field, 422
UPt_3, 323, 349

vacuum polarization, 384
valence
    bands, 132
    bandwidth, 129

van der Waals
    correlations, 86, 97
    interactions, 4
vanadates, 314
Vandermonde determinant, 372
variational methods, 193
Varma-Yafet trial-wavefunction, 172
Verwey transition, 356, 360
virtual crystal, 146
vortex-antivortex pair, 386

Wannier orbitals, 69, 83, 125, 475
Wick's theorem, 109
Wigner
    crystal, 4, 25, 34
    lattice, 221
    liquid, 37

X-ray scattering, 308, 313, 467

$(YSc)Mn_2$, 355
Yb^{3+}, 157
Yb_4As_3, 5, 222, 234, 341, 347
$YBa_2Cu_3O_7$, 282
$YBa_2Cu_3O_{6.6}$, 462
YMn_2, 323

c-ZnS, 135
Zeeman
    effect, 6
    splitting, 176
    term, 423–425, 427
zero-point fluctuations, 78
Zhang-Rice singlet, 288, 289, 450
$ZrSn_2$, 264